Summary of equations for testing hypotheses and for estimation

Sample size	Test of Hypothesis	Hypothesized Parameter	Sample Statistic	Standard Error	Calculated Test statistic	σ is	Confidence Interval for Estimation
Large Sample $n \geq 30$	**MEANS** **One sample**	$H_0: \mu = 80$; $H_1: \mu \neq 80$	$\bar{x}$	$\sigma_{\bar{x}} = \frac{\sigma}{\sqrt{n}}$	$z_c = \frac{\bar{x} - \mu_H}{\sigma_{\bar{x}}}$	Known	$\bar{x} \pm z\,\sigma_{\bar{x}}$
		or $\mu < 80$; > 80		$s_{\bar{x}} = \frac{s}{\sqrt{n}}$	$z_c = \frac{\bar{x} - \mu_H}{s_{\bar{x}}}$	Unknown	$\bar{x} \pm z\,s_{\bar{x}}$
	Two samples	$H_0: \mu_1 - \mu_2 = 0$; $H_1: \mu_1 - \mu_2 \neq 0$	$\bar{x}_1 - \bar{x}_2$	$\sigma_{\bar{x}_1 - \bar{x}_2} = \sqrt{\frac{\sigma_1^2}{n_1} + \frac{\sigma_2^2}{n_2}}$	$z_c = \frac{(\bar{x}_1 - \bar{x}_2) - (\mu_1 - \mu_2)}{\sigma_{\bar{x}_1 - \bar{x}_2}}$	Known	$(\bar{x}_1 - \bar{x}_2) \pm z\,\sigma_{\bar{x}_1 - \bar{x}_2}$
		or $H_0: \mu_1 = \mu_2$; $H_1: \mu_1 \neq \mu_2$		$s_{\bar{x}_1 - \bar{x}_2} = \sqrt{\frac{s_1^2}{n_1} + \frac{s_2^2}{n_2}}$	$z_c = \frac{(\bar{x}_1 - \bar{x}_2) - (\mu_1 - \mu_2)}{s_{\bar{x}_1 - \bar{x}_2}}$	Unknown	$(\bar{x}_1 - \bar{x}_2) \pm z\,s_{\bar{x}_1 - \bar{x}_2}$
	PROPORTIONS **One sample**	$H_0: \pi = .20$; $H_1: \pi \neq .20$	p	$\sigma_p = \sqrt{\frac{\pi_H(1 - \pi_H)}{n}}$	(no test)	Known or (assumed)	(no estimate)
		or $\pi < .20$; $\pi > .20$		$s_p = \sqrt{\frac{p\,q}{n}}$	$z_c = \frac{p - \pi_H}{\sigma_p}$	Unknown	$p \pm z\,s_p$
	Two samples				(no test)	Known	(no estimate)
		$H_0: \pi_1 - \pi_2 = 0$; $H_1: \pi_1 - \pi_2 \neq 0$	$p_1 - p_2$	$s_{\hat{p}_1 - \hat{p}_2} = \sqrt{\frac{\hat{p}\,\hat{q}}{n_1} + \frac{\hat{p}\,\hat{q}}{n_2}}$ where $\hat{p} = \frac{x_1 + x_2}{n_1 + n_2}$	$z_c = \frac{(p_1 - p_2) - (\pi_1 - \pi_2)}{s_{\hat{p}_1 - \hat{p}_2}}$	Unknown	$(p_1 - p_2) \pm z\,s_{p_1 - p_2}$ where $s_{p_1 - p_2} = \sqrt{\frac{p_1\,q_1}{n_1} + \frac{p_2\,q_2}{n_2}}$
Small Sample $n < 30$	**MEAN** **One sample**	$H_0: \mu = 70$; $H_1: \mu < 70$	$\bar{x}$	$s_{\bar{x}} = \frac{s}{\sqrt{n}}$	$t_c = \frac{\bar{x} - \mu}{s_{\bar{x}}}$	Unknown	$\bar{x} \pm t\,s_{\bar{x}}$
	Two samples	$H_0: \mu_1 - \mu_2 = 0$; $H_1: \mu_1 - \mu_2 \neq 0$	$\bar{x}_1 - \bar{x}_2$	$s_{\overline{pl}} = \sqrt{\frac{s_{pl}^2}{n_1} + \frac{s_{pl}^2}{n_2}}$ where $s_{pl}^2 = \frac{s_1^2(n_1 - 1) + s_2^2(n_2 - 1)}{n_1 + n_2 - 2}$	$t_c = \frac{(\bar{x}_1 - \bar{x}_2) - (\mu_1 - \mu_2)}{s_{\overline{pl}}}$	Unknown	$(\bar{x}_1 - \bar{x}_2) \pm t\,s_{\overline{pl}}$
	PAIRED DIFFERENCE	$H_0: \mu_1 - \mu_2 = 0$; $H_1: \mu_1 - \mu_2 \neq 0$	$\bar{d}$	$s_{\bar{d}} = \frac{s_d}{\sqrt{n}}$	$t_c = \frac{\bar{d} - \mu_d}{s_{\bar{d}}}$	Unknown	$\bar{d} \pm t\,s_{\bar{d}}$

Second Edition

STATISTICS FOR BUSINESS

Second Edition

STATISTICS FOR BUSINESS

Joseph G. Monks
Gonzaga University

Byron L. Newton

SCIENCE RESEARCH ASSOCIATES, INC.
Chicago, Henley-on-Thames, Sydney, Toronto

An IBM Company

Acquisition Editor — Roger Ross
Development Editor — Molly Gardiner
Designer — Kirk George Panikis
Composition/Illustrations — Graphics West

Library of Congress Cataloging-in-Publication Data

Monks, Joseph G.
Statistics for business / Joseph G. Monks, Bryon L. Newton.
p. cm.
Includes index.
ISBN 0-574-19585-8
1. Social sciences—Statistical methods. 2. Commercial statistics. 3. Statistics. I. Newton, Byron L. II. Newton, Byron L. Statistics for business. III. Title.
HA29.M7465 1988
519.5′024658—dc19 87-31173
CIP

Printed in the United States of America.

10 9 8 7 6 5 4 3 2 1

Contents

PREFACE X

CHAPTER 1 Introduction 1

Introduction: The Role of Statistics 2
- Why Study Statistics? **2**
- Managerial Statistics **3**
- Statistical Terminology **7**
- Steps in a Statistical Study **10**
- Sources and Types of Business Data **12**
- Using a Computerized Database **15**
- **Summary 16**
- **CASE: Statistics in the Supermarket 17**
- **Questions and Problems 18**
- Appendix A: Corporate Statistical Database **20**

CHAPTER 2 Organizing and Displaying Data 24

Introduction 26
- Types of Data **26**
- Raw Data and Arrays **27**
- Statistical Tables **28**
- Frequency Distributions **30**
- Graphic Presentation of Data **36**
- Graphs for Other Types of Data **42**
- Using Microcomputers to Display Data **44**
- **Summary 46**
- **CASE: Midland Distribution Company 46**
- **Questions and Problems 49**

CHAPTER 3 Summary Measures of Data 54

Introduction 56
- Mathematical Notation **58**
- Measures of Central Tendency (Ungrouped Data) **61**
- Measures of Dispersion (Ungrouped Data) **71**
- Summary Measures of Grouped Data **77**
- Shapes of Distributions: Skewness and Kurtosis **82**
- Comparison Measures: The Coefficient of Variation **85**
- Finding Numerical Measures on the Computer **88**
- Summary **88**
- **CASE: Pfafner, Jeffrey and Jones Investment Company 91**
- **Questions and Problems 94**

CHAPTER 4 Probability: Concepts and Rules 100
Introduction 102
Meaning of Probability **102**
Set Theory and Venn Diagrams **105**
Marginal, Joint, and Conditional Probabilities **112**
Probability Rules **115**
Bayes' Theorem **124**
Counting Procedures **130**
Using Computers to Find Probabilities **137**
Summary 137
CASE: Alliance National Bank 138
Questions and Problems 140

CHAPTER 5 Probability Distributions 146
Introduction 148
The Binomial Distribution **155**
The Poisson Distribution **164**
The Normal Distribution **169**
Other Characteristics of Statistical Distributions **186**
Summary 189
CASE: Detroit Lakes Power Company 190
Questions and Problems 193

CHAPTER 6 Sampling and Sampling Distributions 196
Introduction 198
Using Samples to Learn about Populations **198**
Using a Random Number Table **202**
Judgment and Probability Sampling **204**
Sampling Distributions **209**
Sampling Distribution Theorems **214**
Mean and Standard Error of Sampling Distribution **216**
Finite Population Correction Factor **218**
Inference When σ is Unknown: The *t*-Distribution **221**
Standard Errors for Means and Proportions **225**
Summary 230
CASE: Debate at The City Council 231
Questions and Problems 232

CHAPTER 7 Estimation 234
Introduction: Interval Estimation 236
Confidence Interval Estimates of the Mean: Large Samples, σ Known **239**
Confidence Interval Estimates of the Mean: Small Samples, σ Unknown **240**
Finding the Sample Size for Estimating the Mean **243**

Confidence Interval Estimates of Proportion **246**
Finding the Sample Size for Estimating the Population Proportion **250**
Summary 252
CASE: Northwestern Grain Cooperative 253
Questions and Problems 255

CHAPTER 8 Testing Hypotheses: One-Sample Tests 258
Introduction 260
Null and Alternate Hypotheses **264**
Types of Errors **266**
Level of Significance **268**
Two-Tailed Versus One-Tailed Tests **271**
A Standardized Testing Procedure **275**
Testing the Population Mean: σ Known **277**
Testing the Population Mean: σ Unknown **280**
Testing Population Proportion **284**
Computing the Type II Error **290**
Power Curves and Operating Characteristic Curves **294**
Summary 299
CASE: International Printing and Publishing Company 300
Questions and Problems 302

CHAPTER 9 Testing Hypotheses: Two-Sample Tests 306
Introduction 308
Testing the Difference Between Means from Two Independent Samples **312**
Testing the Difference Between Means of Two Paired Samples **317**
Testing the Difference Between Proportions from Two Independent Samples **324**
Decision Making Rules and Sample Size for Quality Control **329**
Summary 332
CASE: Transcontinental Communications Company 336
Questions and Problems 338

CHAPTER 10 Chi-Square and Analysis of Variance 342
Introduction 344
The Chi-Square Distribution **344**
Tests of Goodness of Fit **346**
Tests for Independence of Classification **354**
Additional Applications of Chi-Square **357**
The *F*-Distribution **360**
One-Way Analysis of Variance **363**
Using the ANOVA Table **367**
Two-Way ANOVA Tests **368**

Summary **371**
CASE: Aerospace Industries **373**
Questions and Problems **375**

CHAPTER 11 **Regression and Correlation Analysis** **380**
Introduction: What is Regression and Correlation? **382**
Finding the Linear Regression Equation **386**
Using the Regression Model for Inference **393**
Simple Linear Correlation **406**
Summary **415**
CASE: Kathy DuBois, Legislative Assistant **416**
Questions and Problems **418**

CHAPTER 12 **Multiple Regression and Correlation** **422**
Introduction: Multiple Regression and Correlation **424**
Finding the Multiple Linear Regression Equation **427**
Using the Multiple Regression Model for Inference **433**
The Coefficient of Multiple Determination **440**
Computer Approaches to Multivariate Analysis **449**
Summary **451**
CASE: Forecasting at Sun Belt Equipment Company **452**
Questions and Problems **454**

CHAPTER 13 **Time Series, Index Numbers, and Forecasting** **458**
Introduction: Time Series and Index Numbers **460**
Classical Components of a Time Series **461**
Trend Estimation **463**
Seasonal, Cyclical, and Irregular Components **471**
Index Numbers: Purpose and Types **486**
Simple Price, Quantity, and Value Indexes **488**
Composite Index Numbers **490**
Selecting and Changing the Base Period **495**
Using Indexes in Business **496**
Forecasting Methods **501**
Summary **503**
CASE: Winter Coats Inventory at Fashionworld **505**
Questions and Problems **508**

CHAPTER 14 **Nonparametric Statistical Methods** **514**
Introduction **516**
Runs Test for Randomness **517**
Sign Tests for Central Tendency and Differences **520**
Rank Sum Test of Means **527**

Rank Correlation **533**
Summary 538
CASE: Globe Personnel Services—We'll Find You a Job! 539
Questions and Problems 541

APPENDIXES 545
A Corporate Statistical Data Base **546**
B Areas Under the Normal Curve **550**
C The Binomial Distribution **552**
D The Poisson Distribution **559**
E Student's *t*-Distribution **565**
F The Chi-Square Distribution **566**
G The *F*-Distribution **568**
H Answers to Odd Numbered Problems **574**

INDEX 599

Preface To Instructors

We have tried to offer a theoretically sound introduction to statistics that is appropriate for either the undergraduate or introductory MBA level. We hope you'll find the writing style clear and concise. Each chapter begins with learning objectives and an illustration or business situation that introduces the topic. Within the chapters, abundant figures, charts, and drawings have been designed to enhance the learning process. Key concepts and definitions are highlighted by boxes, examples are clearly set out, and exercises and problems of varying difficulty are provided. Most answers to the odd-numbered problems (Appendix H) offer limited guidance toward the solution, in addition to the final answer.

We have included two features not always found in texts of this size: (1) *computer-based problems* and (2) *cases*. Both computer materials and cases are presented in a way that enables you to include or omit them without any loss of continuity. They are truly supplemental materials designed to give you more flexibility in structuring your course

Computer problems The computer-based problems appear at the end of the Questions and Problems section of most chapters and can be assigned or skipped depending upon whether time and facilities are available for computer work. They are not software-specific, so almost any (generic) statistical software program for a mainframe or microcomputer can be used to solve them. The problems use a database included in Appendix A, which contains statistics such as sales, earnings, and employment for 100 firms.

Cases Each chapter concludes with a case study that includes discussion questions. The cases reinforce chapter materials and require deeper thought than ordinary problems. However, they do not introduce new statistical concepts, so their use is also at your discretion.

In addition to the computer problems and cases, some other materials are noted as optional. These allow you to extend coverage beyond what is generally included in a more basic course.

Acknowledgments

Our grateful appreciation must be expressed to the reviewers who contributed so significantly to this work. Our thanks to Larry Claypool of Oklahoma State University, Richard M. Smith of Byrant College, Thomas

Wedel of California State University, Northridge, Ron Dattero of Texas A&M University, Bulent Uyar of University of North Dakota, David Krueger of St. Cloud State University, Roger C. Gledhill of Eastern Michigan State University, Robert E. Meier of Eastern Illinois University, Jeff Mock of Diablo Valley College, Kris Moore of Baylor University, and Bernard "Chuck" W. Taylor III of Virginia Polytechnic Institute and State University. Special thanks to Mildred Massey of California State University, Los Angeles, who so meticulously scrutinized the theory and to Lynda Borucki at Northwestern University for her thorough review of the exercises.

The team at Science Research Associates has been exceptionally cooperative and helpful, and their suggestions, assistance, and encouragement are greatly appreciated. Phil Gerould, Dave McEttrick, and John Levstik were all instrumental in structuring the project from the start. Roger Ross provided the real professional management needed throughout the project, and Molly Gardiner handled the multitude of day-to-day editorial decisions in a most organized and skillful manner. Others at SRA who helped in a special way include Timothy Taylor in production and Steve Leonardo in design.

Many thanks also to Dean Bud Barnes and to Terry Coombes for facilitating the project at Gonzaga, to our students for class-testing the material, and to Lorie Allen for her work on the manuscript.

I am grateful to the Literary Executor of the late Sir Ronald A. Fisher, F.R.S., to Dr. Frank Yates, F.R.S., and the Longman Group Ltd, London, for permission to reprint Table III and Table IV from their book *Statistical Tables for Biological, Agricultural, and Medical Research* (6th Edition, 1974).

Finally, we would like to express our thanks to our wives and families who were so patient and understood our need to spend countless hours working on "the book." To them our efforts are dedicated.

Joseph G. Monks
Byron L. Newton

Preface To Students

Statistics is a fascinating subject—but it can be somewhat difficult. The purpose of this book is to increase your interest in statistics, while at the same time easing the learning process by providing you with a clear, organized, and intuitively understandable presentation of the material. As a first objective, we'd like you to appreciate the tremendous role of statistics in the analysis of business problems. Beyond that, our aim is to help you develop your analytical skills to a point where you will feel confident using statistical methods in decision making situations. Thousands of managers already use a broad spectrum of statistical methods every day. These same statistical skills could be useful to you as well.

Having each taught statistics for many years, we recognize the difficulties of a formal mathematical approach to the subject. We believe that you can learn the essential statistical concepts without embroiling yourself in mathematical notation and calculus formulas aimed at proving statistical theorems. Ours is a more applied orientation. Our approach has been to make this text as clear and understandable to beginning students as possible—not to impress those who want proofs or already understand the material. Mathematical symbols are used where necessary but are omitted when a concept can be presented verbally or schematically with equal clarity and conciseness. If you have a reasonable understanding of algebra, you should have no difficulty with the mathematics of this text.

However, statistical methods require analytical thought patterns best developed through consistent practice over time. You may have a fine instructor and a good book, but that is not enough. Nor will sporadic study followed by long periods of rest do the trick. Independent problem solving on a regular basis is essential. It will help determine whether you understand the concepts well enough to use them in situations that are slightly different from the examples given in class.

We have found that one successful way of learning this material is to (1) read the assigned materials before class, (2) take notes of key points from class, (3) do a good selection of (assigned) problems, and (4) summarize all your study notes in condensed form. The last step may be the most important. Your summary should be a well-organized condensation of the most important points from each chapter; it should include the items especially emphasized by your instructor, as well as the cogent points from the homework (not all the problems—just the key ideas that you struggled with or that turned you on to the correct solution). Scan the existing materials for three or four minutes before you add to this summary sheet each week. After a while, the *rules of probability* and other items on your sheet will become an intellectual part of you. This systematic and repeated review can help you assimilate the material and perform well on exams. And for those exams—good luck!

Joseph G. Monks
Byron L. Newton

CHAPTER 1

Introduction to Statistics

INTRODUCTION: THE ROLE OF STATISTICS

WHY STUDY STATISTICS?

MANAGERIAL STATISTICS
- Descriptive Statistics
- Inferential Statistics
- Statistics and Mathematics

STATISTICAL TERMINOLOGY
- Census and Sample Data
- Point and Interval Estimates
- Parameters and Statistics
- Sampling Error
- Bias and Other Error

STEPS IN A STATISTICAL STUDY

SOURCES AND TYPES OF BUSINESS DATA
- Internal Data
- External Data
- Surveys and Field Studies

USING A COMPUTERIZED DATABASE

SUMMARY

CASE: STATISTICS IN THE SUPERMARKET

QUESTIONS AND PROBLEMS

APPENDIX A: CORPORATE STATISTICAL DATABASE

Chapter 1 Objectives

1. Define the terms statistics, population, sample, and estimate.
2. Distinguish descriptive vs inferential statistics.
3. Explain the concept of sampling error.
4. Describe the steps in a statistical study.
5. Explain the role of data in statistics.

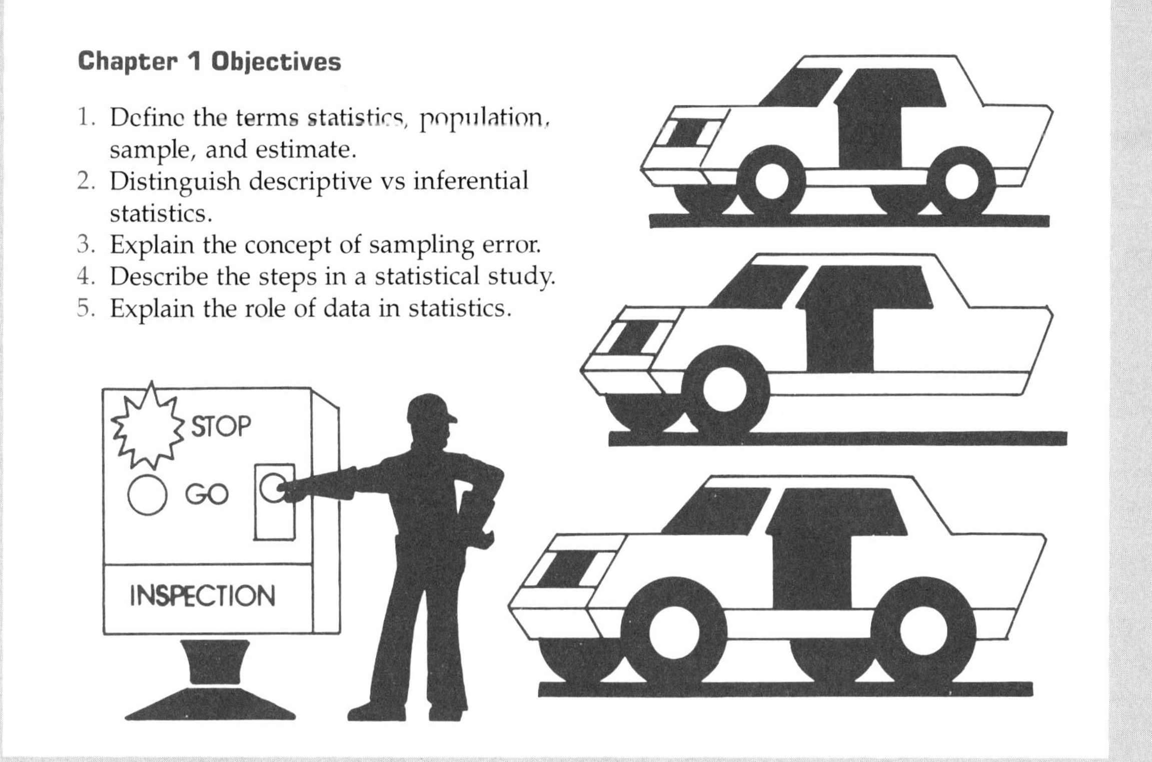

In the middle of the Toyota engine plant in Japan is a large scoreboard-like sign. If a worker notices a defect in one of the auto parts coming along on the assembly line, the worker pushes a button and stops the entire assembly line. The sign then identifies which worker is responsible for the defect. When this happens, everyone's productivity drops to zero. However, one third of each worker's income is a bonus based upon the worker's productivity. So defects don't occur very often in Toyota engines!

Welcome to statistics—which plays a vital role at Toyota and at thousands of other firms throughout the world. Statistics can be a challenging subject, but it is also one of the most useful topics you will encounter in a business curriculum. We'll try to help you feel comfortable with it by avoiding unnecessary mathematical complexity, while at the same time presenting the theory and applications in a way that will convince you of its enormous value to managers.

In this first chapter we will verify the need for statistics and introduce some essential terminology. Then, after reviewing the steps involved in a statistical study, we identify some sources and types of statistical data. The end of the chapter describes a data set (i.e., database) which we will use for both manual and computerized exercises throughout the text. When you finish this chapter, you should be able to define statistics and respond to the learning objectives mentioned above.

INTRODUCTION: THE ROLE OF STATISTICS

Toyota cars are only one of many successful Japanese products. An individual largely responsible for the quality level and competitiveness of Japanese products is Dr. W. Edwards Deming. His task, as a representative of the U.S. Government, was to help rebuild the Japanese economy following World War II.

Early in his assignment, Dr. Deming observed that Japan, with its skilled managers, and proficient engineers, mathematicians, and statisticians, should be able to build a reputation for quality. He concluded that statistical methods could help. In fact, he felt that it would be impossible to raise quality to a sufficiently high level without using statistical methods on a broad scale.

> I predicted, at an assembly of Japanese manufacturers in Tokyo, in July 1950 that in five years, manufacturers in other industrial nations would be on the defensive, and that in ten years the reputation for top quality in Japanese products would be firmly established the world over. Statistical techniques became a living, vital, and essential force in all stages of Japanese industry.[1]

American car manufacturers, steel companies, and electronics firms know only too well how the Japanese manufacturers met the predicted timetable. Dr. Deming started a *statistical revolution* in Japan. With statistical techniques workers were able to detect variation in quality characteristics and test for the presence (and responsibility) of special causes. They applied the fundamentals of probability to design, process control, and reliability problems. That was followed by the practical use of estimation, tests of hypotheses, and other topics we shall study in this book. Dr. Deming's philosophy is concisely summarized in the admonition to top management contained in his book, *Quality/Productivity: A Competitive Position,* "Make maximum use of statistical knowledge and talent in your company."

WHY STUDY STATISTICS?

As a student of business, statistics is as essential to you as it was (and still is) to the Japanese—maybe more so. It has become a professional language of modern business. Suppose your manager suggests you "do a simple regression of advertising on sales" to justify your budget request for next year. Wouldn't it be nice to know what the term "regression" means before he sits

[1] W. Edwards Deming, "What Happened in Japan?" **Industrial Quality Control,** based upon an article in SANKHYA (Calcutta), series B, vol.28: 1966.

down beside you at your computer terminal? (Don't worry—we'll delve into computer outputs too!)

Another reason for studying statistics is to learn how to accurately describe factual data. Statistical information is vital to nearly every facet of business operations. It describes personnel records, productivity, financial conditions, supplier performance, production schedules, inventory levels, customer characteristics, and more. But statistics can also deceive us, so it must be handled carefully to avoid disastrous results.

The volume of data is, however, increasing at an exponential rate. It is already so extensive that we use more than ten million computers to help keep track of data. Businesses have invested heavily in computers because they can assemble, manipulate, and describe data so fast and efficiently. Statistical methods, used in conjunction with these computers, help condense and summarize these data to bring them into a form that managers can comprehend and use.

A final reason for studying statistics relates to the decisions you may have to make as a practicing manager. Statistical methodologies help managers identify problems, make comparisons and tests, and decide what action to take. After decisions are implemented, statistical analysis is also used to evaluate results.

In summary, these are some important benefits from statistics:

- **Statistics is the factual language of business** and is required to communicate within business.
- **Statistics are used to describe data** which relate to the status and relationships that exist in all aspects of operations.
- **Statistical methodologies support conclusions, inferences, and decision making** processes, which are the prime responsibility of management.

Organizations (and people) that make good use of statistical methods are likely to be more successful than those that do not. It's as simple as that!

MANAGERIAL STATISTICS

Every organization needs managers to plan and control its operations. Managers must make the decisions and take the actions necessary to direct the day-to-day activities toward the organization's goals.

Years ago, management was largely a subjective art. Today, in the age of computers, we speak of *management as a science*. This is primarily because

(a) organized principles of management now exist, (b) objective (business) data can be obtained, (c) mathematical and statistical methods can be used to analyze the data, and (d) the results are repeatable.

As with other sciences, experimentation in management is now possible. This is largely because business situations can be *simulated* on computers through the use of mathematical and statistical models. For example, a computer model was used to simulate the (underwater) phone line capacity connecting Seattle with an island in Puget Sound. Different levels of projected growth in demand for the next ten years were then "tried out" on the model to help determine when more equipment would be needed.

Managerial success today depends heavily upon the availability of valid, timely, and relevant information. However, each manager has unique requirements. A marketing manager may want to analyze charts of sales by product line, whereas a quality control inspector may need to estimate the number of defective units in an outgoing shipment. A good information system can supply much of the needed data. Statistical theory and procedures help ensure that these data are in the proper form and are most effectively used.

Statistics is the body of methods for the collection, analysis, presentation, and interpretation of quantitative data, and for the use of such data.

In statistical analysis, a **population** is the whole collection of data under study, whereas a **sample** is any part of the population. (Note, however, that a *good* sample must be representative of the population.) The term *statistics* may refer to either of two accepted methods of presenting or analyzing data: (1) descriptive and (2) inferential.

DESCRIPTIVE STATISTICS

Descriptive statistics are concerned with describing data by collecting, classifying, summarizing, and displaying either population or sample data. In Chapter 2 we will see how data can be summarized by charts and graphs, and in Chapter 3 we turn to numerical measures of the central tendency and the amount of spread in the data. For example, suppose a personnel manager of an airline company is faced with negotiating a new wage contract for various workers of the firm. Charts and summary measures of the current wages paid to pilots, mechanics, stewardesses, and other employee groups would give the negotiators a composite overview, or summary, of wage levels and differences among workers. It's true—a picture is often worth a thousand words!

INFERENTIAL STATISTICS

Statistical inference takes one of two forms: (1) deduction or (2) induction. Figure 1–1 illustrates the difference.

FIGURE 1–1 Statistical inferential procedures

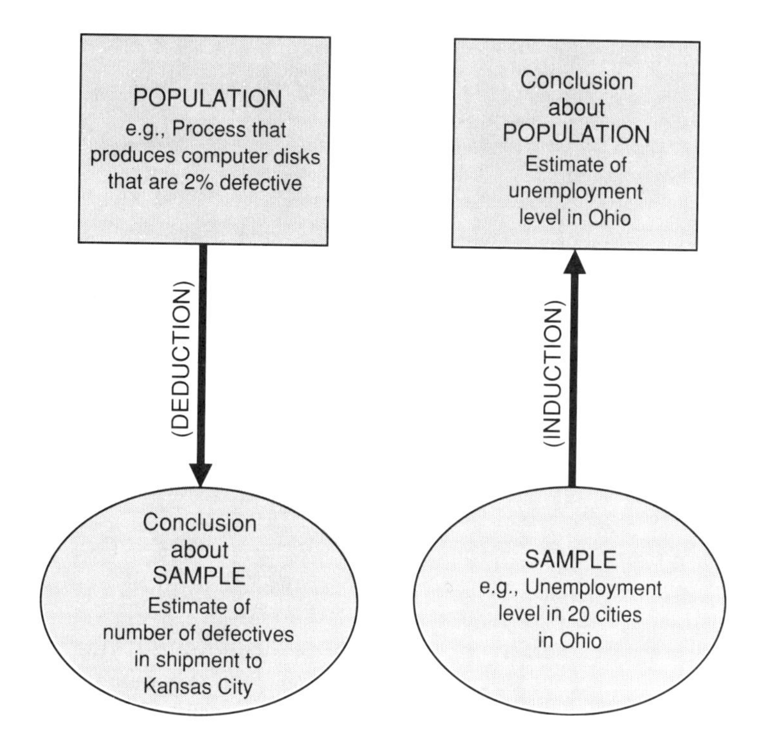

Deduction is the logical process of using general knowledge of a population to draw a conclusion about a specific element or sample of that population. As we will see in Chapter 4, some probability statements use deductive logic. For example, assume a manufacturing process (i.e., our population) produces computer disks, and a small proportion of them are defective. If we knew the defective rate was 2%, we could estimate (deduce) the likelihood of having one or more defects in a box of 50 disks sent to Kansas City.

Induction is the logical process of using specific knowledge of a sample of a population to infer some generalization about the population. We will use inference procedures extensively in Chapters 7, 8, and 9 to estimate population values and test whether assumptions about populations are supported by sample data. For example, a bank economist may use a sample of the unemployment levels in 20 cities to estimate the statewide unemployment level in Ohio.

STATISTICS AND MATHEMATICS

Statistics and mathematics obviously have some similarities. Both deal with quantitative data and equations—some of which can get quite complex. However, statistics differs from mathematics in the type of data being analyzed and the models or equations used in that analysis.

FIGURE 1–2 Continuum of data

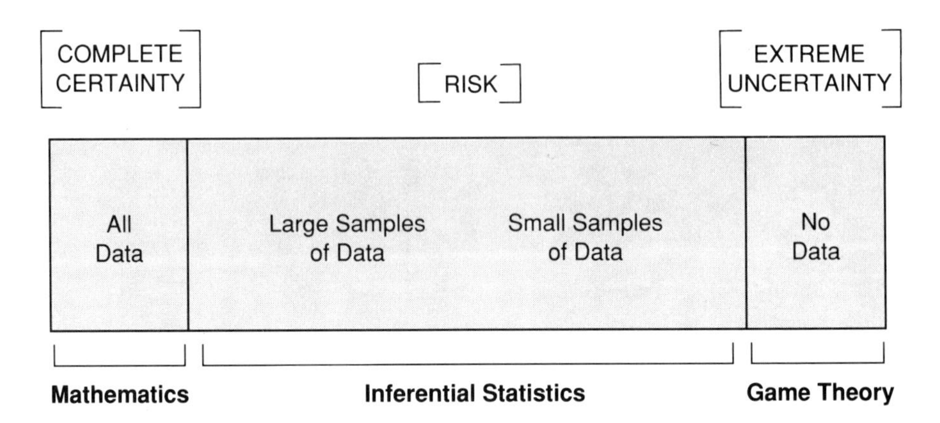

Mathematics is a precise science whose equations yield predictable results. As suggested in Figure 1–2, mathematics lies on the "complete certainty" end of the data continuum where there is no question about the accuracy or completeness of the data being analyzed. If the data are questionable or incomplete, we simply *assume* they are correct, and proceed as if they were. For example, accountants may have to assume production costs are accurate in order to complete the calculations of profit or loss for a firm's annual report. Once all questions about the validity of the data are dispensed with (or assumed away), the methods of algebra, calculus, linear programming, and other mathematical techniques yield well-defined, deterministic results.

Game theory lies on the "extreme uncertainty" end of the continuum—where we have *no information* about what conditions prevail. Although game theory has been the subject of considerable academic inquiry, it is not a commonly used approach to decision making and is outside the scope of our text.

Statistical techniques apply to situations where the database is uncertain or incomplete. As illustrated in Figure 1–2, we must rely upon sample information in this case—and large samples give us more certainty than small ones. But they also cost more! So, much of our analysis involves balancing the benefits of more information against the costs of larger samples.

STATISTICAL TERMINOLOGY

Suppose you are a management trainee for a grocery distribution firm. Your manager has just learned that a competitive warehouse of canned goods is being auctioned off next week by a bank because the competitor is bankrupt. He has asked you to review the data and recommend whether your firm should bid on the warehouse goods. The advertisement says there are 420,810 cases of food items in the warehouse, but it doesn't specifically say what they are and you don't know their total value. However, the bank conducting the sale has used invoices to collect a representative sample of 50 cases of goods and placed them in a truck for public inspection. The contents include an assortment of foods such as peaches, canned milk, dog food, olives, and much more. You price the truckload of 50 items and estimate their value at $385. Figure 1–3 illustrates the situation.

FIGURE 1–3 Population and sample data

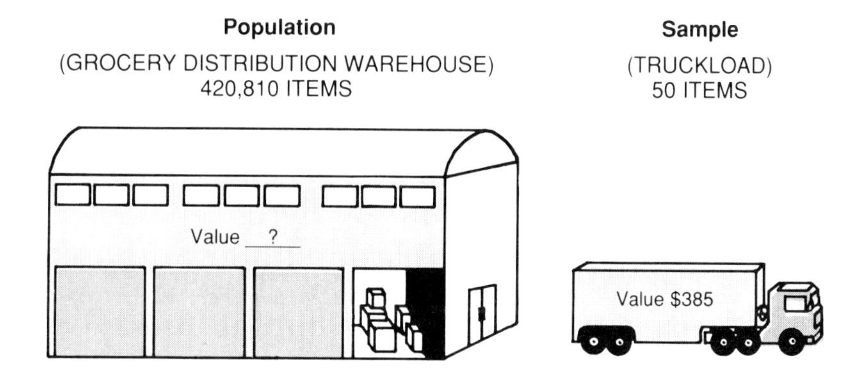

CENSUS DATA AND SAMPLE DATA

A **census** is a survey or collection of data on every element in the statistical universe, whereas a **sample** describes only part of the population. In Figure 1–3, you would have to take a census to determine the precise value of the 420,810 items in the warehouse. However, that information is not available, so the sample data will have to be used to make the bid decision. You need an *estimate* of the population value.

POINT AND INTERVAL ESTIMATES

A **point estimate** is a single value, obtained from a sample, that is used to estimate some characteristic of a population.

Example 1–1 Make a point estimate of the total value of the groceries in the warehouse of Figure 1–3.

Solution

Assuming the 50 items in the sample are representative of all items in the warehouse:

From Sample

$$\text{Estimated unit value} = \$385.00 \div 50 \text{ items} = \$7.70/\text{item}$$

For Population

$$\text{Estimated total value} = (\$7.70/\text{item})\,(420{,}810 \text{ items}) = \$3{,}240{,}237$$

This example uses a form of induction. We reason from a knowledge of the sample (truckload) to a point estimate of the population value. However, point estimates, though close, are almost always incorrect. Statisticians frequently use intervals to express their estimates.

Interval estimates show the bounds within which the point estimate is believed to lie. For example, an interval estimate around the $7.70 sample average might be from $7.40 to $8.00 per item. This would yield an estimated total value in the range from $3,113,994 to $3,366,480. Using statistical theory (from Chapter 7), you could also state the likelihood, or probability, that this range really does include the true but unknown population value.

PARAMETERS AND STATISTICS

A **population parameter** is a characteristic of a population, whereas a **sample statistic** is a characteristic of a sample. Parameters are usually designated by Greek symbols, while we use the standard English symbols for statistics. In Figure 1–3, the population parameter is unknown (which is often the case), but one sample statistic (i.e., its mean) is known to be $7.70/item. We designate the sample mean as *X-bar,* so we would say $\overline{X} = \$7.70$.

The word "statistic" has another meaning, however. People often refer to any single item of quantitative data as *a statistic,* and to collections of such items as *statistics*. We could refer to the age of an individual truckdriver as a statistic, although some statisticians might prefer to reserve that term for the average age (the $\overline{X}$) of a sample of truckdrivers.

SAMPLING ERROR

Firms apply statistical sampling to everything from inspecting raw materials to auditing accounts or assessing customer satisfaction with their products. In each case, only part of the population is inspected or measured. But that limited information is used to make some decision about the whole population, such as whether the warehouse of groceries is worth $3,240,237. There is some risk of making a mistake because the decision must be made on the basis of only partial information. In Figure 1–1, the bank economist estimating statewide unemployment might happen to get several cities with unusually high unemployment in her sample of 20 cities. She then might estimate state unemployment at a higher rate than it really is.

Sampling error is the deviation that exists whenever a sample statistic and population parameter differ because of the *chance* failure of a sample to perfectly represent the population from which it is taken.

Sampling error is due to the "luck of the draw," and is characteristic of statistical analysis. So we expect it and account for it in our analysis. In fact, the analysis of error is a trademark of inferential statistics. If it were not for the analysis of sampling error, our text would end after Chapter 3 (but unfortunately, you must face a few more chapters than that!).

BIAS AND OTHER ERROR

BIAS Statistical investigations do, however, sometimes contain errors that are not due to sampling. These errors arise from unrecognized or systematic causes, such as unskilled interviewers or inaccurate measuring devices. For example, if the sample of grocery products in Figure 1–3 was taken from only one type of canned goods (e.g., peaches and other canned fruits), or from easy-to-load items, or from "impulse" items, the sample would be *biased*. Biased sample statistics do not accurately represent their population values because they are consistently off the mark, much like a speedometer that always registers too low. If you don't know about the bias, you may be in for a surprise, but if you do know the amount (and direction) of bias, you can sometimes make a correction for it.

Systematic error can also bias results. For example, suppose a company has 10 molding machines making parts that are fed onto an assembly line at regular intervals. If a quality control inspector samples every 10th item (or multiples of 10), he may (unwittingly) be getting all the observations from one machine.

OTHER ERROR Errors can also arise from administrative procedures associated with taking, recording, or reporting data. However, these errors can occur in both census and sample data. In fact, sample statistics may be more accurate than census data, if large masses of data have to be handled. On the other hand, "accuracy" is not the only criterion of a good sample. Thus a sample of two items can be very "accurate," but may be too small to provide a reliable estimate of the population parameter. And, although a sample of 1,000 may have some individual inaccuracies, the decision maker will probably feel more confident about using it as a representative of the population.

STEPS IN A STATISTICAL STUDY

Statistical studies are one of the more "scientific" and systematic activities of business. We find them used more frequently in large, rather than small, organizations. Although they are not always conducted in the same way, these studies generally encompass the steps shown in the flowchart of Figure 1–4.

Careful *identification of the problem* ensures that the question or hypothesis is clearly stated in a form that allows us to draw valid conclusions. Care must be taken to abstract the relevant variables from the problem environment, and address the *real* question (rather than a *symptom* of the problem). The question is often stated in such a way that the results of the study clearly support (or fail to support) some decision to be made. Expected costs and benefits of conducting the study may also be relevant considerations in this initial step.

The *criteria for solution* are the standards by which the results will be judged. They include the type of statistical test to be performed, computational procedures to follow, and the amount of precision expected in the result. These are considerations we shall take up throughout the text. We shall also be concerned with maintaining a control over the amount and type of data to be used in the analysis.

The *data collection* phase refers to all activities related to acquisition of data to solve the problem. This may include an investigation of related studies to determine what has already been done in the problem area. The next section of this chapter addresses the sources and types of business data. Data may come from internal company sources, such as from accounting or operating records, or from external reference books, periodicals, trade associations, and government agencies.

Extreme care must be taken when using any data from other sources that do not explain the terms used or how the data were collected, classified, and recorded. For example, if a university enrollment is reported to be 6,500 students, does this mean 6,500 full-time students, or is it the equivalent of 6,500 full-time students? How many credit hours constitute a full-time student? Are executives who attend a two-week conference counted as students? Is the number exactly 6,500, or has it been rounded to the nearest hundred?

Analysis and testing of the data is the computational phase, and is often done with the aid of a computer. Both descriptive and inferential procedures are used to discover the meaning of the data and explore significant relationships.

Conclusions may be stated verbally or in tabular, graphic, or numerical form. Chapter 2 deals with tables and graphic presentations of data, and Chapter 3 takes up numerical summaries of data. The conclusions should be qualified by any assumptions or limitations that could influence the findings. This last step can also include recommended actions and guidelines for making appropriate decisions.

FIGURE 1–4 Steps in a statistical investigation

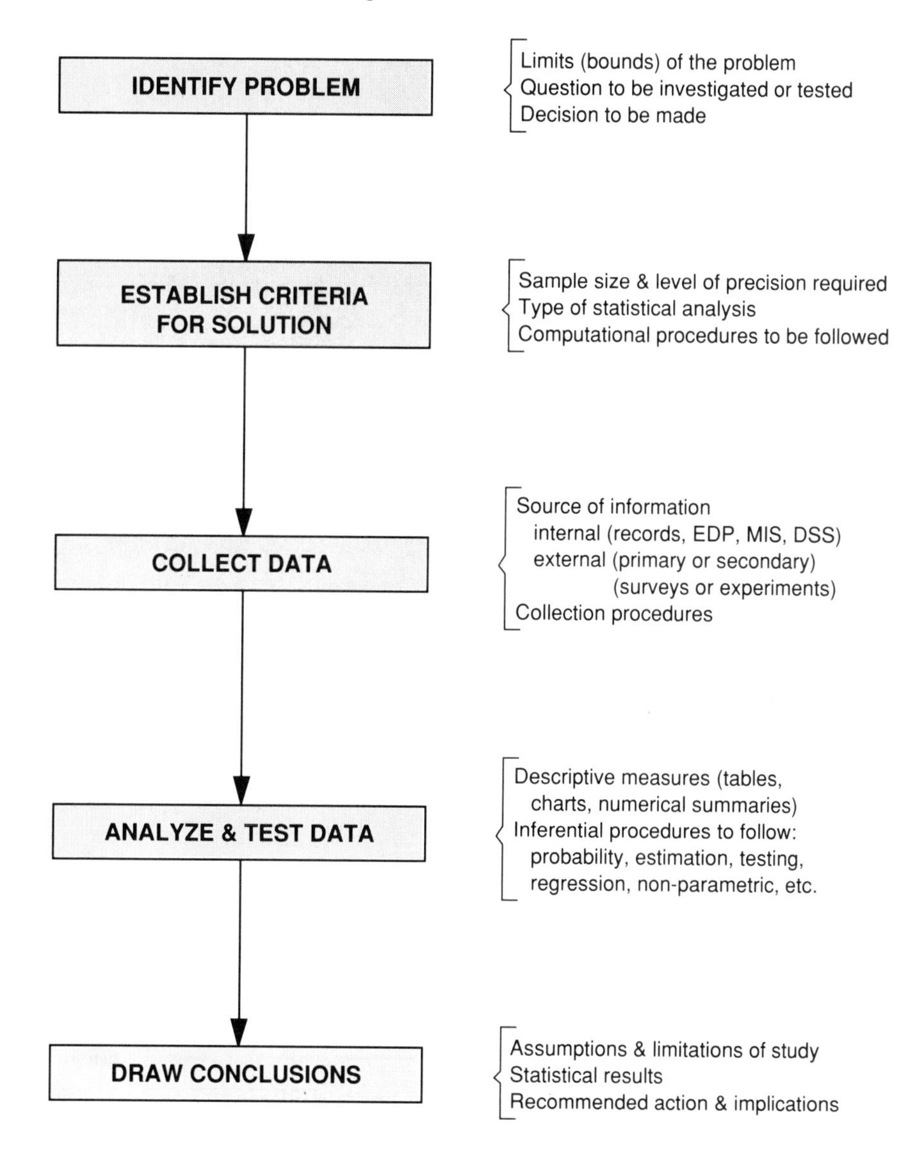

SOURCES AND TYPES OF BUSINESS DATA

A convenient classification is to distinguish between data sources that are (a) *internal* or (b) *external* to the organization.

INTERNAL DATA

Internally, the firm's own records and data files are usually its most useful source of information. For example, the grocery warehouse of Figure 1–3 would likely have invoices and inventory records that show the types and quantities of items on hand, as well as costs, selling prices, customer lists, sales volumes, and inventory turnover rates. Other internal data might include personnel records and salary and budget information, as well as market research and other studies.

Today, many organizations also have well-designed *management information systems* or *decision support systems* that can supply a variety of statistical data about past and present operations.

A **Management Information System (MIS)** is a computer-based set of data and procedures used in common by different departments of an organization to provide information for planning, controlling, and evaluating its activities.

A **Decision Support System (DSS)** is also a computer-based system with data and procedures, except it is designed to facilitate less-structured (as well as more specific) decision processes by incorporating appropriate analytical methodology.

EXTERNAL DATA

External data come from other organizations or from studies conducted outside the organization using the data. They are sometimes classified as either (1) *primary data* or (2) *secondary data*.

Primary data are data obtained from the organization that collected and published the data. Such data are typically related to a specific study or research project. **Secondary data** are data from a (secondary) source that did not originally collect them. Primary sources are preferred because they often contain more complete tables, explanations of how the data were collected, and definitions of terms, as well as uses and limitations of the data. Secondary data are often more readily available, but they are more likely to have been incorrectly transcribed or have qualifying footnotes omitted.

Some popular sources of data are (a) from national and regional *government agencies* and (b) from *trade and periodical publications*.

GOVERNMENTAL SOURCES

The Statistical Abstract of the United States, published annually by the U.S. Bureau of the Census.

Survey of Current Business, published monthly by the U. S. Bureau of the Census, Office of Business Economics.

Federal Reserve Bulletin, published monthly by the Board of Governors of the Federal Reserve System.

Monthly Labor Review, published monthly by the U. S. Department of Labor.

Bureau of the Census Catalog, published quarterly by the U. S. Bureau of the Census.

The U.S. Bureau of the Census also publishes a useful *County and City Data Book.* It contains information on income, housing, labor, banking, vital statistics, population, and other facts.

TRADE AND PERIODICAL DATA Trade associations publish data on lumber, steel, automotive, insurance, savings and loan, and numerous other industries. Most reference libraries have such references as the *Business Conditions Digest, Federal Reserve Bulletin,* and the *Life Insurance Fact Book.* In addition, current information of a statistical nature can be obtained from *The Wall Street Journal, U.S.A. Today, New York Times, Newsweek, Time, The Economist, U.S. News and World Report, Fortune, Business Week,* and other publications. A number of guides and indexes are also useful references although they contain no data themselves. Examples of these include the *Business Periodical Index* and *The Wall Street Journal Index.*

SURVEYS AND FIELD STUDIES

Surveys and field studies are questioning and investigative processes that yield data about the status or elements of a population. They are concerned primarily with observation and opinion as it exists in the subject environment. This differs from the more carefully controlled statistical experiments. In *statistically controlled experiments,* control is exercised over factors that affect the variable of interest. For example, in a taste test to discover preference among different brands of soft drinks, all brand identifying marks would be carefully hidden. Such controls are not always used in surveys and field studies.

Figure 1–5 illustrates three of the most common sources of data from surveys and field studies. These include (1) *observation,* (2) *personal interviews,* and (c) *mail questionnaires.*

Observation is the personal witnessing of an ongoing activity, such as a furniture assembly operation or an advertising campaign. Care must be taken to ensure the observer is objective and that the subjects' behavior typifies their normal operations.

FIGURE 1–5 Data sources from surveys and field studies

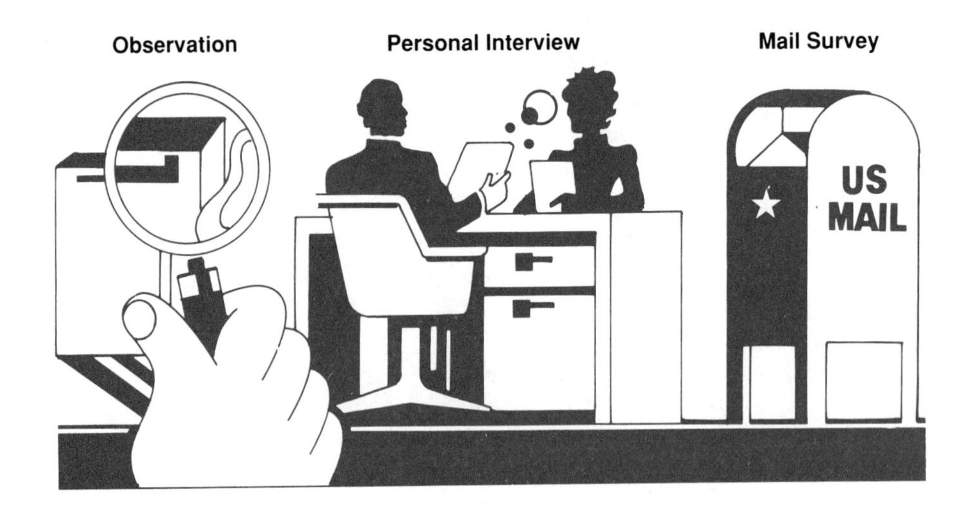

Personal interviews entail having a trained interviewer ask prescribed questions of a respondent and record the response on a standard form. The response rate is usually higher than the response from mail questionnaires. In addition, more detailed information can be obtained because questions can be explained and clarified as needed. But personal interviews are expensive, sometimes costing as much as $40 or $50 per respondent. Phone interviews are a less-expensive and widely used method of obtaining data. However, the credibility of phone interviews has been jeopardized by two factors: (1) some firms claim to be legitimate surveys when they are really just disguised sales campaigns, and (2) some organizations try to use prerecorded (computerized) messages that are not really all that "personal."

Mail questionnaires are one of the chief means of collecting external statistical data. Great care must be exercised designing mail questionnaires to avoid misunderstandings, to eliminate questions that can yield biased results, and to ensure that the needed data can be obtained and tabulated. Inexperienced analysts frequently mail a lengthy questionnaire to several hundred potential respondents, only to find that they have failed to ask for vital information (e.g., income level, product preferences, etc.) or have unusable responses. For example, suppose you are surveying the production managers of oil companies in an attempt to forecast the price of oil two years from now. Assume you have used the categories of:

- Under $10/barrel
- $10 to under $20/barrel
- $20 to under $30/barrel
- $30 to under $40/barrel
- $40/barrel or above

If 85% of your respondents checked the category of $40/barrel or above, you may still not have adequately narrowed down the price question. Maybe the majority feel it will be over $100/barrel! Open-ended intervals such as this are of limited usefulness because they do not allow the researcher to compute the principal measures used to summarize data. Pretesting the questionnaire and "dry runs" of the data analysis can help avoid some of these potentially disastrous problems.

Well-designed questions are generally phrased to elicit a short response that can be numerically coded for easy data processing on computer. Multiple-choice questions should be clear, reasonable, easy to answer, and contain sufficient alternatives to cover all possibilities. Open-ended questions often require special attention to avoid biasing the respondent or suggesting a desired response.

USING A COMPUTERIZED DATABASE

When computers were first introduced into businesses, they were used to automate well-established data processing procedures, such as wage and salary calculations. Different departments of the organization gradually automated their own records, e.g., marketing computerized their sales records, production put their inventory records on computer, etc. However, the corporate files were not linked together or widely accessible. Files were often duplicated in two or three departments, and they were not always consistent with each other.

Today, management information systems usually operate from a common database. That is, information is centrally stored in a location that is accurately maintained and is accessible via computer terminals throughout the organization. This means that the same current (on-line) information is available almost instantaneously to anyone in the organization who needs that data. Of course, not all data need be in the corporate information system.

We cannot duplicate a corporate database here, nor would it be worthwhile to do so. However, there is some value in using a data set to practice some of the statistical procedures we will be studying in this text. Therefore, we have included some abbreviated data in Appendix A. We will use these data for problems as we work through the chapters of the text.

Appendix A is a small (disguised and modified) data set, and most problems we will use it for can be solved manually. However, the data set can also be used to practice solving statistical problems on a computer if you have one available. Problems which lend themselves to solution via standard statistical computer programs are marked with a . Because our data set is relatively small, most of the problems can be done quite efficiently by hand or calculator. Your instructor may require manual calculations because they focus more attention on the logic of the solution.

So a computer solution to the problems should be considered optional, unless it is required by your instructor. In some cases, however, you may want to use the computer solution as a quick check of your manual calculations. The problems are stated in sufficiently general terms that commonly available statistical software packages should provide satisfactory solutions. Most of the problems can be solved either on mainframe computers (using packages like SAS, Minitab, or SPSS) or on IBM, Apple, or other microcomputers, depending upon what is available to you. Some solutions illustrated in the text were done on StatView, a microcomputer program available from BrainPower, Inc.

SUMMARY

Statistics is the body of methods for the collection, analysis, presentation, and interpretation of quantitative data and for the use of such data. As a language of business, statistics enables us to describe data and to make inferences for decision making purposes. Deductive inference proceeds from general to specific knowledge, whereas inductive inference proceeds from specific knowledge to general conclusions. Mathematics is more certain than inferential statistics. It yields predictable results, whereas statistics applies where the database is uncertain or incomplete.

A population is a total collection of elements and is described by parameters, whereas a sample is a part of a population and is described by statistics. We use point and interval estimates from samples to make inferences about population values. However, sample statistics are not expected to perfectly represent population parameters, so we allow for sampling error. But we do try to eliminate bias through the use of proper sampling techniques.

Statistical studies entail (1) identifying the problem, (2) establishing the criteria for solution, (3) collecting data, (4) analyzing and testing the data, and (5) drawing conclusions. Every step is important. Data are either from internal sources (corporate files, MIS, and DSS) or external sources, such as national, regional, trade, and periodical data, or the organization's own surveys and field studies. Field studies generally include either direct observations, personal interviews, or mail questionnaires.

As societies become more affluent and knowledge-oriented, information systems and databases take on increasing importance. Nearly every large organization has some form of computerized database. Appendix A is a small data set that we will use to practice some of the statistical methods in this book.

CASE: STATISTICS IN THE SUPERMARKET

Albertsons needed something dramatic to change their image! Rosauers was the premier supermarket for the discriminating housewife, while Tidyman's Warehouse Foods was the acknowledged discount center. Albertsons was always right in there competing with Safeway—but that wasn't good enough.

On the Wednesday before Thanksgiving, Albertsons locked their doors, covered their windows, and began marking down virtually every item in their stores. When they reopened on Sunday, they advertised "**No Games, No Gimmicks, No Double Coupons that cost you money—just low prices**". Their newspaper advertisement compared the prices of 119 items, as verified by an independent auditor.

COMPARE AND SAVE	ALBERTSONS	SAFEWAY	ROSAUERS	TIDYMAN'S
Strawberry Preserves	1.85	2.09	1.98	1.85
Jif Peanut Butter	2.58	2.99	2.68	2.58
Banquet Dills	1.65	1.89	2.68	1.67
Best Foods Mayonnaise	1.88	2.09	1.92	1.89
–	–	–	–	–
–	–	–	–	–
–	–	–	–	–
French Hard Rolls	.06	.13	.17	.13

In the center of the advertisement they showed the total cost of the items from Albertsons of $216.55, from Safeway of $257.98, from Rosauers of $252.62, and from Tidyman's of $225.77. Albertsons' highest priced item was Pampers Diapers ($9.29), and the lowest was the French Hard Rolls ($.06).

Case Questions

(a) In what way is Albertsons employing the term *statistics* as defined in this chapter?

(b) Using Figure 1–4, reconstruct (briefly) how Albertsons' management used statistical methodology to help solve their problem.

(c) Why do you suppose Albertsons listed the comparative prices of so many (119) items in their advertisement?

(d) Is the $216.55 a good point estimate for reflecting the savings a customer can expect? Why or why not?

(e) What type of information would be required to ensure the continued success of this marketing campaign?

QUESTIONS AND PROBLEMS

1. Define (a) statistics, (b) population, (c) sample.
2. A friend of yours from high school has gone directly to work for a steel company and claims you are wasting your time studying statistics in college. How might you respond to him to convince him that your time (hopefully) is well spent?
3. What justification is there for referring to management as a science?
4. Distinguish between the following:
   (a) descriptive and inferential statistics
   (b) deduction and induction
   (c) mathematics and statistics
   (d) census data and sample data
5. Indicate whether each of the following represents deductive or inductive reasoning:
   (a) You receive a recall notice to return your 1985 model car to an authorized shop for correction of a possible steering problem.
   (b) A marketing representative in a supermarket encourages you to taste a new cheese that is on special today.
   (c) You decide to purchase a state lottery ticket because you believe you have a good chance of winning $1.5 million.
   (d) First National Bank sends you a questionnaire explaining that they are trying to obtain a regional estimate of student income and loan amounts.
6. Give an example of (a) a point estimate and (b) an interval estimate.
7. Explain the difference between sampling error and bias.
8. Nationwide Car Rental Co. has just announced that a survey of 100 customer records revealed that their customers drove an average of 22.6 miles per day for work transportation, and from 6,100 to 6,500 miles per year total. Indicate whether this is (a) sample or population data, (b) internal or external data, (c) a point or interval estimate. (d) Why would Nationwide want this information and what type of inference would they be likely to use it for? (e) Are Nationwide's conclusions likely to be accurate? Explain.
9. Allfresh Baking Co. supplies a variable number of loaves of bread to supermarkets in Missouri. They use the previous week's data, along with some "intuition," to decide how many loaves to deliver. However, some supermarket managers have complained that the number of loaves is "usually too many, sometimes not enough, and rarely just right!" Use the concepts of uncertainty, sampling error, and bias to explain what might be the problem.

10. Suppose you had to conduct a study to determine the cost of producing plywood in the United States. How would you go about it?
11. A competitor claims (on TV) that his product is superior to your firm's product. Explain how you might conduct a study to determine whether your competitor's claim is justified by the facts or is merely sales propaganda.
12. Suppose you wanted to locate recent data concerning the United States money supply and the unemployment status. What national data sources would yield this type of information?
13. Refer to the database in Appendix A. What percent of the organizations in the database (a) are utilities, (b) have profit sharing?
14. Explain, by example, how you might use the data from Appendix A as (a) descriptive statistics and (b) inferential statistics.

APPENDIX A: Corporate Statistical Database

This database provides information on 100 business organizations as follows:

U = 20 utilities
T = 5 transportation firms
R = 15 retail firms
C = 20 chemical, drug firms
B = 10 banking, insurance firms
E = 30 electronics firms

The data have been disguised and/or modified and are not intended to reflect actual corporate statistics, although they are representative of a cross section of firms. The content of each column of the database is as listed below:

Col. No.	Designation	Content
1	No	Number assigned to firm
2	Corp. Name	Corporate Name
3	I	Industry Code (as designated above)
4	OP	Years of Operation
5	Sales	Annual Sales ($m)
6	Exp	Exports as % of Sales
7	Sh	Price per Share ($)
8	EPS2	Earnings per Share 2 years ago
9	EPS1	Earnings per Share 1 year ago
10	EPSC	Earnings per Share current year
11	PE	Price/Earnings Ratio = Sh ÷ EPSC
12	Div$	Annual Dividend ($)
13	P	Profit Sharing? (yes or no)
14	Ad Ex	Advertising and Public Relations Expense ($m)
15	Slry	Average Annual Salary of Employees ($000)
16	AG	Average Employee Age
17	No Emp	Number of Employees
18	HQ	Corporate State Headquarters
19	ST $	State's per Capita Income ($ annual)

No	Corp Name	I	OP	Sales	Exp	Sh	EPS2	EPS1	EPSC	PE	Div$	P	Ad Ex	Slry	AG	No Emp	HQ	ST $
1	Mountain Gas	U	22	399	0	29	4.07	3.90	4.04	7	2.34	N	2.79	20	51	2577	ND	6400
2	Mohawk Power	U	44	489	0	27	3.97	4.22	3.13	9	2.60	N	4.89	22	38	2575	MT	6600
3	Mountain P & L	U	60	567	0	24	2.78	3.04	2.98	8	1.50	N	4.82	20	45	3040	UT	6300
4	Portland Electric	U	54	585	0	15	2.54	2.31	2.21	7	1.80	N	2.93	22	41	3240	OR	7600
5	Puget Electric	U	24	519	0	15	2.29	1.69	1.92	8	1.50	N	3.12	25	49	2380	WA	8100
6	Utah Gas & Elec	U	56	855	0	23	2.38	2.46	2.26	10	0.90	N	9.49	20	45	4575	UT	6300
7	Centre Electric	U	61	715	0	17	1.81	2.33	1.95	9	1.61	N	5.72	25	42	2900	IL	8100
8	Cleveland Gas	U	47	1450	0	15	2.75	2.94	2.35	6	2.16	N	13.05	22	47	4800	OH	7300
9	Common Electric	U	71	4900	0	27	3.75	4.39	4.30	6	3.00	Y	54.88	25	41	17757	IL	8100
10	Day Power & Lt	U	73	1040	0	16	2.65	2.80	2.20	7	2.00	N	6.80	23	39	3000	OH	7300
11	Denton Electric	U	83	2530	0	15	1.75	2.21	2.15	7	1.68	N	50.60	22	36	11152	MI	7700
12	Great States Util	U	59	1560	0	13	1.95	2.13	2.30	6	1.60	N	12.48	31	38	4958	TX	7200
13	Hector Industries	U	8	4200	0	21	3.77	3.54	3.42	6	2.25	N	46.20	26	29	10700	TX	7200
14	Interlake Power	U	61	1290	0	22	3.04	3.80	4.25	5	2.50	N	14.95	23	34	3969	IL	8100
15	Isoquant Power	U	52	460	0	31	3.42	4.17	4.60	7	3.00	N	4.58	24	46	2300	IN	7100
16	Kanton City Gas	U	75	430	0	27	3.00	3.08	3.20	8	2.27	N	5.16	22	31	2015	KS	7400
17	Lakeville Gas	U	71	700	0	35	2.21	2.87	2.95	12	2.32	N	7.20	21	35	3499	KY	6000
18	Northeast Power	U	62	1800	0	42	4.79	5.60	5.85	7	4.00	N	25.20	23	47	6450	MN	7500
19	Oklahoma P & L	U	82	1040	0	28	2.57	2.62	2.55	11	1.85	N	10.38	20	34	3850	OK	6900
20	Texas Group	U	19	3900	0	26	3.85	3.90	3.21	8	2.85	N	43.30	21	39	16240	TX	7200
21	Alkan Airlines	T	50	4110	0	28	–0.99	4.79	4.00	7	0.00	Y	120.87	21	41	35500	TX	7200
22	Overland Airways	T	41	480	0	9	0.77	0.17	1.45	6	0.00	Y	9.12	23	29	3980	MO	6900
23	Northern Air Lines	T	4	2050	0	38	0.23	–2.00	5.00	8	0.90	N	34.76	21	37	11225	MN	7500
24	Continental Truck	T	55	1200	0	23	2.04	2.43	2.70	9	0.88	N	24.13	22	35	24400	CA	8300
25	Burgandy Rail	T	23	4508	0	21	2.28	2.08	3.70	6	1.10	Y	85.65	23	47	35721	WA	8100
26	Norsteads	R	45	788	0	35	1.50	1.51	2.15	16	0.50	Y	32.31	17	36	8400	WA	8100
27	Pacific Stores	R	53	833	0	21	1.54	1.07	0.81	26	0.10	N	30.80	17	35	5900	WA	8100
28	Pay Little Stores	R	79	1202	0	22	1.52	2.02	1.25	18	0.55	N	57.69	16	37	11543	WA	8100
29	Albi Dept Stores	R	55	4040	0	48	4.41	6.15	6.90	7	1.80	Y	161.60	17	38	62540	NY	7500
30	Angus Stores	R	22	800	0	26	1.00	1.54	2.00	13	0.14	Y	29.20	18	40	9200	CT	8500
31	The Trend Setter	R	12	525	0	20	2.33	2.52	2.30	9	0.40	N	26.25	17	31	8700	CA	8300
32	Hadrians	R	25	480	0	10	0.60	1.03	0.95	11	0.37	N	30.65	17	33	6800	WV	6100
33	Jasper Fashions	R	18	475	22	42	1.20	1.77	2.00	21	0.00	Y	19.47	18	35	5300	NJ	8100
34	Magic-City Inc	R	74	4700	0	40	3.25	4.32	4.95	8	1.30	Y	202.15	19	35	58000	MO	6900
35	Merchants Leigh	R	65	1765	0	53	4.74	5.65	6.20	9	1.00	N	72.36	16	39	21100	NY	7500
36	Lucy May Stores	R	65	4065	0	49	2.74	3.72	4.10	12	1.00	N	154.00	17	43	49760	NY	7500
37	Pepperbox Stores	R	52	920	0	35	2.10	2.29	2.65	13	1.10	N	3.55	17	31	9010	NJ	8100
38	Pier Five Imports	R	6	255	0	9	0.67	2.82	2.25	4	0.00	Y	10.46	16	33	2530	TX	7200

No	Corp Name	I	OP	Sales	Exp	Sh	EPS2	EPS1	EPSC	PE	Div$	P	Ad Ex	Slry	AG	No Emp	HQ	ST $
39	Stop & Go Shops	R	24	3200	0	50	3.29	4.10	5.00	10	1.30	N	121.60	17	28	35470	MA	7500
40	Xron Stores	R	25	2985	0	47	2.05	3.19	4.05	12	2.00	N	120.70	16	31	38580	MA	7500
41	Allied Products	C	58	4750	30	51	3.59	4.00	4.30	12	2.64	N	140.92	16	46	54680	NY	7500
42	Boundary Mining	C	84	4200	30	50	2.59	3.00	3.45	14	1.50	Y	130.20	22	37	14560	NY	7500
43	Eastern Chem	C	83	3100	33	63	5.42	6.13	9.65	7	3.00	N	96.10	24	39	29200	IN	7100
44	Millwood Ltd	C	23	3535	43	89	5.61	6.10	6.75	13	3.00	Y	98.98	23	40	32620	NJ	8100
45	Periodic Chem Co	C	42	3900	52	40	2.13	2.73	3.10	13	1.35	Y	128.76	21	37	40750	NY	7500
46	Rainland Mines	C	4	1282	30	29	2.74	2.10	2.69	11	1.50	Y	39.90	25	40	11123	CT	8500
47	Rome & Richards	C	16	530	28	29	1.82	2.02	2.05	14	1.00	Y	20.14	21	42	4580	PA	7100
48	Sand Petroleum	C	76	1235	0	59	2.76	2.39	3.25	18	0.50	Y	46.93	23	45	10200	IL	8100
49	Shoshone Falls Co	C	17	1885	38	52	3.01	3.31	3.70	14	2.40	Y	60.32	21	37	24100	NY	7500
50	Union City Co	C	82	2170	15	66	4.18	5.28	5.40	12	4.00	Y	71.61	20	36	21425	NJ	8100
51	Warner-Massey	C	64	3180	33	35	2.05	2.51	2.85	12	1.50	Y	98.58	22	39	42380	NJ	8100
52	Hammer Powder	C	72	2600	35	34	1.97	2.76	4.90	7	1.46	N	8.32	21	43	24200	DE	7500
53	Mocha Chemical	C	51	6700	27	45	4.24	4.72	5.75	8	2.18	Y	194.31	20	44	42330	MO	6900
54	Ortega Products	C	95	2060	40	34	2.66	3.01	3.75	9	1.40	N	63.58	24	47	18705	CT	8500
55	Stone Mtn Chem	C	14	1506	12	17	3.06	-0.03	1.12	15	0.00	Y	28.62	28	42	9752	CT	8500
56	Regency Group	C	7	2035	35	61	2.92	5.33	6.86	9	5.00	N	50.87	23	43	11450	PA	7100
57	Bauer Chemical	C	27	310	15	31	1.93	2.08	2.35	13	1.10	N	8.68	21	44	3890	PA	7100
58	Chem Corp Am	C	14	350	13	29	1.79	2.15	2.45	12	1.48	Y	10.01	20	36	4252	OH	7300
59	Crater Lake Chem	C	36	245	19	20	1.05	1.81	2.40	8	1.00	Y	0.65	28	38	2400	NY	7500
60	Dewey Labs	C	60	704	26	25	2.44	2.40	2.69	9	1.25	N	18.58	21	39	6713	MN	7500
61	Chem Trust Co	B	17	1350	0	49	4.02	4.56	3.70	13	2.14	N	14.85	19	29	7600	NJ	8100
62	Cork Trust Co	B	16	2600	0	32	2.04	-0.07	1.45	22	2.60	Y	31.20	18	31	19200	NY	7500
63	Safeguard Insur	B	55	153	0	34	2.86	3.56	2.86	12	1.10	N	3.21	19	44	10305	WA	8100
64	Radium Bank	B	57	47	0	31	4.12	4.90	5.75	5	2.20	N	0.42	17	34	5422	WA	8100
65	Penny Bank	B	95	6	0	20	3.33	1.56	1.50	13	1.00	N	0.07	18	35	2209	WA	8100
66	Fire Insur Am	B	50	873	0	63	3.67	4.48	6.05	10	0.75	N	5.98	20	45	4743	DC	9000
67	Ohio Insurance	B	15	842	0	42	4.91	4.85	2.50	17	2.50	N	7.57	18	41	5300	OH	7300
68	St. Louis Bank	B	95	1744	0	48	9.23	6.03	2.80	17	2.00	N	6.97	19	43	9870	MN	7500
69	Central Bank Corp	B	65	41	0	23	3.04	2.67	3.00	8	1.80	N	0.49	17	32	3518	MO	6900
70	American Trust	B	12	120	0	51	6.05	5.49	7.85	6	2.88	N	1.22	16	39	3906	OH	7500
71	Adv Computer	E	12	235	16	19	0.43	0.98	1.60	12	0.32	Y	2.35	23	29	4300	NY	7500
72	Wesland Labs	E	18	360	42	13	-0.45	1.05	1.50	9	0.00	N	4.32	25	31	1965	CA	8300
73	Advtech	E	15	1025	0	32	0.39	1.23	2.85	11	0.00	N	9.15	25	33	10680	CA	8300
74	Antigua Lazer	E	19	313	40	12	0.41	0.73	1.35	9	0.00	N	2.96	23	37	1050	MA	7500
75	Arrow Edwards	E	38	765	0	15	-1.20	0.85	2.10	7	0.20	N	9.18	30	40	1985	CT	8500

No	Corp Name	I	OP	Sales	Exp	Sh	EPS2	EPS1	EPSC	PE	Div$	P	Ad Ex	Slry	AG	No Emp	HQ	ST $
76	Ashton Software	E	19	155	11	22	0.60	0.63	0.90	24	0.00	Y	1.24	28	31	2230	CA	8300
77	Battle Creek Inc	E	32	312	0	24	1.70	1.50	2.21	11	0.35	N	1.25	23	42	4005	CA	8300
78	Fort Wayne Instr	E	61	1060	16	22	3.34	1.16	1.40	16	1.00	Y	4.51	21	31	26700	NY	7500
79	Genesee Ind	E	69	260	24	16	0.63	1.25	1.15	14	0.40	Y	0.08	21	41	3059	MA	7500
80	Geologic Systems	E	36	1550	0	23	2.10	1.75	1.90	12	0.60	Y	10.82	25	43	20550	IL	8100
81	International Chip	E	14	1650	28	58	-0.33	1.02	2.15	27	0.00	Y	14.85	22	41	21500	CA	8300
82	National Chip Co	E	7	1655	30	12	-0.20	-0.16	0.66	18	0.00	N	24.82	25	33	32700	CA	8300
83	Techdyne	E	24	385	35	26	0.24	1.01	1.75	15	0.00	Y	2.86	24	43	3900	MA	7500
84	Thomas Electric	E	60	330	0	36	1.52	1.74	2.50	14	1.00	N	39.60	22	39	3420	NJ	8100
85	Vinyard Mtn Ltd	E	3	929	20	39	1.48	2.01	2.72	14	0.00	Y	8.73	21	29	13670	CA	8300
86	Zag Electric	E	61	1685	0	24	-1.10	2.11	2.45	10	1.00	N	18.54	23	40	30242	IL	8100
87	Tarawa Electron	E	7	1331	55	65	4.25	2.57	4.44	15	3.50	N	0.88	22	38	20693	OR	7600
88	Apolo Electronics	E	17	598	10	17	2.23	0.72	0.91	19	0.27	N	6.15	21	35	2700	IL	8100
89	Badger Electric	E	64	185	5	19	1.33	1.20	1.55	12	0.30	N	2.04	19	36	3985	AR	5600
90	Cosmos Systems	E	48	1750	60	62	1.92	3.50	4.95	13	2.00	Y	24.50	18	36	27600	NY	7500
91	Designtronics	E	39	625	0	10	1.37	1.00	0.95	11	0.00	N	0.85	21	31	10560	VA	7500
92	Ellington	E	94	4200	24	70	4.37	4.42	6.10	11	2.50	Y	54.60	21	34	48760	MO	6900
93	Erromatics	E	15	260	0	14	1.42	0.78	1.20	12	0.80	N	2.86	23	39	3118	IL	8100
94	Federal Signal Co	E	31	210	23	26	1.85	3.45	5.34	5	0.00	N	0.95	31	36	1850	WA	8100
95	General Service	E	80	1810	10	47	3.85	3.16	3.85	12	2.10	N	15.80	25	31	23530	CT	8500
96	Grandview Ltd	E	56	1065	25	57	3.50	3.56	4.80	12	1.25	N	12.78	24	29	5970	IL	8100
97	Guild Industries	E	16	150	15	15	-1.20	1.33	1.85	8	0.60	Y	0.62	23	33	2669	NJ	8100
98	Jordan Controls	E	32	1425	30	42	3.83	4.17	4.71	9	1.80	Y	13.68	23	43	20700	WI	7200
99	Mayfield Electric	E	58	2315	0	35	3.09	1.92	3.75	9	2.00	N	19.68	24	39	28000	IL	8100
100	Scotch Electric	E	81	1395	9	38	2.61	2.12	3.80	10	1.00	Y	12.55	24	41	21320	IL	8100

CHAPTER 2

Organizing and Displaying Data

INTRODUCTION

TYPES OF DATA

Data Points and Data Sets
Discrete and Continuous Variables

RAW DATA AND ARRAYS

STATISTICAL TABLES

FREQUENCY DISTRIBUTIONS

Steps in Constructing a Frequency Distribution
Number of Intervals
Class Limits and Class Boundaries
Class Midpoints
Relative and Cumulative Frequency Distributions

GRAPHIC PRESENTATION OF DATA

Histogram
Frequency Polygon
Cumulative Frequency Distribution (Ogive)

GRAPHS FOR OTHER TYPES OF DATA

USING MICROCOMPUTERS TO DISPLAY DATA

SUMMARY

CASE: MIDLAND DISTRIBUTION COMPANY

QUESTIONS AND PROBLEMS

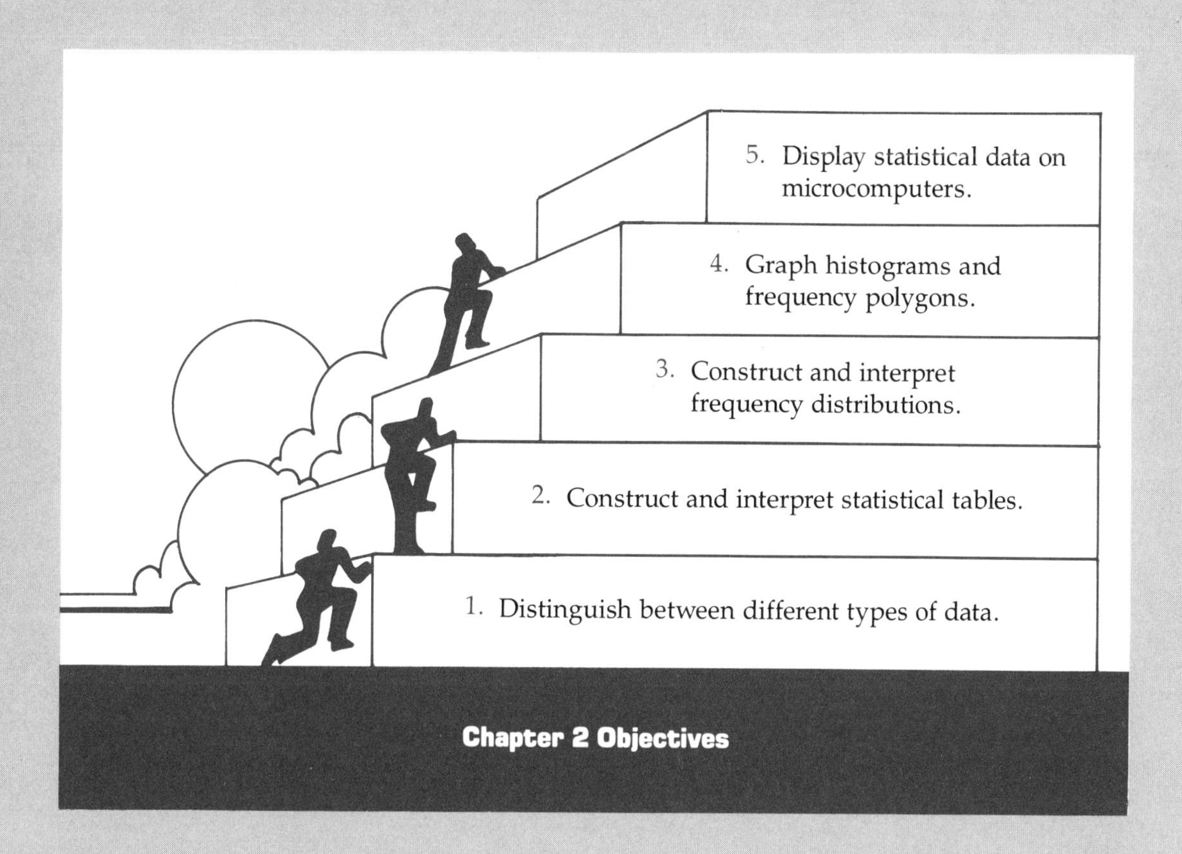

The United States is now an information society. *Forty years ago most workers were associated with manufacturing activities. Today, more than two thirds of the jobs are in information or "knowledge" activities. They range from collecting market research data to conveying legal, investment, and health advice. The chances that your work will involve generating, controlling, or using information are now very high. For accounting, finance, marketing, and other managers, this information frequently takes the form of numbers (or data) that describe the flow of economic activities. And what is better for storing and handling numbers than a computer!* Conclusion*: If you're a manager today, you probably have to work with data (and your nearest friend may be a computer)!*

INTRODUCTION

Data are numbers that convey information. They are used to describe nearly every facet of business and government operations—from the number of olives produced in California to the Department of Defense budget at the Pentagon. But without some sort of organization and summarization, our understanding of data seems to be inversely related to the amount of data we are trying to comprehend.

In this chapter we will first draw some distinctions in the types of data you are likely to encounter in business. Then we will review two (visual) forms of organizing and presenting data: tables and graphs. The graphs often take the form of frequency distributions, which we will use in the chapters that follow. The material in this chapter is largely descriptive, and not too difficult—but very essential for working with *any* statistical data.

TYPES OF DATA

There are many ways of classifying data. Some widely used bases for grouping data are by size, type, time, and place. For example, sales reports of a publishing company might reveal data classified as below:

CLASSIFICATION BASIS	DATA IN REPORT
Quantitatively (size)	Total yearly sales ($) of each product line
Qualitatively (type)	Classification of books sold as texts, novels, and periodicals
Chronologically (time)	Sales during each month of the year
Geographically (place)	Sales in east, midwest, west, and south

Most publishers use two or more bases of classification to help discover relationships that might be useful in activities such as planning production or advertising or setting inventory levels. For example, sales data by product line, which are also grouped into geographical regions, would help managers anticipate where new inventories might be needed.

DATA POINTS AND DATA SETS

In each of the above classifications, single observations, or *data points,* are grouped into collections, or *data sets.* For example, the number of periodicals sold in the Midwest in June (a data point) is recorded along with sales values at other locations (and other times) to constitute a data set that can eventually yield a national, annual sales figure. We shall deal first with *quantitatively* classified data.

DISCRETE AND CONTINUOUS VARIABLES

Insofar as the value of each data point can vary from one location (and time period) to the next, the data points are referred to as values of *variables*. A **variable** is a common attribute (e.g., number of books sold) that can differ from one observation to the next. In statistics, we often let the letter **X**, or some other letter, represent the (unspecified) value of a variable. Furthermore, it is frequently necessary to distinguish between *discrete* and *continuous* variables.

Discrete Variables are variables that can assume only separate and distinct (countable) values.

Continuous Variables are variables that can take on any values in a range of (measurable) values.

You can usually differentiate between the variables by asking yourself whether they can be counted (i.e., are discrete) or measured. Figure 2–1 gives some examples.

FIGURE 2–1 Some constants and types of variables

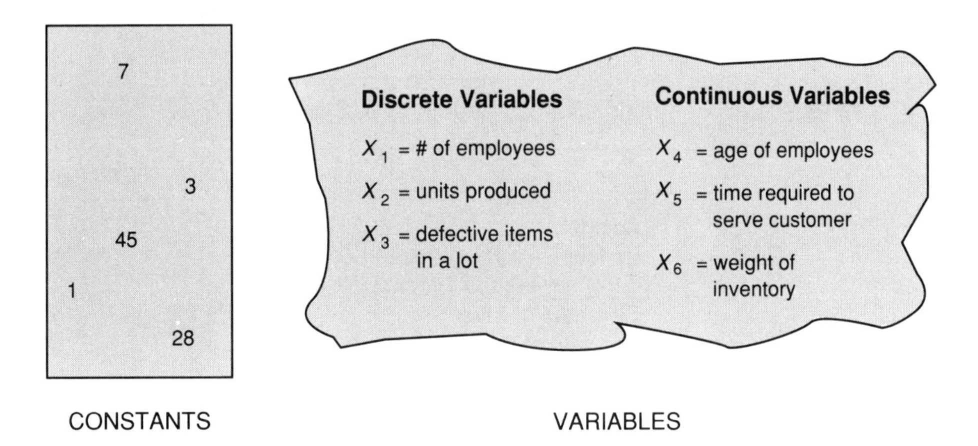

RAW DATA AND ARRAYS

Raw data is quantitative information that is presented in the order in which it is collected (i.e., with only one classification scheme). For example, suppose you asked ten employees at a utility company how many years they had worked for their employer. Your listing of the responses in the order received (e.g., 3, 12, 7, 1, 15, 7, 2, 13, 5, and 2) would be raw data. With more observations, you might have more difficulty grasping "the picture." However, arranging the data in an array would help. An **array** is a sequential ordering of data according to their value. An array of the employment data would be the listing 1, 2, 2, 3, 5, 7, 7, 12, 13, 15 (or the reverse order of this).

STATISTICAL TABLES

Tables are often a clear and forceful method of presenting data in a row and column format. *Analytical* (or text) tables are arranged to show comparisons or relationships in a form that lends itself to analysis. They should convey the writer's message in as simple, concise, and effective a manner as possible. *Reference* (e.g., appendix) tables typically contain larger masses of data and should be arranged for ease of reference and completeness. Tables are usually numbered and should acknowledge the source as a means of giving proper credit and confirming the reliability (or bias) of the data.

In our society, managers in the hospital, insurance, and health care industries use research data on mortality to help guide their managerial decisions. Table 2–1 is an analytical table that reveals the ten leading causes of death among the age group of persons 15 through 34 years. Note that these data are *cross classified* by type of mortality (on the rows) and sex (columns). The table quickly makes it apparent that accidents are the prime cause of mortality in this age group, with homicide, suicide, and cancer following at significantly lower levels. It also facilitates some interesting comparisons between male and female regarding causes of death.

TABLE 2–1 Leading causes of death in U. S. for ages 15–34

	MALE	FEMALE	TOTAL
Accidents	34,309	8,717	43,026
Cancer	3,933	3,434	7,367
Cerebrovascular Disease	687	639	1,326
Cirrhosis of Liver	903	415	1,318
Congenital Anomalies	613	378	991
Diabetes	368	315	683
Heart Disease	2,758	1,392	4,150
Homicide	10,195	2,506	12,701
Pneumonia, Influenza	511	354	865
Suicide	8,750	2,353	11,103
TOTALS	63,027	20,503	83,530

Source: Vital Statistics of the United States, 1979, as reproduced in publication of the American Cancer Society, *CANCER STATISTICS*, 1984.

EXERCISES I

1. What basis of classification (qualitative, quantitative, chronological, or geographical) would most likely be used to report each of the following?
   (a) monthly production
   (b) auto sales by territory
   (c) alternative investment opportunities
   (d) retirement program costs
2. Indicate whether the following are discrete or continuous variables:
   (a) dimensions of a product
   (b) mileage on company car
   (c) number of books in library
   (d) flight time to Boston
   (e) number of unionized employees
   (f) tons of wheat in a warehouse
3. The times required by ten word processing applicants to complete a manual dexterity test were as follows (seconds):

32	27
54	65
38	46
45	37
32	42

   (a) Are the data discrete or continuous?
   (b) Arrange the data in an (ascending ordered) array.
   (c) Using the array, what is the value of X_6?
4. Use the library (or other source) to locate a cross-classified table. Describe the two types of classification and indicate whether they are qualitative, quantitative, chronological, or geographic (or other).

FREQUENCY DISTRIBUTIONS

In addition to tables, one of the most common methods of grouping data is in the form of *frequency distributions*.

> **Frequency distributions** are condensed tables or graphs that show the number (or %) of data points as grouped into mutually exclusive classes.

As with other tables, analysts use frequency distributions to give the reader a clearer impression of the data. Because the frequency distributions "condense" the raw data into only a few classes, the values of the individual observations are lost, but the intuitive benefit to the reader usually outweighs this loss of accuracy.

Example 2–1 The raw data below are from a study designed to help establish the staffing needs (just before closing time) for a chain of convenience stores in Texas. The study encompassed 100 days (data points) and consisted of recording the number of customers who entered a store during the last hour before closing. Use a tally table to group the data, and show the resulting frequency distribution.

RAW DATA: Number of Customers Entering Store Before Closing																			
39	22	35	45	30	14	31	24	22	41	12	30	50	43	47	31	17	27	21	47
32	29	34	32	59	37	21	17	24	31	23	11	45	39	32	30	11	43	20	28
49	38	38	24	44	24	29	13	23	42	53	14	37	54	30	60	37	21	39	33
33	50	41	44	24	32	31	48	42	49	25	17	50	49	25	35	32	12	59	44
21	14	25	48	15	13	43	37	35	52	32	26	28	32	18	64	21	36	63	39

Solution

The raw data show a minimum of 11 customers and and maximum of 64, leaving a range of 53 to be covered by our frequency distribution. Suppose we (arbitrarily) choose to use about seven or eight class intervals of equal width. If we divide the 53 by 7 we arrive at a preliminary interval width of about 8 customers. This can then be rounded up to a convenient interval of 10, so our first interval is 10 to $<$ (less than) 20. We then tally the number of observations in each class, as shown on the left in Figure 2–2. The corresponding frequency distribution is shown on the right. (Although our frequency distribution uses the standard mathematical notation of "$10 < 20$" to express the interval "10 to under 20," you should remember that this condensed notation is not always appropriate for lay readers.)

TALLY TABLE

CLASS INTERVAL (No. Customers)	TALLY (No. Days)
10 to under 20	𝍸 𝍸 IIII
20 to under 30	𝍸 𝍸 𝍸 𝍸 IIII
30 to under 40	𝍸 𝍸 𝍸 𝍸 𝍸 𝍸 II
40 to under 50	𝍸 𝍸 𝍸 IIII
50 to under 60	𝍸 III
60 to under 70	III
TOTAL	100

FREQUENCY DISTRIBUTION

CLASS INTERVAL (No. Customers)	FREQUENCY f
10 < 20	14
20 < 30	24
30 < 40	32
40 < 50	19
50 < 60	8
60 < 70	3
TOTAL	100

FIGURE 2–2 Tally table and frequency distribution of number of convenience store customers

A frequency distribution like Figure 2–2 could help management plan for how many customers to accommodate during the last hour before closing, taking costs into account. In the few instances where the number of customers is more than about 50, they might find it more cost effective to make the customers wait a little, rather than incur the cost of keeping an extra clerk available.

STEPS IN CONSTRUCTING A FREQUENCY DISTRIBUTION

We can summarize the procedure followed by the analyst in the form of some useful guidelines for constructing frequency distributions, as given in Table 2–2.

TABLE 2–2 Steps in Constructing a Frequency Distribution

1. Find the *range* of values from the raw data.
2. Determine a *preliminary class interval* by choosing a reasonable number of classes (i.e., between 5 and 20) and dividing that number into the range.
3. Set the *stated class interval* by adjusting the preliminary value upward to a common or convenient width (e.g., 10 or 20). Avoid fractional or odd-numbered interval widths, if possible.
4. Record, or *tally*, the number of observations that fall into each class.
5. Use the tally amounts as class frequencies and prepare a *frequency table* (i.e., of absolute or relative frequencies).

NUMBER OF INTERVALS

A major purpose of frequency distributions is to reveal the underlying pattern of the data. Thus, the number of classes may be more a function of the actual data than of any specific rules. With fewer than five classes, the patterns may remain hidden. But more than 15 or 20 classes can obscure the benefits of grouping the data. Six to 10 intervals are common and can serve as a good starting point.

Large quantities of data that are significantly spread out may justify more classes than small, concentrated data sets. Also, the intervals of a frequency distribution should, if possible, be of equal size. Otherwise, graphic and numerical summaries are more difficult to prepare. Nevertheless, some distributions (such as income) may have a few extremely high observations that would seem to rule out the use of equal size classes. In such cases, unequal class sizes could be considered, or the top class may be left *open ended,* such as "$100,000 and up."

CLASS LIMITS AND CLASS BOUNDARIES

DISCRETE DATA Class intervals should be stated in such a way that every observation will fit into one, and *only one,* class (i.e., no overlapping). This is usually not a problem when dealing with discrete data as in Figure 2–2. When the interval is from "20 to under 30" (or stated as $20 < 30$), data points of 20 and 29 would definitely fall into that interval, and an observation of 30 would fall into the next. The same would be true if the limits were stated as 20–29 and 30–39. Even so, some analysts prefer to simply express the intervals to one more level of accuracy (19.5–29.5, and 29.5 – 39.5) so as to avoid any question about where data points fall.

Continuous variables demand a little more attention because measurements can fall at any point along a continuum. Suppose we were dealing with a continuous variable such as pounds, and the classes were given as

20–29 pounds and 30–39 pounds. An observation of 29.4 pounds would not appear to fit in either class. For this reason, statisticians distinguish between *stated class limits* (such as 20–29 and 30–39), and *real* class limits, or *class boundaries* (see Figure 2–3). The term "class limits" refers to stated limits unless delineated otherwise.

FIGURE 2–3 Stated Class Limits and Class Boundaries

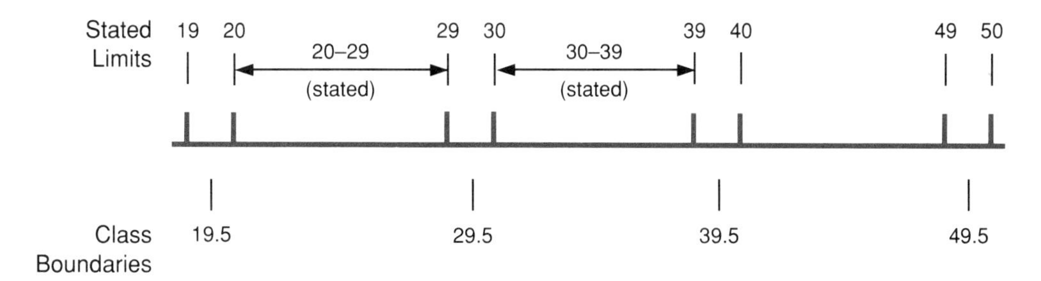

The class boundaries for the first two classes are 19.5–29.5 pounds and 29.5–39.5 pounds, respectively—the width between boundaries is one class interval. Also, the class boundaries actually touch each other, but they are stated in one more level of accuracy than the data points themselves. This avoids potential problems because the data are recorded to the accuracy of the stated limits only. Thus, an observation of 29.4 pounds would have been rounded to 29 pounds and tallied in the 20–29 pound class. If the analyst wished to record data to the nearest tenth of a pound, the stated class limits would be 20.0–29.9 and 30.0–39.9 pounds. Then the class boundaries would be 19.95–29.95 and 29.95–39.95 pounds. Again, every observation would fall into only one class.

CLASS MIDPOINTS

Some of the equations and problems that follow in the next chapter require the use of class boundaries and class midpoints, so be sure you have a clear understanding of them. In the future we will sometimes use the letters LCB for *lower class boundary* and UCB for *upper class boundary*.

The **class midpoint** is simply the center of the class, and can be found by adding either the stated or real class limits (i.e., boundaries), and dividing by 2. Thus, the midpoint of the 20–29 pound class is 49/2 or 24.5, and the midpoint of a 20.0–29.9 class is 49.9/2 or 24.95. A significant advantage of stating class limits as "20 to under 30" (or 20 < 30 as we did in Figure 2–2), is that the *stated limits and real limits coincide*. This results in a convenient midpoint (e.g., 25) that will be easy to work with later on.

RELATIVE AND CUMULATIVE FREQUENCY DISTRIBUTIONS

The frequency distribution in Figure 2–2 is an absolute frequency distribution because the frequencies for each class tell the actual number of data points, or observations, in that class. **Relative frequency distributions** show the class frequencies expressed as a percent of the total number of observations. They are especially useful for analyzing the relative importance of different classes and for comparing frequency distributions having an unequal number of observations.

Table 2–3 is a *relative frequency distribution* useful to the Human Resources Manager of a hotel chain. This distribution uses class intervals of $8,000 to describe the wages and salaries of 220 employees of the firm. (The actual numbers are also included for reference. For example, the 7.3% was computed by dividing the first class frequency of 16 by the total 220.) Note that 60% of the employees have wages within the $16,000 < $24,000 interval, and less than 10% exceed $40,000.

TABLE 2–3 Relative Frequency Distribution of Wages & Salaries

INCOME CLASS $	Actual No. Employees f	Relative No. Employees %
8,000 < 16,000	16	7.3
16,000 < 24,000	132	60.0
24,000 < 32,000	38	17.3
32,000 < 40,000	14	6.4
40,000 < 48,000	10	4.5
48,000 < 56,000	8	3.6
56,000 < 64,000	2	.9
TOTALS	220	100.0

TABLE 2–4 Cumulative Frequency Distribution of Wages & Salaries

INCOME CLASS $	Actual No. Employees f	Cumulative No. Employees F < UCB
8,000 < 16,000	16	16
16,000 < 24,000	132	148
24,000 < 32,000	38	186
32,000 < 40,000	14	200
40,000 < 48,000	10	210
48,000 < 56,000	8	218
56,000 < 64,000	2	220
TOTALS	220	

Table 2–4 is a cumulative frequency distribution that would give the Human Resources Manager a slightly different insight from the same data. **Cumulative frequency distributions** show the number (absolute) or percentage (relative) of data points falling below (or above) various class boundaries of the distribution. Here we must distinguish the class frequencies (*f*) from the cumulative frequencies (F). Table 2–4 shows the number of employees with salaries *less than* the upper class boundary (UCB), so it is cumulating values "from below." Only 16 employees had incomes below the upper boundary of the first class, so this is also the first cumulative amount. The cumulative frequency of the second class is that cumulative column sum (16) plus the second class frequency of 132. So 148 employees had incomes below $24,000. The last figure tells us that all 220 employees had incomes below $64,000. In summary, a "less than" cumulative fre-

quency table shows the number of points falling below successive class boundaries.

Whereas Table 2–4 is an absolute cumulative frequency distribution, it could easily be converted to a *relative cumulative distribution* by dividing the numbers in the cumulative column by the total frequency of 220. Another option is to show the cumulated frequencies that are *greater than* the lower class boundry of each class. In this case, the cumulative frequency column would read "$F >$ LCB," and the first column value would be 220, with the second 204 (i.e., 220–16), and so on.

EXERCISES II

5. Distinguish between:
   (a) stated class limits and real class limits
   (b) absolute and relative frequency distributions
6. Using the raw data in Example 2–1, construct a tally table and an absolute frequency distribution with classes of $5 < 15$, $15 < 25$, . . . , $55 < 65$.
7. The Everbright Battery Company guarantees their batteries for 30 hrs. They have life-tested some batteries from a competitive manufacturer in Singapore, and classified the data as shown below, with times in hours:

Class Interval	5–9	10–14	15–19	20–24	25–29	30–34	35–39	40–44
Frequency	1	5	13	19	30	17	9	6

   (a) To what level of accuracy were the data recorded (i.e., hrs, .1 hr)?
   (b) For the second class (i.e., 10–14), what are the stated class limits?
   (c) What are the class boundaries of the second and third classes?
   (d) What is the midpoint of the third class?
8. Use the frequency distribution values in the previous problem.
   (a) Construct a cumulative distribution showing absolute values cumulated from below (i.e., a "less than" distribution with values of F < UCB).
   (b) How many batteries lasted < 29.5 hrs?
9. Using the frequency distribution values in Figure 2–2, formulate the following frequency distributions:
   (a) a cumulative distribution showing *absolute* values cumulated from above (i.e., a "greater than" distribution with values of F > LCB)
   (b) a cumulative distribution showing *relative* values cumulated from above (i.e., a "greater than" distribution with values of %F > LCB)

GRAPHIC PRESENTATION OF DATA

Tables, such as the frequency distributions discussed above, are usually preferred if the data presented are to be used for specific reference or for computation of numerical characteristics or summaries. However, graphs and charts have several advantages over tables: (1) graphs are more likely to attract attention of the casual reader than would the same data presented in tabular or paragraph form, (2) graphs can sometimes present the data more compactly (if less exactly) than a table, (3) graphs enable the analyst and the reader to spot trends or relationships which would be less obvious in a table, and (4) graphs may be used as a computational device since values can be read from any point once a curve has been drawn.

As used in statistical work, graphs often take the form of (a) *histograms,* (b) *frequency polygons,* or (c) *ogives.*

HISTOGRAM

A **histogram** is a vertical bar graph of a frequency distribution, where the height of each bar is determined by the number of observations in the class interval. As with frequency distributions, histograms can be used to show either *actual* frequencies or *relative* frequencies. If equal intervals are used, the bars are of equal width and the height of each bar is determined directly by the frequency of the interval. (If unequal intervals are used, the *area* of each bar is proportional to the frequencies.)

Example 2–2 The study of the number of customers entering the convenience store in the hour before closing (Example 2–1) resulted in the frequency distribution repeated below. Use this data to construct a histogram.

Class Interval	$10 < 20$	$20 < 30$	$30 < 40$	$40 < 50$	$50 < 60$	$60 < 70$
Freq. (No. days)	14	24	32	19	8	3

Solution

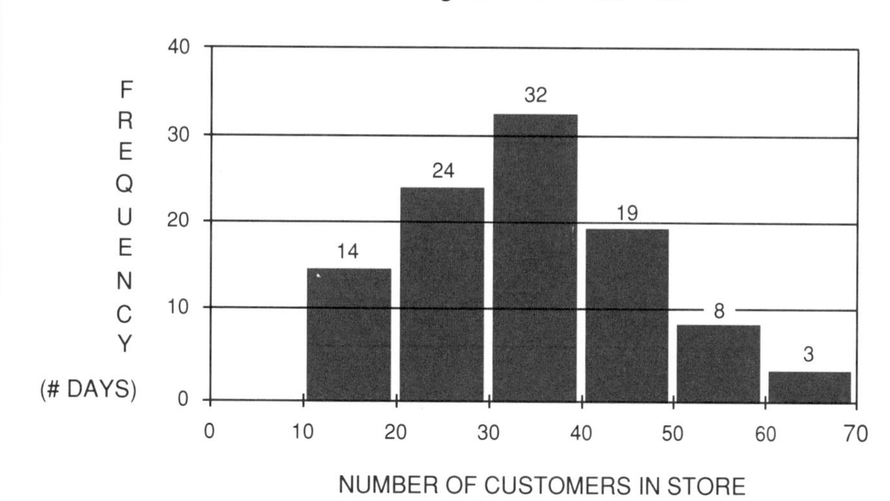

For the frequency distribution graphed in Example 2–2, the stated class limits and the class boundaries coincide, so the horizontal (x-axis) shows these points. When the stated limits and boundaries do not coincide, statistical reports sometimes show the class boundaries on the x-axis. However, for clarity of presentation, the stated limits are often shown, or perhaps only the lower limit of each class is shown. For example, if the classes are \$200–\$299 and \$300–\$399, etc., the x-axis may show only \$200, \$300, \$400, etc., on the horizontal scale. And some business literature, such as corporate annual reports, may even show the rounded class midpoint value in preference to an interval. Thus these same classes might be marked \$250, \$350, etc. Frequency polygons, which we take up next, use the class midpoints as a basis for a line graph of the data.

FREQUENCY POLYGON

A **frequency polygon** is a (many sided) line graph that also portrays the data of a frequency distribution. It has the same height (and approximately the same spread) as a histogram, for the height of each plotted point is again determined by the frequency of the respective class interval. However, the polygon is formed by using straight lines to connect the *midpoints* at the top of each bar. The lines on each end are brought down to the horizontal axis at the midpoints of what "would be" the next lower interval (on the left) and higher interval (on the right). The x-axis scale can show either the class limits or the class midpoints.

Example 2–3 Use the data of Example 2–1 to construct a frequency polygon.

Solution

The sketch below shows the outline of the histogram for reference (only) so you can see how the polygon lines cross the column midpoint values.

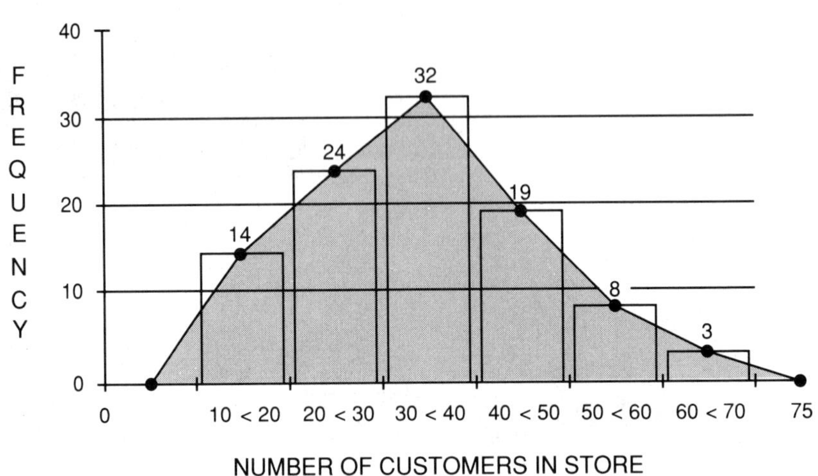

CUMULATIVE FREQUENCY DISTRIBUTION (OGIVE)

The **ogive** (pronounced oh-jive) is a graph of a cumulative frequency distribution with points connected by straight lines. It is constructed by plotting the cumulative frequency (y-axis) that corresponds to each class interval boundary (x-axis). Depending upon whether the frequencies are cumulated from below (i.e., "less than") or above (i.e., "more than"), the cumulative function will start at zero and rise to the top of the graph, or vice versa.

OGIVE TYPE	FORM	START AT	[NEXT X	NEXT Y]	[NEXT X	NEXT Y]
LESS THAN	F < UCB	X = 0, Y = 0	1st UCB	1st Class *f*	2nd UCB	1st *f* + 2nd *f*
GREATER THAN	F > LCB	X = 1st LCB, Y = F	2nd LCB	F–1st Class *f*	3rd LCB	F – (1st + 2nd class *f*)

Example 2–4 (a) Use the data of Example 2–1 to construct an ogive showing the number of days when less than 20, 30, . . . , 70 customers entered the convenience store. (b) What would be required to make the ogive show the relative (i.e., %) values as well?

Solution

(a) The frequency distribution values of Example 2–1 are repeated below along with the cumulative values which are needed for this example. Note that this is to be a "less than" cumulative graph, so the values have been cumulated from below.

Class Interval	10 < 20	20 < 30	30 < 40	40 < 50	50 < 60	60 < 70
Frequency *f*	14	24	32	19	8	3
Cumulative F < UCB	14	38	70	89	97	100

As with the frequency polygon, we will begin on the axis at what would be the upper class boundary of the next lower class (i.e., 10). The first plotted point will then be at a cumulative frequency (Y value) of 14 that corresponds to the upper class boundary (or X value) of 20. Then the next is at $Y = 38$ and $X = 30$. This means that on 38 of the days (observations), there were less than 30 customers in the store during the last hour.

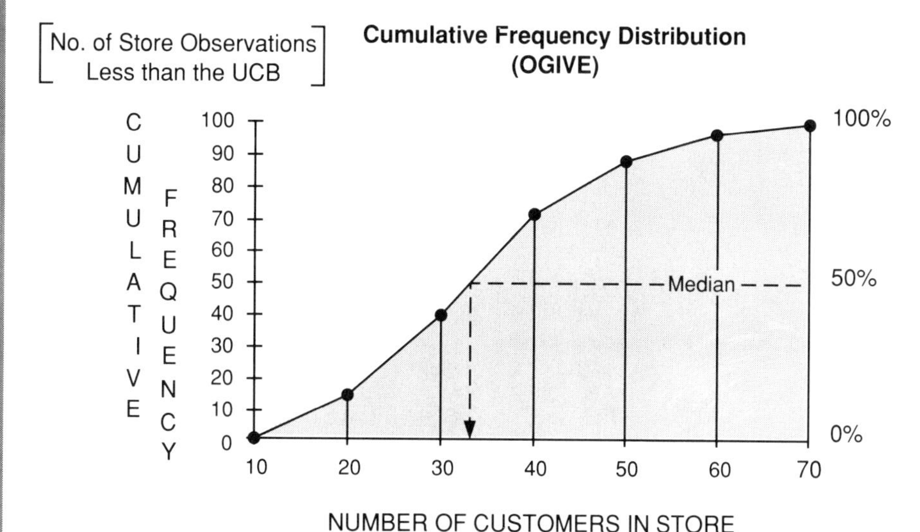

(b) Percentage values are easy to show, for we simply let the total number of observations be 100%, and apportion the other values accordingly. The percentage scale (shown on the right in the graph above) is the same as the absolute scale (on the left) because there happen to be exactly 100 observations—this is an unusual case.

Several useful statistics can be obtained from the ogive. The point on the x-axis which lies directly below the intersection of the curve and the 50 percent line is the value above and below which exactly half of the data points should lie. This point is the **median**. If a percent scale is shown on the chart, the percent of observations lying above and below other points is also easy to determine. The point which separates the bottom one-quarter from the top three-quarters of the observations is called the **first-quartile** (Q_1). Next higher is the median (Q_2), and the point which separates the top one-quarter from the bottom three-quarters is the **third-quartile** (Q_3). The points that divide the scores into ten equal groups (deciles) are also sometimes useful. All of these points can be located mathematically, without the use of the ogive, but the ogive simplifies the procedure and clarifies the meaning of the values.

EXERCISES III

10. Construct a histogram showing the annual wage rates ($000) of the following applicants who applied for loans at a bank during one week in June.

Wage Interval ($000)	0 < 10	10 < 20	20 < 30	30 < 40	40 < 50
Number of Applicants	35	20	90	45	10

11. Using the raw data shown below, construct a histogram with the classes of 5 < 15, 15 < 25, . . . , 55 < 65. [Note: The data were used in Example 2–1. You may already have the frequency distribution computed if you did Problem 6.]

RAW DATA: Number of Customers Entering Store Before Closing

39	22	35	45	30	14	31	24	22	41	12	30	50	43	47	31	17	27	21	47
32	29	34	32	59	37	21	17	24	31	23	11	45	39	32	30	11	43	20	28
49	38	38	24	44	24	29	13	23	42	53	14	37	54	30	60	37	21	39	33
33	50	41	44	24	32	31	48	42	49	25	17	50	49	25	35	32	12	59	44
21	14	25	48	15	13	43	37	35	52	32	26	28	32	18	64	21	36	63	39

12. For the Everbright Battery Company data shown below, construct (a) a histogram and (b) a frequency polygon. [Note: The data are repeated from Problem 7 and show the battery lifetimes in hours.]

Class Interval	5–9	10–14	15–19	20–24	25–29	30–34	35–39	40–44
Frequency	1	5	13	19	30	17	9	6

13. For the Everbright Battery data in the problem above, construct an ogive showing the cumulative number of observations *less than* the upper class boundary (i.e., $F < UCB$).

Use the following data concerning the number of VCR tapes rented per year to answer problems 14 and 15. (Source: Photo Marketing Association International. Data adjusted to total 100%.)

No Tapes Rented	0	1–4	5–9	10–19	20–29	30–39	40–49	50 or more
% of Households	27	5	8	17	16	9	3	15

14. Using the data above, construct a histogram showing the number of VCR rentals. [Note: Let the x-axis be the number of tapes rented, and combine the first three classes.]
15. For the VCR rental data shown above, (a) construct a cumulative ogive showing the cumulative percent of households greater than the lower class boundary. (b) What would be your estimate of the number of tapes rented by the median (50th percentile) household? (c) What is the maximum number of tapes rented by any household?

GRAPHS FOR OTHER TYPES OF DATA

Data which are classified according to *type, time,* and *place* can be shown graphically as well as in tabular form. A graphic presentation is often less exact, but more compact, more likely to be noticed, and more likely to reveal trends, regularities, and relationships. Business organizations often use many different types of multicolored graphs to describe financial, marketing, and other activities because graphs are such an excellent means of communication. Some of the more commonly used displays are (a) area charts, (b) bar graphs, (c) line diagrams, (d) pie charts, and (e) combinations of these. Figure 2–4 illustrates some typical charts and graphs.

Area charts show the relative importance of components. They are especially useful for depicting the changing proportions of some variables, such as energy, over time.

Bar charts (including "column" charts) are frequently used to show comparative values of different types of products, or changes over time, and to make locational comparisons. Horizontal bars are easier to label, but vertical bars are often used for chronological data. They are typically separated by a space about half as wide as the bars themselves. If one of the bars is extremely long in comparison to the others, it is permissible to show a break in the bar, but that break must be clearly indicated.

A glance at the annual reports of major firms will reveal three variations of bar charts in common use. Individual bars are sometimes apportioned to depict the changing composition of the bars (i.e., a *component bar chart*), or sets of bars are grouped together for each period to show the changing relationship between paired bars (i.e., a *grouped bar chart*). Other bar charts (i.e., *duo-directional bar charts*) have a "zero line" in the center and use deviations from that to depict percentage changes or increases/decreases by departments, products, and so on, from one year to the next.

Line diagrams are frequently used to depict data which cover a long period of time, with the time scale being on the horizontal axis. They are effective for comparing one time period (hour, month, year) with the next and for presenting a picture of long-term trends in the series of data. As with all charts, scales that do not begin at zero must be clearly marked so as not to mislead the reader. Although *arithmetic* line diagrams are more widely used, *logarithmic* vertical scales are sometimes better for presenting data that have been increasing (or decreasing) at an almost constant rate or for comparing fluctuations in two series that have vastly different magnitudes.

Pie charts have an intuitive appeal to many people because they show proportional effects very readily. We all understand the meaning of a "large" versus a "small" piece of the pie. The largest component is frequently shown first, starting at the twelve o'clock position, but this is not a universal practice.

FIGURE 2–4 Commonly used charts and graphs

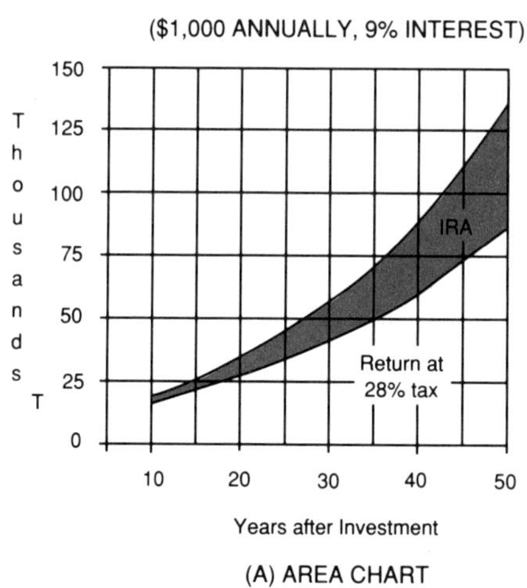

(A) AREA CHART

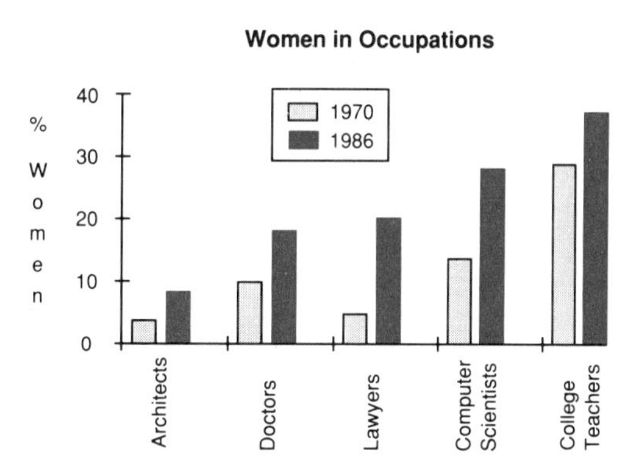

(B) COLUMN CHART

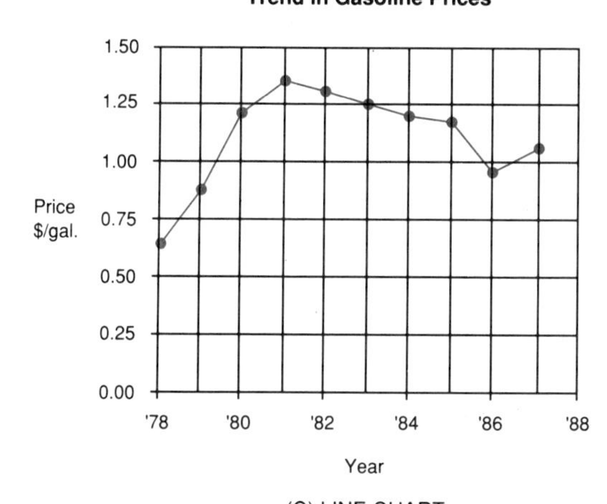

(C) LINE CHART

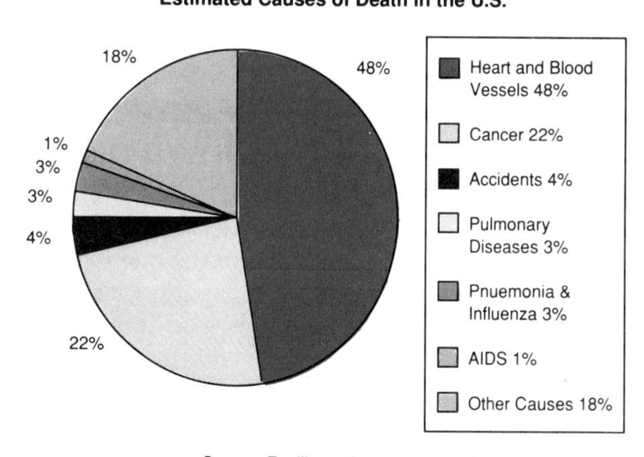

(D) PIE CHART

EXERCISES IV

16. Use an *appropriate graph* to illustrate the number of new business starts in River City since an industrial park was formed:

Years Since Park	0	1	2	3	4	5	6	7
New Businesses	12	7	4	10	18	28	32	25

17. Use the data in Table 2–1 to draw a *pie chart* to depict the five leading causes of death in the United States for females age 15–34.
18. Use the data in Table 2–1 to prepare a *grouped bar chart* comparing the number of male and female deaths in age group 15–34 from accidents, cancer, heart disease, homicide, and suicide. [Note: Show two bars for each category of death, but distinguish them by different shading or crosshatching.]
19. Use your library to find and *graph* (in an appropriate manner) one of the following:
    (a) a representative price of gold over the past 8 years
    (b) the rate of unemployment in the United States over the past 5 years
    (c) the earnings per share of DuPont Company stock for the past 5 years

USING MICROCOMPUTERS TO DISPLAY DATA

Software disks are available for producing tables, charts, and graphs on virtually all the microcomputers used extensively in business today. These are often called (a) *statistical packages* or (b) *chart/graph packages,* and they sometimes work on only the computer system for which they have been written. This means you may need a disk for your particular brand of computer. Beyond that, however, many of the chart/graph packages do basically the same thing, though some are more "user friendly" and/or more versatile than others.

Many of the **statistical packages** contain program options that can be used on problems in nearly every chapter of this book. If you are using one of these packages, you will first need to "boot-up" the disk, and then choose the type of statistical program you want. At this point, our problems and exercises are limited to tables, frequency distributions, and graphs, so you should respond to the computer question prompts by selecting the appropriate chart or graphing program from the menu. You will also need to "input" your data as directed by the program.

Suppose, for example, that you wish to use a microcomputer to construct a frequency distribution. You will boot-up the program disk and request the frequency distribution option. The computer will ask for your (raw) data and may ask how wide the class intervals are to be (or how many). After you type in the data, the computer will group them and offer to print out a copy of the frequency distribution for you. In similar manner, other programs may offer to print a graph of the frequency distribution or of the cumulative distribution.

Chart/graph packages that are designed exclusively for charts and graphs frequently offer options that generic statistical packages do not include. This is because much of the space on the statistical disks is needed for other equations and computational programs. Graphics packages typically offer a menu of subprograms that give the user considerable flexibility in constructing and enhancing graph displays. These include the following:

1. **HELP MENU** This explains everything else and helps you if you need something explained or if you don't know what to do next.
2. **DATA INPUT MENU** This enables you to type in (input) and display (on the screen) your data points.
3. **EDIT MENU** This allows you to change your data or to add titles, labels, legends, etc., to your tables or graphs. Edit menus are sometimes also used to receive tables or graphs from another location (file) or to send them to a location (e.g., to a letter in a word processing file).
4. **SAVE & FILE MENU** This menu typically allows you to store your tables or graphs in the file (storage location) you designate (i.e., on your data disk), or allows you to print the material, or to quit working with it. Sometimes the SAVE/FILE subprogram is also used to bring previously prepared materials back from a file so you can do more work with them.
5. **CHART TYPE** This allows you to choose and draw alternative types of charts or graphs such as area charts, bar graphs, column charts, line graphs, pie charts, scatter diagrams, and combinations of these. The combinations may include "overlays," which enable you to enhance comparisons by overlaying one chart on another.

SUMMARY

Data are numbers that are classified by type, size, time, or place. Discrete data are countable, whereas continuous data are measurable. Frequency distributions are among the most important tabular devices for summarizing or presenting data. They show the number of data points grouped into a few mutually exclusive classes and can be either absolute (numbers) or relative (percentages).

Graphs and charts give a more visual impression of data, with histograms and frequency polygons being among the most commonly used. Cumulative graphs are also useful for showing the number (or proportion) of observations that are greater than or less than designated class boundaries.

Microcomputer programs are available to receive raw data and print virtually any form or chart or graph used in business. These include area charts, bar graphs, column charts, line graphs, pie charts, and various combinations of these.

CASE: MIDLAND DISTRIBUTION COMPANY

Midland Distribution Company was a $140 million/year operation with headquarters in a Chicago suburb. They distributed toasters, irons, and other appliances to retailers in eleven midwestern states. The company's competitive strategy was to emphasize good service to customers, so they had always stressed the importance of having plenty of inventory to meet all their retailer demands. However, with rising interest rates, the cost to buy and hold inventory was beginning to put a serious strain on their financial structure. The comptroller, Brian Kirchfield, insisted that the firm give more attention to their inventory situation. Part of that attention was to channel responsibility to someone who could do some analysis of their inventory records. That is where Joyce Brennan came in.

Joyce Brennan had been with Midland only since last September. She had expected to land a job in June. There she was with a legitimate business degree from a fine state university, but no companies had been standing in line for her expertise. Her disgust had almost turned to despair when Midland Distribution offered her the job in September. The job had a fine title (*Inventory Records Analyst*), but the first six months had been basically clerical work. Most of her time had been spent entering inventory changes on the firm's computer, with an occasional report to the Comptroller's Office. That was until last Friday when her boss, Bufford Emmerson, left to join the transportation department of a trucking firm in Cleveland. Brian Kirchfield had been quick to recommend that Joyce be seriously considered for the vacancy.

Now Joyce finally felt challenged—perhaps overwhelmed was a better word. She had been given the opportunity to take charge of the Materials Supply Assurance (MSA) activities. This meant that she was responsible for ensuring that the firm had an adequate stock of each of the fourteen appliance lines it carried. The challenge in MSA was to stock enough appliances to satisfy the field sales staff, but not so much as to aggravate the financial department. The financial manager exerted tremendous pressure to keep inventories low because the firm didn't have extra cash to tie up in unused inventory. He frequently reminded others that "every dollar of inventory costs us $.30 a year to support."

The problem with setting inventory levels was that the demand from the retailers was very uncertain and uneven, and it differed for each line. Joyce spent considerable time talking to the various managers to try to establish some trade-off between giving customers immediate service (from inventory) versus the cost of stocking extra inventory. She determined that if the company could ship appliances from stock 90% of the time, both marketing and finance would be "acceptably" happy.

By the end of her second week in the new position, Joyce had decided on a plan. She selected one appliance (irons) and dug out demand data for the past three years. Her main concern was the number of irons demanded during the two-week (lead) time between when irons were ordered from the factory until they arrived into stock at the Midland Distribution Center. She entered the demand data into her desktop computer and called for a frequency distribution, which turned out as shown below.

FREQUENCY DISTRIBUTION OF DEMAND
(Standard M–20 Electric Irons, 110 volt)

DEMAND FOR IRONS (During 2 Week Lead Time)	FREQUENCY (*f*)	CUMULATIVE FREQUENCY (F > Lower Class Boundary)
25 < 50 units	4	80
50 < 75 units	8	76
75 < 100 units	14	68
100 < 125 units	14	54
125 < 150 units	20	40
150 < 175 units	12	20
175 < 200 units	5	8
200 < 225 units	1	3
225 < 250 units	2	2
	80	

Joyce's computer program did not automatically graph the cumulative distribution, so she used the data above and developed the following graph by hand.

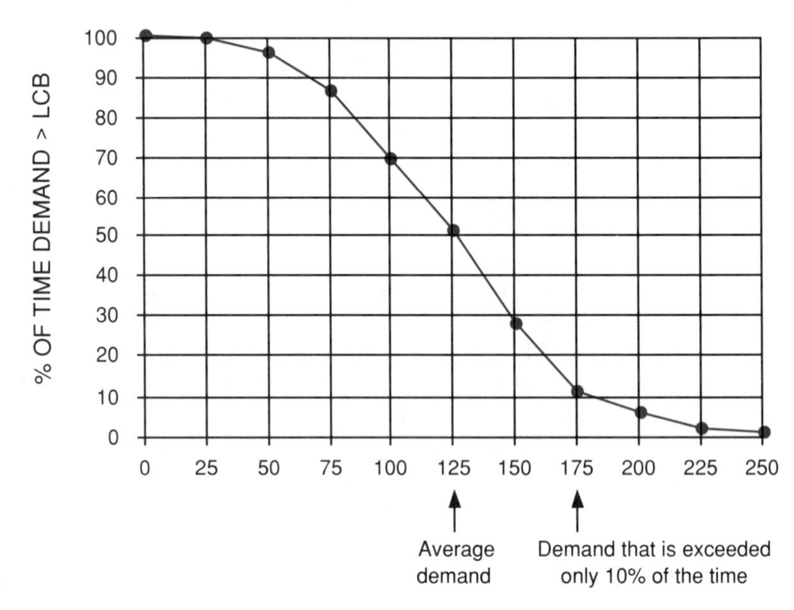

From the cumulative distribution, Joyce took the average demand to be 125 units during the two-week time it took shipments to arrive. She reasoned that if the firm had normally carried 50 units more than this, they would have met 90% of the demands over the past three years. Having done a thorough analysis of the past data, she felt fairly comfortable "issuing" her first directive as Supervisor of MSA. She directed that a two-week supply of irons (i.e., 125 units) be ordered whenever the inventory levels dropped to 175 units. This way they had 125 units available for average use and 50 units as a "safety stock" to offset any unexpectedly large demands. Fifty units was the minimum amount of inventory required to provide the 90% service level the marketing manager would accept.

Case Questions

(a) What type of distribution did Miss Brennan use (i.e., a "less than" or a "greater than")?

(b) Is the cumulative distribution curve she developed an "absolute" or a "relative" cumulative distribution?

(c) The values used to plot the cumulative distribution curve are not shown. Compute the values that correspond to the LCB points of 25, 125, 175, and 225.

(d) What justification did Miss Brennan have for using 125 units as an average demand rate for the two-week period?

(e) What does the decision to hold 50 units of safety stock assume about future demands?

(f) Suppose the marketing manager later decided that they must ship from stock 95% of the time, instead of 90%. Use the graph to estimate how many units of safety stock would be required to provide 95% service.

QUESTIONS AND PROBLEMS

20. Which of the four common bases for analysis or classification of data would be used for a table showing (a) sales by product line, (b) sales by product line by sales district, (c) sales by month by product line, (d) number of telephones installed by service district by month, (e) number of years telephone poles have been in use by kind of wood preservative treatment?

21. Distinguish between the terms:
    (a) discrete and continuous variables
    (b) raw data and arrays
    (c) class limits and class boundaries

22. Compare or contrast:
    (a) a histogram versus a frequency polygon
    (b) a "less than" ogive versus a "greater than" ogive
    (c) a bar chart versus a pie chart

23. Indicate what the appropriate stated class limits would be for frequency distributions designed to present the following data:
    (a) weights of 400 boxes of Spanish oranges to the nearest tenth of a pound, ranging from 126.1 to 132.3 pounds. Use seven or eight class intervals.
    (b) ages of 90 employees of a food processing plant, reported to the last birthday, and ranging from 16 to 70 years. Use no more than ten class intervals.
    (c) for 350 boxes of Florida fruit, the number of pieces of fruit per box, ranging from 220 to 259. Use no more than eight class intervals.

24. For each of the frequency distributions prepared in Problem 23, state:
    (a) whether the data are inherently discrete or continuous
    (b) the lower limit of the first class interval
    (c) the size of the class interval used
    (d) the midpoint of the first class interval

Following are census bureau estimates of selected state populations by age group (data in thousands of people with ages reported to last birthday):

	< 5 yrs	5–17	18–24	25–44	45–64	> 64 yrs
Maine	82	225	142	328	220	149
Mississippi	226	573	336	697	451	303
Wyoming	53	107	63	168	83	40
Hawaii	89	195	142	324	185	89

25. Construct a *histogram* showing the Maine population.
26. What would be the problem in constructing a *frequency polygon* showing the Mississippi population?
27. Construct an *ogive* showing the number of people "greater than" a given age in the state of Hawaii.
28. Construct *some form of chart* that will highlight any proportional differences in the different age groups for Maine and Wyoming.
29. Consider each of the sets of class intervals listed below:

(1)	(2)	(3)	(4)
20–24	20.0–24.9	20 and under 25	20–29
25–29	25.0–29.9	25 and under 30	30–39
30–34	30.0–34.9	30 and under 35	40–49
35–39	35.0–39.9	35 and under 40	50–99
40–44	40.0–44.9	40 and under 45	100–149
45–49	45.0–49.9	45 and under 50	150–199

(a) What is the lower boundary of the first interval for each distribution?

(b) What is the midpoint of the first interval for each distribution?

(c) What is the size of the first interval for each distribution?

(d) In which distributions are the observations recorded to the nearest full unit? nearest tenth of a unit? last full unit?

(e) Are the intervals of equal size in the first three distributions?

(f) Are the intervals in the fourth distribution of equal size?

30. The data below represent the weights (pounds) of boxes of canned goods collected for a food bank by a service club in their Christmas Drive.

Interval	Frequency
10–19	6
20–29	18
30–39	20
40–49	14
50–59	12
60–69	6
70–79	4
Total	80

(a) Are the data discrete or continuous?

(b) Were the observations (i.e., weight measurements) most likely made to the nearest one tenth of a pound? How do you know?

(c) How many class intervals were used to present this data?

(d) What is the size of the class intervals?

(e) What are the stated class limits of the interval having the highest frequency?

(f) What are the real class limits (i.e., boundaries) of the class having the lowest frequency?

(g) What are the midpoints of each class interval?

31. Use the data from Problem 30 to prepare the following graphs:

(a) a histogram (b) a frequency polygon (c) an ogive (F < UCB).

32. Use the data from Problem 30 to prepare

(a) a relative frequency distribution

(b) an ogive showing absolute values of F > LCB

(c) Add a relative scale on the right of your distribution in (b).

(d) Find the X values exceeded by 25%, 50%, and 75% of the observations. What are these values called?

(e) What is the ordinate (Y value) of your ogive at 29.5 on the X scale?

= **Computer Problems** The following data represent the number of years of operation of the 100 firms listed in the Corporate Statistical Database of Chapter 1 Appendix (and repeated here for convenience):

22	44	60	54	24	56	61	47	71	73	83
59	8	61	52	75	71	62	82	19	50	41
4	55	23	45	53	79	55	22	12	25	18
74	65	65	52	6	24	25	58	84	83	23
42	4	16	76	17	82	64	72	51	95	14
7	27	14	36	60	17	16	55	57	95	50
15	95	65	12	12	18	15	19	38	19	32
61	69	36	14	7	24	60	3	61	7	17
64	48	39	94	15	31	80	56	16	32	58
81										

33. Using the *years of operation* data for the 100 companies in the Corporate Statistical Database (repeated above), construct a *frequency distribution* using an interval width of 10 years. Start with class $0 < 10$.
    (a) How many class intervals result?
    (b) Which class interval includes the most number of companies?
    (c) How many of the companies have been in operation for 50 years or more?
    (d) Using the same data, reconstruct the frequency distribution using an interval width of 20 years. Now how many class intervals result?
    (e) Suppose you wished to determine the number of companies that had been in operation for 50 years or more from this new distribution. What problem do you encounter?
    (f) What generalization concerning number of classes versus amount of information can you draw from this comparison?
34. Using the same *years of operation* data (above) and a class interval of 10 years, construct a histogram showing the number of companies in operation for each class interval of the data.
35. Using the same *years of operation* data (above) and a class interval of 10 years, (a) construct a *cumulative distribution* showing the number of companies in operation less than 10 years, less than 20 years, etc., up to less than 100 years. What percent of the companies have been in operation for less than (b) 20 years? (c) 80 years?
36. Using the same *years of operation* data (above), and a class interval of 20 years, construct one or more of the following graphs:
    (a) a *histogram*
    (b) a *frequency polygon*
    (c) an *ogive*

37. Using the Corporate Statistical Database (Chapter 1 Appendix), construct one or more of the following charts to describe the data:
    (a) a vertical *bar chart* (i.e., column chart) showing the exports as a % of sales for firms number 41, 42, 57, 74, 88, and 89.
    (b) a *pie chart* showing the proportion of firms in the various industry classifications (i.e., Banking, Chemical, Electrical, Retail, Transportation, and Utility).

CHAPTER

Summary Measures of Data

INTRODUCTION

MATHEMATICAL NOTATION

MEASURES OF CENTRAL TENDENCY (UNGROUPED DATA)

The Mean and Weighted Mean
Median
Quartiles and Percentiles
Mode
Comparisons and Other Special Purpose Means

MEASURES OF DISPERSION (UNGROUPED DATA)

Range
Mean Absolute Deviation (MAD)
Variance and Standard Deviation

SUMMARY MEASURES OF GROUPED DATA

Measures of Central Tendency (Grouped Data)
Measures of Dispersion (Grouped Data)

SHAPES OF DISTRIBUTIONS: SKEWNESS AND KURTOSIS

COMPARISON MEASURES: THE COEFFICIENT OF VARIATION

FINDING NUMERICAL MEASURES ON THE COMPUTER

SUMMARY

CASE: PFAFNER, JEFFREY AND JONES INVESTMENT COMPANY

QUESTIONS AND PROBLEMS

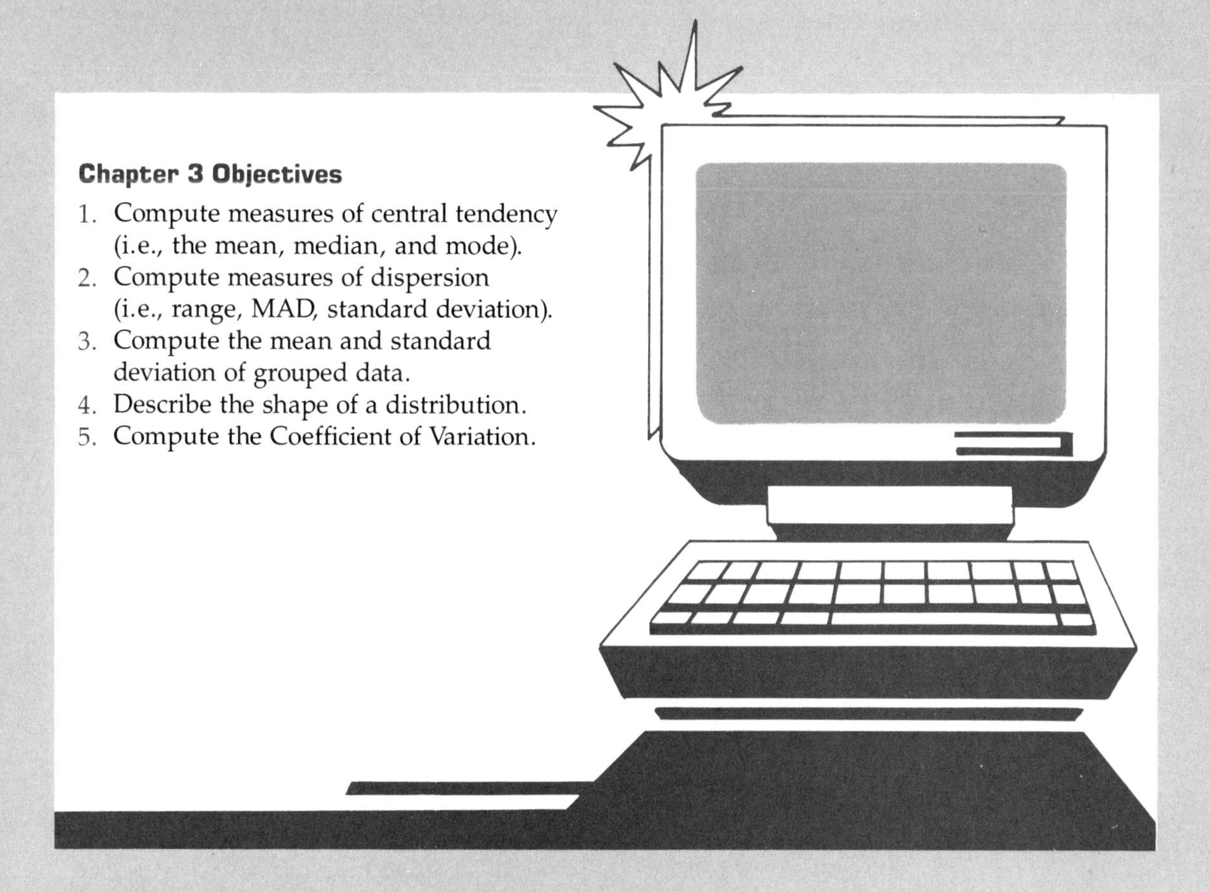

Chapter 3 Objectives

1. Compute measures of central tendency (i.e., the mean, median, and mode).
2. Compute measures of dispersion (i.e., range, MAD, standard deviation).
3. Compute the mean and standard deviation of grouped data.
4. Describe the shape of a distribution.
5. Compute the Coefficient of Variation.

How are we doing—as Americans? Our population is now 250 million with a median age of 30.6 years. Males weigh an average of 173 pounds, and females average 142 pounds. We watch an average of 7 hrs of television per day and consume 18 pounds of ice cream per year. In 1970, only 55% of our population had completed high school; by 1982 this increased to 71%, with nearly 18% completing college. Over 90% of all workers are employed; median family income is $25,735, and the average weekly pay of factory workers is nearing $400 per week. The average value of an acre of farmland ($739) is a little less than that of a personal computer—a market now worth about $18 billion per year. About 15% of the population are below the "official" poverty level. Heart disease and cancer account for more than a million deaths per year (38% and 21% respectively of all deaths), but the trend in AIDS-related deaths is rising rapidly. Annual per capita consumption of beer has leveled off at about 24 gallons per person. About 32% of the adult population in the U.S. smoke, and for those that smoke, the average is 195 packs per year. The divorce rate is still increasing; 22% of all youngsters are living in single parent homes.

INTRODUCTION

We are bombarded with statistics like the above every day! Fortunately many of the statistics we get are already condensed in one way or another. As you already know, numerical values are more condensed than the tables and graphs we worked with in the last chapter. This makes them convenient to convey and use, but it also means that some of the individual characteristics of the data are lost.

Figure 3–1 relates our forthcoming discussion to the material already studied in Chapters 1 and 2. After introducing the concept of data in Chapter 1, we explored the use of tables, frequency distributions, and graphs for presenting data in Chapter 2. Now it is time to go on to the calculation of numerical summaries of data, as in the "statistics" above. As shown in Figure 3–1, data from the business environment are typically first arranged into arrays or tables and can be presented as frequency distributions and graphs or condensed directly by calculation. When we compute numerical descriptions directly from the arrays and tables of data, we are working with *ungrouped data*. This is where we will begin. Later we will see what modifications must be made to obtain the same numerical descriptions from data that have already been *grouped* into frequency distributions.

Suppose you are the manager of the West Virginia Disaster Relief Agency, and you have just received an emergency phone call. There has been an *"explosion at a pharmaceutical plant and poisonous chemicals have been released."* Now the problem is in your hands! Your first two most urgent questions might well be: (1) Where is the plant (**location**) and (2) how widespread is the damage (**dispersion**). Once you find that out, you might also want to look in your records to learn if there were any similar types of emergencies. You can then (3) **compare** the current situation (e.g., contamination level) with that in earlier periods, or compare the dispersion with other chemical spills. These very basic measures of data are the subject of this chapter—only we'll refer to routine business operations rather than emergencies.

We will review some essential mathematical notation before getting into the heart of the chapter. Then we will take up **(1)** three measures of location, or central tendency, (i.e., the *mean, median,* and *mode*) and **(2)** four measures of dispersion (i.e., the *range, absolute deviation, variance,* and *standard deviation*). A related description of location and dispersion is the shape of the distribution. Following this we look briefly at **(3)** the *coefficient of variation*, which is used for comparing the dispersion of two (or more) sets of data. These are all fairly common measures of data, and are not conceptually difficult. In addition, microcomputer programs to compute all these statistics are readily available. By the time you finish the chapter, you should feel comfortable working with any of these measures.

FIGURE 3–1 Collection and description of statistical data

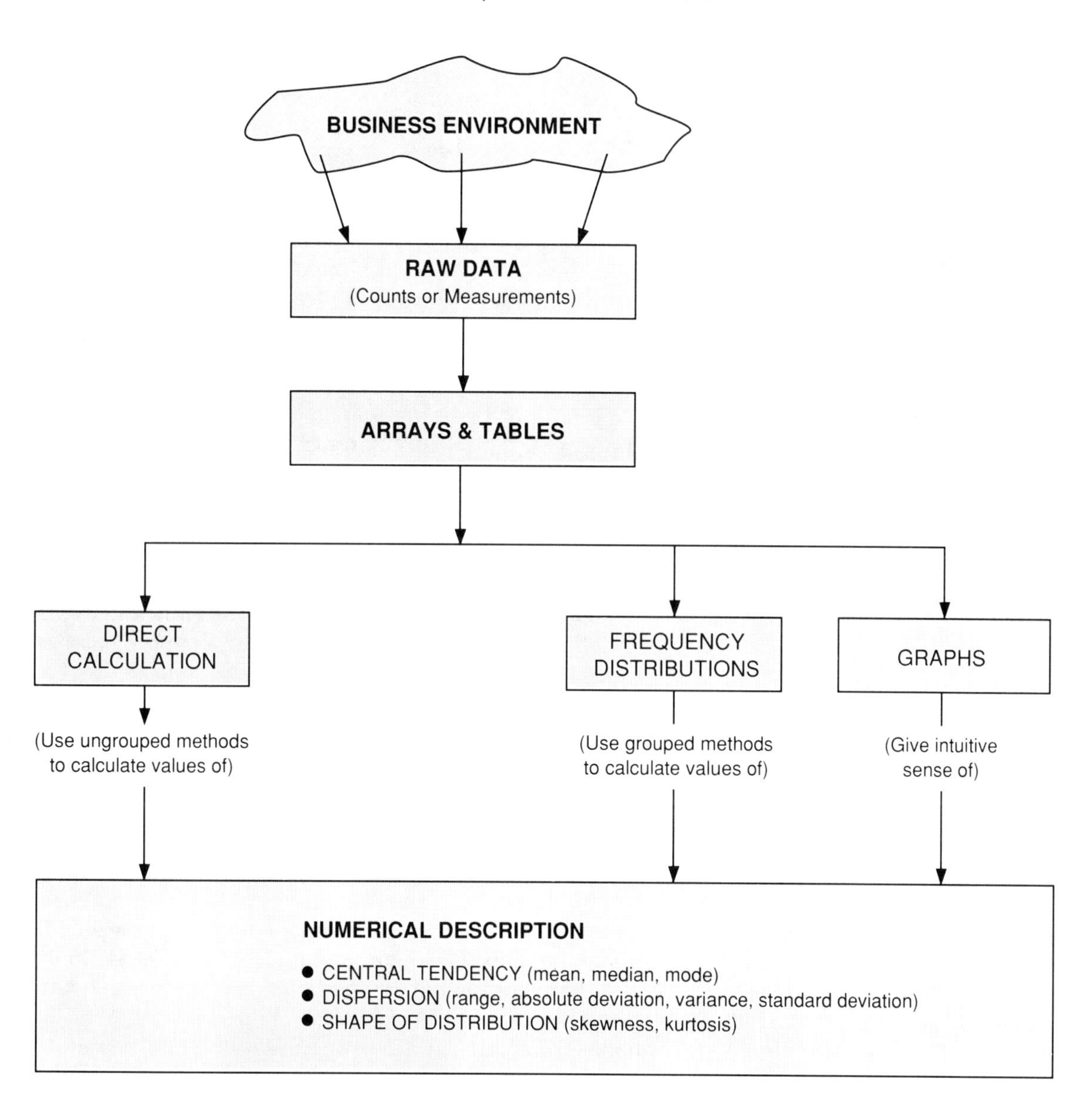

MATHEMATICAL NOTATION

Three mathematical details that merit attention in statistics are *significant digits, rounding,* and *summation notation.*

SIGNIFICANT DIGITS Discrete data takes on integer values only, so the counts are exact numbers. However, continuous data such as weight and distance are recorded to the nearest measurement. Reported values are never as "exact" as they might be. So, how accurate is an answer? How many digits are significant? *Digits that correctly convey the accuracy of the measurements are significant, and digits that merely locate the decimal point are not.* For example, if a shipment of goods was measured to the nearest pound and weighed exactly 24,900 pounds, the number 24,900 would have five significant digits. However, if the weight was measured to the nearest one hundred pounds, the number 24,900 would have only three significant digits. Review the measurements below. The number of significant digits in each is shown in parentheses:

24,900 pounds	(3 to 5)	3.2 million miles	(2)
5,602 pounds	(4)	239,000 miles	(3 to 6)
4.026 pounds	(4)	16.34 miles	(4)
.026 pounds	(2)	.280 miles	(3)

MULTIPLICATION AND DIVISION The number of significant digits resulting from the multiplication or division *should not exceed the number of significant digits in the original number with the fewest significant digits.* For example, the number of significant digits in the answer to each problem below is shown in parentheses:

$3.26 \times 4{,}082$	(3)	$16.04 \div 123.0$	(4)
$1.02 \times .0032$	(2)	$16.04 \div .0123$	(3)
$3{,}465 \times 1.0032$	(4)	$326.0 \div 1.0123$	(4)

ROOTS AND POWERS The rules for extracting roots and raising numbers to powers are the same. For example, the square root of 436.2 cannot have more than four significant digits, no matter how many digits are computed. Thus it is inappropriate to report the square root of 436.2 as 20.885402 just because your calculator happens to report eight digits. Final answers should report only significant digits (e.g., 20.89). However, in intermediate stages of cumulative calculations, extra digits are typically carried to reduce rounding error. Also, in solving a problem with both multiplication and division, it is better to multiply first and divide last to reduce the rounding error.

ADDITION AND SUBTRACTION The rule for the addition and subtraction of numbers is based upon the position of the numbers relative to the decimal. The answer should have *no more decimal places than there are in the number with the fewest significant decimal places.* Thus, if the numbers are placed in a column with their decimal points aligned and a vertical line drawn after the column where the first digit ceases to be significant, the digits to the left of the line are significant and those to the right are not.

Example 3–1 Find:

(a) the sum of 32.06 pounds, 156.2 pounds, and 953.872 pounds
(b) the difference between 1,826.36 pounds and 196.2 pounds

Solution

(a) Add

32.0	6
156.2	
953.8	72
1,142.1	32
significant	not significant

(b) Subtract

1,826.3	6
-196.2	
1,630.1	6
significant	not significant

The answers would be reported as (a) 1,142.1 pounds and (b) 1,630.2 pounds. [Note that the extra digits were used to minimize rounding error.]

ROUNDING CONVENTIONS When rounding digits in the final answer, round up or down to the nearest digit. If the digit to be dropped is a 5, round up to the next even number (or drop the 5 if the last nonzero digit is already even). For example, in (b) of Example 3–1, the rightmost 6 in the answer causes the 1,630.16 to be rounded up to 1,630.2; in (a) the 32, being less than 50, is dropped.

CARRYING EXTRA DIGITS Statistical calculations often necessitate that a series of calculations be performed in sequence. Analysts frequently carry several extra digits in the intermediate computations to avoid the effect of cumulative error. Most calculators do this automatically, even though the extra digits are not displayed. This is especially important when working with the squares and square roots of the differences between values, as we shall be doing throughout the text.

SUMMATION NOTATION Statistical calculations frequently involve adding individual values of a variable. In statistics, the upper case Greek letter sigma (Σ) is used as mathematical shorthand to indicate the "*summation* of" the values of the variables to be added. For example, suppose X represents some variable (e.g., number of days worked). Then X_1 (i.e., "X sub 1") could represent the days worked by one worker, X_2 those worked by the second, and so forth. If there are n (n being the total number) workers, then the total days worked by all n workers would be designated as $\sum_{i=1}^{n} X_i$ where the i designates the variables in sequence. This notation means to sum the values of X beginning with $i = 1$ and continuing through $i = n$. Because we usually sum all the values from the first to the nth observation we will omit the $i = 1$ *to* n designation unless only some of the values are to be summed. Other summations frequently used in statistics are ΣX^2, which designates the summation of the squared X values, and $(\Sigma X)^2$, which means summation of X-the-quantity, squared. Note carefully the difference in these last two terms.

Example 3–2 Five engineers worked the following days on a construction project.

$$X_1 = 4 \quad X_2 = 2 \quad X_3 = 7 \quad X_4 = 9 \quad X_5 = 3$$

Find (a) ΣX, (b) ΣX^2, and (c) $(\Sigma X)^2$ (d) What would be $\sum_{i=1}^{3} X_i$?

Solution

We show the computations in detail here to ensure a good understanding:

i	X	X²
X_1	4	16
X_2	2	4
X_3	7	49
X_4	9	81
X_5	3	9
	25	159

Thus: **(a)** $\Sigma X = 25$
(b) $\Sigma X^2 = 159$
(c) $(\Sigma X)^2 = (25)^2 = 625$
(d) $\sum_{i=1}^{3} X = 4 + 2 + 7 = 13$

Sometimes we will find it necessary to multiply frequencies (f) by the value of a variable (X). This will be designated as ΣfX, which means to first multiply the individual values of f times X and then sum them. Note that this is different from $\Sigma f \, \Sigma X$, which calls for first summing the f values and the X values, and then multiplying the totals. Finally, when a summation instruction involves parentheses, start with the operation inside the innermost parentheses first. For example, to compute the values of $\Sigma (X_i - 5)^2$ for the data in Example 3–2, we would add:

$$(4-5)^2 + (2-5)^2 + \ldots + (3-5)^2$$

to obtain the final answer of 34. The summation sign is a convenient mathematical symbol for addition that is used frequently throughout the text.

EXERCISES I

1. How many significant digits are in each of the following measurements?
   (a) 3.1416
   (b) 2.040
   (c) 2.8 million
   (d) .0046

2. What is the maximum number of significant digits in the answer to each of the following?
   (a) \$126.42 ÷ 1.375
   (b) 1413 × 1.64
   (c) square root of 168.4
   (d) 126.4 + 4.28
3. Find the following for the set of X variables: 1, 4, 5, and 6:

   (a) ΣX (b) $\Sigma (X)^2$ (c) $(\Sigma X)^2$ (d) $\sum_{i=1}^{3} X^2$ (e) $\Sigma (X - 4)^2$

MEASURES OF CENTRAL TENDENCY (UNGROUPED DATA)

Our first concern is for a single number that describes the typical size (or location) of the variable of interest. We seek a representative point on a scale around which the other observations tend to be clustered, and refer to these points as **measures of central tendency**.

THE MEAN AND WEIGHTED MEAN

The **arithmetic mean** is the most widely used and most important of all averages. In fact, the word "average" usually refers to the arithmetic mean unless stated otherwise. It is calculated by summing the individual values of the variable (X), and dividing by the number of observations. If the data are from a sample of size n, the equation for the mean, $\overline{X}$, is:

$$\overline{X} = \frac{X_1 + X_2 + X_3 + \ldots + X_n}{n} = \frac{\sum_{i=1}^{n} X_i}{n}$$

which we will shorten to

$$\overline{X} = \frac{\Sigma X}{n} \tag{3–1}$$

If the X-values constitute a population of size N, the mean is designated as the Greek letter mu (μ), so we have

$$\mu = \frac{\Sigma X}{N} \tag{3–2}$$

Example 3–3 Computer Graphics Ltd. has 300 sales representatives in England. The number of sales calls made by a sample of five representatives last week were 12, 4, 9, 0, and 5. What is the mean number of calls made?

Solution

Because this is a sample, we designate the mean as $\overline{X}$.

$$\overline{X} = \frac{\Sigma X}{n} = \frac{30}{5} = 6 \text{ calls}$$

The mean is stated in the same units as the original data (e.g., number of calls). It is influenced by every value (even the zero), and is essentially the balance point around which the observations center. It is sometimes referred to as the "first moment" of the data, as illustrated in Figure 3–2.

FIGURE 3–2 The mean as a balance point

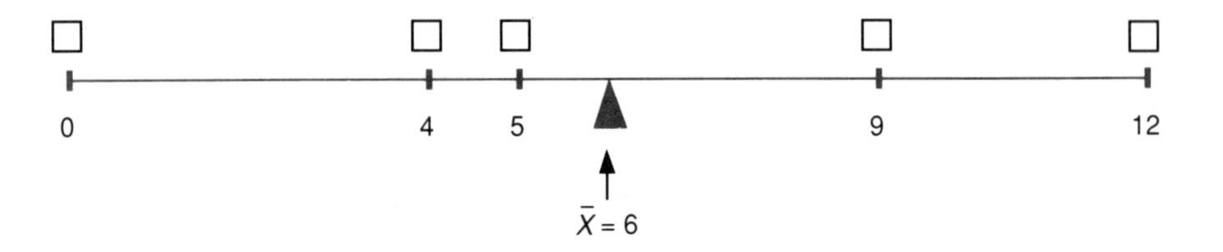

As a point of balance, the sum of the deviations from the mean (of 6) in the positive and negative directions offset each other. For example:

NEGATIVE DEVIATIONS	[offset]	POSITIVE DEVIATIONS
(0 − 6) + (4 − 6) + (5 − 6)	[offset]	(9 − 6) + (12 − 6)
−9	[offsets]	+9

WEIGHTED MEAN The arithmetic mean automatically assigns equal weight to each observation. However, in some instances we may wish to *weight* the different scores by some measure of their individual importance. This is easily accomplished by multiplying the individual (X) values by their assigned weight (w) and dividing by the sum of the weights (instead of the number of observations).

$$\overline{X}_{wt} = \frac{\Sigma (Xw)}{\Sigma w} \qquad (3\text{–}3)$$

Grade point averages are weighted averages with which you may be familiar. In the calculation of grade point averages, the letter grades (A, B, C, D) are assigned points (e.g., 4, 3, 2, 1) and then weighted by the number of credit hours.

Example 3–4 A student received the following term grades. All were three credit hour courses except statistics (4 hr) and P.E. (1 hr):

English	B	Statistics	A	History	C
Logic	A	Psychology	B	P.E.	A

Compute the term g.p.a. letting A = 4, B = 3, C = 2, and D = 1.

Solution

		VALUE	WEIGHT (hrs)	
COURSE	**GRADE**	**X**	**w**	**Xw**
ENGLISH	B	3	3	9
LOGIC	A	4	3	12
STATISTICS	A	4	4	16
PSYCHOLOGY	C	2	3	6
P.E.	A	4	1	4
		Totals	14 (hrs)	47 (points)

$$\overline{X}_{wt} = \frac{\Sigma(Xw)}{\Sigma w} = \frac{47}{14} = 3.36$$

Note that the grade values (X) are discrete, and the weights (w) are accurate to more than two digits beyond the decimal, so the result (3.36) is expressed at whatever level of accuracy is commonly used.

Other weights may be such things as the cost per unit, or a percentage of errors or defectives. However, if the percentages do not add up to 100%, the numbers will not have a common base, and the percentages must be converted to numbers as in Example 3–5.

Example 3–5 Find the average percent defective, given the following data obtained from a survey:

WORKMAN	NO. UNITS PRODUCED	PERCENT DEFECTIVE	[For Solution] NO. DEFECTIVE
A	10,000	6.1%	610
B	20,000	2.2%	440
C	12,000	3.8%	456
	42,000		1,506

Solution

We cannot directly compute the arithmetic mean of the Percent Defective column because the percentages refer to different amounts produced. They do not have a common base. Therefore, we first multiply the number produced by the percent defective to obtain the number defective (shown on right above). Since there were 1,506 defective units out of 42,000 produced, the combined percent defective is equal to:

$$\text{Weighted } \% = \frac{\text{Number Defective}}{\text{Total Number}} = \frac{1{,}506}{42{,}000} = .03586 = 3.6\%$$

MEDIAN

The **median** is that point on the number line such that an equal number of the scores lie above it and below it. It is typically the "halfway" point in the data, for at least half of the scores will have values less than or equal to the median, and half will have values greater than or equal to the median.

To find the median of ungrouped data, simply array the data in ascending or descending order and select the middle item; its value is the median. In equation form, the position of the median in an array of n ungrouped data points is:

$$\text{Position of median} = \frac{(n + 1)}{2}$$

For example, if we arrange the five observations from Example 3–3 into an array, we have: 0, 4, 5, 9, and 12 calls. The median is in the $(n + 1) \div 2$ position which is the $(5 + 1) \div 2$ or 3rd position. So it has the value of 5 calls.

If the number of observations is even, the median is assumed to be halfway between the two middle numbers. Suppose there were six observations in the sample, and the array was 0, 4, 4, 5, 9, and 12. In this case the median would be in the $(6 + 1) \div 2 = 3.5$th position, which would have a value of $(4 + 5) \div 2$ or 4.5.

The median is sometimes called the "position" average by virtue of its position in the center of the data. Because it is not based on the total value of X, it is not affected by the extreme items, and may be a better average than the mean when a few extremely large (or small) values would distort the mean. For example, if the 12 in the above array was replaced by 38, the mean would be 10, i.e., more calls than all of the salespersons except one. The 10 may not be as good a standard for comparing performance of the salespersons as the median, which is still 4.5. Note, however, that the median does not provide as much information as the mean. It is possible to determine the total from the mean and the number of obervations—this is not the case with the median.

QUARTILES AND PERCENTILES

Closely related to the positional concept of the median are the concepts of *quartiles, deciles*, and *percentiles*. Whereas the median is at the "center" of the array and divides the data into halves, quartiles divide it into quarters, deciles into tenths, and percentiles into one hundredths. These measures of "noncentral" location are all special cases of the more general designation called *quantiles*. Quantiles are numbers between 0 and 1 that apportion ordered data. Thus the .50 quantile is the second quartile, the fifth decile, and the fiftieth percentile.

QUARTILES The first quartile, Q_1, is the value in the data set at or below which 25 percent of the data points lie. Next comes the median, or second quartile, Q_2. The third quartile, Q_3, is then the value at or below which 75 percent of the data points lie.

When a small number of discrete data points are involved, the quartiles do not always mark the exact 25%, 50%, and 75% points that would correspond to a continuous scale. This is because it is customary to express quartiles either as integers or as points midway between two integers. For example, if we divide only six observations into four groups we should have $6 \div 4 = 1.5$ observations in each quartile. Using the same sample observations as above, we can identify the quartiles as shown below:

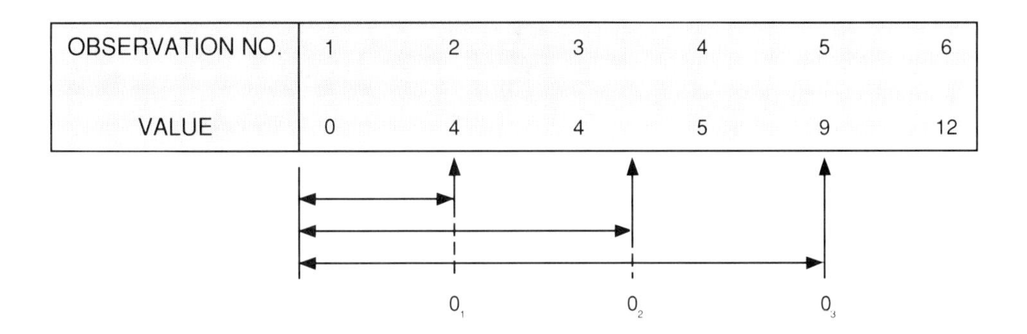

OBSERVATION NO.	1	2	3	4	5	6
VALUE	0	4	4	5	9	12

Note that the first quartile encompasses the first observation and half of the second, the second quartile (the median of 4.5) encompasses 2×1.5 or three observations, and the third quartile encompasses 3×1.5 or 4.5 observations. The following locational equations can be used to find the positional location of quartiles to be determined from ungrouped data in an array.

Quartile	**Location corresponds to value of observation in the**
Q_1	$\frac{(n+1)}{4}$th ordered position
Q_2	$\frac{2(n+1)}{4}\text{ th} = \frac{(n+1)}{2}$ th ordered position
Q_3	$\frac{3(n+1)}{4}$ th ordered position

When the locational equation yields an integer, the observation in that position is accepted as the quartile. If the equation yields a point midway between two positions, the mean of the observed values in those two positions is taken as the quartile. If the result is neither an integer nor halfway between two integers, the quartile is assumed to be the value of the observation in the nearest integer position. For example, fractional positions such as the 6.25 and 14.75 would be assigned to the values of the 6th and 15th observations respectively. We can summarize the rule for finding Q_1 and Q_3 of ungrouped data in an array as follows:

Finding Quartiles To find Q_1, if $(n + 1)/4$ is an integer, then Q_1 is the value of the observation in that position of the array. If it is a point halfway between two positions, then Q_1 is the mean of the values in those two positions. If it is neither an integer nor halfway between two integers, then Q_1 is the value of the observation in the nearest integer position. Follow the same rule for Q_3 except use $3(n + 1)/4$.

Example 3–6 Use the locational equations to find Q_1 and Q_3 for the two sets of data:

(a) $n = 5$ observations
0 4 5 9 12

(b) $n = 6$ observations
0 4 4 5 9 12

Solution

The position of the quartiles for the two sets of data are:

(a)

$$Q_1 \quad \frac{(n+1)}{4} = \frac{(5+1)}{4}$$
$$= 1.5\text{th position}$$

Q_1 value = midpoint of (4 + 5)
= (4 + 5)/2 = 4.5

$$Q_3 \quad \frac{3(n+1)}{4} = \frac{3(5+1)}{4}$$
$$= 4.5\text{th position}$$

Q_3 value = midpoint of (9 + 12)
= (9 + 12)/ 2 = 10.5

(b)

$$Q_1 \quad \frac{(n+1)}{4} = \frac{(6+1)}{4}$$
$$= 1.75\text{th position}$$

Q_1 value = value of 2nd observation
= 4

$$Q_3 \quad \frac{3(n+1)}{4} = \frac{3(6+1)}{4}$$
$$= 5.25\text{th position}$$

Q_3 value = value of 5th observation
= 9

PERCENTILES **Percentiles** are quantiles that divide the data into 100 equal subgroups. They are most appropriate for large data sets where a simple percentage can be applied to the number of observations. For example, a survey of several hundred workers may yield earnings data that show 42 percent of factory workers earn less than $9.50 per hour. Then $9.50 would be the 42nd percentile. Similarly, 85 percent of the observations would be less than (or to the left of) the 85th percentile, etc.

As noted in Chapter 2, cumulative frequency distributions (ogives) often have a relative, or percent, scale on the y-axis. This enables us to estimate the percentile, or percent of observations *"less than"* or *"greater than"* a specified class limit of grouped data.

MODE

The **mode** is the most frequently occurring value in a data set. It is simple to identify after the data have been arrayed or tallied. For example, the mode of the array 0, 4, 4, 5, 9, and 12 is 4. If the array had only one 4, it would not have a mode. And if it had two 9s as well as two 4s, it would have two modes, i.e., it would be *bimodal*.

The mode is the least important of the three measures of central tendency, but it does help identify a prominent feature of a data set. For example, the *modal* (most commonly occurring) PE ratio of the 100 firms in the Corporate Statistical Database (Appendix A) is 12 because the greatest number of firms (14 firms) have a PE ratio of 12.

COMPARISONS AND OTHER SPECIAL PURPOSE MEANS

COMPARISON OF MEAN, MEDIAN, AND MODE Although the mean, median, and mode are all "averages," they do differ. The mean is affected by every data point, and extreme values can even pull it away from the main concentration of data points. For example, if a data set of salaries has a few extremely high salaries, the mean tends to be higher than the median. The median, being a middle value only, is not influenced by the value of other data points. It may be more representative of the center of the observations, but it conveys less information than the mean—and is not as useful as the mean for purposes of statistical inference. The mode is even less useful, but it does indicate where a major cluster of data lies.

In addition to the averages discussed above, some special purpose averages have been developed to deal with rates. Full treatment of them is beyond our scope here, but you should be alerted to situations where the (1) *harmonic* and (2) *geometric* means apply.

The **harmonic mean** is sometimes used to compute an average of quantities per unit of time when the times per unit vary. For example, if workers A, B, and C took 10, 12, and 20 minutes per unit to assemble a product, the average assembly time *would not be* (10 + 12 + 20) ÷ 3, or 14 minutes per unit. This average is incorrect because the number of units produced in a given time varies from one worker to another. In essence, the simple arithmetic average fails to account for the fact that only 10 minutes of worker A's time are incorporated in the average, whereas 20 minutes of worker C's time are included. This is because the times required per unit vary from worker to worker.

More advanced statistical reference texts may include equations for computing the harmonic mean, but that is beyond the scope of our inquiry in this chapter. Alternatively, harmonic mean problems can often be solved simply by restating them in a common time period, such as output per hour. For example, the output per hour of A is 60 min/hr ÷ 10 min/unit, or 6 units per hour. B and C would then be 5 and 3 units per hour respectively. This means that the workers average (6 + 5 + 3) ÷ 3, or 4.67 units per hour. Dividing this into 60 minutes yields the correct average of 12.86 minutes per unit.

The **geometric mean** is used to find the average rate of change in a series of percentages where the individual percentage rates are known. It is the nth root of the product of n factors:

$$G = \sqrt[n]{(X_1)\,(X_2)\,.\,.\,.\,(X_n)} \tag{3–4}$$

The geometric mean is useful for averaging growth rates such as sales or productivity.

Example 3–7 Sales of Video Record Company increased 10% the first year, decreased 5% the second, increased 40% the third, and increased 20% the fourth. What is the average rate of increase?

Solution

Since each rate is based on the preceding year, the percentages do not have a common base. The 10% increase in the first year is 110% of the preceding year, so our first value in the equation is 1.10. Stating each year as a multiple of the preceding year, the geometric mean would be:

$$G = \sqrt[4]{(1.10)\,(.95)\,(1.40)\,(1.20)} = 1.15$$

Therefore the average growth rate is .15 or 15%. [Note: Taking the 4th root is equivalent to taking the square root twice.]

The geometric mean is a very effective way of averaging rates of change in percentages where each period is a function of the previous period.

EXERCISES II

4. What type of average would be most appropriate for determining
   (a) the average income of employees of Sacred Heart Hospital
   (b) a course grade where projects count 15%, quizzes 30%, and exams 55%
   (c) the average growth rate in productivity over a five-year period
5. A telephone service man installed the following number of phones on successive days: 4, 7, 5, 8, 5, 6, 2, 7, 4, 5. Find the (a) mean, (b) median, and (c) mode.
6. For the phone installation data above (i.e., 4, 7, 5, 8, 5, 6, 2, 7, 4, 5) find (a) Q_1 and (b) Q_3.

7. In a summer month, the number of passengers denied boarding (bumped) by major air travel carriers was as shown below.[1]

American	165	Piedmont	184
Delta	55	Trans World	252
Eastern	172	United	436
Northwest	117	USAir	207
Pan American	280	Western	408
People Express	396		

For this data, determine (a) the median number of passengers denied boarding per 100,000 passengers flown and (b) the first and third quartiles. (c) Explain the meaning of Q_3.

8. Evergreen Nursery purchased 1,100 fruit trees from an Iowa supplier at the costs shown below. Find the (weighted) average cost per tree.

DATE	NO. OF TREES	COST PER TREE	INVOICE COST
Mar. 10	200	$5.00	$1,000
April 12	500	4.00	2,000
May 17	100	6.00	600
Oct. 30	300	4.50	1,350

9. Data were collected on the time required by three employees to handle complaints at a large department store. The times required by customer service representatives Doris, Jim, and Randy were 6, 15, and 10 minutes per customer, respectively. Assuming this *rate* is maintained over a long period, how many customers per hour can be serviced on average?

10. During the past four years, a New York mutual fund specializing in international investments has experienced the following growth in their share price from the prior year:

First year = 18% increase	Third year = 4% decrease
Second year = 12% increase	Fourth year = 35% increase

What is the average rate of change in the share price of this mutual fund?

[1] *The Wall Street Journal*, Nov 10, 1986, p. 23. Data are for each 100,000 passengers flown, as collected from the Department of Transportation and individual airlines, and include volunteers who left the plane and received compensation.

MEASURES OF DISPERSION (UNGROUPED DATA)

Practically all business data varies. Bicycle rim diameters vary from one to the next, cables are not uniformly strong, cans and bottles contain slightly different amounts, and light bulbs do not burn for exactly the same number of hours. Although all such differences are caused by something, we usually do not know (or don't bother with) why the differences exist. We simply accept them as an "inherent variability" in the product, process, or situation.

But some measure of variability is a useful way to characterize a set of data. This is because the level of variability in the data affects the accuracy of the inference we make from the data. For example, suppose you are managing a project that involves construction of a 10-story office building as shown in Figure 3–3. Your company has hired a contractor to do the work, but you still have overall responsibility for the project. Your contractor has proposed to lift a 100,000 lb roof section with a cable that has an average lift strength of 120,000 lbs. The cable is made of hundreds of small strands of steel wire twisted into a ropelike cord. You would probably feel a little more comfortable knowing the strength of the cables varied between only plus or minus 5,000 lbs, rather than plus or minus 25,000 lbs.

FIGURE 3–3 Cables with same mean and different dispersion

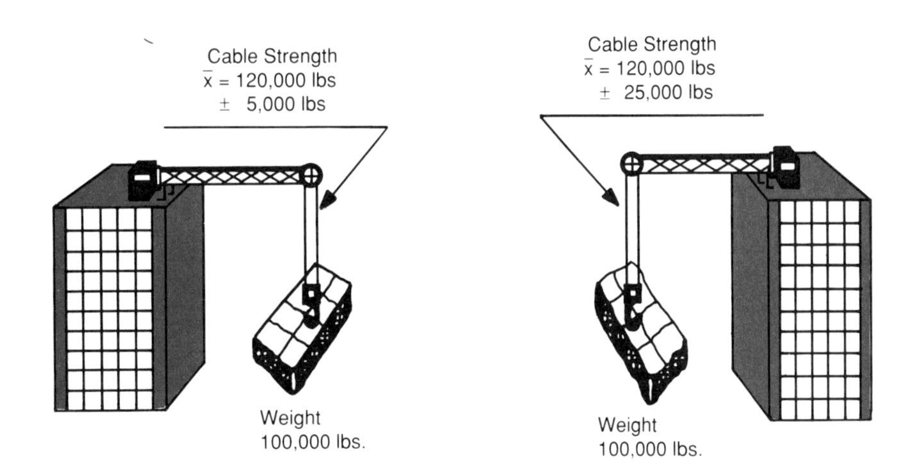

In this section, we consider four commonly used measures of dispersion : (1) *range*, (2) *mean absolute deviation*, (3) *variance*, and (4) *standard deviation*. For now, our discussion will utilize the (ungrouped) data in Table 3–1 which represents the individual breaking strength (lbs) of a *sample* of 10 wire strands used in a steel cable, such as in Figure 3–3. The strength measurements on the strands have been arranged in an array.

TABLE 3–1 Strength of sample of ten strands of steel wire

STRAND NUMBER	1	2	3	4	5	6	7	8	9	10
STRENGTH (LBS)	54	62	63	65	68	71	73	78	82	84
			↑ Q_1		↑ Q_2			↑ Q_3		

RANGE

The **range** is the difference between the largest and smallest values in the data set. In Table 3–1 the range is 84–54, or 30 lbs. Ranges are used in financial analysis, quality control work, and personnel reports because they are so easy to find. However, a single extreme data point can cause a distorted picture of the dispersion. For example, suppose a weakened strand of wire exhibited a test strength of only 4 pounds. The range of 80 pounds would not be representative of the bulk of the data in Table 3–1. Although the range is simple to calculate, it fails to make full use of the data (e.g., by taking account of any clusters of data points).

The **interquartile range** eliminates the extremes by using the difference between the first and third quartiles of the data, rather than the difference in end points. That is, it starts at the 25th percentile, extends only over the middle 50% of the data, and ends at the 75th percentile. For example, from Table 3–1, the first quartile is at $Q_1 = 63$ and the third at $Q_3 = 78$. The interquartile range, IR, is then:

$$\text{Interquartile Range} = Q_3 - Q_1 = 78 - 63 = 15 \text{ lbs}$$

MEAN ABSOLUTE DEVIATION (MAD)

A measure of the average deviation of the scores from some central point makes better use of the information available by taking every value into account. The *"central point"* is normally the arithmetic mean, although a case can be made for using the median. However, as noted earlier, the mean is a *"balance point."* If we were to sum the positive and negative deviations from the mean, the total would be zero. We could not compute an arithmetic average deviation. Fortunately, we are able to circumvent this by summing the *absolute* deviations about the mean instead.

The **Mean Absolute Deviation (MAD)** is the average deviation of values from their mean, and is obtained by summing the absolute deviations and dividing by the number of values, *n*.

$$\text{MAD} = \frac{\Sigma|\overline{X} - X|}{n} \tag{3–5}$$

Example 3–8 Find MAD for the data of Table 3–1.

Solution

First we must determine the mean of the data.

$$\bar{X} = \frac{\Sigma X}{n} = \frac{700}{10} = 70 \text{ lbs}$$

Then, the absolute deviations from the mean are:

	SCORE X	DEVIATION FROM $\bar{X}$ $X - \bar{X}$	ABSOLUTE DEVIATION $\lvert X - \bar{X}\rvert$
	54	−16	16
	62	−8	8
	63	−7	7
	65	−5	5
	68	−2	2
	71	1	1
	73	3	3
	78	8	8
	82	12	12
	84	14	14
TOTALS	700	0	76

$$\text{MAD} = \frac{\Sigma\lvert X - \bar{X}\rvert}{n} = \frac{76}{10} = 7.6 \text{ lbs}$$

MAD is used frequently in forecasting and inventory control activities to assess the extent to which current demand (X) for a product is deviating from past average demand ($\bar{X}$). If MAD is too large, it means the firm's forecasting or control systems are not functioning adequately and costs may be getting out of line.

VARIANCE AND STANDARD DEVIATION

Two of the most useful of all statistical measures are the *variance* and the *standard deviation*. They provide information about the uniformity of a data set and about the reliability of other statistics obtained from samples. We will use them throughout the text, so a very good understanding of them is vital. Fortunately, they are simple to calculate (but tedious—hence the advantage of a calculator or computer). And if you compute one, it is easy to obtain the other.

The **variance** is the average of the squared deviations of values from their mean, and the **standard deviation** is the square root of the variance.

There is a slight difference between population and sample values. A population variance is designated by the lowercase Greek letter sigma, σ^2. It is computed by subtracting the mean (μ) from each score (X), squaring the difference, adding the squared values, and dividing by the number of observations, N.

$$\text{Population Variance} \qquad \sigma^2 = \frac{\Sigma (X - \mu)^2}{N} \tag{3–6}$$

The population standard deviation, σ, is then simply the square root of the population variance.

$$\text{Population Standard Deviation} \qquad \sigma = \sqrt{\frac{\Sigma (X - \mu)^2}{N}} \tag{3–7}$$

We seldom work with the population variance and standard deviation because they can be obtained only if a complete census is taken of the whole population. That is, we need to know the true mean, μ, before either of these parameters can be calculated. Nevertheless, even though we rarely know or calculate either μ or σ, they do exist for every population. Much of the inference we use in the latter chapters of the text relates to estimating their values or making statements about them.

The more common situation in business is to work with only partial (or sample) information. However, sample information is always incomplete, so we must make a minor adjustment to the above formulas when applying them to sample data. Otherwise, the results would have a slight downward bias. This is partly because the deviations of data points about a sample mean ($\overline{X}$) tend to be smaller than if the deviations were measured about the true (but unknown) mean (μ). We correct for this bias by using $n - 1$ in the denominator, rather than n. This yields an *unbiased* estimate of the population variance. The sample variance, S^2, is then:

$$s^2 = \frac{\Sigma (X - \overline{X})^2}{n - 1} \tag{3–8}$$

And the sample standard deviation is:

$$s = \sqrt{\frac{\Sigma (X - \overline{X})^2}{n - 1}} \tag{3–9}$$

By taking the square root of the variance, the equation for standard deviation returns the measure of variability back into the same units as those of the original variable. So, for example, our measure of dispersion is in more easily grasped units such as dollars or pounds, instead of $(\text{dollars})^2$ or $(\text{pounds})^2$.

Example 3–9 Compute the (a) variance and (b) standard deviation of the sample data given in Table 3–1.

Solution

We have previously found the mean ($\overline{X} = 70$) and deviations. We must now square the deviations, sum them, and divide by $n - 1$ to obtain the sample variance.

	SCORE X	DEVIATION FROM $\overline{X}$ $(X - \overline{X})$	DEVIATION SQUARED $(X - \overline{X})^2$
	54	−16	256
	62	−8	64
	63	−7	49
	65	−5	25
	68	−2	4
	71	1	1
	73	3	9
	78	8	64
	82	12	144
	84	14	196
TOTALS	700	0	812

(a) $$s^2 = \frac{\Sigma (X - \overline{X})^2}{n - 1} = \frac{812}{10 - 1} = 90.22 \text{ lbs}^2$$

(b) $$s = \sqrt{\frac{\Sigma (X - \overline{X})^2}{n - 1}} = \sqrt{\frac{812}{10 - 1}} = 9.50 \text{ lbs}$$

The sample standard deviation is a widely known (and used) measure of dispersion, and in business you are likely to use it more than the variance. However, variances are used extensively for comparing characteristics of two or more data sets, and variances are additive, whereas standard deviations are not.

Although Equation 3-9 is an accurate "definitional" formula for the sample standard deviation, the computation can be rather tedious—especially if the mean is not some nice round number. In practice, an alternative equation is used to simplify the computations. This "computational formula" yields the same result, but does not require that each deviation be individually computed and squared:

$$s = \sqrt{\frac{\Sigma X^2 - [(\Sigma X)^2 / n]}{n - 1}} \qquad (3\text{–}10)$$

Example 3–10 Use the data of Table 3–1 to compute the sample standard deviation using the computational Equation 3-10.

Solution

The Table 3–1 values of X, along with the X^2 values, are given below:

											TOTALS
STRENGTH X	54	62	63	65	68	71	73	78	82	84	$\Sigma X = 700$
X^2	2916	3844	3969	4225	4624	5041	5329	6084	6724	7056	$\Sigma X^2 = 49{,}812$

$$s = \sqrt{\frac{\Sigma X^2 - [(\Sigma X)^2/n]}{n-1}} = \sqrt{\frac{49{,}812 - (700)^2/10}{10-1}} = \sqrt{90.22} = 9.50 \text{ lbs}$$

Equation 3-10 could, of course, be squared to yield the sample variance. As a matter of fact, the sample variance of 90.22 pounds squared was calculated as an interim step in the example above.

The standard deviation is, perhaps, more meaningful to us because it is expressed in the same units as the data (i.e., pounds). Moreover, we shall use it extensively for analysis of statistical data throughout the text. Like MAD, it is a measure of average deviation, except the deviations are first squared, then averaged, and then "unsquared." We shall use it to express the amount of dispersion in a set of data, and also as a reference or standard to locate a specific data point relative to the mean of the data set. For example, the standard deviation of the data set above is 9.5 lbs. The largest value in the data set (84 lbs) is approximately 1.5 standard deviations above its mean of 70 lbs. And all of the points happen to be within plus or minus two standard deviations of the mean for this set of numbers.

EXERCISES III

11. Boxes of bulk candy are supposed to weigh within 4 ounces of 28 pounds. (a) Which measures of dispersion would tell, on average, how much weight deviation exists? (b) What would be signified by a sample standard deviation of zero?
12. A population of $N = 425$ pieces of airline luggage has a mean weight of $\mu = 43$ lbs and $\sigma = 12$ lbs. What would be the effect on μ and σ of adding a constant 5 lbs of weight to each piece of luggage?
13. Compute the range and interquartile range for the following *population* data: 5, 9, 8, 3, 6, 10, 4, 7, 2, 6.
14. Compute (a) MAD, (b) the standard deviation, and (c) the variance for the population in the previous problem.

15. Compute (a) MAD and (b) the standard deviation for the following *sample* data: 7, 12, 5, 7, 9.
16. The numbers of sales calls made by the sample of five marketing representatives (Example 3–3) were 12, 4, 9, 0, and 5. For these data compute the (a) range, (b) mean absolute deviation, (c) variance, and (d) standard deviation.
17. Suppose the population in the problem above included 300 marketing representatives and had a true (population) mean of $\mu = 7$.
    (a) What would be the population standard deviation (σ) if $\Sigma (X - \mu)^2 = 6{,}350$?
    (b) Use the five sample values to compute $\Sigma (X - 7)^2$ and $\Sigma (X - 6)^2$.

    How does the summation of the deviations squared about $\mu = 7$ compare with the summation of deviations around the sample mean $\overline{X} = 6$? Explain.

SUMMARY MEASURES OF GROUPED DATA

Computers can sum, square, and otherwise manipulate large blocks of data so rapidly that most data can be summarized as ungrouped data. However, we sometimes need to compute summary measures from data that have already been grouped or presented in the form of a frequency distribution. We will consider some measures of (1) *central tendency* and (2) *dispersion* for grouped data in this section. Insofar as grouping always condenses the individual data points into classes, these summary measures for grouped data are only close approximations of the more accurate measures that would result from using the ungrouped data directly.

MEASURES OF CENTRAL TENDENCY (GROUPED DATA)

MEAN Before finding the mean, median, and mode of grouped data, we must make some assumptions about how the data are distributed within each class of the distribution. To find the mean, we assume that the midpoint of each class is a representative (or average) value for all data points in that class. Although this assumption is not always true, the errors tend to offset each other, so the results are generally satisfactory.

The mean of a frequency distribution is computed by weighting each class midpoint (X) by the frequency of the class (f), summing the products, and dividing by the total number of frequencies, n (i.e., which is the Σf).

$$\overline{X} = \frac{\Sigma fX}{n} \tag{3–11}$$

Example 3–11 The data below represent the scores achieved on a personality test given by a personnel department to 100 applicants for a sales position. The columns on the right of the frequency distribution have been added to facilitate calculation of the mean. Those columns show the midpoint of the class interval (X), and the weighted value of each class (fX).

FREQUENCY DISTRIBUTION OF TEST SCORES		[For calculation of $\overline{X}$]	
TEST SCORE INTERVAL	FREQUENCY f	MIDPOINT X	CLASS VALUE fX
50–54	1	52	52
55–59	5	57	285
60–64	17	62	1054
65–69	36	67	2412
70–74	25	72	1800
75–79	11	77	847
80–84	4	82	328
85–89	1	87	87
Total	100	Total	6865

Determine the mean of the frequency distribution of test scores.

Solution

The arithmetic mean is easily obtained by substituting the column totals into Equation 3–11:

$$\overline{X} = \frac{\Sigma fX}{n} = \frac{6865}{100} = 68.6$$

MEDIAN The median is the point where half the scores are on one side and half on the other. However, when the original data are unavailable, we must again settle for an approximation. This time we assume that the data are evenly spaced throughout the interval that contains the median. To compute the median, we start with the lower boundary of the median class (L_m) and advance just far enough into the class (proportionally) so as to include 50% of the observations. If we let f_i = the number of observations (distance) *into the class* needed to encompass 50%, and f_m = the number *already in the median class*, then f_i/f_m is the proportion of the class interval (i) needed to reach the median. In equation form, this is equivalent to:

$$\text{Median [grouped]} = L_m + \frac{f_i}{f_m}(i) \tag{3–12}$$

Example 3–12 Estimate the median for the grouped data in the previous example.

Solution

There are 100 data points, so we will want to accumulate data points until we include 50 (half) of the observations. This means we must include the first three class intervals (with frequencies of 1 + 5 + 17 = 23) and move into the next or median class which has a lower boundary of 64.5.

FIGURE 3–4 Locating the median in grouped data

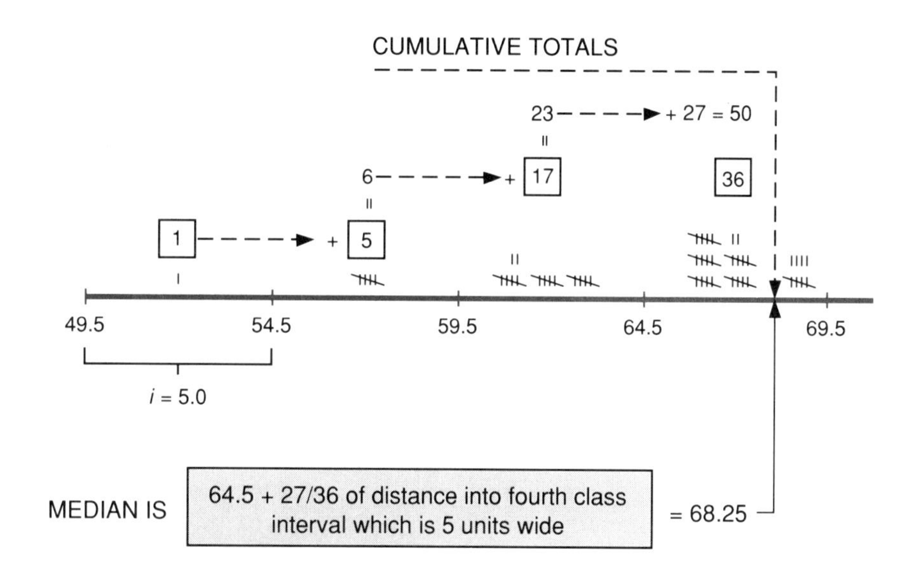

At this point, 27 more observations are required to bring the total up to 50. Since there are 36 observations in that interval, the median is a point located 27/36 of the interval width above the lower boundary of the interval as illustrated in Figure 3–3. Using Equation 3–12, and for $i = 5$:

$$\text{Median [grouped]} = L_m + \frac{f_i}{f_m}(i) = 64.5 + \frac{27}{36}(5) = 68.25$$

Note that the median could also be found by working from the "top down" just as easily as working from the "bottom up." Remember also, that the median constitutes a "position" in the class interval. If there are an even number of observations such as the above, we simply incorporate half of them into our cumulative total. If there had been an odd number of data points, such as 101, the median would have to incorporate half of the central value. If, for example, the median class contained 37 instead of 36 observations, we would have had to advance 27.5/37 of the way instead of 27/36.

MODE The mode can be (geometrically) estimated from the histogram of a frequency distribution, but the procedure is inexact and not often used. However, the class interval with the highest frequency is often referred to as the *modal class*. The modal class in the distribution of test scores (Example 3–11) is the class interval 65-69, which has a frequency of 36. It is a *unimodal* distribution. If the data evidenced a second separate and identifiable peak of roughly the same magnitude, we would refer to the distribution as being *bimodal*.

MEASURES OF DISPERSION [GROUPED DATA]

We turn now to methods of determining the range, variance, and standard deviation of grouped data.

RANGE As with ungrouped data, the range computed from grouped data is of limited value because it tells only the extremes of the data. For grouped data, the range is the difference between the upper boundary of the highest class and the lower boundary of the lowest class. Thus, if the class intervals are $20 < 30$, $30 < 40$, . . . , $60 < 70$, the range would be $70 - 20$, or 50 units.

VARIANCE AND STANDARD DEVIATION We shall, again, be working with sample data unless stated otherwise. The difference is, of course, that we use N in the denominator for population data and $n - 1$ for sample data. All the grouped data computations assume that the class midpoints adequately represent all values in the class.

As before, we can calculate these statistics using a "definitional" equation, or a more efficient "computational" formula. The definitional equation for the standard deviation of grouped data is very similar to the ungrouped equation, except the deviations of each class midpoint (X) from the mean of the data ($\overline{X}$) are weighted by the frequency (f) of each class.

Sample
Standard Deviation
[grouped data]

$$s = \sqrt{\frac{\Sigma f(X - \overline{X})^2}{n - 1}} \qquad (3\text{–}13)$$

Example 3–13 The frequency distribution below shows the number of hospital beds needed in a community on 60 randomly selected days. Find the standard deviation and variance of this sample data.

CLASS INTERVAL [No. Beds]	MIDPOINT X	FREQUENCY f	[For Calculation of s			]
			CLASS VALUE fX	DEVIATION $(X - \bar{X})$	DEV SQ $(X - \bar{X})^2$	WEIGHTED DEV SQ $f(X - \bar{X})^2$
0 < 10	5	16	80	−10	100	1600
10 < 20	15	30	450	0	0	0
20 < 30	25	12	300	10	100	1200
30 < 40	35	2	70	20	400	800
TOTALS		**60**	**900**			**3600**

Solution

This method requires that we first determine the mean of the distribution, which is $\Sigma fX/n, = 900/60 = 15$. Then the deviation of each midpoint from this mean of 15 is computed, squared, and weighted by the frequency of the class. The standard deviation is then

$$s = \sqrt{\frac{\Sigma f(X - \bar{X})^2}{n - 1}} = \sqrt{\frac{3600}{60 - 1}} = \sqrt{61.02} = 7.81 \text{ Beds}$$

The variance is again already calculated as part of the standard deviation calculation, and is 61.02 beds squared.

We will use the same data now to illustrate the more computationally oriented formula for the standard deviation for grouped data which is:

$$s = \sqrt{\frac{\Sigma fX^2 - [(\Sigma fX)^2 / n]}{n - 1}} \qquad (3\text{–}14)$$

Example 3–14 Use the grouped data of Example 3-13 and compute the sample standard deviation using the computational Equation 3-14.

Solution

For this approach, we do not need the deviation columns, but we need to have columns for fX and fX^2. Note this latter value is $(fX)(X)$.

CLASS INTERVAL [No. beds]	MIDPOINT X	FREQUENCY f	[For Calculation of s] CLASS fX	VALUE fX^2
0 < 10	5	16	80	400
10 < 20	15	30	450	6750
20 < 30	25	12	300	7500
30 < 40	35	2	70	2450
TOTALS		**60**	**900**	**17,100**

$$s = \sqrt{\frac{\Sigma fX^2 - [(\Sigma fX)^2/n]}{n - 1}}$$

$$= \sqrt{\frac{17{,}100 - [(900)^2/60]}{60 - 1}} = \sqrt{\frac{3600}{59}} = 7.81 \text{ Beds}$$

SHAPES OF DISTRIBUTIONS: SKEWNESS AND KURTOSIS

Suppose you wished to convey some information about the average income of residents in a locality. The arithmetic mean alone may not be a satisfactory measure, because a few persons with high incomes can distort the mean. You may wish to convey something about the shape of the income distribution as well.

Two terms sometimes used to describe the shape of distributions are *skewness* and *kurtosis*. **Skewness** is defined as a departure from symmetry. Thus, skewed distributions appear lopsided and do not have a symmetrical concentration of data. This *asymmetry* stems from the presence of extreme scores on either the high or low end of the data that tend to pull the mean away from the center of the data.

Figure 3–5 shows the effects of skewness on the relative location of the mean, median, and mode of the data. The mean is affected most, the median next, and the mode is not affected. Note the difference between the mean and the median is sufficient to provide us with an indication of the shape of the distribution. When $\overline{X} >$ the median, the data is right or positively skewed, whereas if $\overline{X} <$ the median, the data is left or negatively skewed. Specific measures of skewness are available in advanced descriptive texts.

FIGURE 3–5 Effect of skewness on relative location of the mean, median, and mode

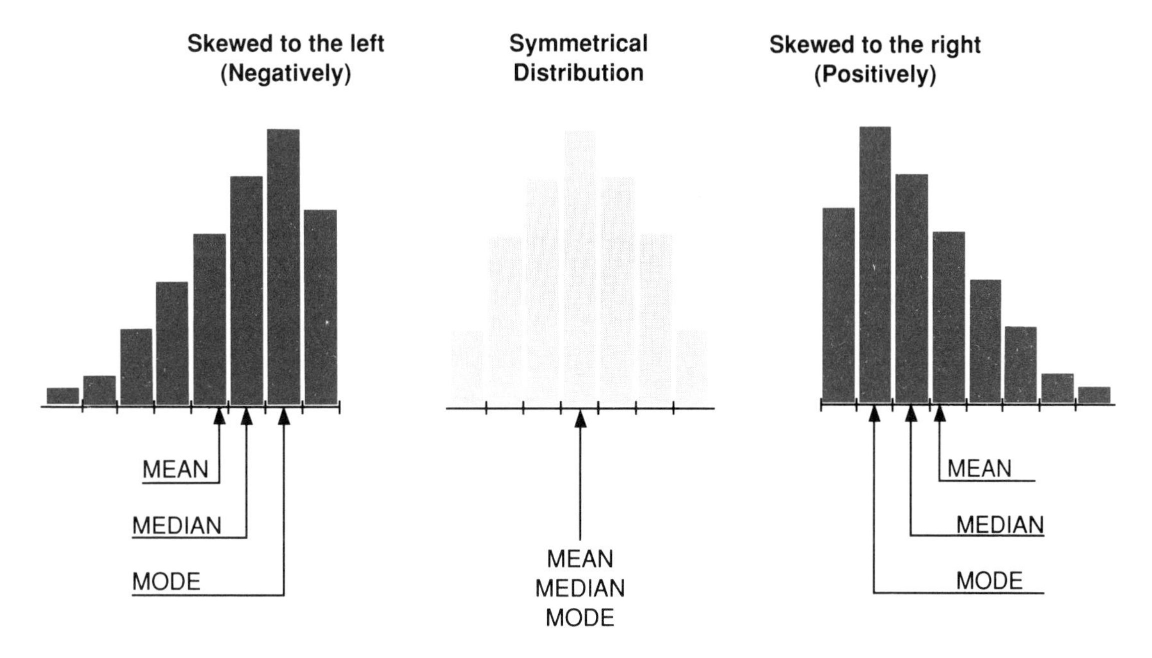

Kurtosis refers to the peakedness (or flatness) of a distribution. A highly peaked distribution is said to be *leptokurtic,* a moderate distribution *mesokurtic,* and a flat distribution is termed *platykurtic.*

EXERCISES IV

18. Explain why calculations of the mean and standard deviation from grouped data are not as accurate as if ungrouped data were used.
19. Find the (a) mean, (b) median, and (c) mode of the hospital bed demand distribution of Example 3–13.
20. Find the (a) range, (b) variance, and (c) standard deviation for the personality test data of Example 3–11.

[GROUPED DATA: EXERCISES 21–25 REFER TO THIS DATA] The production analyst at Donegal Textiles Ltd. has made some studies (using a stopwatch) to learn how much time is required for a sewing activity used in the production of Irish wool jackets. Sample data from 50 observations are as follows, with times recorded in minutes:

TIME INTERVAL	NO. OBSERVATIONS
7 < 9 min	3
9 < 11 min	27
11 < 13 min	14
13 < 15 min	4
15 < 17 min	2

21. What is the average (mean) time for the sewing activity?
22. Identify the median class interval and estimate the value of the median.
23. Describe the distribution of sewing times in terms of (a) the range, (b) the mode, (c) the shape of the distribution (i.e., its skewness).
24. Compute the standard deviation of the sewing times distribution.
25. What is the variance of the distribution?
26. A shipment (population) of 500 pieces of concrete sewer pipe has the following weight distribution (weights measured to nearest 1 lb).

CLASS INTERVAL	FREQUENCY
120.0–121.9	40
122.0–123.9	70
124.0–125.9	95
126.0–127.9	115
128.0–129.9	85
130.0–131.9	60
132.0–133.9	35

Estimate (a) the mean, μ, and (b) the standard deviation, σ, of this population distribution.

COMPARISON MEASURES: THE COEFFICIENT OF VARIATION

We will use statistical techniques to compare sets of data throughout the text. Thus this section is not intended to substitute for later chapters. However, before finishing this chapter, we should mention two measures of comparison that relate rather closely to the concepts of central tendency and dispersion that we have worked with in the chapter. These are the use of (a) *index numbers* for comparing the relative magnitude (or location) of a data set at different points in time and (b) the *coefficient of variation* for comparing the relative amount of dispersion in sets of data. Both of these are *relative measures*, i.e., their value lies in making comparisons with other sets of data.

INDEX NUMBERS FOR COMPARISONS OF CENTRAL TENDENCY

You have already been introduced to index numbers if you have heard of the consumer price index, the index of industrial production, or any of hundreds of other indexes. They are one of the most widely used techniques for summarizing economic information, and we shall study them in detail in a later chapter. For now, we will *simply note that index numbers are a way of summarizing and comparing two data sets.*

Index numbers describe the relationship of two sets of data, such as the prices of a common set of items at two different points in time. The condensation of the two summary measures into one number is accomplished by expressing the two as a percentage ratio of a numerical measure of the items in a given time period to the same measure in a reference or base period.

For example, a car rental firm may use an index number to monitor its automobile repair costs. If the cost of parts necessary to rebuild an automobile engine is $700 today and the cost of the same parts four years ago (base period) was only $580, the price index of these parts would be [$700 ÷ $580] = 1.21. Converting this to a percentage we would express the index as 121% of the base year price.

Several forms of index numbers are in use. They range from simple ratios of single prices or quantities to weighted aggregates, where prices or quantities are used as weights—in order to make the data more meaningful. But index numbers represent a substantial area unto itself, and we shall take them up in more detail later in the text. For now, simply note that index numbers are one of the more widely used methods of expressing the central tendency of economic data, such as wages or prices over time.

COEFFICIENT OF VARIATION FOR COMPARISONS OF DISPERSION

Suppose you were a recruiter for your college basketball team. The coach has asked you to search especially hard for a good, consistently high-scoring guard. You collect lots of statistics, attend numerous high school tournaments, and end up with the three players depicted in Figure 3–6.

FIGURE 3–6 Potential recruits for college team

Scoring

	MIKE	JIM	LESTER
Average Points/Game	22 points	27 points	23 points
Standard Deviation	4 points	9 points	3 points

Which player most closely satisfies the coach's criteria? It's difficult to say, but if the coach is serious about consistency, he may concentrate on recruiting Lester, because of the low standard deviation associated with his scoring data. However, Jim does offer a higher average. The question is, does his higher average offset the variability in his performance?

It is difficult to compare absolute measures of dispersion unless both series are in the same terms and have means which are approximately equal. The coefficient of variation offers us a nondimensional measure.

The **coefficient of variation** (V) is a relative measure which facilitates comparison of the dispersion of sets of data by expressing each standard deviation relative to the mean of its set. It is frequently expressed as a percentage.

$$V = \frac{s}{\overline{X}}(100) \tag{3–15}$$

Example 3–15 Compute the coefficient of variation as a measure of the scoring consistency for the three players depicted in Figure 3–6. What can you conclude?

Solution

Mike: $V = \frac{S}{\overline{X}}(100) = \frac{4}{22}(100) = 18.2\%$

Jim: $V = \frac{S}{\overline{X}}(100) = \frac{9}{27}(100) = 33.3\%$

Lester: $V = \frac{S}{\overline{X}}(100) = \frac{3}{23}(100) = 13.0\%$

In this case, Lester has the lowest coefficient of variation and is the most consistent scorer relative to his average scoring history.

A significant advantage of using the coefficient of variation is that it is a dimensionless measurement. Thus, it can be used to compare dispersions stated in different units of measurement. The following example illustrates how this characteristic can be useful in an international business situation.

Example 3–16 Office Products International produces white typing fluid (to cover typing mistakes) in two plants. One plant is in Brussels (18 ml bottle with $\sigma = .9$ ml), and the other is in Cleveland (3.5 oz bottle with $\sigma = .3$ oz). The Quality Control Manager is concerned that the filling machine controls in the Brussels plant are not "tight" enough. Compare the relative dispersions at the two locations, and comment.

Solution

In this case one measure is in ml and the other in oz, but the coefficient of variation can accommodate this. Moreover, because we apparently have population values, we can use σ and μ instead of S and $\overline{X}$.

BRUSSELS: $V = \frac{\sigma}{\mu}(100) = \frac{.9}{18}(100) = 5.0\%$

CLEVELAND: $V = \frac{\sigma}{\mu}(100) = \frac{.3}{3.5}(100) = 8.6\%$

The dispersions are relatively close, but the Brussels plant is actually under closer control. It has less variation in the amount of fill relative to the amount of fluid in the bottle.

FINDING NUMERICAL MEASURES ON THE COMPUTER

Most of the statistical measures we have discussed in this chapter can be readily computed on small calculators. If your calculator has a statistical feature, it may even compute the mean, variance, and standard deviation directly—you simply enter the data, and press the desired button. But check your instruction book. When calculating the standard deviation, some calculators assume you are working with sample data (i.e., they use $n - 1$ in the denominator), and others offer both population and sample standard deviations. [This may not be too critical if you are working with large samples because the effect of the (-1) in the denominator becomes less and less as the sample size increases.]

There are also numerous software packages for making these computations on personal and mainframe computers as well. For ungrouped data, the personal computer programs typically ask you to enter the raw data. The numbers are then sorted for you, and the mean, median, and mode are displayed on the screen. Dispersion programs just as readily give you the range, variance, and standard deviation. (*MAD* is not as commonly provided.) Grouped data programs may ask you to enter the class limits (and frequencies), and then they provide you with the grouped data measures.

Programs are also available for other measures, such as the coefficient of variation, and lesser used statistics, such as the geometric mean. Some coefficient of variation programs allow you to enter two sets of data (i.e., a two-population case), and the programs automatically compute means, standard deviations, and coefficients of variation for both sets of data.

SUMMARY

Numerical summaries should be stated with the correct number of significant digits (by not counting those used simply to locate the decimal point), and in properly rounded form (to the nearest digit). Some of the key characteristics of the measures we studied in the chapter are listed in Table 3–2.

TABLE 3–2 Characteristics of summary measures

CENTRAL TENDENCY		
Measure	**Advantages**	**Disadvantages**
Mean (arithmetic)	▪ Most widely used & easily computed ▪ Influenced by every value of the data set ▪ Good statistical properties (balances points so $\Sigma(X - \overline{X}) = 0$) ▪ Can be weighted	▪ Cannot be computed for open-ended frequency distribution ▪ Affected by extreme values
(harmonic)	▪ Applies to quantities when times per unit vary	▪ Formula is more difficult to remember and use
(geometric)	▪ Applies to percentages that are a function of previous period	▪ Need calculator that can take *n*th root
Median	▪ Widely understood & easily computed ▪ Influenced by number (as opposed to value) of the observations ▪ Can be used on open-ended distributions ▪ Good measure for highly skewed distributions	▪ Identifies center of data only ▪ Doesn't reflect any importance of individual value ▪ Not highly useful for statistical inference
Mode	▪ Easy to find (if it exists) ▪ Identifies locational concentration of data ▪ Can be used on open-ended distribution ▪ Can be used on highly skewed distribution	▪ Not always unique ▪ May be more than one ▪ Not widely used ▪ Lacks good statistical properties

TABLE 3–2 Continued

Measure	DISPERSION Advantages	Disadvantages
Range	▪ Widely used and understood ▪ Easily computed	▪ Cannot be determined for open-ended data ▪ Uses extreme end points of data only ▪ Doesn't reflect dispersion of individual values ▪ Has limited value for statistical inference
Interquartile Range	▪ Avoids influence of extreme values ▪ Can be used on open-ended distributions ▪ Can be used on highly skewed distributions	▪ Still doesn't reflect dispersion of individual values ▪ Has limited value for statistical inference
MAD	▪ Best intuitive measure of "average" deviation ▪ Easily computed, understood, and used ▪ Influenced by every value of the data set	▪ Requires more calculation than range ▪ Has limited value for statistical inference
Standard Deviation	▪ Widely used measure of dispersion ▪ Very good properties for statistical inference ▪ Influenced by every value of the data set ▪ Yields a "geometric" average of deviations ▪ In same units as mean	▪ Not as easily understood as MAD ▪ Extreme values can distort it because deviations are squared ▪ Reasons for using *n* versus $n - 1$ frequently not understood ▪ Requires more calculation than MAD
Variance	▪ Similar to standard deviation ▪ Excellent properties for inference (Variances are additive whereas standard deviations are not.)	▪ Similar to standard deviation ▪ Units not same as original variable

Two other topics considered were the concepts of *skewness* and *kurtosis*. Skewness reflects the presence of extreme scores that pull the mean (most) and the median (next) away from the mode of the data. Kurtosis refers to the peakedness of a distribution.

Finally, two measures for comparing data are *index numbers* and the *coefficient of variation*. Index numbers tell the ratio of two summary measures of the same data at different points in time, and are widely used for economic data. The coefficient of variation is a ratio of the standard deviation to the mean of a data set, and it is a method of comparing the relative dispersion in two sets of data.

CASE: PFAFNER, JEFFREY AND JONES INVESTMENT COMPANY

Pfafner, Jeffrey & Jones Investment Co. (PJ&J) has been in business for nearly 60 years. Although they now have 12 offices nationwide, they have never been outstandingly successful in attracting new business. That's why one of the partners, Phillip Jeffrey, was surprised to see the Dallas office setting new records each month. Upon checking, Mr. Jeffrey learned that much of the new business was attributed to Ann Sanders. Ann was a relatively new account executive, who had joined the firm a year earlier in San Francisco. Mr. Jeffrey decided it was worth a trip from Boston to Dallas to learn Ann's secret. After a brief introduction, they got down to a discussion of Ann's work.

MR. JEFFREY: You've done well, Ann. The people in the Boston office think you must be using some kind of magic.

ANN SANDERS: I appreciate the compliment, but I really don't know what you mean. And I have no secret. I've just been trying to construct the type of investment packages my customers want.

MR. JEFREY: That's just it. You must be doing an outstanding job of that because your follow-on business is triple the company average. Could we just review some of your customer files—as an example?

ANN SANDERS: Sure, how about Mable Thornton? I happen to have her file here on my terminal right now. She's a widow, you know—well off financially, but her investments are her only livelihood. She wants a steady income, and minimum risk.

Ann went over Mrs. Thornton's investment portfolio with Mr. Jeffrey. She had presented Mrs. Thornton with three alternative investment plans. Each had a group of stocks listed, along with their prices, as shown below:

COMMON STOCK	PRICE	COMMON STOCK	PRICE
Mid States Insurance	17	Medical Systems Ltd.	44
U.S. Manufacturing Co.	44	Aerospace Enterprises	55
Falls River Electric.	60	Southern Telephone Co.	84
American Bank & Trust Co.	34	Learning Systems Inc.	25

Ann explained that she had collected at least two years' data on prices and earnings for several stocks. She also had the daily closing prices of each stock for the past month (22 days). For example, the daily closing price of two stocks she had investigated is shown below:

FALLS RIVER ELECTRIC

60 1/4	59 3/8	62 1/4	61 1/4	60	61 1/8	58 5/8	58 3/4
62 1/8	62 1/2	60 3/8	60 1/2	60 1/4	59 3/4	60 3/8	61
58 1/4	60 7/8	57 1/2	59 3/4	59 7/8	60 1/2		

MICROTECH INC.

15 3/4	15 1/2	16 7/8	16 1/2	16 7/8	14 1/4	16 3/4	14 1/8
14 1/8	14 1/4	14 1/8	14 1/4	14 1/8	14 1/2	15	14 1/4
14 1/4	16 3/8	14 1/8	14 1/8	16 5/8	16 7/8		

Ann had included Falls River Electric as a potential stock in Mrs. Thornton's portfolio, but not Microtech. She explained that she had dropped Microtech "on the first cut" because her first criteria was that the stocks must exhibit relatively good price stability. Mr. Jeffrey started to question her on this, but decided to let her continue. She went on to point out that she had, of course, used other criteria such as growth, dividends, and rate of return considerations, as well as price stability. But Mrs. Thornton mostly needed to know her investment was secure from wild price fluctuations.

Ann's strategy was to research the potential stocks, compute a few statistics on each one (e.g., means, standard deviations, etc.), and identify the stocks suitable for her particular client. Then she used only those stocks to develop three alternative packages. She explained that the alternatives looked quite different because the industries represented in each alternative were different. But the alternatives all had an underlying characteristic of price stability. Then, even though her customer "made her own choice" of which alternative, Ann felt quite confident that the stocks would give her clients the main objectives they wanted. She noted that some of her other customers were more willing to take some risks (with the hope of higher profits), and she used stocks with higher coefficients of variation for their portfolios.

MR. JEFFREY: It makes good sense, Ann. Where did you get the idea of using the coefficient of variation?

ANN SANDERS: Oh, I'm afraid that goes back 5 or 6 years now. Coefficients of variation have been one of my favorite animals since I was a sophomore in college.

MR. JEFFREY: Why's that?

ANN SANDERS: Well, you see, I got a lower grade than I should have on an exam once—because I missed a coefficient of variation problem. I told myself that would never happen again. And to make sure, I began calculating the coefficient of variation of just about every group of numbers I came across, and continued that for quite a while. I used it to compare the variabilities in everything from accounting costs, to production times, to product demands in marketing. Using it to compare the variability of price in investment portfolios just came naturally. I guess I wouldn't know what else to do now!

Case Questions

(a) From Mrs. Thornton's standpoint, which do you feel is a better measure of average stock price over a month, the mean, median, or mode? Why?

(b) Upon comparing the stock prices for Falls River Electric and Microtech Inc., Mr. Jeffreys started to question Ann's statement that Falls River Electric exhibited better price stability. Using the range as a criterion, which of the two sets of stock prices shows more dispersion?

(c) What is the mean and (sample) standard deviation for the Falls River Electric stock?

(d) For the Microtech Inc. stock, the mean price is $15.16 and the standard deviation is $1.16. What is its coefficient of variation?

(e) Using Ann's criteria (of the coefficient of variation), which of the two stocks (Falls River Electric or Microtech Inc.) would be more preferable for Mrs. Thornton's investment portfolio? Why?

(f) What problems (or lost opportunities) might result from using the coefficient of variation alone, as the only criteria for recommending a stock?

QUESTIONS AND PROBLEMS

27. State the rule for significant digits with respect to (a) multiplication and division, and (b) addition and subtraction.
28. How many significant digits are in the answer to the following?
    (a) 27.4×2.3075 (c) $\sqrt{244}$
    (b) $588 \div .0006$ (d) $(3.112 + 570.05 + .004) \div 256$
29. Identify (a) three measures of central tendency, (b) four measures of dispersion, and (c) two terms that describe the shape of a distribution.
30. What are the major differences between calculating means, medians, and standard deviations directly from ungrouped data versus using grouped methods?
31. Explain the difference between data that is skewed positively versus data skewed negatively. How are the mean, median, and mode affected?
32. Differentiate between index numbers and the coefficient of variation in terms of their use.
33. For the following values, find the: (a) mean, (b) median, and (c) mode.
    23, 13, 17, 19, 20, 21, 16, 18, 17, 20
34. Valley Lumber Company management is concerned about the financial viability of the independently owned retail lumber yards it serves. A survey of the current ratios (i.e., current assets ÷ current liabilities) of fifteen yards resulted in the following data.

COMPANY	CURRENT RATIO
A	1.6
B	1.7
C	1.9
D	1.9
E	1.9
F	2.0
G	2.0
H	2.1
I	2.2
J	2.4
K	2.9
L	3.1
M	3.8
N	4.6
O	7.2

Find the (a) range of current ratios, (b) mode, (c) median, (d) arithmetic mean, and (e) mean of the middle nine scores; that is, the mean if the highest three scores and the lowest three scores are discarded. (f) Which of these measures are averages, and which are measures of dispersion?

35. A producer of fine lead crystal in Germany has collected the following data on the number of flaws per unit in 1,000 pieces of crystal:

NUMBER OF FLAWS	0	1	2	3	4	5	TOTAL
FREQUENCY (no. units)	280	360	190	80	60	30	1,000

Find the mean number of defects per unit using the equation $\overline{X} = \frac{\Sigma fX}{n}$

36. This problem calls for a comparison of the arithmetic mean, the median, and the mode as measures of central tendency. Values of X are:
12 13 17 18 19 19 20 21 22 22 24 27 30 31 31 31
(a) Compare the arithmetic mean, the median, and the mode.
(b) Why is the mode said to be an erratic measure of central tendency?
(c) Why is the median called a position average?
(d) If the sixteen observations are weights (in lbs.), explain why the mean may be a better average than the median.

37. Given the following *sample* values: 4, 8, 5, 3. Find the (a) mean, (b) median, (c) standard deviation, and (d) variance.

[DATA FOR PROBLEMS 38–40] The following represents the stock price and the total shares held (i.e., a *population*) by an investor on the closing day of the year:

COMPANY DESIGNATION	A	B	C	D	E	F
STOCK PRICE (nearest $)	16	9	20	14	4	9
NUMBER OF SHARES HELD	100	100	400	100	200	100

38. For the stock prices only, find the (a) mean, (b) median, and (c) mode.

39. Calculate the *population* (a) variance and (b) standard deviation of stock prices.

40. Using the number of shares as weights, compute the weighted mean value of stock price for shares held by this investor.

41. The frequency distribution of the number of customers in a sample of convenience stores during the hour before closing (from Chapter 2) was as follows:

NUMBER CUSTOMERS	FREQUENCY
10 < 20	14
20 < 30	24
30 < 40	32
40 < 50	19
50 < 60	8
60 < 70	3
Total	100

Find the (a) mean and (b) standard deviation of this distribution.

42. For the sample data shown, find the:

(a) arithmetic mean
(b) median
(c) modal class
(d) range
(e) variance
(f) standard deviation
(g) coefficient of variation

Class Interval	*f*
130–134	3
135–139	12
140–144	21
145–149	28
150–154	19
155–159	12
160–164	5
Total	100

43. Zag Auto Parts Centers has several parts stores, and their accounting department bills each store on the basis of the average cost per unit of inventory acquired during the month. During the month of October, the Boone Avenue Store received the following number of quarts of motor oil.

Date Received	Number of Quarts	Price	Total
October 3	500	@ $1.26	$630
October 18	300	@ 1.32	396
October 26	200	@ 1.36	272

(a) What is the average cost per quart that should be used for accounting purposes? (b) Suppose each store is also billed a "loading" charge of 8% of the total purchase price to cover freight, sales tax, and handling costs. What total would the Boone Avenue Store be billed for the 1,000 quarts?

44. Keytop Corporation has three plastic molding machines producing keys for computer keyboards. The machines operate at different speeds, but the output from them flows onto a common conveyor belt. The output rates and percent defectives produced are as follows:

Machine	Keys/Hour Produced	Percent Defective
A	16,000	4.0
B	12,000	5.2
C	18,000	3.6
Totals	46,000	

(a) Find the average number of defects per 1,000 keys produced. (b) What kind of mean did you use? (c) Why?

45. Management Services Corporation provides cleaning and laundry facilities to office buildings in Atlanta. Over the past four years, their sales, as a percent of the preceding year, have been 110%, 106%, 92%, and 124%. (a) What is their average annual rate of increase in sales? (b) Why can you not take an arithmetic mean?

46. Coast-to-Coast Transport Co. has a fleet of trucks and uses a large number of RD2 tires each year. An operations analyst has collected some sample data on the number of miles an RD2 tire had been driven before it had to be replaced (i.e., when the center tread depth was worn to 1/16 inch).

Miles Driven	*f*
12,000 < 14,000	16
14,000 < 16,000	30
16,000 < 18,000	46
18,000 < 20,000	50
20,000 < 22,000	40
22,000 < 24,000	26
24,000 < 26,000	12
Total	220

(a) How many observations were recorded?
(b) What is the midpoint of the interval with a frequency of 50?
(c) Compute the average that would be most useful if the firm wants to compare mileage of these tires with the mileage obtained from a competitive brand.
(d) Compute the standard deviation.
(e) Competitive Brand X has a mean of 31,500 and standard deviation of 2,000 miles. Which tires give relatively more uniform mileage?

[COMPUTER AND DATABASE PROBLEMS] The following data are the price earnings ratios of the 100 firms listed in the Corporate Statistical Database of Appendix A (and repeated here for convenience). Consider them a sample. [Note: If your software package will not accommodate 100 data points, use the maximum number it will handle up to 100. Use the data row by row.]

7	9	8	7	8	10	9	6	6	7	7	6	6	5	7	8	12	7	11	8	7	6
8	9	6	16	26	18	7	13	9	11	21	8	9	12	13	4	10	12	12	14	7	13
13	11	14	18	14	12	12	7	8	9	15	9	13	12	8	9	13	22	12	5	13	10
17	17	8	6	12	9	11	9	7	24	11	16	14	12	27	18	15	14	14	10	15	19
12	13	11	11	12	5	12	12	8	9	9	10										

47. Using the price/earnings ratio for the 100 companies in the Corporate Statistical Database (repeated above), find the:
(a) mean, (b) median, and (c) mode.

48. Using the same price/earnings data (above), find the:
(a) range, (b) variance, and (c) standard deviation

49. Using the same price/earnings data (above), (a) construct a frequency distribution using approximately 6 class intervals (e.g., with widths of $4 < 8$, $8 < 12$, $12 < 16$, $16 < 20$, $20 < 24$, $24 < 28$). From your distribution find the (b) mean and (c) standard deviation of the grouped data.

50. Using the first 20 companies in the Corporate Statistical Database (Appendix A), find the (a) mean, (b) median, and (c) mode of the average annual salary of employees. (d) [OPTIONAL], Weight the average salaries of each of the 20 companies by the number of employees in the company, and find the weighted mean salary of the 20 companies.

51. Using the first 20 companies in the Corporate Statistical Database (Appendix A), find the (a) range and (b) standard deviation of the average annual salary of employees.

52. We wish to compare the relative dispersion in the price/earnings ratio of the 20 Chemical and Drug companies in the Corporate Statistical Database (Appendix A) with the 20 Electrical Utilities. Compute and state the coefficient of variation for each. Which group of companies has a greater relative dispersion?

CHAPTER 4

Probability: Concepts and Rules

INTRODUCTION

MEANING OF PROBABILITY

SET THEORY AND VENN DIAGRAMS
- Sets, Events, and Sample Spaces
- Types of Probabilistic Events
- Using Venn Diagrams to Describe Probability Situations

MARGINAL, JOINT, AND CONDITIONAL PROBABILITIES
- Marginal Probabilities
- Joint Probabilities
- Conditional Probabilities

PROBABILITY RULES
- Complement Rule
- Multiplication Rule
- Addition Rule
- Using Cross-Classification Tables to Compute Probabilities

BAYES' THEOREM
- Revising Prior Probabilities
- Using Tree Diagrams to Describe Probability Situations

COUNTING PROCEDURES
- The Fundamental Rule
- Multiple Choices
- Permutations
- Combinations

USING COMPUTERS TO FIND PROBABILITIES

SUMMARY

CASE: ALLIANCE NATIONAL BANK

QUESTIONS AND PROBLEMS

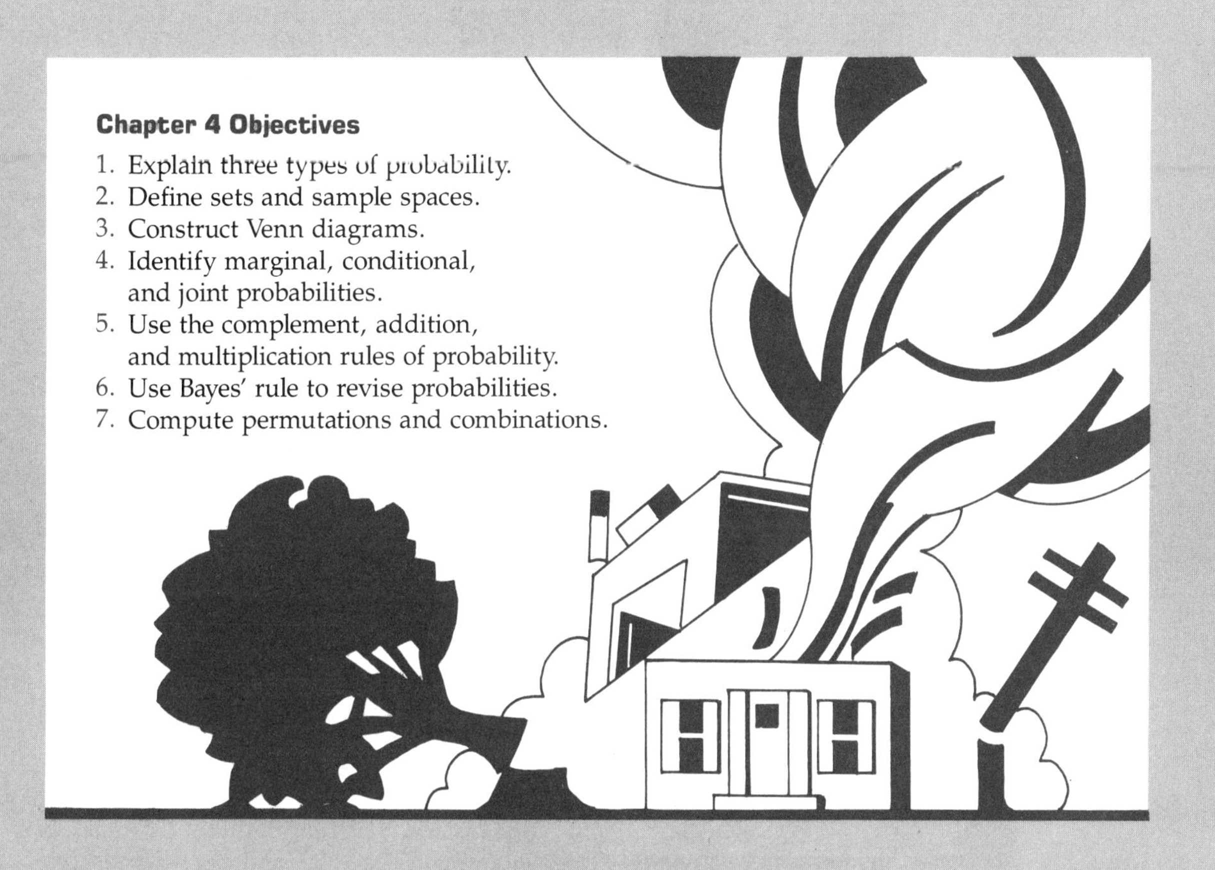

Chapter 4 Objectives

1. Explain three types of probability.
2. Define sets and sample spaces.
3. Construct Venn diagrams.
4. Identify marginal, conditional, and joint probabilities.
5. Use the complement, addition, and multiplication rules of probability.
6. Use Bayes' rule to revise probabilities.
7. Compute permutations and combinations.

On the night of September 8, 1900, a hurricane swept ashore at Galveston, Texas, with a wall of water 20 feet high and 50 miles wide. It killed about 6,000 people and struck a devastating blow to the economic viability of the city. The weather bulletin had called for rain and "brisk to high northerly winds."

But, you may say, "That was long ago, and weather predictions have improved." According to one source,[1] *predictions of hurricanes today are not much better than they were years ago. When a hurricane is 72 hours offshore, the weather service "has only a 10% chance of predicting accurately where it will finally come ashore—if it comes ashore at all." Even at just 24 hours offshore, predictions are "wrong more than half the time." Hurricanes are so unpredictable that even the latest electronic technology can't do much to improve forecasts. With the increasingly dense population centers on coastal areas, some scientists are now warning that each year we are one step closer to what could be an even worse disaster than that of 1900.*

---

[1] *The Wall Street Journal*, October 14, 1983, p.1.

INTRODUCTION

People say nothing is as uncertain or as unpredictable as the weather. That certainly appears to be true with respect to hurricanes. The narrative above verifies that they are quite unpredictable. It also suggests that if we can't predict what will happen, we should be prepared to live with the uncertainty. These two points constitute a good lesson for managers who must work in a competitive business environment. First, uncertainty exists. Second, it is best to assess uncertainty and take it into account—ignoring it could have disastrous consequences.

Chapters 1–3 were concerned with describing and summarizing data. We now move on to making inferences when these data are uncertain or limited. This requires an understanding of *probability* because *probabilities are numbers we use to measure and describe the degree of uncertainty of an event*. You are already somewhat familiar with the concept of probability simply from talking about the weather or the flip of a coin. We'll try to lend a little more structure to your understanding and develop some rules that enable you to derive numerical values for probabilities.

We begin the chapter by distinguishing among some different meanings of the term "probability." Then we will introduce some theory and terms that will be useful for working with probabilities. The main body of the chapter then explains and illustrates four rules that allow us to do almost any probability calculation. Many students feel that committing those rules to memory early is time well spent.

We end the chapter with three shortcut methods (formulas) of counting the number of ways events can occur. These "counting procedures" are included because many probability calculations rely upon counts of the number of ways events can occur. Thus, they serve as a valuable aid in probability calculations. When you finish the chapter, you should have a good grasp of the principles of probability needed for the inference and statistical decision making procedures used in the remainder of the text.

MEANING OF PROBABILITY

We spoke earlier of probability as a means of enabling us to describe a level of uncertainty of an "event." More specifically, probabilities are expressed as numbers, and the event can be any uncertain occurrence such as a 60% chance of the event "rain."

Probabilities are numbers (decimals, fractions, or percentages) used to express the likelihood of the occurrence of an uncertain or chance event. They range from zero (event will not occur) to one (event will occur for certain).

Following the development of mathematical probability in the 1700s, three methods of assigning probabilities have emerged. The first two methods are considered "objective," whereas the third is simply classified as "subjective."

(1) Classical Probabilities are probabilities based upon equally likely outcomes that can be calculated prior to (i.e., *a priori*) an event on the basis of mathematical logic. Classical probability calculations thus yield theoretically correct probabilities, as opposed to historical or subjective probabilities.

$$P(X) = \frac{\text{Number of ways to obtain outcome } X}{\text{Total number of possible outcomes}} \tag{4–1}$$

Example 4–1 We know one lightbulb in a package of eight is burned out (B), but don't know which one it is. If a bulb is randomly selected from the package, what is the probability of it being burned out?

Solution

$$P(B) = \frac{\text{1 burned out bulb}}{\text{8 bulbs total}} = \frac{1}{8} = .125$$

(2) Relative Frequency Probabilities are probabilities based upon historical frequencies or empirical evidence or experiments.

$$P(X) = \frac{\text{Number of times outcome } X \text{ occurred}}{\text{Total number of observations}} \tag{4–2}$$

Example 4–2 Records show that 360 of the last 500 customers at a fast food restaurant ordered some type of burger (B). Estimate the probability of the next customer ordering a burger.

Solution

$$P(B) = \frac{\text{360 burgers ordered}}{\text{500 customers total}} = .72$$

(3) Subjective Probabilities are probabilities based upon personal beliefs or judgment. For example, a personnel manager may assess the probability of a labor strike at 40%, or a financial manager may feel there is a .90 probability of making a profit on an investment.

Classical probabilities concerned with gambling situations (rolling dice, tossing coins, and playing cards) were the first to receive much rigorous investigation. They represented a logical situation and could be calculated before the action (or bet) was taken. For example, the probability of rolling a 5 on a six-sided die is $1/6$, and the probability of drawing any one of four aces from a deck of 52 playing cards is $4/52$.

Since these early beginnings, classical probabilities have been applied to business and social situations where all the possible outcomes of an event have an equally likely chance of occurring. For example, suppose a midwest supplier of plants to nurseries wanted to forecast the last weekday in the spring in which the temperature would dip to 32°F. Assuming all days have an equal likelihood, the probability for every day, P(Sunday), P(Monday), etc., would be the same. Thus for any day, X, we could compute $P(X) = 1/7$.

The *relative frequency* approach assumes that what happened in the past will continue into the future. It relies upon the notion that in the long run, the sample will take on the same characteristics as the population. In this case, a larger sample tends to yield a better estimate than a smaller sample. This is why an auto insurance company with several years of experience is in a better position to estimate the probability you will have an auto accident, than is a newly formed company.

Subjective probabilities are unique, as they allow the decision maker to inject his or her own professional judgment into the situation. They do not necessarily rely upon long history and can even be applied to one-time situations. They can be used to describe the likelihood of success of a new product or the chance of having the low bid in sealed bid competition for a construction project.

Insofar as business decisions frequently relate to specific one-time situations, some managers make fairly extensive use of subjective probabilities. Nevertheless, it is well to remember that subjective probabilities often represent only one person's assessment—and quantifying such values does not automatically make them correct.

GENERAL PROBABILITY EQUATION At the risk of some oversimplification, we can combine the probability concepts above into one operational definition of probability, expressed in terms of successful outcomes as a percent of total outcomes:

$$P(X) = \frac{\text{Successful Outcomes}}{\text{Total outcomes}} = \frac{\text{Successes}}{\text{Successes} + \text{Failures}} = \frac{S}{S + F} \quad (4\text{–}3)$$

In this expression, a "successful outcome" is any particular outcome we are interested in assessing. Thus, it may either be the probability of finding a good product or a defective product, depending upon what we are attempting to evaluate.

SET THEORY AND VENN DIAGRAMS

Probability concepts can perhaps best be explained through the use of set theory and Venn diagrams, which present set relationships in a simple graphic form.

SETS, EVENTS, AND SAMPLE SPACES

A **set** is any clearly defined group of objects. It can be either a listing of specific objects or a criteria that establishes membership in the set. For example, some sets are

- the letters a, b, c, d
- accounts 01–04
- employees living within a 5-mile radius
- all cast parts with missing end brackets

Sets are usually designated by capital letters, with their elements enclosed in braces. For example, the set of letters above might be designated as:

$$A = \{a, b, c, d\}$$

Or we might designate the set of employees above as:

$$R = \{x \mid x \text{ is an employee living within a 5-mile radius}\}$$

which means, R is the set of all employees x, given that x lives within a 5-mile radius. Notice that the vertical line (|) is read as "given." Note also that R could be a *subset* of a larger set of all employees (E). The set of all employees of the company would be referred to as the *universal* set.

An **event** is any distinguishable outcome (or combination of outcomes) of an experiment. Suppose a production lot contains 4,000 parts, and 800 have missing end-brackets. The *event* of randomly selecting one part with a missing end-bracket would have a probability of:

$$P(X) = \frac{S}{S+F} = \frac{\text{800 parts with missing end-brackets}}{\text{4,000 parts total}} = .2$$

A **sample space** is a collection of all the possible outcomes of a statistical experiment. We use the concept of a sample space to help us determine the total number of outcomes for computing probabilities (e.g., as in Equation 4–3 above). For example, the sample space for the roll of a die contains the six possibilities depicted in Figure 4–1.

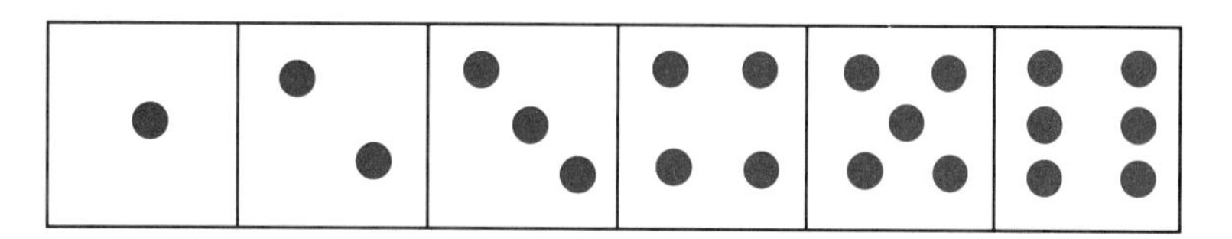

FIGURE 4–1 Sample space for one roll of a die

Example 4–3 What is the probability of getting the five on one roll of the die?

Solution

In this case, the five is the successful event among the six subsets of the sample space.

$$P(X = 5) = \frac{S}{S + F} = \frac{1 \text{ five}}{1 \text{ five} + 5 \text{ others}} = \frac{1}{6}$$

As a second example, suppose you are an accountant checking the validity of four accounts. Suppose further that you *know* that three accounts are correct **(C)** and one is incorrect **(I)**, and you're interested in the probability of randomly selecting the incorrect account. So a "success" is choosing the incorrect account from the four accounts.

Figure 4–2 shows the simplified sample space of the four accounts. The logic we developed above tells us that the probability of selecting the incorrect account is one in four, or .25.

FIGURE 4–2 Simplified sample space of four accounts

C	C
C	I

Further reflection might suggest that the sample space consists of more than four possible outcomes, depending upon the sequence in which the incorrect account is arranged. For example, assume the accounts could be in any one of the four arrangements #1 through #4 as shown in Figure 4–3. If the incorrect account is in the first position, as in #1, then the other three accounts in positions #2, #3, and #4 are correct. But given this same arrangement, one might randomly select the account in #3 position, and find it is correct. That would be a valid outcome. It wouldn't tell us whether the account in the #1, or #2, or #4 position was the incorrect account, but it would be a valid point on a sample space. If similar reasoning were applied to the other three arrangements, we would have 4 possibilities of selecting an incorrect account, 12 possibilities of selecting good accounts, and 16 possible outcomes in total. (The ratio of successful ways to total ways is still $4/16$, or .25.)

FIGURE 4–3 Sample space showing four possible outcomes

Possible Arrangements			
1	**2**	**3**	**4**
I	C	C	C
C	I	C	C
C	C	I	C
C	C	C	I

Note however, that only one of these possible arrangements can exist at any one time. Moreover, each possible arrangement has the same number of correct and incorrect accounts. Thus, the problem is no more difficult than originally depicted in Figure 4–2. The uncertainty is that we don't know which of the possible arrangements actually exists. But regardless of which uncertainty exists, our chance of randomly selecting an incorrect account is still one in four.

In this example of four accounts, it is not necessary for us to allow for other arrangements because we were seeking the probability of finding *the incorrect* account from the (already selected) sample of four accounts. Only one alternative actually exists, and we do not have to specifically allow for more than that one alternative. This approach will be adequate for much of our work. But as we go on to calculate probabilities of some events, we may have occasion to take into account the different order in the number of ways each possible outcome can occur. Fortunately, there are some easily applied counting rules that enable us to do this in a rather simple way. We will examine them toward the end of this chapter.

TYPES OF PROBABILISTIC EVENTS

Before proceeding further, we must note carefully some terms used to describe events:

- complement
- mutually exclusive
- collectively exhaustive
- statistically independent

The **complement** of an event (A) is all the outcomes other than A and is designated as $\overline{A}$, or A'. We shall interpret $\overline{A}$ as meaning "not A." For example, if the probability that a parachute opens (A) is .98, then the complement (i.e., that it won't open!) is $P(\overline{A}) = .02$.

As suggested above, events are **mutually exclusive** if they have no elements in common or cannot occur at the same time. For example, suppose you are projecting the level of employment in your firm for one year from now. Your categories are that employment will be (a) down, (b) the same, or (c) up. These categories are mutually exclusive because the existence of any one of the states precludes the existence of the others. On the other hand, the events "unemployment up" and "recession" are not mutually exclusive since both can occur simultaneously.

Events are **collectively exhaustive** when they include all possible outcomes. For example, the employment categories of "down," "the same," and "up" exhaust all possibilities, and are collectively exhaustive. The sum of the individual probabilities of these categories must equal 1.0.

Events are **statistically independent** if the occurrence (or non-occurrence) of one in no way affects the probability of occurrence of the other. Thus, the likelihood of an error in an account in the office is independent of the probability of receiving defective material—unless the account in some way influenced the material that was ordered.

USING VENN DIAGRAMS TO DESCRIBE PROBABILITY SITUATIONS

Now that we have introduced a number of new terms, let's try to relate them in a little more visual, or intuitive, form. *Venn diagrams* are a very useful technique for this. They depict the sample space as a rectangle and set relationships as circles inside the rectangle. For example, Figure 4–4 illustrates two concepts used in probability calculations: unions and intersections.

FIGURE 4–4 Venn diagrams of union and intersection of sets

UNION

A F

(a) $A \cup F$ is shaded

INTERSECTION

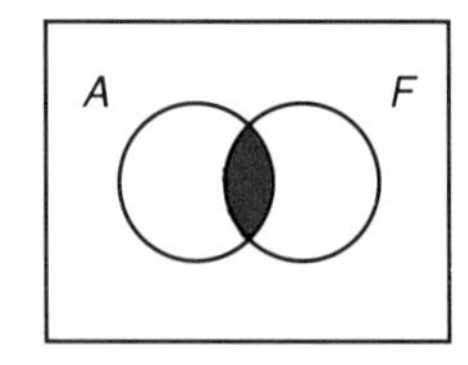

(b) $A \cap F$ is shaded

The figure represents the assembly workers of an electronics plant (A), and the female employees of the company (F). Figure 4–4 (a) represents all employees who are either assembly workers, or female, *or both*. Its shaded area depicts a **union** of the two subsets (designated $A \cup F$). Figure 4–4 (b) highlights only those assembly workers who are female. The shaded area depicts an **intersection** of the subsets (designated $A \cap F$, where the $\cap$ symbol means "and"). The subsets $\{A\}$ and $\{F\}$ are not mutually exclusive because some assembly workers are female. We can also use Venn diagrams to demonstrate the meaning of complement and mutually exclusive events. An example will illustrate.

Example 4–4 The table below shows the composition of Cork Manufacturing Company work force with data cross-classified by location of work and sex of employee. Use a Venn diagram to depict

(a) the sample space as employees (E), and the complement of E ($\overline{E}$).

(b) the classification by sex as male (M) and female (F) into mutually exclusive categories.

(c) the classification of males (M) and office employees (O) as non-mutually exclusive categories.

TABLE 4–1 Employees of Cork Manufacturing Company

WORK LOCATION	SEX OF EMPLOYEE		TOTAL
	Male	Female	
Plant	1,600	200	1,800
Office	20	140	160
Sales	40	0	40
Total	1,660	340	2,000

Solution

(a) Event $\overline{E}$ is the complement of Event E.

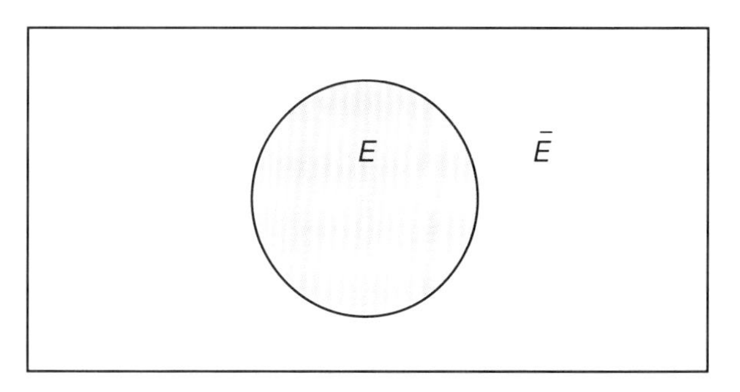

(b) Events M and F are mutually exclusive.

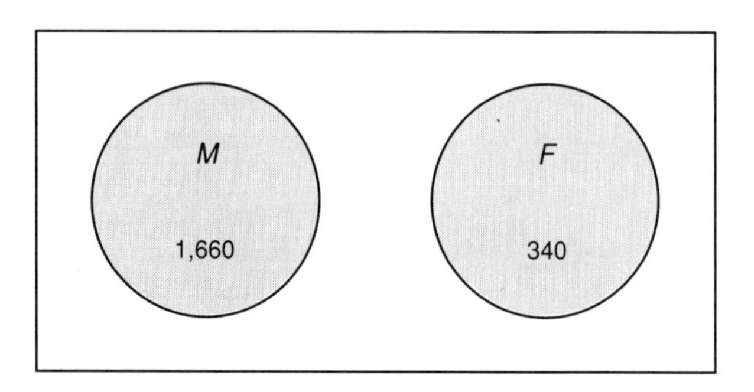

(c) Events M and O are not mutually exclusive because they have 20 elements in common.

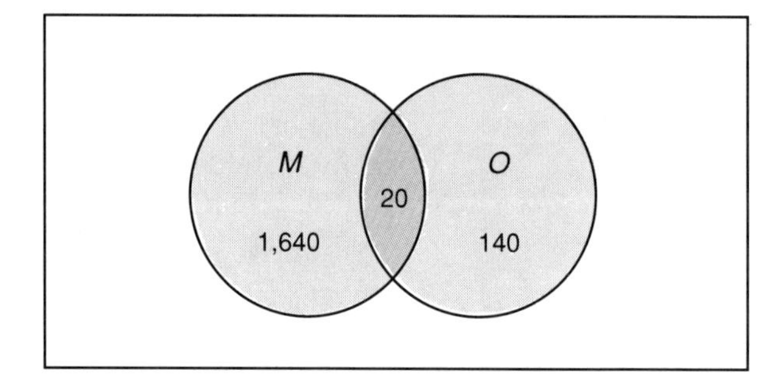

This completes our introduction to Venn diagrams, but we will be using them as we continue on through the chapter. Although we have used them to depict frequencies, they can also be used to depict probability values directly (i.e., in decimal form).

EXERCISES I

1. Distinguish between classical, relative frequency, and subjective probabilities.
2. Which of the methods of assigning probabilities (i.e., classical, relative frequency, or subjective) applies to each of the following situations?
   (a) Using mortality tables and claims history to develop probabilities of survival that will allow an insurance company to make a profit.
   (b) Using field tests of tire mileage to develop probabilities that the company's tire mileage guarantees will not exceed the average wear-out period of their tires.
   (c) Estimating the likelihood that the initial expansion of marketing programs into foreign markets will be successful.
   (d) Determining the probability of winning the state lottery.

3. What is the probability of:
   (a) getting a six on one roll of a die?
   (b) drawing a red card from a deck of 52 cards (half of which are red)?
   (c) receiving the next shipment on time if only 18 of the last 24 shipments have been on time?
   (d) having a strike if 340 of 400 members vote not to strike during a "strike vote."
4. Provide a one-sentence description of the following terms:
   (a) a set (b) an event (c) a sample space
5. Use set theory designation to describe:
   (a) the set Y which consists of the five weekdays of a week
   (b) the set Z of company cars, c, but only those that are < 2 years old
6. How many outcomes are in each of the following sample spaces?
   (a) A space describing which one of five management trainees will be randomly selected for a one-week program in Phoenix.
   (b) A space describing the probability of finding one defect in any of three computer boards. Let D = defect and G = good.
7. Distinguish between mutually exclusive and collectively exhaustive events.
8. What is the complement of each of the following?
   (a) 75% of the trainees pass the test
   (b) getting a five on the roll of a die
   (c) no customers ordered less than $5,000 of equipment
9. Which of the following sets of events are mutually exclusive?
   (a) marketing software domestically and marketing it internationally
   (b) shipping a computer to a utility on schedule and shipping it late
   (c) talking on the phone (at the same time) to someone in a factory in Maine and to a customer in Florida

10. Given the data in the table:

EMPLOYEES OF WEXFORD COMPANY

Where Employed	Male (M)	Female (F)
Plant (P)	120	10
Office (O)	10	20
Sales (S)	30	10
TOTAL	160	40

(a) Draw a Venn diagram to present the data regarding the sex of the employees.

(b) Are the subsets M and F mutually exclusive? Are they collectively exhaustive?

(c) Draw a Venn diagram showing only the subsets P and M of the (universal) set of all employees of Wexford Company.

(d) Are the subsets P and M mutually exclusive? Are they collectively exhaustive?

(e) How many elements of the universal set are not in P and not in M? Is this the complement of the number in P or M?

MARGINAL, JOINT, AND CONDITIONAL PROBABILITIES

We have examined the concept of probability in terms of sets and sample spaces, which can be depicted as Venn diagrams. However, it is often more convenient to work with probabilities directly from the tables in which the frequencies are presented. We can use the employee data of the Cork Manufacturing Company (repeated as Table 4–2 below) to illustrate this. Note that the data are cross-classified (by location of work and sex) and the classes are mutually exclusive and collectively exhaustive.

TABLE 4–2 Employees of Cork Manufacturing Company

WORK LOCATION	SEX OF EMPLOYEE Male	Female	TOTAL
Plant	1,600	200	1,800
Office	20	140	160
Sales	40	0	40
Total	1,660	340	2,000

Before focusing specifically on the rules of probability, we must, however, define three terms commonly used to refer to probabilities in a tabular format. We can use Table 4–2 to illustrate (a) marginal, (b) joint, and (c) conditional probabilities.

MARGINAL PROBABILITIES

Marginal probabilities express the frequency of one event or category of a variable relative to the total frequency of the table. For example, if a person is selected at random from the complete list of Cork Manufacturing Company employees, the probability that the person will be in a particular one of the five categories (three location and two sex categories) is equal to the relative frequency of that category. These probabilities are called *marginal probabilities*, possibly because they are computed from the column or row totals located in the margins of the table.

Example 4–5 Using Table 4–2, what are the marginal probabilities of being (a) male, (b) female, and (c) employed in the plant?

Solution

For all of these problems we will let M = number of males, P = number of plant employees. Because we will be working with relative frequency probabilities, we can use the general form of

$$P(X) = \frac{\text{Number of times outcome } X \text{ occurred}}{\text{Total number of observations}}$$

$$\text{(a)}\quad P(M) = \frac{\text{Male}}{\text{Total}} = \frac{M}{T} = \frac{1{,}660}{2{,}000} = .83$$

$$\text{(b)}\quad P(F) = \frac{\text{Female}}{\text{Total}} = \frac{F}{T} = \frac{340}{2{,}000} = .17$$

Note that the two categories of male and female completely exhaust the sex classification, and the sum of their two probabilities equals 1.00. Another way of expressing this would be to say that the two probabilities *partition* the sample space, and their union probability would be: $P(M \cup F) = 1.00$.

$$\text{(c)}\quad P(P) = \frac{\text{Plant}}{\text{Total}} = \frac{P}{T} = \frac{1{,}800}{2{,}000} = .90$$

Marginal probabilities express the probabilities of individual events. They are also called *unconditional* probabilities because no conditions restrict them; that is, no information is given which reduces the size of the sample space upon which the probability is based.

JOINT PROBABILITIES

Joint probabilities express the frequency at the intersection of two cross-classified events, or categories, relative to the total frequency of the table. For example, Table 4–2 would yield six joint probabilities, one for each of the six cells in the body of the table.

Example 4–6 Using Table 4–2, what are the joint probabilities of randomly selecting an employee that is (a) a plant worker and male, (b) a plant worker and female, (c) a sales employee and male?

Solution

The joint probabilities are computed from the frequencies of the intersections of the categories.

$$\text{(a)}\quad P(P \cap M) = \frac{\text{Plant and Male}}{\text{Total}} = \frac{P \cap M}{T} = \frac{1{,}600}{2{,}000} = .80$$

$$\text{(b)}\quad P(P \cap F) = \frac{\text{Plant and Female}}{\text{Total}} = \frac{P \cap F}{T} = \frac{200}{2{,}000} = .10$$

Note that the sum of the two joint probabilities for the Plant category (i.e., $P \cap M$ and $P \cap F$) add up to .90, which is the marginal probability of randomly selecting a Plant employee (.90).

$$\text{(c)}\quad P(S \cap M) = \frac{\text{Sales and Male}}{\text{Total}} = \frac{S \cap M}{T} = \frac{40}{2{,}000} = .02$$

The joint probabilities express the probability that two events will occur together. If we were to compute the joint probabilities for all six of the mutually exclusive and collectively exhaustive categories, the sum of them would be 1.0.

CONDITIONAL PROBABILITIES

Conditional probabilities express the frequency at the intersection of two cross-classified events or categories, relative to the total frequency of one of the categories. Conditional probabilities always involve a reduced sample space because the element of interest is limited to one of the major categories. For example, if the subset of interest is limited to males (i.e., given male), then the number of males (1,660) would be considered the total in the denominator of the probability expression. The probability of a plant worker, given selection is restricted to males, would be designated $P(P|M)$.

Example 4–7 Using Table 4–2, what are the conditional probabilities of (a) a plant worker, given selection is restricted to males; (b) an office worker, given male; and (c) a female, given office worker.

Solution

(a) $P(P|M) = \dfrac{\text{Plant and Male}}{\text{Male}} = \dfrac{P \cap M}{M} = \dfrac{1{,}600}{1{,}660} = .96+$

(b) $P(O|M) = \dfrac{\text{Office and Male}}{\text{Male}} = \dfrac{O \cap M}{M} = \dfrac{20}{1{,}660} = .01+$

Note that if the conditional probability of $P(S|M)$ were also computed, the three conditional probabilities, given male would total 1.00.

(c) $P(F|O) = \dfrac{\text{Female and Office}}{\text{Office}} = \dfrac{F \cap O}{O} = \dfrac{140}{160} = .88$

PROBABILITY RULES

In working with probabilities, we have ensured that our computations satisfy certain properties or *axioms* of probability:

1. The probability of each outcome in the sample space must be a number between zero and one.
2. The sum of the probabilities of all possible outcomes in the sample space must equal 1.0.
3. If two events in a sample space are mutually exclusive, the total probability of the two is the probability of the first plus the probability of the second.

Some well-developed rules, or theorems, for computing probabilities will now be introduced that will enable us to deal with most any probability problem. The first rule (*Complement Rule*) stems from the first and third properties above. The other two rules (*Addition Rule* and *Multiplication Rule*) utilize mathematical extensions of the axioms and allow us to calculate the probabilities of unions and intersections of events in a sample space.

COMPLEMENT RULE

The **complement rule** states that the probability of the event is 1.0 minus the probability of its complement. It applies to mutually exclusive events where the sample space allows only for an event to occur or not occur.

$$P(A) = 1 - P(\overline{A}) \qquad (4\text{–}4)$$

Example 4–8 A firm estimates their probability of losing a contract at .25. What is their probability of winning the contract?

Solution

Because winning and losing the contract are mutually exclusive events:

$$P(W) = 1.00 - P(\overline{W}) = 1.00 - .25 = .75$$

MULTIPLICATION RULE

The **multiplication rule** states that the joint probability of two events is equal to the marginal probability of one event, multiplied by the conditional probability of the other. The general equation, which holds true for all events A and B, is:

$$P(A \cap B) = P(A)P(B|A) \tag{4–5}$$

The left side of the equation designates the probability of *A and B* occurring together. The right side could be stated either as $P(A)P(B|A)$ or as $P(B)\ P(A|B)$—both are equally valid. Note that the multiplication rule is basically saying that the intersection (or common space of two events) is the probability of the first, times the probability of the second, given that the first has occurred. It does not matter which event is classified "first" or "second," so long as the same designation is maintained throughout.

Example 4–9 The Venn diagram below describes the cars in the Dublin Used Car Lot. The lot contains 100 cars, 30 of which are red (R) and 10 of which are Datsuns (D). Suppose a buyer selects a car at random. What is the probability that the car is a red Datsun?

Solution

$$\begin{aligned} P(R \cap D) &= P(R)\ P(D|R) \\ &= \left(\frac{30}{100}\right)\left(\frac{3}{30}\right) \\ &= .03 \end{aligned}$$

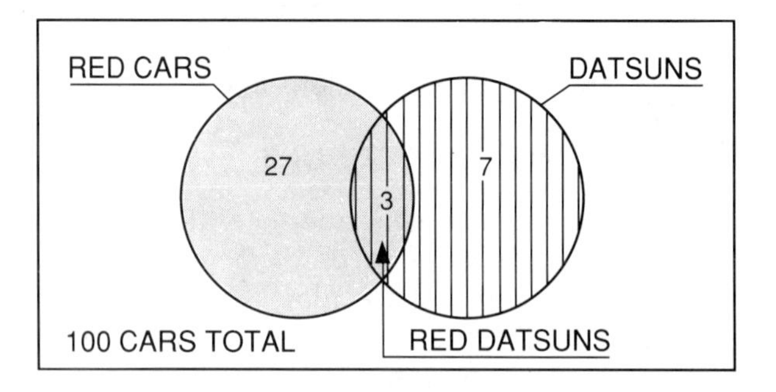

When the events are *statistically independent*, the multiplication rule reduces to:

$$P(A \cap B) = P(A)\,P(B) \qquad (4\text{–}5a)$$

Events are **statistically independent** when the probability of the second is not "conditioned" by whether or not the first occurs. That is, the "given condition" at the end of the multiplication rule does not affect the resultant probability and can be omitted. Therefore, under statistical independence, the joint probability $[P(A \cap B)]$ equals the product of the two marginal probabilities $[P(A)\,P(B)]$.

The probabilities in the car lot example above happen to be independent because the probability of getting a red Datsun is not conditioned by whether the selection is restricted to red cars. Datsuns make up 30% of the cars and they also make up 30% of the red cars. Had we recognized the statistical independence, our calculation could have used the marginal probabilities:

$$P(R \cap D) = P(R)\,P(D) = \left(\frac{30}{100}\right)\left(\frac{10}{100}\right) = .03$$

As a second example of statistical independence, consider the probability of drawing an ace (one of four) from a deck of 52 playing cards. The probability of drawing an ace is $4/52$ or $1/13$. If one card is drawn, and then a second card is drawn without replacing the first card, the second probability is not statistically independent of the first. This is because the second draw is "conditioned" by what happened on the first draw. If an ace happened to be drawn on the first draw, the second probability of an ace is $3/51$. However, if the first card is *replaced* and the deck thoroughly shuffled before the second card is drawn, then the second probability is again $4/52$.

ADDITION RULE

The **addition rule** states that the probability of either event A **or** event B is equal to the sum of their marginal probabilities minus their joint probability. The **or** in the rule is inclusive and means A or B or both, and designates the union of A and B. In equation form, this rule is

$$P(A \cup B) = P(A) + P(B) - P(A \cap B) \qquad (4\text{–}6)$$

Example 4–10 Use the car lot data from the previous example where there are 100 cars, 30 of which are red, and 10 of which are Datsuns. Find the probability of randomly selecting a red car or a Datsun.

Solution

$$P(R \cup D) = P(R) + P(D) - P(R \cap D)$$
$$= \frac{30}{100} + \frac{10}{100} - \frac{3}{100}$$
$$= .37$$

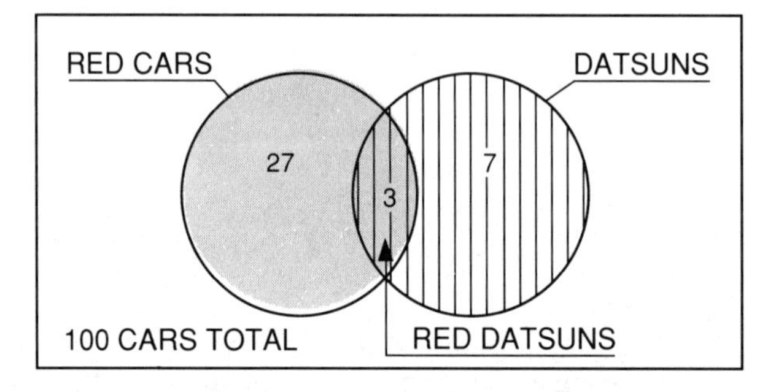

Note that the simple addition of the $P(R)$ and $P(D)$ would have counted the 3 red Datsuns twice, but the addition rule compensates for that by subtracting off the intersection $P(R \cap D)$. If the two events were mutually exclusive, there would, of course, be no intersection. For mutually exclusive events the addition rule simplifies to

$$P(A \cup B) = P(A) + P(B) \qquad \text{(4–6a)}$$

For example, suppose 15 of the cars were Fiats (from Italy). If you wanted to calculate the probability of randomly selecting a Datsun or a Fiat, the probability could be calculated

$$P(D \cup F) = P(D) + P(F) - P(D \cap F)$$
$$= \frac{10}{100} + \frac{15}{100} - 0 = .25$$

Selecting a Datsun or a Fiat are mutually exclusive events. The addition rule can still be used; the joint $P(D \cap F)$ simply takes on a value of zero.

The addition rule can be extended to the union of three probabilities, and we present that equation here for reference only (i.e., without examples).

$$P(A \cup B \cup C) = P(A) + P(B) + P(C) - P(A \cap B) - P(A \cap C) - P(B \cap C) + P(A \cap B \cap C) \qquad (4\text{–}7)$$

Figure 4–5 summarizes the probability rules we have been working with, and adds a fourth, Bayes' Rule.

FIGURE 4–5 Probability rules

Complement Rule

$$P(A) = 1 - P(\bar{A}) \qquad (4\text{-}4)$$

Multiplication Rule

$$P(A \text{ and } B) = P(A)\,P(B|A) \qquad (4\text{-}5)$$
$$= P(A)\,P(B) \quad \text{(if independent)} \qquad (4\text{-}5a)$$

Addition Rule

$$P(A \text{ or } B) = P(A) + P(B) - P(A \text{ and } B) \qquad (4\text{-}6)$$
$$= P(A) + P(B) \quad \text{(if mutually exclusive)} \qquad (4\text{-}6a)$$

Bayes' Rule

$$P(A|B) = \frac{P(A \text{ and } B)}{P(B)} = \frac{P(A)\,P(B|A)}{P(A)P(B|A) + P(\bar{A})P(B|\bar{A})} \qquad (4\text{-}8a)$$

As a final example of the rules of probability, let us revisit the data of the Cork Manufacturing Company to see how readily the rules can be applied to tabular data.

Example 4–11 Given the employment data for the Cork Manufacturing Company below:

WORK	SEX OF EMPLOYEE		TOTAL
LOCATION	Male	Female	
Plant	1,600	200	1,800
Office	20	140	160
Sales	40	0	40
Total	1,660	340	2,000

Assume an employee is selected at random, and use the rules specified to find the probabilities requested:

(a) Use the ***complement rule*** to find the probability of selecting a male.

(b) Use the ***multiplication rule*** to find the probability of selecting a female who is an office worker.

(c) Use the ***addition rule*** to find the probability of selecting either a female or a plant worker (or one who is both).

(d) Are the categories employment and sex ***statistically independent***?

Solution

(a) $P(M) = 1 - P(\overline{M}) = 1 - 340/2{,}000 = 1 - .17 = .83$

(b) $P(F \cap O) = P(F)\,P(O|F) \quad \textit{where } P(O|F) = 140/340$

$$= \left(\frac{340}{2{,}000}\right)\left(\frac{140}{340}\right) = .07$$

(c) $P(F \cup P) = P(F) + P(P) - P(F \cap P)$

$$\textit{where } P(F \cap P) = P(F)\,P(P|F) = \left(\frac{340}{2{,}000}\right)\left(\frac{200}{340}\right) = .10$$

$$= \left(\frac{340}{2{,}000}\right) + \left(\frac{1{,}800}{2{,}000}\right) - \left(\frac{200}{2{,}000}\right)$$

$$= .17 + .90 - .10 = .97$$

(d) The two variables are statistically independent if the percentage breakdown among plant, office, and sales workers in the male and female columns is the same as the percentage breakdown in the total column. In other words, the conditional *P*(Plant | Male) should equal *P*(Plant | Female) and these should equal the marginal *P*(Plant). To test this:

	***P*(Plant | Male)**		***P*(Plant | Female)**		***P*(Plant)**
	$\frac{1{,}600}{1{,}660}$		$\frac{200}{340}$		$\frac{1{,}800}{2{,}000}$
But:	.96	$\neq$	.59	$\neq$	.90

Any one inequality would void the statistical independence, so they are *not* statistically independent.

Another approach is to note that the joint probabilities [i.e., $P(P \cap M)$] do not equal the product of the marginal probabilities [i.e., $P(P)\,P(M)$]. That is, the multiplication rule for $P(P \cap M)$ cannot be shortened to $P(P)\,P(M)$.

USING CROSS-CLASSIFICATION TABLES TO COMPUTE PROBABILITIES

Tables are extremely useful techniques for visualizing (and solving) probability problems involving two variables of classification. The values in the body of the table can be either frequency amounts or probability values. Probability values work very well because the joint probabilities across rows and down columns are additive, and the sum of the marginal probabilities in the table must equal 1.0.

Example 4–12 Given $P(A) = .65$, $P(B) = .45$, and $P(A|B) = .40$. Find $P(A \cup B)$.

Solution

We can consider A and its complement as one variable of classification, and B and its complement as the other. Then the two steps in using a table to help solve the problem are as shown below:

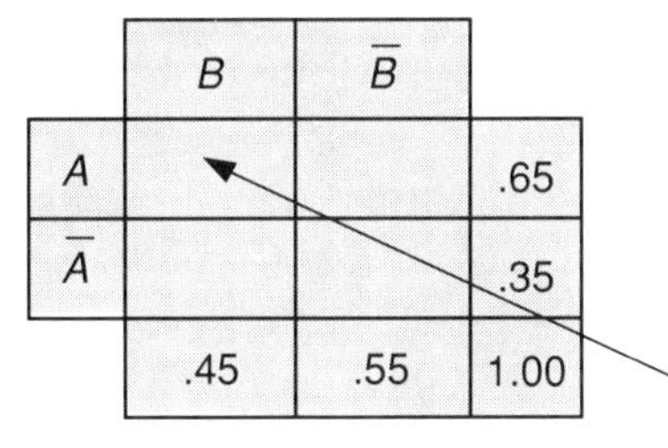

Knowing P(A|B) = .40, we can express this as the intersection A ∩ B ÷ P(B):

$$P(A|B) = \frac{P(A \cap B)}{P(B)} = \frac{P(A \cap B)}{.45}$$

$\therefore$ $P(A \cap B) = (.40)\,(.45) = .18$

	B	$\bar{B}$	
A	.18	.47	.65
$\bar{A}$	.27	.08	.35
	.45	.55	1.00

And once the A ∩ B cell is determined, the other cell values must satisfy the row and column totals. Then;

$$P(A \cup B) = P(A) + P(B) - P(A \cap B)$$
$$= .65 + .45 - .18 = .92$$

This completes our discussion of the rules of probability. They are so widely used in statistics that you may find it worthwhile to simply *memorize them*. The next section covers Bayes' Rule, which is in one sense a restatement or manipulation of the above rules.

EXERCISES II

[Note: Use the Cork Manufacturing Company data, repeated below, for problems 11–16.]

WORK LOCATION	SEX OF EMPLOYEE Male	Female	TOTAL
Plant	1,600	200	1,800
Office	20	140	160
Sales	40	0	40
Total	1,660	340	2,000

11. What is the *marginal probability* of being employed in (a) the office and (b) sales?
12. What are the *joint probabilities* of randomly selecting an employee that is (a) an office worker and female, (b) a sales worker and male, and (c) a sales worker and female?
13. Find the *conditional probabilities* of (a) a male given selection is restricted to office workers, (b) a sales worker given male, and (c) a female given office worker.
14. Use the *multiplication rule* to find the probability of selecting a male who works in the plant.
15. Use the *addition rule* to find the probability of selecting a person who is either a male or an office worker (or both).
16. Use the necessary expressions to compute the following *marginal, conditional,* and *joint* probabilities:

    (a) $P(M)$ (c) $P(\overline{M})$ (e) $P(\overline{O})$ (g) $P(M|S)$

    (b) $P(O|M)$ (d) $P(\overline{M} \cup \overline{O})$ (f) $P(O \cap S)$ (h) $P(S|M)$
17. North State University Business College has two majors, Accounting and Economics. They have 500 students with 220 majoring in Accounting. The student body is 60% female, and of these, 100 are in Accounting.
    (a) Show the data as a two-way cross classification (sex/major) table.
    (b) Draw a Venn diagram that depicts the number of Accounting (A) and Female (F) students.

    Assuming random selection of a student, find the:

    (c) $P(E)$ (g) $P(E \cap F)$

    (d) $P(A|F)$ (h) $P(E \cup F)$

    (e) $P(A \cap F)$ (i) What categories are mutually exclusive?

    (f) $P(F)$ (j) Are the variables sex and major statistically independent?

18. Given that $P(A) = .30$, $P(B) = .60$, and $P(A \cap B) = .15$. Find:
   (a) $P(A \cup B)$
   (b) $P(A|B)$
   (c) Are A and B mutually exclusive?
   (d) Are A and B statistically independent?
19. Suppose $P(A) = .70$, $P(B) = .20$, and $P(B|A) = .10$. Find (a) $P(B \cap A)$, and (b) $P(A \cup B)$.
20. The Clonmel Crystal Company sometimes produces "seconds" because of defects in the glass. The probability of an air bubble, $P(A) = .15$, and the probability of a scratch, $P(S) = .10$. The probability of a scratch, given an air bubble is present, $P(S|A) = .20$. What is the probability that a randomly selected piece of crystal (a) has both an air bubble and a scratch and (b) has either an air bubble or a scratch?
21. Taiwan Shoe Company urgently needs two raw materials, on time, if they are to finish an order of running shoes by March 1. The supplier of the leather parts (in Germany) is typically late 20% of the time, and the supplier of rubber materials (in Malaysia) is late 40% of the time.
   (a) Would you think the two marginal probabilities are independent? Why?
   (b) Assuming they are independent, what is the probability that the shipment can be finished by March 1?
22. A market research firm surveyed 200 business leaders to learn whether they read the *Management News and Digest* regularly. Of those surveyed, 140 had graduated from college. The survey showed that 50% of those surveyed read the periodical, and given that the person was a college graduate, the probability he or she read the periodical was .60. What is the probability that one of the randomly selected leaders either read the periodical or was a college graduate?
23. The Safety Officer of Trans Atlantic Air Cargo Co. has been asked to recommend whether the airline could safely fly two-engine planes on an overseas route where they currently use three-engine planes. The probabilities of engine failure are independent of each other and, for illustration purposes, we will assume they are 2%. (a) For the two-engine plane, what is the probability both engines (A and B) will fail? (b) For the two-engine plane, what is the probability that either Engine A or Engine B will fail? (c) For the three-engine plane, what is the probability that all three engines will fail on the same flight? (d) Is the two-engine plane sufficiently safe?

BAYES' THEOREM

Bayes' theorem is a statement of conditional probabilities developed by the Reverend Thomas Bayes (1702–1761), a Presbyterian minister in England. Bayes attempted to reason from an effect (evidence in the world) back to a cause (i.e., that God exists), and in doing so became the founder of a branch of modern statistics known as Bayesian Decision Theory.

The basic focus of Bayesian statistics today is that prior probability statements can be revised on the basis of empirical evidence that is obtained later. *Prior probabilities* are frequently subjective probabilities that have been assigned to events by the decision maker. The empirical data may come from market research or database studies that make use of historical information describing the success rate of past studies. The resultant *posterior* probabilities then provide a better basis for decision making than the prior estimate.

In business, analysts frequently use Bayes' theorem to *revise probabilities* of market demand (e.g., of sales, successful products, etc.) based upon sample or subsequent information. But it is also used in other analysis. For example, a Bayesian analysis of the booster engines that powered the Challenger space shuttle was done by the Air Force prior to the tragic shuttle launch in 1986. The Bayesian analysis yielded a booster failure probability of 1 in 35 flights. The disastrous launch would have been the 25th flight (see *Discover*, April 1986, p.56).

In its simplest form, Bayes' theorem is little more than a restatement of the multiplication rule, $P(A \cap B) = P(A)\,P(B|A)$. You may recall, we noted that it could also be written as $P(A \cap B) = P(B)\,P(A|B)$. If we solve for $P(A|B)$ we have Bayes' theorem:

$$P(A|B) = \frac{P(A \cap B)}{P(B)} \tag{4–8}$$

Notice that the conditional probability $P(A|B)$ is equal to the joint probability of A and B, divided by the marginal probability of B. The numerator, $P(A \cap B)$, is computed from $P(A)\,P(B|A)$. The denominator, $P(B)$, is the sum of all the mutually exclusive joint probabilities in which B occurs; for example, $P(B)$ is equal to $P(A \cap B) + P(\overline{A} \cap B)$. Hence, a more general statement of the revised probability applied to n events is:

$$P(A_1|B) = \frac{P(A_1)\,P(B|A_1)}{P(A_1)\,P(B|A_1) + P(A_2)\,P(B|A_2) + \ldots + P(A_n)\,P(B|A_n)} \tag{4–8a}$$

REVISING PRIOR PROBABILITIES

To illustrate the use of Bayes' theorem, we will first solve a problem using the basic equation and then show how a tabular approach might be useful. Following this, we will illustrate the same problem on a tree diagram.

Example 4–13 The Champion Chocolate Company marketing manager feels there is a .70 probability that marketing ten-pound boxes of chocolates rather than five-pound boxes will result in a sales increase (S) (rather than a decrease, $\bar{S}$). A market research firm has agreed to test her strategy in selected cities, and inform the company if the test results are favorable (F) or unfavorable ($\bar{F}$).

Both firms recognize that such tests are not always accurate and that the results may be misleading. Based on previous market tests, the research company says there is a .60 probability of getting **F**avorable (F) results **given** the **S**ales (S) would increase [i.e., $P(F|S) = .60$]. Also, the probability of getting Favorable results followed by a Sales decrease is .20 [i.e., $P(F|\bar{S}) = .20$]. A test is made, and the results are Favorable. Given the test information, revise the probability of a sales increase.

Solution

Summarizing the data we have

Prior Probability: $P(\text{Sales increase}) = P(S) = .70$. Thus $P(\bar{S}) = .30$

Conditional Probabilities:

$P(F|S) = .60$ (This also means $P(\bar{F}|S) = 1.00 - .60 = .40$)

$P(F|\bar{S}) = .20$ (This also means $P(\bar{F}|\bar{S}) = 1.00 - .20 = .80$)

The probability we seek is: $P(\text{Sales increase}|\text{Favorable test result}) = ?$
Applying Bayes' theorem:

$$P(S|F) = \frac{P(S \cap F)}{P(F)} = \frac{.42}{.48} = .875$$

Where:

$$P(S \cap F) = (\text{prior})\,(\text{conditional}) = P(S)\,P(F|S) = (.70)\,(.60) = .42$$

$$P(F) = P(S)\,P(F|S) + P(\bar{S})\,P(F|\bar{S}) = (.70)\,(.60) + (.30)\,(.20) = .48$$

The "tricky" part of solving most Bayesian problems is stating the conditional probabilities correctly, and computing the marginal probabilities. Using a table may help answer the computational questions.

	F	$\bar{F}$	TOTAL
S			.70
$\bar{S}$			.30
TOTAL			1.00

	F	$\bar{F}$	TOTAL
S	.42	.28	.70
$\bar{S}$	.06	.24	.30
TOTAL	.48	.52	1.00

The value of $P(S)$ is given as .70, so the complement, $P(\bar{S})$, is .30. The conditional probability, $P(F|S) = .60$, means that in the row marked S, the value under F must be 60% of the total for that row. Since $.60 \times .70 = .42$, enter the .42 under F in the row marked S. Then subtract .42 from .70 to find the amount to be entered in row S under $\bar{F}$, and enter the resulting .28 as shown on the right above. Next, consider the given conditional probability $P(F|\bar{S}) = .20$. This means that the figure entered at the intersection of row S and column F must be 20% of the total for the given row, that is $.20 \times .30 = .06$. Enter the .06 under F and the remainder, $.30 - .06 = .24$, in the column marked $\bar{F}$. Total the F and $\bar{F}$ columns. All the data required are now available to revise the probabilities if the results are either favorable or unfavorable:

IF TEST RESULTS FAVORABLE:

$$P(S|F) = \frac{P(S \cap F)}{P(F)} = \frac{.42}{.48} = .875$$

IF TEST RESULTS UNFAVORABLE:

$$P(S|\bar{F}) = \frac{P(S \cap \bar{F})}{P(\bar{F})} = \frac{.28}{.52} = .54$$

Table 4–3 presents the analysis of the previous example in a form that has become identified with Bayesian problems. This form simplifies the calculations for problems with several events. Column 1 identifies the basic events and column 2 gives the prior probability of each. Column 3 gives the conditional probabilities (or "likelihoods"), and column 4 lists the joint probabilities obtained from multiplying the priors and conditionals. The sum of the joint probabilities in column 4 is the marginal probability that is called for in the denominator of Bayes' formula. Dividing the respective joint probabilities in column 4 by this marginal probability yields the revised probabilities of column 5. The .875 shown in this column is the probability of **S**ales increase, given **F**avorable test results [i.e., $P(S|F)$] requested in the example.

TABLE 4–3 Bayes' theorem calculations for chocolate company example

(Calculations based upon having received a **F**avorable test market report)

EVENTS S	PRIOR PROBABILITIES $P(S)$	CONDITIONAL PROBABILITIES $P(F\|S)$	JOINT PROBABILITIES $P(S)\ P(S\|F)$	REVISED PROBABILITIES $P(S\|F)$
S Sales Increase	.70	.60	.42	.42 ÷ .48 = .875
$\bar{S}$ No Increase	.30	.20	.06	.06 ÷ .48 = .125
	1.00		.48	1.000

USING TREE DIAGRAMS TO DESCRIBE PROBABILITY SITUATIONS

Tree diagrams of probability situations are another helpful way of visualizing problems, provided the number of outcomes (branches) is not too large. Figure 4–6 presents a complete tree diagram for the chocolate company example. The diagram is constructed as follows:

FIGURE 4–6 Tree diagram for Bayes' Theorem problem

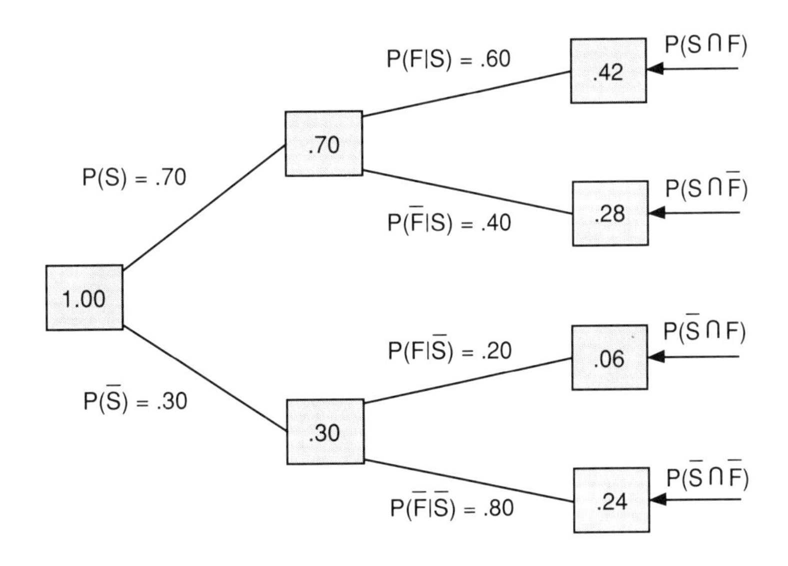

Begin at the left side of the chart with the full probability 1.00. The prior marginal probabilities are entered first. They must be known or given. For our example, the $P(S) = .70$ value and its complement, $P(\bar{S}) = .30$, are entered in the boxes as shown.

The next step is to enter the given conditional probabilities, which are $P(F|S) = .60$ and $P(F|\bar{S}) = .20$. The complements of the given conditional probabilities, $P(\bar{F}|S) = .40$ and $P(\bar{F}|\bar{S}) = .80$, respectively, are then entered so that the two original branches of the tree are subdivided to include all outcomes of the sample space.

When the conditional probabilities entered in the preceding step are multiplied by the marginal probabilities which are in the boxes of the two major branches, the required joint probabilities are obtained. These are entered in the final boxes and labeled for ease of identification. The problem originally posed can now be solved by using values from the tree.

$$P(S|F) = \frac{P(S \cap F)}{P(F)} = \frac{.42}{.48} = .875$$

Note that the numerator, $P(S \cap F)$, is equal to $P(S)\ P(F|S)$, and that the denominator, $P(F)$, is the sum of the various joint probabilities containing F, which in this case are $P(S \cap F)$ and $P(\bar{S} \cap F)$. Since F and $\bar{F}$ are complements, $P(S \cap F) + P(\bar{S} \cap F)$ represent the total probability of F. It is given that F occurred. The $P(S|F)$ is merely the proportion of the time that both S and F occurred when F occurred.

The meaning of the Bayesian result is important. Note that the marketing manager's prior probability of the success of the 10-pound box was .70. Now the test market information has come in and supports her opinion, and the revised probability of an increase in sales is raised to .875. If the test results were unfavorable, the estimate of an increase in sales would be revised downward to .54, as was calculated previously (and could also be taken from the tree diagram).

This completes the journey into Bayesian statistics for now. The final section of this chapter concerns some quick methods of counting the number of outcomes of a sample space.

EXERCISES III

24. State Bayes' Rule and identify the kinds of probabilities used in it.
25. Distinguish between *prior* and *posterior* probabilities. Explain how Bayes' theorem is used to revise prior probabilities on the basis of experimental data.
26. If $P(A) = .60$, and $P(B) = .40$, and $P(B|A) = .20$, what is the $P(A|B)$?

27. Suppose the marketing manager of Example 4–12 feels there is a .90 probability that 10-pound boxes will increase sales (instead of the .70 prior probability in the example). Using the other data from the example, use Bayes' Rule to compute the revised probability of a sales increase, assuming test results are favorable.

28. Suppose the marketing manager of Example 4–12 has a .85 prior probability (instead of .70) and $P(F|S) = .65$ and $P(F|\bar{S}) = .25$. Using the other data from the problem, compute the revised probability of a sales increase, assuming that test results are unfavorable.

29. An Allied Insurance Company representative has been asked to investigate a recent accident at a major construction project in the city, and to estimate the likelihood that it was caused by faulty equipment. Let θ represent the existence of defective equipment, and A represent the occurrence of an accident. Past examinations of the construction equipment have revealed that 15% of it is defective (i.e., $P(\theta) = .15$). Given that equipment is defective, the probability of an accident occurring on a project is .75 (i.e., $P(A|\theta) = .75$), and if the equipment is not defective, the chance of an accident is reduced to .05 (i.e., $P(A|\bar{\theta}) = .05$). Use Bayes' Rule to estimate the likelihood that the recent accident was caused by faulty equipment.

30. Because of high packaging costs, a company is considering marketing a 5-pound package of their product for $60 rather than a 3-pound box for $40. (Assume the gross margin per pound is the same in either case.) Management assigns a probability of .90 to the likelihood of national success. Market research evidence suggests that the probability of getting unfavorable results in a test made in one city is .30 if the national outcome would be favorable and about .80 if the national outcome would be unfavorable. The prior probability of success is to be revised on the basis of the results in the one test city. Use the symbols:

    T = Test Results Favorable $\qquad N$ = Nationally Successful
    $\bar{T}$ = Test Results Unfavorable $\qquad \bar{N}$ = Nationally Unsuccessful

    (a) Prepare a tree diagram and find:

    $$P(N \cap T) \qquad P(T) \qquad P(N|T)$$
    $$P(N \cap \bar{T}) \qquad P(\bar{T}) \qquad P(N|\bar{T})$$

    (b) Is it logical that a failure in the test city would decrease management's estimate of the probability of national success? Does your answer agree with this?

    (c) Prepare a table with columns labeled T and $\bar{T}$, and rows labeled N and $\bar{N}$. Compute all joint and marginal probabilities for the table using the data given. Now compute $P(N|T)$ and $P(N|\bar{T})$ from your table. Compare these answers with those obtained using the tree diagram.

COUNTING PROCEDURES

The Old Holland Cheese Co. is a small family-owned business near Amsterdam. They produce the four types of cheese depicted in Figure 4–7: Cheddar, Gouda, Roquefort, and Swiss. Suppose they wished to design a holiday gift pack that would offer customers different packages of the cheeses. How many different arrangements might they come up with?

FIGURE 4–7 Cheeses produced by Old Holland Cheese Co.

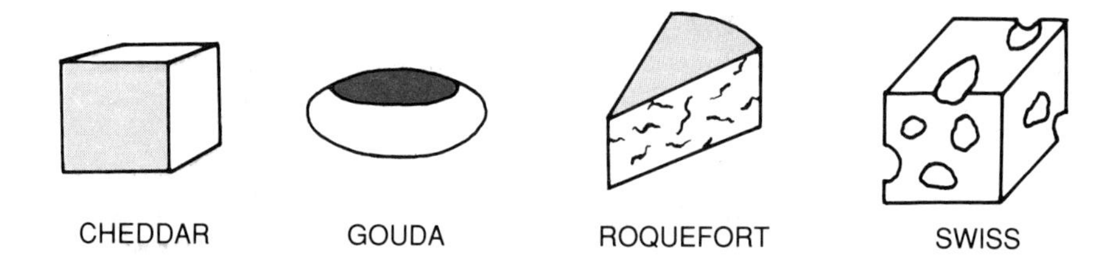

The answer depends, of course, upon what one considers a "different arrangement." Three ways of counting arrangements that you should be familiar with for statistical purposes are: (1) multiple choices, (2) permutations, and (3) combinations.

In discussing these counting methods, we will be referring to n as the total number of items and using x to refer to the group size that is used, such as one item at a time, or two at a time, etc. For example, if each gift box contained only two pieces of cheese, we would be looking for the number of arrangements of $n = 4$ items, taken $x = 2$ at a time.

FACTORIALS Another "housekeeping" detail to note before we discuss the counting methods is to point out that some of them use factorial notation. A *factorial* is a number followed by an exclamation mark, such as 3!, which means that the number is to be multiplied by all smaller positive integers. (By definition, 0! is equal to 1.) Thus, $4! = 4 \cdot 3 \cdot 2 \cdot 1 = 24$. Since $4! = 4 \cdot 3!$, to divide 4! by 3! write

$$\frac{4!}{3!} = \frac{4 \cdot \not{3}!}{\not{3}!} = 4$$

The fact that factorial division can be accomplished by cancellation is an important time saver in combinatorial computations. Note also that many calculators have factorial keys to simplify these calculations.

THE FUNDAMENTAL RULE

The most basic rule for counting outcomes of a sample space is simply a way of stating that the number of outcomes from combining two or more items is the product of the number of choices associated with each item.

RULE: If there are k_1 items of the first type, k_2 of the second and k_n of the nth, then the total number of possible outcomes is $(k_1)(k_2)(k_3) \ldots (k_n)$.

For example, suppose you have just landed a job (with IBM) where you feel you should dress appropriately each day. If you have $k_1 = 2$ suits, $k_2 = 4$ shirts or blouses, and $k_3 = 3$ ties, you could come up with as many as $2 \times 4 \times 3 = 24$ different "outfits."

Example 4–14 A toy store can purchase a wagon in (a) light, standard, or heavy duty, with (b) rubber or plastic wheels, and (c) in any of ten colors. Also, it can be ordered with or without a silver stripe. How many different end items are possible?

Solution

There are 3 models, 2 types of wheels, 10 colors, and 2 choices for the stripe (i.e., with or without). The number of end items possible is thus:

(NO. MODELS) (3)	×	(NO. WHEEL TYPES) (2)	×	(NO. COLORS) (10)	×	(STRIPE/NO STRIPE) (2)	= 120

The fundamental rule has been extended to apply to more restrictive situations. In particular, it enables us to very quickly calculate what we shall refer to as the number of multiple choices, permutations, and combinations. The differences that result from counting multiple choices, permutations, and combinations stem from (1) whether *duplication* is permitted, and (2) whether a *different order* counts. With multiple choices, both are allowed, with permutations no duplication is allowed, and with combinations neither duplication nor different order counts as another arrangement.

The description that follows includes some summary equations (provided without derivation) that may be useful to you later on in the text. We'll present a formula and an example to illustrate each counting method. Then we will return to the cheese company example to compare all three of the counting techniques in one figure to make the distinctions quite apparent.

MULTIPLE CHOICES

When each choice mentioned in the fundamental rule involves the same number of alternatives, we have what is sometimes referred to as a *multiple choice situation*.

Multiple Choices (M), are the number of different ways that n items can be arranged or selected X units at a time if duplications are permitted and different orders of the same items count as separate choices.

$$_nM_x = n^x \qquad (4\text{–}9)$$

Example 4–15 How many different three-letter computer security codes could be made from the 26 letters of the alphabet if letters can be duplicated (i.e., *AAA* is okay) and different order counts (i.e., *ABC*, *ACB*, *BAC*, *BCA*, *CAB*, and *CBA* all count)?

Solution

This is a problem of $n = 26$ items taken $x = 3$ at a time. There are 26 ways to choose the first letter. Then, for each one of these ways, there are 26 ways to choose the second, and another 26 ways to choose the third.

$$_nM_x = n^x = 26^3 = 17{,}576$$

PERMUTATIONS

If the order of selection is important (i.e., counts as an arrangement) but duplications are not permitted, then the problem is one of *permutations*.

Permutations (P), are the number of different ways that n items can be selected x units at a time if **no** duplications are permitted, but different order counts.

$$_nP_x = \frac{n!}{(n - x)!} \qquad (4\text{–}10)$$

Example 4–16 A fashion expert is presented with ten neckties and asked to rank the five she likes best from first to fifth. How many possible outcomes exist?

Solution

Order is important, and each tie can occupy only one position, so no duplication is possible. We have a *permutation* of $n = 10$ items being taken $x = 5$ at a time.

$$_nP_x = \frac{n!}{(n - x)!} = \frac{10!}{(10 - 5)!} = \frac{10 \cdot 9 \cdot 8 \cdot 7 \cdot 6 \cdot 5!}{5!} = 30{,}240$$

If all ten of the ties were to be ranked, the equation $_nP_x$ would be used with $x = 10$. This would result in $10! \div 0!$, which is $10! \div 1$, or simply 10! The answer is 3,628,800 different orders. This results from having 10 ways to select the first tie, 9 ways to select the second, and so forth.

Optional Another feature of permutations allows us to compute the number of arrangements when some elements are indistinguishable. For example, suppose there are sixty electrical resistors in a box (and assume thirty are color coded red, twenty are green, and ten are blue). If the resistors are distinguishable only by color, then the number of permutations of all resistors which can be distinguished is given by the equation:

$$_nP_{(x_1, x_2, \ldots, x_k)} = \frac{n!}{x_1!\, x_2! \ldots x_k!} \qquad (4\text{–}11)$$

This equation can be used *only when all n elements* are selected. Using it, the number of distinguishable permutations of resistors is:

$$_{60}P_{30,\, 20,\, 10} = \frac{60!}{30!\, 20!\, 10!}$$

COMBINATIONS

If the order of selection is irrelevant, and duplications are not permitted, the problem is one of *combinations*. The symbol for combinations is variously written $_nC_x$, C_x^n, or $\binom{n}{x}$.

Combinations (C), are the number of different ways that n items can be selected X units at a time if **no** duplications are permitted and **no** different order counts.

$$_nC_x = \frac{n!}{x!(n - x)!} \qquad (4\text{–}12)$$

Example 4–17 Each of 50 states has sent one delegate to a convention. Five persons are to be selected to draft a resolution to be voted upon. How many different *ad hoc* committees could be selected from the fifty delegates?

Solution

$$_nC_x = \frac{n!}{x!(n-x)!} = \frac{50!}{5!(50-5)!} = \frac{50\cdot 49\cdot 48\cdot 47\cdot 46\cdot 45!}{5\cdot 4\cdot 3\cdot 2\cdot 1\cdot 45!} = 2{,}118{,}760$$

The number of combinations will always be less than the number of permutations, which will be less than the number of multiple choices. Now let us return to the Old Holland Cheese Co. to compare the three.

Example 4–18 A gift box of cheese will hold two pieces. How many different arrangements can be made of four types of cheese if (a) duplication and different order count, (b) no duplication is allowed, but different order counts, and (c) no duplication or no different order counts?

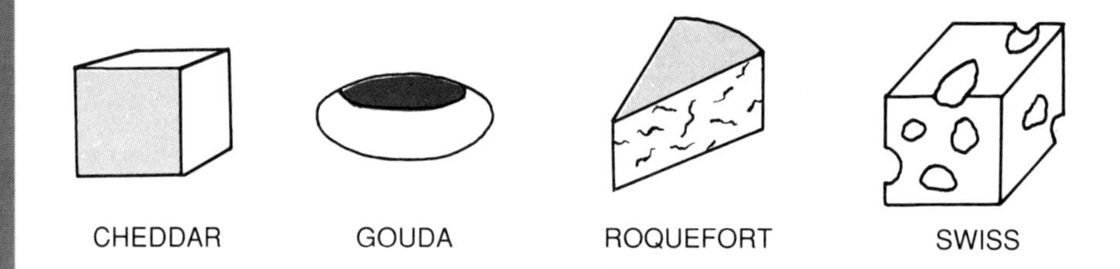

Solution

The problem is asking for (a) multiple choices, (b) permutations, and (c) combinations. Letting C, G, R, and S represent the types of cheeses, the gift boxes could be arranged as follows:

MULTIPLE CHOICES (duplication OK) (diff. order OK)	PERMUTATIONS (**no** duplication) (diff. order OK)	COMBINATIONS (**no** duplication) (**no** diff. order)
C C --------	------------	
C G	C G	C G
C R	C R	C R
C S	C S	C S
G C	G C --------	---
G G --------	---	---
G R	G R	G R
G S	G S	G S
R C	R C	---
R G	R G	---
R R --------	---	---
R S	R S	R S
S C	S C	---
S G	S G	---
S R	S R	---
S S --------	---	---
Total 16	12	6

(a) Multiple Choices

$${}_4M_2 = n^x = 4^2 = 16$$

(b) Permutations

$${}_4P_2 = \frac{4!}{(4-2)!} = 12$$

(c) Combinations

$${}_4C_2 = \frac{4!}{2!(4 = 2)!} = 6$$

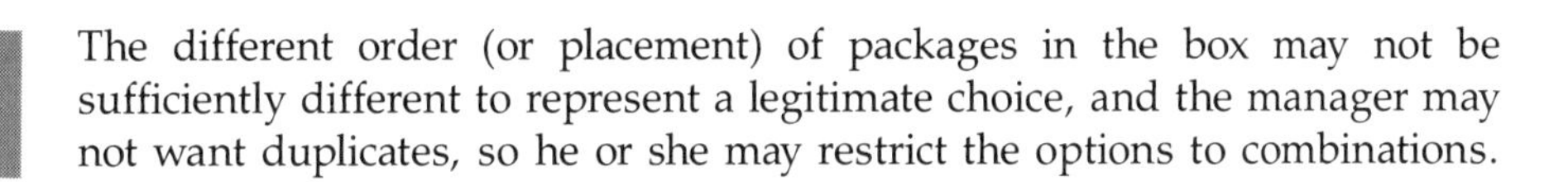

The different order (or placement) of packages in the box may not be sufficiently different to represent a legitimate choice, and the manager may not want duplicates, so he or she may restrict the options to combinations.

EXERCISES IV

31. What is the value of (a) 4!, (b) 1!, (c) 0!, (d) ${}_{12}M_2$, (e) ${}_5M_4$?
32. A business student must select a literature, history, and philosophy class from among five literature classes, six history classes, and two philosophy classes. Assuming that half of the selections result in conflicts, how many possible choices of the three classes would the student have? (Use the fundamental rule.)
33. The Horizon Appliance Company offers customers the following options on their electric washing machines:

    Standard or Deluxe Model

    Choice of six colors (white, golden, tan, cream, yellow, avocado)

    Any model can come with or without a cold rinse feature

    Choice of three frame designs (compact, stand-alone, or apartment)

    How many different end-products can be made from combining these options?

34. What are the chief distinctions among (a) multiple choices, (b) permutations, and (c) combinations?
35. Compute values for the following:
(a) ${}_4P_3$, (b) ${}_6P_1$, (c) ${}_{10}P_2$, (d) ${}_5C_3$, (e) ${}_{12}C_{10}$, (f) ${}_6C_6$
36. A state lottery inspector has been asked to compute how many different four-digit tickets could be derived from the digits 0–9 inclusive. What should his response be?
37. A city transit manager has six bus drivers that must be assigned to six bus routes. How many different combinations of assignments of drivers to routes exist?
38. A quality control committee is to be made up by selecting four employees from ten that have been nominated by their departments. In how many different ways could the committee be structured?
39. Paris Perfume International produces five perfumes in two-ounce bottles: (1) Apri, (2) Ballet, (3) Chanille, (4) DuVant, and (5) Empre. If a gift box is to be made up of three bottles, how many different package arrangements could be made if (a) duplication and different order count, (b) different order counts, but no duplication is permitted, (c) no duplication or different order is permitted?
40. The scheduling nurse at Bay City Hospital must schedule 20 operations in the hospital's three operating rooms, and the sequence in which the operations are performed is significant. Assuming each operation took a similar amount of time, how many different sequences could be considered?
41. A marketing consultant has developed a product packaging design that entails using a different color on each of the six sides of a box that will contain an electronic product. The consultant has identified nine possible colors that he feels are acceptable and has presented the design to his client. How many different color groupings does this leave for the client to choose from? Assume that the box will contain six different colors, so no duplications need be considered.
42. An assembly shop has eight jobs (A, B, C, D, E, F, G, H) that are due to be shipped tomorrow, but only enough test cell capacity to complete two of the jobs. (a) In how many ways can 2 jobs be chosen from the 8 that should be completed? (b) Assuming the two jobs to be completed are selected at random, what is the probability that one of those jobs is Job A? (Hint: You may wish to use a tree diagram to describe this situation.) (c) List all the possible combinations of two jobs and confirm your answer in (b) by showing the ratio of combinations that contain A (i.e., successes) to all the combinations possible (i.e., the total).

USING COMPUTERS TO FIND PROBABILITIES

Numerous software programs exist for computing probabilities, although many probability problems are so short they don't justify using a computer. You are asked to enter values for the variables and whether or not they are (a) mutually exclusive, or (b) independent. The programs then calculate the probabilities. Some programs also provide a two-way classification table, and the probabilities can be obtained from a table. Programs are also available for Bayes' Theorem. You must, of course, supply the conditional probabilities that are used for the probability revision. In addition, probability software often includes counting programs that provide answers to permutation and combination problems.

SUMMARY

Probabilities express the likelihood of chance events. *Classical* probabilities are based upon equally likely outcomes, *relative frequency* probabilities on empirical evidence, and *subjective* probabilities on personal judgment. *Set theory* helps us understand probabilities by depicting a sample space as a collection of all possible outcomes of an experiment. Then the proportion of "successes" relative to the total outcomes represents the probability of an event. *Venn diagrams*, which depict set relationships (as circles) inside the sample space (a rectangle) are also useful, as are *tree diagrams*.

Events are *mutually exclusive* if they have no elements in common and are *statistically independent* if the occurrence of one in no way affects the other. Three other terms used to distinguish probabilities are marginal [e.g., $P(A)$], *joint* [$P(A \cap B)$], and *conditional* [$P(A|B)$].

The *complement rule* [$P(A) = 1 - P(\bar{A})$], *multiplication rule* [$P(A \cap B) = P(A)P(B|A)$], and *addition rule* [$P(A \text{ or } B) = P(A) + P(B) - P(A \cap B)$] are sufficient to solve most probability problems. *Bayes' rule* extends them slightly to allow us to revise prior probabilities on the basis of more data. Bayes' rule allows us to compute the conditional $P(A|B)$ by dividing the joint $P(A \cap B)$ by the marginal $P(B)$.

Counting equations are useful shortcuts for computing the number of possible outcomes. With *multiple choices* both duplications and different orders count as separate outcomes. With *permutations* no duplications are permitted, and with *combinations* neither duplication nor different orders are permitted to count.

Microcomputer *software* is available for computing all the probabilities discussed in the chapter, as well as permutations and combinations.

CASE: ALLIANCE NATIONAL BANK

Alliance National Bank's board-of-directors were elated! They had just pulled off a strategic acquisition that gave them an excellent *entree* into a whole new financial market. The board, by a vote of 7 to 2, had approved the acquisition of Multinational Insurance Corporation (MIC) at a cost of $225 million. Although MIC had some dubious looking assets, the bank's public accountant had tagged it a "reasonably safe" investment, with significant upside potential. Moreover, Alliance was the country's fourth largest bank—it could handle the acquisition.[2]

Because of the technical nature of the discussions, the Alliance Bank's internal auditor had been allowed to sit in on the board-of-directors deliberations as a nonvoting member. Although the bank's outside accountant (Merst and Nagey Inc.) had given a qualified go-ahead on the purchase, the internal auditor had asked for a more definitive statement. The public accounting firm had said they didn't see any advantage in quantifying a risk to the point where it made their recommendations appear more accurate than they really were. They felt it was a "reasonably safe" investment with "slightly more than average risk." On the other hand, it offered better-than-average profit potential if the insurance claims situation leveled out. (There had been some unusually large claims due to tropical storms.) During negotiations, the internal auditor kept pressing for more specifics, but was put off again and again. Finally, when the merger was practically a sure thing, the public accounting firm issued a blanket assessment saying the odds were at least 5 to 1 that the investment would pay off.

Alliance National's directors had voted 7 to 2 to go ahead with the acquisition. The two directors opposed to the merger asked for an independent assessment by an outside consultant, and received approval to have this done. Meanwhile, however, the other members of the board felt they must proceed with the acquisition or lose the opportunity. The outside consultant's assessment came in later citing a 70% chance that a major outstanding (international) loan of ($102m) due to the insurance company would be paid and an 80% chance that the claims situation would improve, "given that the global weather conditions held fairly constant." However, if the weather were to continue about the same, or worsen, "that probability could drop to .65."

Six months later, the auditor and two directors obtained an independent evaluation by a second (merger) consultant who had a record of having correctly predicted collections 85% of the time when the debt really was collected, and correctly predicting a loss 95% of the time when a

[2] Although this case has some similarities to a situation that occurred in a European country during the mid 1980s these facts are not intended to describe that case and should be considered hypothetical.

loss was really incurred. In their opinion, the international loan would not be collected. They also observed that the continuing bad weather was causing the firm to have to pay off on some unexpectedly high risk claims.

As a result of this information and at the urging of the company's internal auditor, the two directors recommended that the bank divest itself of the MIC immediately. However, the chairman and other directors remained adamant in rejecting what they felt were "obviously subjective" estimates, "highly distorted by the current recession." They pointed out that their managerial image would look very poor to the financial community if they were to attempt to back out on the insurance company investment now. Besides, things would probably improve within a short time.

Six months after that, a New York newspaper leaked the news. The legislature was having an emergency (secret) meeting. No one was supposed to know what it was about, but the newspaper claimed a "reliable source" said it had to do with the government takeover of Multinational Insurance Co. The source claimed MIC was totally bankrupt, and if Alliance National had to absorb the loss, it would pull the bank under as well. Furthermore, this might have a cascading effect on other large banks.

Two days later the official news story broke. Yes, MIC was bankrupt. Alliance National Bank's survival had also been endangered temporarily, but an "arrangement" had been worked out with the government so that the losses would be sustained by the national treasury. The bank remained secure, and the Chairman of the Board confidently announced that its stockholder dividends would be paid on schedule. Unfortunately the Chairman of the Board was unavailable for comment at the moment, because he was "in Bermuda on holiday" (as of yesterday).

The news story left many unanswered questions, but the one that seemed to surface most often in the news media was, "Why should we taxpayers pay over $200m when the loss was suffered because of the inept management by a private bank?" The apparent response was that it was in the "best national interest" to do so, but it was clear this issue was not yet put to bed. All Bermuda holidays must end someday.

Case Questions

(a) What kind of probabilities were involved in this case (classical, relative frequency, or subjective)?

(b) In what ways did the bank directors fail to use good judgment? Can a deficiency in one area justify a claim that management of the bank was "inept" or incompetent?

(c) How might the directors have used probability concepts to combine the two major risks and arrive at some average level of risk?

(d) What would have been the revised probability of collecting the loan if the directors had been willing to accept the two consultants' estimates?

QUESTIONS AND PROBLEMS

[Note: Most of the following problems can be done on computer, but probability calculations are frequently so short that it may be faster to do them on a small calculator.]

43. Which one type of probability is most likely to (a) incorporate a manager's own experience and judgment, (b) be calculated prior to an event occurring, and (c) be based primarily upon historical data?
44. Identify the type of probability (*classical*, *relative frequency*, or *subjective*) used in each of the following:
    (a) Randomly selecting a defective computer chip from a box that contains 200 chips, only 3 of which are defective.
    (b) Determining the likelihood of producing a "hit" record by pre-release interviews with radio disk jockeys.
    (c) Having an auto fatality in Syracuse on a holiday weekend, based upon data from 30 previous holiday weekends.
    (d) Selecting a student who is both female *and* a business major by random choice from the registrar's list of all university students. [Note: Thirty percent of all the students are in business and 55% of the business students are female.]
45. Compute the probabilities of events (a) and (d) in the previous problem.
46. Explain why the following general probability rules are simplified.
    (a) Multiplication Rule—when events are *independent*.
    (b) Addition Rule—when events are *mutually exclusive*.
47. The land-based weather forecasting system in the United States yields forecasts that have a 65% chance of being accurate. (a) Illustrate this in terms of a *Venn diagram*, where the universal set is all forecasts $\{F\}$, the accurate set is designated $\{A\}$, and the complement of $\{A\}$ represents the inaccurate forecasts. (b) What is the complement of $\{A\}$?
48. Use a *Venn diagram* to depict the mutually exclusive subsets A, B, and C of the set U. Assume that A has 40 elements, and that B has 100 elements.
    (a) How many elements are there in C if there are 200 elements in total?
    (b) How many elements are there in the intersection of A and B?
    (c) What are two characteristics of the partition of this sample space?

49. The table below is derived from the database in Appendix A and shows the corporate dividend amounts cross-classified by industry code.

	NUMBER OF FIRMS PAYING ANNUAL DIVIDENDS OF			TOTAL NUMBER OF FIRMS
INDUSTRY	**$ 00**	**$.01 ≤ 1.00**	**> $1.00**	
B Banking	—	2	8	10
C Chemical	1	3	16	20
E Electrical	10	13	7	30
R Retail	2	8	5	15
T Transportation	2	2	1	5
U Utility	—	1	19	20
TOTAL	**15**	**29**	**56**	**100**

Assuming a random selection of firms, find the probability of selecting: (a) a chemical firm, (b) a firm that pays no dividend, (c) a retail firm, given that selection is restricted to firms paying dividends of more than $1.00, (d) an electrical firm that pays a dividend of more than $1.00, (e) either an electrical firm or a firm that pays a dividend of from $ .01 to $ 1.00.

50. Using the same data as in problem 49, find:
(a) $P(T)$, (b) $P(T|\$0.00)$, (c) $P(B \cap > \$1.00)$,
(d) $P(C \cap \leq \$1.00)$, (e) $P(U \cap \$0.00)$, (f) $P(R \cup > \$1.00)$.

51. Using the same data as in problem 49, find:
(a) $P(B|\$0.00)$, $P(B|\$.01 \leq \$1.00)$, and $P(B| > \$1.00)$.
(b) Are the industry class and dividend rates independent? Explain.

52. Using the same data as in problem 49, are the events $(C \cup \$0.00)$ and $(C \cap > \$1.00)$ mutually exclusive in this data set? Explain.

53. The probability tree below depicts the probability of an insurance stock having a low (L) or high (H) trading volume, followed by the conditional probability of the price going up (U) or down (D), given the volume. Compute whatever values are missing to answer the following:

(a) What is the unconditional probability of having a high volume?

(b) What is the conditional probability of the price going up, given that the volume is high?

(c) What is the joint probability of the stock experiencing a low volume and the price going up?

(d) What is the probability of the price going down?

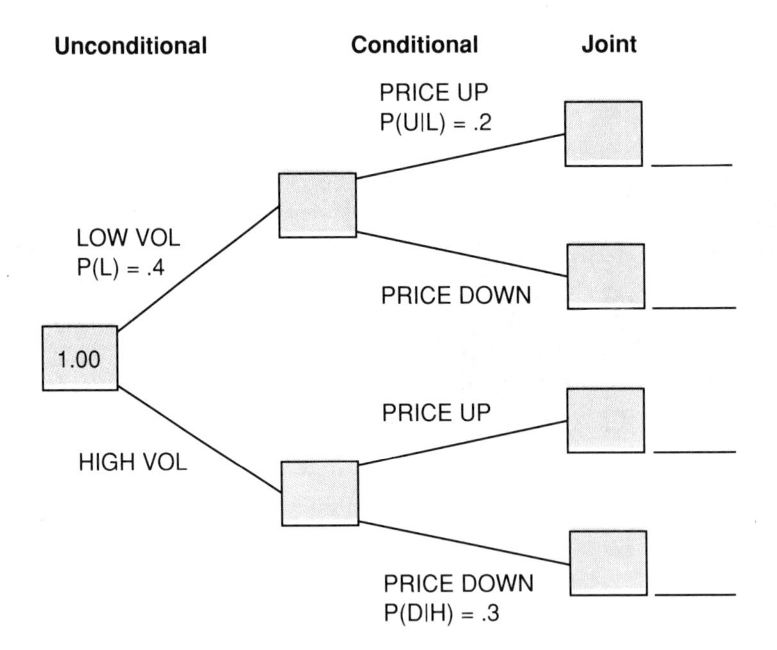

54. Suppose you wished to use Bayes' rule to compute the $P(L|U)$ for the problem above. What would be the notation and numerical values for the:

(a) numerator of Bayes' equation? (b) denominator of Bayes' equation?

55. Assume $P(X) = .20$ and $P(Y) = .40$ and $P(Y|X) = .40$.

(a) What is $P(X \cap Y)$? (b) What is $P(X \cup Y)$?

(c) Are X and Y independent? Explain.

56. Given: $P(A) = .4$ $\quad P(B) = .6$ $\quad P(A \cap B) = .1$ $\quad$ Find $P(A|B)$.

57. Given: $P(A) = .80$ $\quad P(B|A) = .70$ $\quad P(B|\overline{A}) = .10$ $\quad$ Find $P(A|B)$

58. A data processing center has two large computers, an MXII and a BMI20. About 55% of the work requires a PASCAL processor, which can be loaded onto either machine. Twenty percent of the work requires PASCAL *and* must be done on the MXII computer. What is the probability a randomly selected job was done on the MXII, assuming we know it required PASCAL? [*Hint*: Find *P*(MXII *given* PASCAL).]

59. The prior probability of a power outage during March is .06. If there is (i.e., given) a power outage, the probability of losing some computer data is .45. Without a power outage, the probability of losing computer data is .02. An information systems analyst has just discovered that some vital computer data was lost in March. What is the revised probability there was a power outage, i.e., knowing now that the computer data was lost?

60. The Emerald Isle Fertilizer Company has applied to a county council for permission to construct a lime quarry near a coastal area. However, the council members are concerned that the plant will pollute the bay and damage marine life. The firm says the probability of an unacceptable pollution is only .02 and the council accepts this as a starting point. Then, an independent consultant is called in and says the plant is likely to pollute the bay. The consultant has a history of correctly predicting that pollution will occur when in fact it does later occur on 85% of their studies. Also, they have predicted no (or NOT) pollution when pollution did later occur on 10% of their studies. What should be the revised probability of pollution?

61. Machines A and B each produce 50% of the precision castings of a company, but A's output is 90% **G**ood, whereas B's is only 80% **G**ood. A casting is inspected and found to be **N**ot **G**ood. (a) Draw a tree diagram to show the probabilities associated with obtaining *good* or *not good* castings, depending upon the machine on which they are made. (b) Use the tree to compute the probability that the casting came from machine A.

62. A sweater manufacturer in Korea has prepared a display of 50 different ski sweaters for a visit from a large department store buyer from Vienna. The buyer is asked to select five sweaters which she believes will be most popular in Austria.

    (a) In how many ways can she make her selection of the top five sweaters?

    (b) In how many ways can she make her selection if she ranks the winners from first to fifth?

    (c) What is the essential difference between combinations and permutations?

63. Nationwide Distributors has ten regional sales contest winners and must select five finalists for a trip to Hawaii. (a) In how many ways can the five finalists be selected from the ten? (b) Suppose cash prizes are awarded with the most to the top selection, next to the second, and so on. In how many ways can the five finalists be rank ordered from one to five?

64. The hotel key card shown has five columns and ten rows and must be inserted into the electronic lock to open your hotel door in London. Every different arrangement of holes constitutes a new lock.

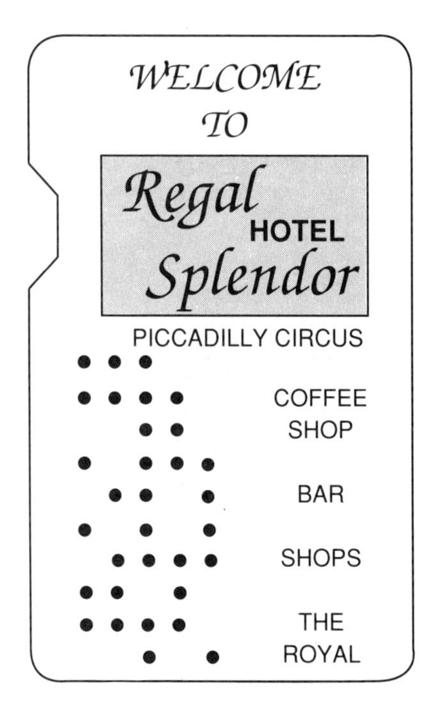

(a) How many different locks could be arranged from the five holes in **one row** only of the card

(b) Assuming the same number of locks can be constructed from the second row of the card, how many different arrangements can be derived from **two rows** of the card?

(c) Comment on the security afforded by this card with ten rows.

CHAPTER 5

Probability Distributions

INTRODUCTION

Random Variables
Probability Distributions and Expected Values
Overview of Probability Distributions

THE BINOMIAL DISTRIBUTION

Using the Binomial Equation
Using Binomial Tables
Binomial Distribution Graphs
Mean and Standard Deviation of the Binomial Distribution

THE POISSON DISTRIBUTION

Using the Poisson Equation
Poisson Tables and Graphs
Poisson Approximation to the Binomial Distribution

THE NORMAL DISTRIBUTION

Characteristics of the Normal Distribution
Finding Areas Under the Normal Curve
The Normal as an Approximation to the Binomial

OTHER CHARACTERISTICS OF STATISTICAL DISTRIBUTIONS

Mean and Standard Deviation of Discrete Distributions
Chebyshev's Inequality

SUMMARY

CASE: DETROIT LAKES POWER COMPANY

QUESTIONS AND PROBLEMS

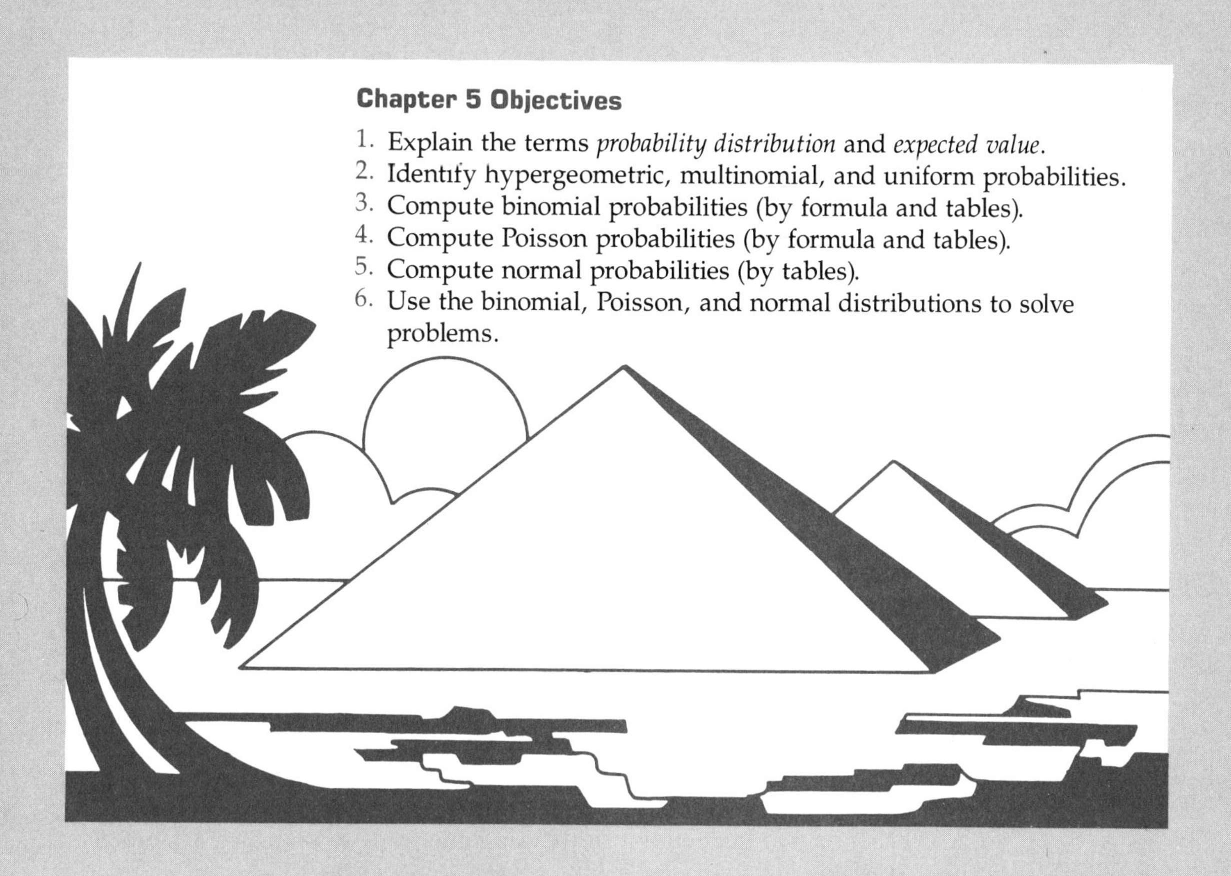

Chapter 5 Objectives

1. Explain the terms *probability distribution* and *expected value*.
2. Identify hypergeometric, multinomial, and uniform probabilities.
3. Compute binomial probabilities (by formula and tables).
4. Compute Poisson probabilities (by formula and tables).
5. Compute normal probabilities (by tables).
6. Use the binomial, Poisson, and normal distributions to solve problems.

On the upper floor of the British Museum in London is a glass case containing the Rhind Mathematical Papyrus that dates back to 1875 BC. The papyrus describes in detail how to multiply and divide fractions and compute volumes of cylinders, areas of squares, circles, and triangles. Yes, it even tells how to determine the slope of the sides of one of the first types of geometrical distributions used by man—a pyramid. The basic geometrical form of the pyramid is the triangle, and planners had to calculate how much labor and materials were required to build them. Some of the blocks in the Great Pyramid in Egypt weigh up to 200 tons, and as many as 900 workers may have been required to drag them into very precise locations.

Triangular and other distributions have extensive use in business today. For example, the planners of large construction projects use a distribution very similar to the triangular distribution to estimate the time required to complete their projects. Consider building a large hydroelectric dam in Idaho or a skyscraper in New York. These projects require excavation, massive concrete foundations, mechanical equipment, electrical work, and so forth. There are so many interactive activities to coordinate that the final completion date can be quite uncertain. One statistical distribution (the beta) allows planners to combine different estimates of completion times for various phases of the project, and another (the normal) enables them to compute the probability that the project will be completed on schedule or by a certain date (e.g., 24 months, 30 months).

INTRODUCTION

The probability rules of Chapter 4 provided us with the underlying logic for drawing conclusions about the probability of specific events when some associated probabilities were given. Now it is time to extend our concern to the source of the probability information and see how the type of variable affects our calculation of probabilities. In particular, probability calculations can differ depending upon whether the variable is *discrete* or *continuous*, and whether it is *dependent* or *independent*.

Once the variable has been characterized, it is often most convenient to compute probabilities using a model (or theoretical probability distribution) of that variable that closely resembles the actual situation. This chapter introduces several probability distributions and shows how three can be used by business managers. We begin with a definition of a *probability distribution*, and then go on to consider the *binomial*, *Poisson*, and *normal distributions*.

RANDOM VARIABLES

We have been using the term *random* since Chapter 1, and have referred to it as an outcome determined by chance. Let us now give the term more specific meaning in the context of probability.

A **random variable** is a numerical function whose value is determined by chance. We frequently use the capital letter X to represent a random variable.

Although a random variable is a specific *function* that assigns a numerical value to an "experiment," we usually focus more upon the value than upon the function. For example, suppose a state lottery provides that first round winners of a lottery shall receive $1,000 times the value that surfaces from the roll of a die. By the classical rule of probability, each of the six sides of the die has an equal chance (1/6) of ending up on top. We cannot predict for certain which value will emerge, so the outcome is truly a *chance event*. The *numerical value* assigned by the random variable X would be one of the sample space outcomes of either $1,000, $1,100, $1,200, $1,300, $1,400, $1,500 or $1,600. Thus we have:

Symbol	Random variable (function that assigns values)	All possible values in the sample space of outcomes
X	$X = \$1{,}000$ (# dots on die)	$1,000, 2,000, 3,000, 4,000, 5,000, 6,000

Although the outcomes are really the values (as opposed to the function that generates those values), they are sometimes referred to as the random variable (value), or simply the random variable itself.

Many of the variables affecting marketing, production, and other business decisions can be analyzed as random variables. For example, the number of customers entering a store may be considered a random variable, as might be the number of defects from a production line, or the time required to handle a banking transaction. Thus, we shall have numerous occasions to work with random variables throughout the text.

PROBABILITY DISTRIBUTIONS AND EXPECTED VALUES

Earlier we saw that relative frequency distributions showed the proportion of times that various outcomes of a random variable occurred. Probability distributions simply describe these outcomes as *theoretical* outcomes.

A **Probability Distribution** of a random variable shows the values the variable can take on and the probability of each value.

EXPECTED VALUE As in frequency distributions, probability distributions also have a mean. The mean of a probability distribution is called its *expected value*, $E(X)$, and is the weighted average of the values the random variable can take on. It is computed by multiplying each possible value of the random variable by its probability of occurrence and summing these products.

$$E(X) = \Sigma[X \cdot P(X)] \tag{5-1}$$

Example 5–1 For the roll of a six-sided die, (a) show the probability distribution, and (b) compute the expected value of the distribution.

Solution

(a)

SAMPLE SPACE	PROBABILITY DISTRIBUTION: RANDOM VARIABLE X	PROBABILITY DISTRIBUTION: PROBABILITY $P(X)$	EXP. VALUE: $X \cdot P(X)$
⚀	1	1/6	1/6
⚁	2	1/6	2/6
⚂	3	1/6	3/6
⚃	4	1/6	4/6
⚄	5	1/6	5/6
⚅	6	1/6	6/6
	TOTAL	6/6	21/6

FIGURE 5–1 Sample space, probability distribution, and expected value

(b) $E(X) = \Sigma[X \cdot P(X)] = 21/6 = 3.5$

Let us now extend the above concepts into an industrial situation where a firm uses these probability concepts to evaluate the reliability of a motor.

Example 5–2 An electrical motor will run if either (or both) of two relays are operable as shown. Each relay has a probability $P(W)$ of .9 of working satisfactorily.

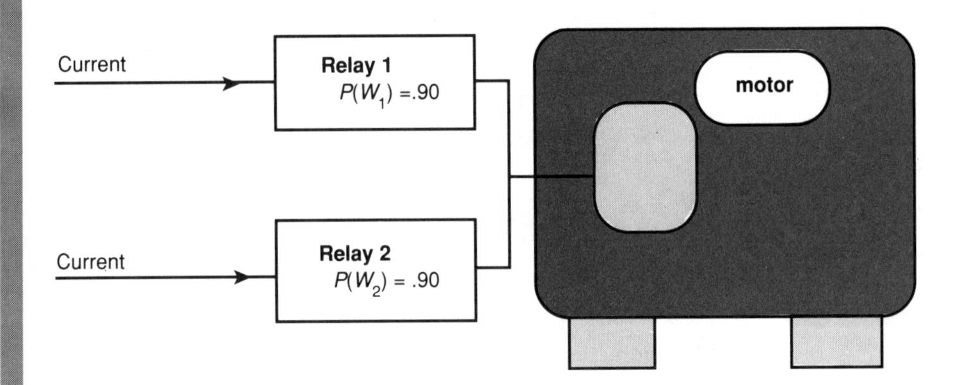

(a) List the elementary events that make up the sample space.

(b) State the probability distribution that describes the likelihood the motor will operate.

Solution

(a) There are four elementary events in the sample space. Letting W_1 = relay 1 works and $\overline{W}_1$ = relay 1 does not work, the sample space is made up of the set of four ordered pairs:

$$S = \{(W_1, W_2), (W_1, \overline{W}_2), (\overline{W}_1, W_2), (\overline{W}_1, \overline{W}_2)\}$$

(b) We shall assume the relays function independently, so the probability of any pair (i.e., via the multiplication rule) is the simple product of the two probabilities. Then let the (random variable) rule that assigns probability values to the events in the sample space be:

Chance that both relays work:	let probability value be $.9 \times .9$
Chance that W_1 works and W_2 does not:	let probability value be $.9 \times .1$
Chance that W_1 does not and W_2 does work:	let probability value be $.1 \times .9$
Chance that neither relay works:	let probability value be $.1 \times .1$

Letting X = the number of relays that work, the resulting probability distribution follows from assigning probability values to X in accordance with the (probability) rule stated above.

For these elements in the sample space	The random variable is (X) [No. Relays that Work]	And the resulting probability of the elementary event is
W_1, W_2	2	.90 × .90 = .81
$W_1, \overline{W}_2$	1	.90 × .10 = .09
$\overline{W}_1, W_2$	1	.10 × .90 = .09
$\overline{W}_1, \overline{W}_2$	0	.10 × .10 = .01

We can then restate the probability distribution a little more succinctly:

PROBABILITY DISTRIBUTION FOR RELAYS

No. Relays that Work X	Probability P(X)
0	.01
1	.18
2	.81

The motor in Example 5–2 thus has a .99 probability of operating satisfactorily. It will not be necessary to go into this much detail on most of our distributions in this chapter; we simply used this example to relate the earlier material to our present discussion.

OVERVIEW OF PROBABILITY DISTRIBUTIONS

All probability distributions satisfy the requirements that (1) individual event probabilities are ≤ 1, and (2) the sum of the probabilities equals 1. However, the calculation of individual probabilities of the distributions differs depending upon whether the data is *discrete* or *continuous*.

Discrete probability distributions are based upon *counts*, and show the probability of each discrete event in the sample space. (Examples of discrete distributions are the *BINOMIAL* and the *POISSON*.)

Continuous probability distributions are based upon *measurements*, and offer the probability of events that occupy some interval in the continuum of the sample space. (An example of a continuous distribution is the *NORMAL*.)

For example, the number of customers purchasing a new car from a dealer today is a discrete random variable; the shipping weight of all the cars purchased is a continuous random variable.

A second major determinant of the probability of an event is statistical independence. Events are considered *statistically independent* if the sample is taken either from an infinite population, or "with replacement" from a finite population. The "with replacement" means elements of the sample are returned before a second sample is drawn.

Example 5–3 A card is to be drawn from a standard deck of 52 cards (i.e., of 26 red and 26 black). What is the probability of (a) getting a red card on the first draw, (b) getting a red card on the second draw, assuming the first card was replaced, and (c) getting a red card on the second draw, assuming the first card was not replaced?

Solution

Let R = red card, B = black card, and R_1 = red card on draw one, etc.

(a) $P(R) = \frac{S}{S + F} = \frac{26}{52} = .50$

(b) With replacement (and assuming the cards are reshuffled), the probability of a red card on the second draw is the same (.50). The probabilities are *independent*.

(c) Without replacement, the probability of red on the second draw depends upon what happened on the first draw (i.e., is *dependent*).

Assuming first card was red: $P(R_2 \mid R_1) = 25/51 = .49$
Assuming first card was black: $P(R_2 \mid B_1) = 26/51 = .51$

Figure 5–2 depicts some differences in probability distributions by distinguishing between *discrete* and *continuous* distributions

FIGURE 5–2 Characteristics of probability distributions

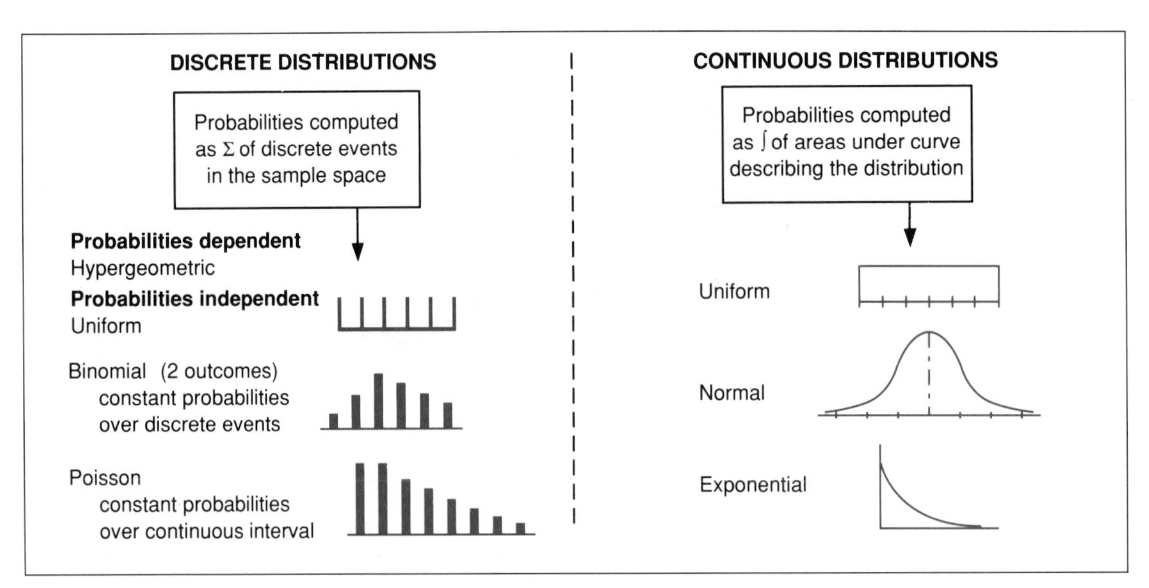

Our major concern in this chapter will be with the *binomial, Poisson,* and *normal distributions* (shown above). However, we have already introduced three of the other distributions (without formally referencing them), so we shall briefly note their characteristics before proceeding.

HYPERGEOMETRIC DISTRIBUTION The previous example uses the *hypergeometric distribution*. It describes the probability of X successes in random samples (of size n) that are drawn without replacement. In Example 5–3 we found that the probability of a red card on the second draw was dependent upon whether a red card had been obtained on the first draw. The hypergeometric distribution applies to two category data, where the probability changes with each trial.

MULTINOMIAL DISTRIBUTION Example 5–2 illustrates the *multinomial distribution*. It describes situations where the random variable can take on ≥ 3 mutually exclusive outcomes. The probabilities of each outcome are constant from trial to trial, and the trials are independent. Our example contains three independent categories (i.e., zero, one, or both relays working), and we have to allow for the fact that there are two ways of having only one relay work.

UNIFORM DISTRIBUTION The uniform distribution describes situations where each value that the random variable can assume has the same probability as every other value. There are two types: the *discrete uniform* and the *continuous uniform*.

Example 5–1 was of a discrete uniform distribution because only a discrete number of dots could come up on the die, and each number had the same probability of occurrence (i.e., 1/6). If the probabilities had related to the time interval of an event (e.g., the time between machine breakdowns) and time was equally likely, then we would be dealing with a continuous uniform distribution.

The parameters of the uniform distribution differ for discrete and continuous variables. Although we shall not explore them in any depth, equations for the mean and variance are provided below for reference.

For a *discrete uniform* random variable consisting of n values:

$$\text{Mean: } \mu = \frac{n+1}{2} \qquad \text{Variance: } \sigma^2 = \frac{n^2-1}{12} \qquad (5\text{–}2)$$

For a *continuous uniform* distribution having values ranging from a low of a to a high of b:

$$\text{Mean: } \mu = \frac{a+b}{2} \qquad \text{Variance: } \sigma^2 = \frac{(b-a)^2}{12} \qquad (5\text{–}3)$$

Discrete The mean number of dots resulting from the roll of a die would be a discrete uniform random variable, with mean $\mu = (6 + 1) \div 2 = 3.5$ dots, and a variance of $\sigma^2 = (6^2 - 1) \div 12 = 2.9$. Note that the mean is the expected value of the distribution, $E(X)$, and the expected value itself need not be a feasible event. That is, the long run average value from the roll of a die is 3.5 points, but you could never actually get 3.5 points on any one roll.

Continuous Assume the time between successive arrivals of cars at a car wash is uniformly distributed, with a minimum interarrival time of 2 minutes, and a maximum interarrival time of 10 minutes. Then the mean interarrival time would be $\mu = (2 + 12) \div 2 = 7$ min, and the variance $\sigma^2 = (10 - 2) \div 12 = .67$.

EXERCISES I

1. Distinguish between a random variable and a probability distribution.
2. Would these probability distributions be discrete or continuous?
   (a) Number of forest fires on a timber company's land each year
   (b) Acres of trees burned in forest fires each year
   (c) Ounces per bottle in cases of dishwashing liquid
   (d) Number of service calls made by a repair man on Tuesdays
3. An oil well drilling firm is equally likely to strike oil on any of the five workdays of the week. The random variable is the day on which oil is struck. (a) What type of probability distribution would describe this situation? (b) Would the distribution be discrete or continuous? (c) What values of probability would apply?
4. A production supervisor has some assembly jobs that take the following times (hrs) to complete:

ASSEMBLY JOB	A	B	C	D	E	F	G
TIME REQUIRED	5	4	6	6	9	5	3

   The supervisor wants to be fair to everyone so plans to assign the jobs to the workers on a random basis. Let the random variable be the time required to do the job. (a) What type of probability distribution would apply? (b) Construct the probability distribution showing the random variable (X) and $P(X)$. (c) Compute $E(X)$.
5. For Example 5–2, assume relays #1 and #2 have probabilities of operating satisfactorily of .80 and .70, respectively. State the resulting probability distribution.

6. A box of a dozen computer disks contains two defective disks, and disks are selected at random. (a) What is the probability of getting a good disk on the first draw? (b) Assuming the first disk was good, what is the probability of getting a defective disk on the second draw? (c) Assuming the first disk was defective, what is the probability of getting another defective disk on the second draw? (d) What type of probability distribution would apply to this situation?
7. Distinguish between two types of uniform distributions with respect to (a) type of variable, and (b) expected value of the distribution.
8. Using the appropriate expressions, find the mean and variance of the following probability distributions:
   (a) discrete uniform where $n = 8$ shipments
   (b) continuous uniform where the range of ages is from 18 to 30 years.
9. A computer is being used to simulate the number of trucks on a bridge at one time. If the computer assigns 1 to 12 trucks on the bridge on an equally likely basis, compute the (a) mean and (b) variance of the resulting probability distribution.
10. A car rental agency has the following cars in stock, which are assigned to governmental employees on a random basis, without regard to model:

MODEL	AX10	BX11	CX12	DX13
NUMBER OF CARS	4	10	2	4
MILES PER GALLON	36	50	28	30

Suppose we let the random variable be X = miles per gallon, and our interest lies in the average mileage from the inventory of the 20 cars.
(a) Construct the resulting probability distribution. (b) What is $E(X)$?

THE BINOMIAL DISTRIBUTION

BINOMIAL DISTRIBUTION The *binomial distribution* is probably the most widely used discrete distribution in business applications. It enables us to compute the probabilities of independent events that are categorized "dichotomously" (i.e., into two mutually exclusive and collectively exhaustive classes). For example, employees may be classified into those who favor a new union contract and those who do not, products may be classified as good or defective, and sales calls by marketing representatives as successful or unsuccessful. In each of these cases, only two categories exist. Moreover, the populations of employees, products, and sales calls are analyzed on the basis of those two proportions that together represent 100% of the items in the data set.

POPULATION PROPORTIONS Statisticians typically denote the population proportion of a binomial distribution with the Greek letter pi (π). Thus, we shall use the letter π to designate the *proportion of successes,* and $(1 - \pi)$ to designate the *proportion of failures* in a population. So, if the probability of a successful sales call is $\pi = .8$, then the probability of an unsuccessful one or a failure is $(1 - \pi)$ or $(1 - .8) = .2$. We are using the notion of "success" here in a generic sense; it really does not matter which category we designate as success. The important point is to categorize the data into two classes, and to maintain that distinction between the two throughout our analysis.

SAMPLE PROPORTIONS Consistent with other parts of the text, we shall reserve the Greek letter π for population data. However, as we use proportions to describe sample data later on in the text, we will use the English letter p to represent the proportion of successes in a sample. The letter q will be used to represent the proportion of failures, so $q = (1 - p)$. Thus the p and q of sample data will correspond with the π, and $(1 - \pi)$ of the population data we are using in this chapter. However, this designation is not universal. Even some computer packages, such as MINITAB, use P to designate the population parameter π.

BINOMIAL PROBABILITIES Binomial probabilities arise from what is known as a *Bernoulli process* (named after a Swiss mathematician). This process (i.e., experiment) generates observations (i.e., trials) that satisfy the following:

Binomial Probability Distribution Conditions:

1. Each of n trials has only *two* possible outcomes (e.g., success or failure).
2. The probabilities of success and failure remain *constant* from trial to trial.
3. Each trial is *independent* of any previous trials.

The binomial distribution will thus show the probabilities of $X = 0, 1, 2,$ up to n successes in n trials, when the probability of success on any one trial is equal to some known (or assumed) proportion, π. It is theoretically correct for describing any process that satisfies these conditions. In addition, it offers a useful approximation of probabilities that change only slightly as each trial is made, and for some continuous distributions that are not highly skewed.

As a preface to calculating binomial probabilities via an equation, let us use a quality control situation to gain an intuitive understanding of what the binomial probabilities represent. Although we are dealing with discrete values that possess greater accuracy, we shall (arbitrarily) limit our final answer to four significant digits, which is consistent with the accuracy reported in many tables.

Example 5–4 A crystal manufacturing process consistently produces 10% defective wine glasses because of unavoidable dust particles in an acid bath. Suppose a sample of 4 glasses is shipped (without inspection) to a customer. Find the probability that one of the glasses is defective.

For Each Glass:

P(Defective) = .10
P(Good) = 1.00 − .10
= .90

Solution

We shall assume that the probability of getting a defective (i.e., success) is *constant* for each trial (glass selected), and that each trial is *independent* of the previous trial. Using the multiplication theorem, the probability of 1 defective (D) and 3 good (G) glasses can then be computed by multiplying (.1)(.9)(.9)(.9) to get .0729. However, from Chapter 4, recall that there are four ways (i.e., 4 combinations) of selecting 1 defective and 3 good glasses (see Figure 5–3). So, the resultant probability is (4)(.0729) = .2916.

Ways of Getting One Defective in Four Glasses	#1	#2	#3	#4	Probability of Each Way
#1 Defective					$(.1)(.9)(.9)(.9) = (.1)^1(.9)^3 = .0729$
#2 Defective					$(.9)(.1)(.9)(.9) = (.1)^1(.9)^3 = .0729$
#3 Defective					$(.9)(.9)(.1)(.9) = (.1)^1(.9)^3 = .0729$
#4 Defective					$(.9)(.9)(.9)(.1) = (.1)^1(.9)^3 = .0729$
					Total .2916

FIGURE 5–3 Binomial probability of 1 defective in sample of 4, when proportion defective = .10

In Example 5–4, we computed the probability of getting *only 1* defective in the sample of 4 glasses, when the proportion in the population π was 10% defective. In more concise notation, this could be expressed as $P(X = 1 \mid n = 4, \pi = .1)$, which we found to equal .2916.

USING THE BINOMIAL EQUATION

The calculation of these binomial probabilities can be simplified by using the binomial equation.

$$P(X|n, \pi) = \frac{n!}{x!(n - x)!} \pi^x (1 - \pi)^{(n-x)} \tag{5–4}$$

Equation (5–4) gives the probability of *exactly* X successes in n trials, given the population proportion value π. Note that the first part of the equation is simply an expression of the number of combinations of n items taken x at a time, or ${}_nC_x$. So the equation can also be written:

$$P(X \mid n, \pi) = {}_nC_x \, \pi^x (1 - \pi)^{(n-x)} \tag{5–4a}$$

Example 5–5 Use the binomial equation to compute the probability of 1 defective in a sample of 4 glasses, when the proportion defective in the population is $\pi = .10$.

Solution

$$P(X = 1 \mid n = 4, \pi = .10) = \frac{n!}{x!(n - x)!} \pi^x (1 - \pi)^{(n-x)}$$

$$= \frac{4!}{1!(4 - 1)!} (.10)^1 (.90)^3 = .2916$$

Recall that a probability distribution must show all the values that a random variable can take on, and the probability of each. A complete binomial probability distribution would also include the probability of 0, 1, 2, 3, and all 4 glasses being defective as well. Table 5–1 shows the complete binomial distribution for $P(X \mid n = 4, \pi = .10)$.

TABLE 5–1 Binomial distribution for $n = 4$, $\pi = .10$

No. Defectives X	No. Combinations ${}_nC_x$	Defective π^x	Good $(1-\pi)^{(n-x)}$	Probability $P(X)$
0	${}_4C_0$ or 1	$.10^0$	$.90^4$	.6561
1	${}_4C_1$ or 4	$.10^1$	$.90^3$	.2916
2	${}_4C_2$ or 6	$.10^2$	$.90^2$	.0486
3	${}_4C_3$ or 4	$.10^3$	$.90^1$	.0036
4	${}_4C_4$ or 1	$.10^4$	$.90^0$	.0001
			TOTAL	1.0000

Note that the probability distribution accounts for all combinations of events, so that $\Sigma P(X)$ equals one. But it is usually unnecessary to compute the entire probability distribution for most problems.

Example 5–6 A work scheduler at the Royal Tea Company Ltd. must plan ahead to reduce labor hours if raw materials are not available on time. Thirty percent of the shipments of tea from India arrive late. If 8 shipments are now en route, what is the probability that (a) the sixth shipment will arrive late? (b) ≥ 6 of the shipments will arrive late? (c) 5 or less of the shipments will arrive on time?

Solution

(a) The population proportion is $\pi = .30$, and the sample size is $n = 8$.

$$P(X = 6 \mid n = 8, \pi = .30) = \frac{n!}{x!(n-x)!}\pi^x (1-\pi)^{(n-x)}$$

$$= \frac{8!}{6!(8-6)!}(.30)^6 (.70)^2 = (28)(.00073)(.49000) = .0100$$

Thus .0100 represents the probability of exactly 6 late shipments in 8 shipments, when 30% of the population of shipments arrive late.

(b) To find the probability of ≥ 6, we must sum the discrete probabilities of 6, 7, and 8 shipments arriving late. The cumulative total is:

$$P(X=6 \mid n=8, \pi=.30) = {}_6C_8\,(.30)^6\,(.70)^2 = .0100$$
$$P(X=7 \mid n=8, \pi=.30) = {}_7C_8\,(.30)^7\,(.70)^1 = .0012$$
$$P(X=8 \mid n=8, \pi=.30) = {}_8C_8\,(.30)^8\,(.70)^0 = .0001$$
$$\text{TOTAL} \quad .0113$$

(c) The probability of ≤ 5 of the shipments arriving on time is the complement of ≥ 6 shipments arriving late, so:

$$P(X \leq 5) = 1.0000 - .0113 = .9887.$$

USING BINOMIAL TABLES

Calculating binomial probabilities can be tedious—even if you have a calculator handy. Fortunately, binomial tables can simplify the calculations a great deal. Tables for individual values are arranged by sample size (n), and show the probability of each specified number of successes (X) for various values of the proportion π. Table 5–2 repeats a portion of the binomial table given in Appendix C that could have been used to solve Example 5–6 above.

TABLE 5–2 A portion of the binomial distribution table

		VALUES OF π									
n	X	.05	.10	.15	.20	.25	.30	.35	.40	.45	.50
8	0	.6634	.4305	.2725	.1678	.1002	.0576	.0319	.0168	.0084	.0039
	1	.2793	.3826	.3847	.3355	.2670	.1977	.1374	.0896	.0548	.0312
	2	.0515	.1488	.2376	.2936	.3115	.2065	.2587	.2090	.1569	.1094
	3	.0054	.0331	.0839	.1468	.2078	.2541	.2786	.2787	.2568	.2188
	4	.0004	.0046	.0185	.0459	.0865	.1361	.1875	.2322	.2627	.2734
	5	.0000	.0004	.0026	.0092	.0231	.0467	.0808	.1239	.1719	.2188
	6	.0000	.0000	.0002	.0011	.0038	.0100	.0217	.0413	.0403	.1094
	7	.0000	.0000	.0000	.0001	.0004	.0012	.0033	.0079	.0164	.0312
	8	.0000	.0000	.0000	.0000	.0000	.0001	.0002	.0007	.0017	.0039

To use the table, first locate the appropriate sample size in the leftmost column. Then go down the X column until you locate the desired number of successes. Then read across the columns until you come to the proportion, π, that applies to the problem.

Appendix C is a table of *individual* binomial probabilities. You should be able to locate the values of .0100, .0012, and .0001 for the $P(X = 6)$, $P(X = 7)$, and $P(X = 8)$ for the distribution $P(X \mid n = 8, \pi = .30)$ in that portion of it shown above. The cumulative value of $P(X \geq 6)$ is thus the sum, .0113. Some statistical references also have *cumulative* binomial tables (of $X \geq$ the number of successes) or ($X \leq$ the number of successes). However, these probabilities can usually be summed up fairly easily, so these tables are not always available. Also, for values of $\pi > .5$ you can find $P(X)$ by substituting $(1-\pi)$ for π and looking up the $P(n-X)$ instead of the $P(X)$.

Example 5–7 Use the binomial table above to find the following probabilities:

(a) $P(X = 4 \mid n = 8, \pi = .30)$

(b) $P(X = 0 \mid n = 8, \pi = .10)$

(c) $P(X \leq 3 \mid n = 8, \pi = .25)$

(d) $P(X \geq 7 \mid n = 8, \pi = .50)$

(e) Use the complement to find the $P(X \geq 2 \mid n = 8, \pi = .40)$.

(f) $P(X = 4 \mid n = 10, \pi = .70)$

Solution

(a) At the intersection of $X = 4$ and $\pi = .30$ in the table is .1361.

(b) On the top row under $\pi = .10$ is the value .4305.

(c) $P(X = 3 \mid n = 8, \pi = .25)$
$= P(X = 0) + P(X = 1) + P(X = 2) + P(X = 3)$
$= .1002 + .2670 + .3115 + .2076 = .8863$

(d) $P(X \geqslant 7 \mid n = 8, \pi = .50) = P(X = 7) + P(X = 8)$
$= .0312 + .0039 = .0351$

(e) $P(X \geqslant 2 \mid n = 8, \pi = .40) = 1.0000 - [P(X = 0) + P(X = 1)]$
$= 1.0000 - [.0168 + .0896] = .8936$

(f) $P(X = 4 \mid n = 10, \pi = .70) = P[(n - X) \mid n = 10, \pi = .30]$
$= P(X = 6 \mid n = 10, \pi = .30) = .0368$

[Note: this is equivalent to recognizing that the probability of 4 successes in 10, where the probability of success is .70, is the same as the probability of 6 failures in 10, where the probability of failure is .30.]

BINOMIAL DISTRIBUTION GRAPHS

The graph of a binomial distribution can be readily constructed from a table of binomial probabilities. Figure 5–4 compares the graphs for sample sizes of $n = 8$ for the different proportions of π equal to .10, .25, .50, and .90. Note that (a) all graphs are *unimodal*, and (b) they are *symmetric when* $\pi = .50$. As π deviates from .50, the distribution is skewed positively (when $\pi < .50$) or negatively (when $\pi > .50$). (Also, not all values are graphed.)

FIGURE 5–4 Graphs of binomial distributions for $n = 8$

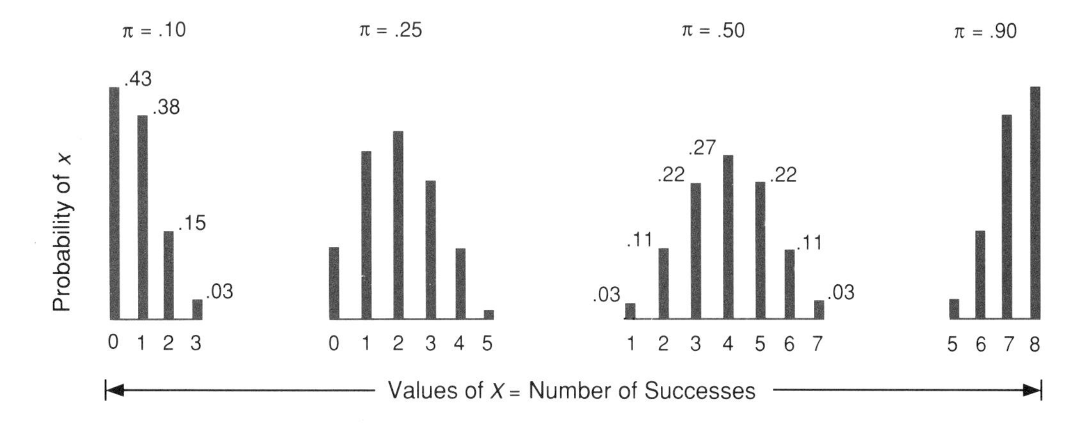

MEAN AND STANDARD DEVIATION OF THE BINOMIAL DISTRIBUTION

The two most important summary characteristics of the binomial distribution are its mean (μ) and its standard deviation (σ). They are quite easy to compute, and can be expressed either in terms of the *proportion* of successes or of the *number* of successes. (Because if you multiply proportion values, π, by the number of items, n, you get the number in the proportion!)

Expressed as a proportion:

Mean: $$\mu = \pi \tag{5–5}$$

Standard Deviation: $$\sigma = \sqrt{\frac{\pi(1 - \pi)}{n}} \tag{5–6}$$

Expressed as numbers:

Mean: $$\mu = n\pi \tag{5–7}$$

Standard Deviation: $$\sigma = \sqrt{n\pi(1 - \pi)} \tag{5–8}$$

Example 5–8 A shipment of 500 crystal wine glasses is to be produced on equipment that generates 10% defectives in the population. Estimate the mean and standard deviation of the binomial distribution for both (a) the *proportion* of successes in total, and (b) the *number* of successes in the shipment.

Solution

(a) *As a proportion:*

Mean: $\mu = \pi = .10$

Std Dev: $$\sigma = \sqrt{\frac{\pi(1 - \pi)}{n}} = \sqrt{\frac{(.10)(.90)}{500}} = .013$$

Thus, the expected (or mean) proportion of defectives is .10, and the standard deviation is .013, or 1.3% defectives.

(b) *As numbers:*

Mean: $\mu = n\pi = (500)(.10) = 50$ glasses

Std Dev: $\sigma = \sqrt{n\pi(1 - \pi)} = \sqrt{(500)(.10)(.90)} = 6.7$ glasses

In terms of numbers, one could expect 50 defective glasses in a sample of 500, when the population is 10% defective.

Note that proportions are long-run (dimensionless) expected values. If they are multiplied by the sample size (500 in this case), we obtain the specific number amounts (of 50 and 6.7 glasses allowing for rounding error). So the values are readily convertible from one to another. Moreover, the standard deviation for any given binomial distribution is completely determined by the proportion of successes and the sample size. And, as the equation and graphs suggest, it is the largest at $\pi = .50$.

EXERCISES II

11. What conditions must prevail to use the binomial distribution?
12. Use the binomial equation to compute the following probabilities:
    (a) $P(X = 2 \mid n = 6, \pi = .20)$
    (b) $P(X = 5 \mid n = 10, \pi = .40)$
13. Use the binomial equation to compute the following probabilities:
    (a) $P(X \leqslant 1 \mid n = 8, \pi = .10)$
    (b) $P(X \geqslant 9 \mid n = 10, \pi = .50)$
14. Determine the following probabilities:
    (a) probability of 3 or more successes in $n = 12$, when $\pi = .30$
    (b) probability of 2 or 3 successes in a sample of 10, when $\pi = .40$
15. European Insurance Company has purchased six new Zolta automobiles at a good price. They are also aware that 25% of the Zolta cars require some factory attention during the first year. Construct a binomial distribution graph to show the probability that none, one . . . up to all six of their cars will require factory attention.
16. Compute the mean (μ) and standard deviation (σ) of the probability distribution in the previous problem, where $n = 6$ and $\pi = .25$. Express *your* answer as a proportion.
17. Employee absences can seriously affect productivity. In one country, employees are absent from work an average of 2.6 weeks of the 52-week work-year. Assuming this average holds for a small factory that employs 12 workers, what is the probability that during any given week:
    (a) no workers will be absent?
    (b) one worker will be absent?
    (c) two or more workers will be absent?
18. Cancer statistics show that approximately 20% of all deaths in the United States are attributed to cancer. What is the probability that from a group of 6 workers, (a) none will die of cancer, (b) all will die of cancer?
19. A Space Lab experiment requires 12 acceptable solar cells. However, past experience has shown that approximately 10% of all cells are not satisfactory. If a shipment of 15 cells is received, what is the probability that it will contain enough satisfactory cells to do the experiment?
20. Given the following binomial probability distributions:
    (I) $n = 100, \pi = .40$; (II) $n = 20, \pi = .05$

    For both distributions, compute the mean and standard deviation for: (a) the proportion of successes and (b) the number of successes.

THE POISSON DISTRIBUTION

The Poisson distribution is named after Simeon Poisson (1781–1840), who is said to have been studying the number of personnel kicked to death by mules in army divisions employing mules for transportation. It is a discrete distribution used to describe the pattern of *discrete* events over a continuous dimension of measurement, such as time or space. And it applies where the probabilities of the events, $P(X)$, are based upon some prior knowledge about the average number of occurrences in the time or space interval. For example, the Poisson distribution may describe the probability of a specific number of patient arrivals at a hospital (based upon the mean number of arrivals per hour), or a specific number of defects in a sheet of plywood (based upon the average number of defects per sheet).

Insofar as Poisson probabilities are completely defined by the average or mean number of occurrences, the Poisson is referred to as a *single parameter* distribution. The possible number of occurrences, such as the potential number of accidents or defects, may not even be a countable number. We assume the random variable X can take on nonnegative integer values ranging from zero to infinity. But the probability of occurrence of an event during a small interval of the time or space is usually very low. And, as we move away from its average, these probabilities decay at a logarithmic rate. Hence, the Poisson is also referred to as a *rare event* distribution.

The parameter that completely defines the Poisson distribution is its mean, which is designated by the Greek letter lambda (λ). Moreover, the variance of a Poisson distribution also equals λ, so the standard deviation is $\sqrt{\lambda}$.

Poisson probability distribution conditons are:

1. There are *many* possible (discrete) outcomes over a continuous space.
2. The probability of any single outcome is *low and constant* from trial to trial.
3. Each trial is *independent* of any previous trial.

Whereas the Poisson is theoretically correct for describing situations that satisfy the above conditions, it also offers a useful approximation to binomial probabilities, where π values are small and n values are large. We'll return to this after we illustrate some basic Poisson probability calculations.

USING THE POISSON EQUATION

For random variables that are distributed according to the Poisson distribution with mean λ, the probability of any given number of occurrences, X, is:

$$P(X \mid \lambda) = \frac{\lambda^x e^{-\lambda}}{X!} \tag{5–9}$$

where: λ = mean or expected number of occurrences
e = 2.7183 (the base of the natural logarithms)

Recall that $e^{-\lambda}$ is equivalent to $\frac{1}{e^{\lambda}}$. So, for example, $e^{-2} = \frac{1}{(2.7183)^2} = .1353$

Example 5–9 Telephone calls arrive at an emergency (911) number at an average rate of 2 per hour. Assuming the call probabilities can be described by a Poisson distribution, what is the probability of (a) no calls during any given hour, and (b) three calls during any given hour?

Solution

(a)

$$P(X=0 \mid \lambda=2) = \frac{\lambda^x e^{-\lambda}}{X!} = \frac{(2)^0 e^{-2}}{0!} = \frac{(1)(2.7183)^{-2}}{1} = \frac{(1)}{(2.7183)^2} = .1353$$

(b)

$$P(X=3 \mid \lambda=2) = \frac{\lambda^x e^{-\lambda}}{X!} = \frac{(2)^3 e^{-2}}{3!} = \frac{(8)(2.7183)^{-2}}{6} = \frac{(8)}{6(2.7183)^2} = .1804$$

Example 5–10 A truck assembly plant in Michigan uses welding robots that produce truck body welds with only 4 minor defects per 100 feet of weld. If one of the robots does 2.5 feet of welding on each truck body, what is the probability of that robot working on 10 truck bodies without producing any defects?

Solution

This is a "defect per unit of distance" problem, and we must first clarify the X and λ values. The 2.5 feet of welding on 10 truck bodies equals 25 feet of weld, so we are concerned with the probability of zero defects in 25 feet of weld. Next, we must establish λ as the mean number of defects in 25 feet of weld. The rate of 4 defects per 100 feet can be stated in a smaller common unit rate of $4/100$ or .04 defects/ft. Then in 25 feet, the mean rate would be (.04) (25) = 1 defect.

$$P(X = 0 \mid \lambda = 1) = \frac{\lambda^x e^{-\lambda}}{X!} = \frac{(2)^0 e^{-1}}{0!} = \frac{(1)(2.7183)^{-1}}{1} = \frac{(1)}{(2.7183)} = .3679$$

POISSON TABLES AND GRAPHS

A simpler method of determining Poisson probabilities is to use tables, which can take several forms. Appendix D shows the individual Poisson probability values at the intersection of the λ values (across the top) and X values (down the margin columns). For example, we could have used the table for Example 5–9 to find $P(X = 0 \mid \lambda = 2)$ by locating the $\lambda = 2$ value at the head of the rightmost column in the second set, and then going down the column to $X = 0$ (for .1353) and $X = 3$ (for .1804). A complete probability distribution for $P(X \mid \lambda = 2)$ would have the distribution and graph shown in Figure 5–5.

FIGURE 5–5 Values and graph for Poisson distribution

Poisson Table Values $\lambda = 2.00$

X	$P(X)$
0	.1353
1	.2707
2	.2707
3	.1804
4	.0902
5	.0361
6	.0120
7	.0034
8	.0009
9	.0002

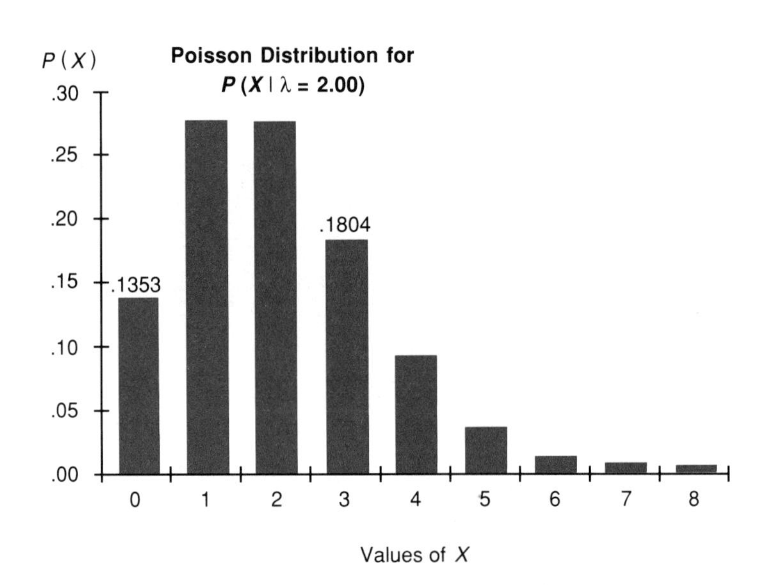

Some Poisson tables show the cumulative probabilities of $\leq X$ for selected values of the mean, λ. However, cumulative totals can also be obtained quite readily from Appendix D by adding the individual values. For example:

$$P(X \leq 2 \mid \lambda = 2) = P(X = 0) + P(X = 1) + P(X = 2)$$
$$= .1353 + .2707 + .2707 = .6767$$

Cumulative tables are especially useful in quality control work because the analysis of defects is often analyzed via the Poisson distribution.

POISSON APPROXIMATION TO THE BINOMIAL DISTRIBUTION

Although the binomial distribution is theoretically correct for specific situations, the binomial tables in textbooks seldom extend much beyond sample sizes of 20. However, when the proportion of successes is small (say $\pi < .10$) and the sample size is more than 20, the Poisson distribution may provide a satisfactory approximation to the binomial solution. The approximation improves as π approaches zero, and as n gets larger. The mean and standard deviation of the Poisson distribution are then estimated as:

$$\text{Mean:} \quad \lambda = n\pi \qquad (5\text{–}10)$$

$$\text{Standard Deviation:} \quad \sigma = \sqrt{n\pi} \qquad (5\text{–}11)$$

The following example compares the values obtained from a binomial probability calculation and a Poisson approximation to the binomial.

Example 5–11 A sportswear manufacturing plant has 20 sewing machines used for stitching jogging suits. On some days all machines work satisfactorily, but on other days, one, two, three, or even more of the machines may be out of service. On average, experience has shown that on any given day, there is a .10 chance that a machine will be inoperable. Find the probability that two or more of the machines are inoperable using (a) the binomial distribution, and (b) the Poisson approximation.

Solution

(a) If we consider a "success" as a machine out of service, we are looking for $P(X \geq 2 \mid n = 20, \pi = .10)$. Rather than calculating the probability of 3, 4, 5, on up to 20 machines being out of service, we can calculate the $P(X < 2)$, which is $P(X = 0) + P(X = 1)$. Then we can use the complement, $1 - [P(0) + P(1)]$. Using the *binomial equation*, this is:

$$P(X \geq 2) = 1.0 - \left[\left[\frac{20!}{0!20!}(.10)^0(.90)^{20}\right] + \left[\frac{20!}{1!\,19!}(.10)^1(.90)^{19}\right]\right]$$

$$= 1.0 - [.1216 + .2702] = .6082$$

The same answer can be obtained from a *binomial distribution table* of $n = 20$, $\pi = .10$, where the $P(X = 0)$ is .1216 and the $P(X = 1)$ is .2702.

(b) Using the *Poisson distribution* in Appendix D:

First let $\lambda = n\pi = (20)(.10) = 2.0$ Then, from the table we have:

$$P(X \geq 2 \mid \lambda = 2.0) = 1.0 - [P(X = 0) + P(X = 1)]$$
$$= 1.0 - [.1353 + .2707] = .5940$$

The cumulative Poisson distribution yields the answer even more directly because it gives the $P(X \leq 1 \mid \lambda = 2)$ of .4060. Subtracting this from 1.0 gives the .5940. Although .5940 is only an approximation, it is reasonably close to the correct answer of .6082.

EXERCISES III

21. Under what conditions does the Poisson distribution apply?
22. Use the *Poisson equation* to compute the following probabilities:
    (a) $P(X = 2 \mid \lambda = 4)$ (c) $P(X \leqslant 1 \mid \lambda = 1)$
    (b) $P(X > 0 \mid \lambda = 2)$
23. Use the *Poisson tables* to find the following probabilities:
    (a) $P(X = 9 \mid \lambda = 6.8)$ (c) $P(X \geqslant 3 \mid \lambda = 6.6)$
    (b) $P(X < 4 \mid \lambda = 8.8)$
24. An oil refinery has averaged three pump failures per year. Using the *Poisson equation*, determine the probability of fewer than an average number of pump failures this year. Check your figures with table values.
25. For an aluminum pot line that averages four emergency shutdowns per year, find the probability of having (a) no shutdowns in a year, and (b) two shutdowns during a six-month period. Assume the Poisson distribution applies.
26. Wheat trucks arrive at Midwest Grain Elevators at an average rate of 24 per hour. (a) Use the *Poisson equation* to estimate the probability of receiving no trucks in a 5-minute period, and (b) no trucks in a 10-minute period. (c) Use the *Poisson tables* to estimate the probability of receiving more than 4 trucks in a 10-minute period.
27. International Glass Co. supplies hospitals with glass tubing that has, on average, one scratch or other defect in every 20 feet. All sales are in 60-foot-length packages.
    (a) *Graph* the Poisson probability distribution to show the probability of zero, one, two, etc., defects in the 60-foot-length packages.
    (b) What is the *mean* and *standard deviation* of the distribution?
    (c) What is the probability of $\leqslant 2$ defects in a 60-foot-length package?
28. The probability of a stockbroker quoting an incorrect price to a customer is .04. If the broker handles 60 calls on a given day, what is the (Poisson) probability he will quote the wrong price three times?
29. A public accounting firm has found that 3% of the accounts that it audits are incorrect. If 40 accounts are audited, what is the probability that (a) 2 accounts are incorrect, (b) > 2 accounts are incorrect?
30. The personnel manager of a movie theater chain has data to show that employees fail to show up for work about 10% of the time. He would like to do more follow-up studies in a theater where ten people are employed. (a) Use the *Poisson distribution* to estimate the

probability of one or more employees being absent at any given time. (b) Compare this with the *binomial probability* of one or more employees being absent.

31. A manufacturer of a numerically controlled machine has included a box of 200 small bulbs to be used on a control panel. If the bulbs average 3% defective and 190 bulbs are required to complete the assembly, what is the probability that the customer will have a sufficient number of good bulbs to complete the assembly?

THE NORMAL DISTRIBUTION

DISCRETE Both the binomial and Poisson distributions are *discrete*. They assign probabilities to events on the basis of a *count* of the number of occurrences of some specified outcome. As we have seen, discrete probability computations often require some form of summation (Σ) activity.

CONTINUOUS When a random variable is assigned a value on the basis of a *measurement* rather than a count, the variable is said to be *continuous*. A continuous scale has an infinite number of possible values—depending upon how precise the measurement is. Any point along the measurement scale represents a potential value for the random variable to take on. However, insofar as a "mathematical point" has no width, the probability at any point is undefined for continuous variables. Thus, the probability of any one point is, for analytical purposes, equal to zero.

Nevertheless, probabilities for continuous variables can be determined from the number of occurrences in an *interval* along the real number line. This, of course, relies upon a unification (or integration $\int$) over more than one potential value. For example, the probability of getting through a supermarket checkout counter in exactly 2.25 minutes (i.e., 2.25000) may be virtually nil. But the probability of getting through the line between 2.00 and 3.00 minutes (or even 2.20 to 2.30) can be estimated—despite the fact that the probability of any one of the infinite number of points between these two values is virtually zero.

Thus, we use intervals to work with continuous variables. The mathematical functions used to describe these intervals in terms of the area under the continuous probability distributions are sometimes referred to as *density functions*. In statistics, the most important such function is the *normal density function*, which describes the normal probability distribution.[1]

---

[1] The normal density function equation is:

$$f(X) = \frac{1}{\sigma\sqrt{2\pi}} e^{-\frac{(X-\mu)^2}{2\sigma^2}}$$

where $f(X)$ is the ordinate of the curve at point X, π is the constant 3.1416, e is the constant 2.7183, μ is the true mean, and σ is the standard deviation of the X values. This is presented for reference only, for we will not be working directly with this function.

CHARACTERISTICS OF THE NORMAL DISTRIBUTION

The **normal distribution** is described by a continuous, unimodal, symmetrical, bell-shaped curve. As illustrated in Figure 5–6, the probability (area) under the curve is equal to 1.00. And the curve approaches the x-axis in both directions, although it (theoretically) never actually touches it. This is because the range of the normal density function is from minus infinity to plus infinity. For practical purposes, however, the area under the curve outside of plus and minus three or more standard deviations away from the mean is extremely small.

FIGURE 5–6 Graphic presentation of a normal distribution

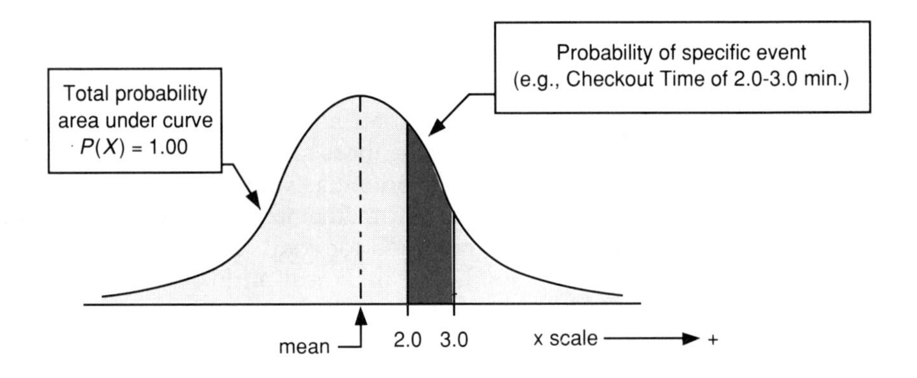

The normal distribution is completely defined by its mean (μ) and standard deviation (σ). Thus, there are as many normal distributions as there are values of μ and σ. Fortunately, however, any set of normally distributed data can be expressed in a standardized (Z) format, where the area under the curve represents the total probability (i.e., of 1.00) of the random variable X, and the spread of the distribution is expressed in terms of standard deviations. Then the probability of any interval along the X-scale equals the proportion of area lying between the upper and lower limits of the interval.

The proportion of area under a normal curve between a mean and any X-value is a function of the number of standard deviations the X-value is located from the mean. Figure 5–7 shows the approximate proportions for *any* normal distribution. Thus, about 68.3% of the area lies within $\pm$ 1 standard deviation of the mean, 95.5% within two σ's, and 99.7% within three σ's.

FIGURE 5–7 Areas under the normal curve

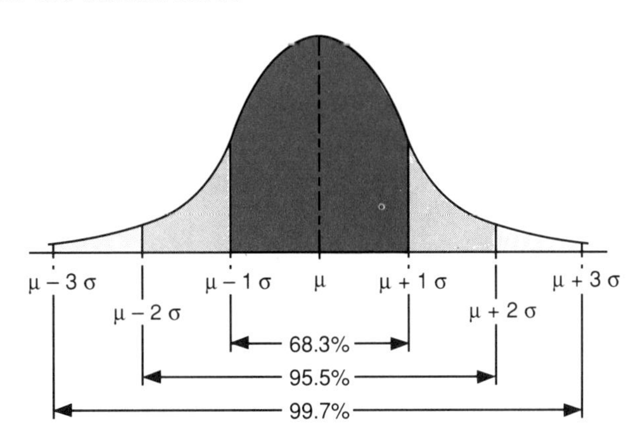

IMPORTANCE OF NORMAL DISTRIBUTION The normal distribution is sometimes called the Gaussian distribution after Carl Gauss (1777–1855), who was one of the first persons to describe its mathematical properties. During his time, scientists observed that repeated measurements of some physical quantities, such as the distance to the moon, did not result in exactly the same value each time. However, the measurements tended to cluster around their mean, and the farther a value was from the mean, the less likely that value was to occur. Because the distribution represented the normal error in repeated measurements, the graph of the distribution was called the *normal curve of error* and later became known simply as the *normal curve*.

Today, the normal distribution is probably the most important statistical distribution for several reasons. *First*, a large number of natural measurements are approximately normal, from the weights of cereal boxes to the gasoline mileage estimates for the newest cars. *Second*, and of even greater significance, it forms the basis for much of our statistical inference. This is because the probability distribution of the means of large random samples from the same population tends to be normally distributed about the true population mean. *Finally*, the normal is useful as an approximation to discrete distributions such as the binomial and Poisson, especially as the sample size becomes large. We shall return to this later in this section.

FINDING AREAS UNDER THE NORMAL CURVE

To illustrate the use of the normal distribution, let's assume we have a normally distributed variable with a known mean and standard deviation. Suppose we know that the weight of boxes of tea shipped from Kenya to Dublin averages $\mu = 300$ pounds, and the standard deviation is $\sigma = 20$ pounds. We wish to use the normal distribution to make probability statements about the weight of individual boxes selected at random from a large shipment. For example, suppose we're interested in the probability that a box weighs between 300 and 320 pounds.

FIGURE 5–8 A population known to be normally distributed

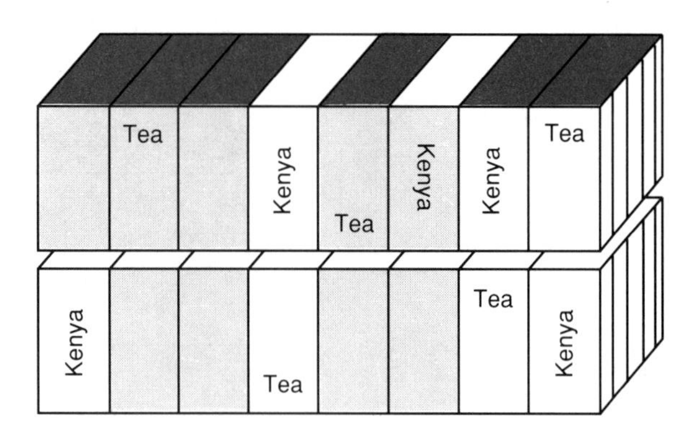

The distribution with a mean of 300 and standard deviation of 20 is only one of many possible normal distributions. We could even have many distributions with the same mean, but with different standard deviations, as in Figure 5–9.

FIGURE 5–9 Normal distributions with the same mean

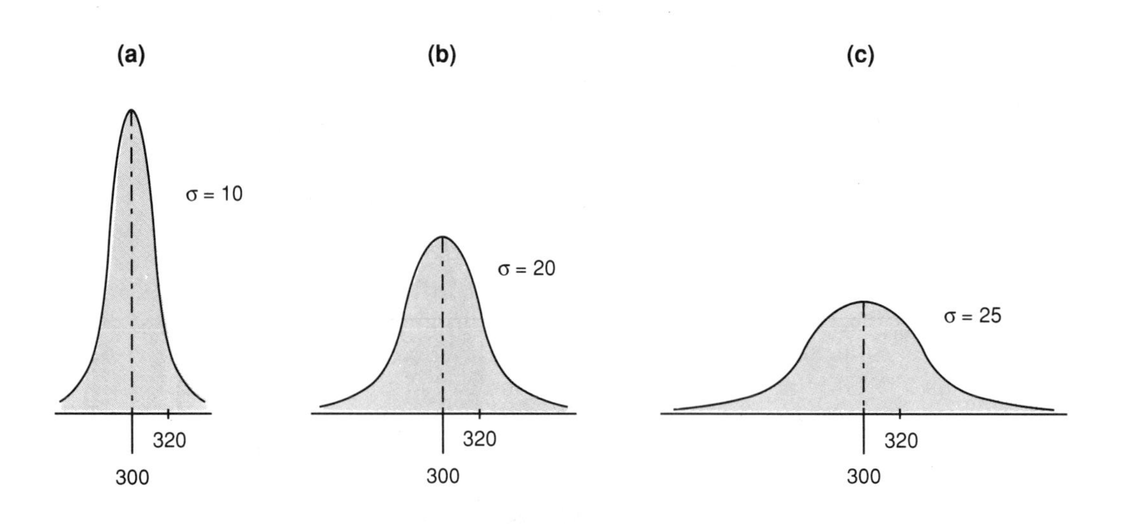

However, we need not use the normal curve equation to make probability statements about a particular distribution. This is because the standardized (Z) scale allows us to express any point in the distribution in terms of its distance (in standard deviation units) from the mean of the distribution. For example, in Figure 5–9 (b), $\sigma = 20$ pounds. So the 320-pound point is one standard deviation unit from the mean of the distribution. [Note: If our distribution had been the "tighter" one in (a), the 320 pounds would be two standard deviations away from μ. On the other hand, the distribution in (c) is more dispersed, and 320 pounds is less than one standard deviation away from the mean there.]

Z VALUES The standardized (Z) value of an observation is simply an expression of the *number of standard deviations* a value is from the mean of the distribution. It is computed by taking the difference between the mean, μ, and the subject, X value, and dividing by the amount of one standard deviation.

$$Z = \frac{X - \mu}{\sigma} \qquad (5\text{–}10)$$

FIGURE 5–10 Normal distribution with $\mu = 300$ and $\sigma = 20$

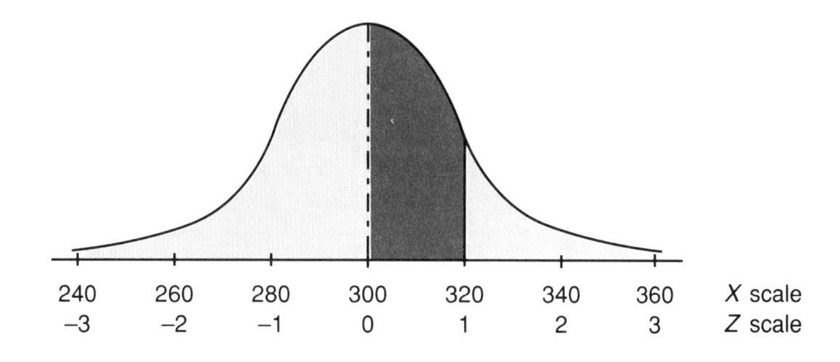

Figure 5–10 shows both the X scale and the standardized Z scale for the distribution of boxes of tea, where $\mu = 300$ and $\sigma = 20$.

Note that the mean of the distribution in Figure 5–10 is at 300 on the X scale, and the Z value for 300 is zero because the Z-scale measures the deviation from the mean. The X scale value of 320 has a corresponding Z scale value of 1.00, because $Z = (X - \mu)/\sigma = (320 - 300)/20 = 1.00$, and the X value at 340 is equivalent to a Z value of 2.00, since $(340 - 300)/20 = 2.00$. All other Z scale values are obtained in a similar manner.

NORMAL DISTRIBUTION TABLE The proportion of area under the normal curve between the mean and any given number of standard deviations is fixed and can be computed by integrating the area under the curve. Fortunately, these values are also readily accessible in table form as a Normal Distribution Table, one of which is included in Appendix B. These tables make it possible to compute the area for an interval under any normal curve from a single, one-page table.

The areas in the table in Appendix B correspond with the shaded portion of Figure 5–10. That is, they reflect the proportion of area contained in a segment bounded by a perpendicular line at the mean, and another line raised at a positive distance of Z standard deviation units. The following examples illustrate the use of the table. Each example first raises a question in a general form, and then as a specific example of that general type of problem. Do not attempt to memorize the numbered steps of the solutions to the problems—just use the problems and diagrams to be sure you understand how the normal curve is used.

Example 5–12 What is the probability that an element selected at random will have a score between the mean and a point X lying above (to the right of) the mean; for example, what is the probability of a weight between 300 and 330?

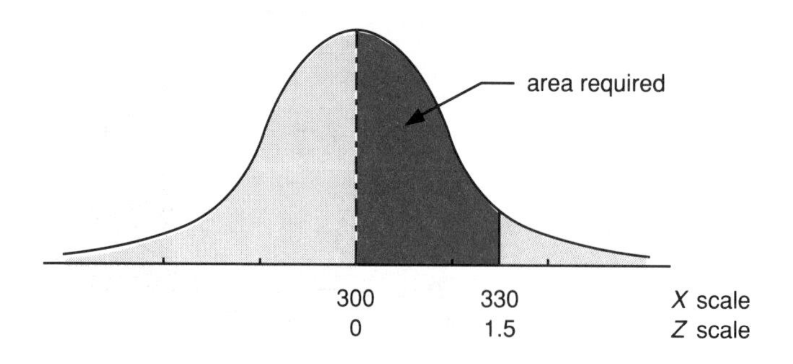

Solution

1. Compute the Z value for the point above the mean:

$$Z = \frac{(X - \mu)}{\sigma} = \frac{330 - 300}{20} = 1.50$$

2. Determine from Appendix B the area under the normal curve between the mean and the point where $Z = 1.50$. Read down the left-hand column marked Z to 1.5, then look at the column headings to find the second decimal digit. Since the next digit desired in this case is a zero, look in the column headed .00. The figure .4332 is the desired area and represents the desired probability.

Example 5–13 What is the probability that an element selected at random will have a score greater than some specified X value; for example, what is the probability that the weight will be more than 355?

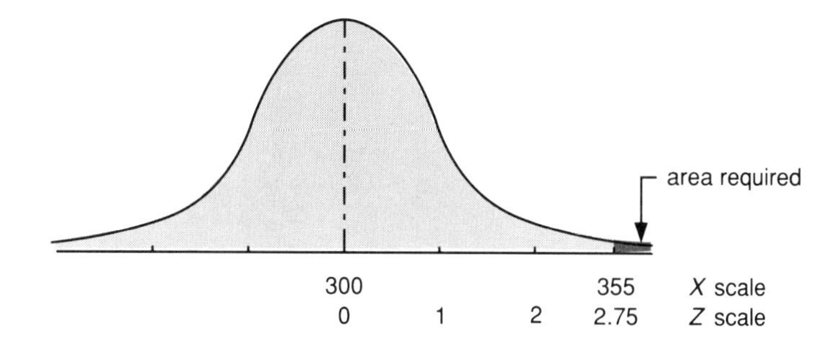

Solution

1. Compute the Z value corresponding to an X value of 355:

$$Z = \frac{(X - \mu)}{\sigma} = \frac{355 - 300}{20} = 2.75$$

2. Find the area between the mean and the point which has a Z value of 2.75. Look under the left-hand column to 2.7, then read the probability score under the column headed .05; the probability for 300 to 355 is .4970.
3. Since one half of the total area under the curve is on each side of the mean, and the area between 300 and 355 is equal to .4970, the remaining .0030 represents the probability that a box selected at random will lie above the point 355.

Example 5–14 What is the probability that an element selected at random will have a score between two specified points, one of which is less than the mean, and the other greater than the mean; for example, what is the probability that a box will weigh between 250 and 350?

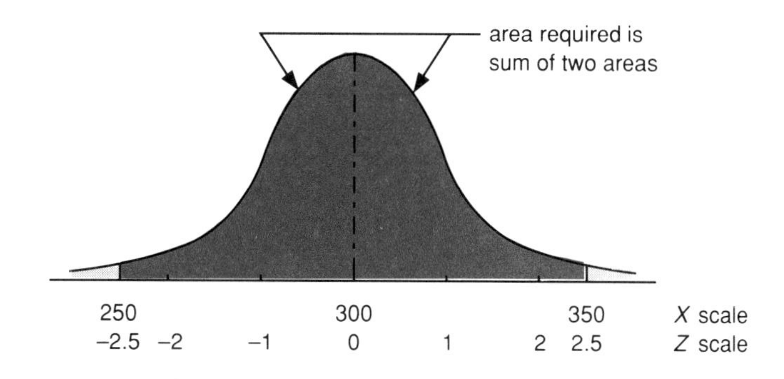

Solution

1. It is necessary to find two Z scores because the table shows the area between the mean and a single point lying Z standard deviations from the mean. Call the Z scores Z_1 and Z_2 and compute both:

$$Z_1 = \frac{250 - 300}{20} = -2.50 \qquad Z_2 = \frac{350 - 300}{20} = 2.50$$

2. Since the curve is symmetrical, the area lying between the mean and $Z_1 = -2.50$ is the same as the area lying between the mean and $Z_2 = +2.50$. From the table the area is found to be .4938, and .4938 + .4938 is equal to .9876, which is the area between 250 and 350. This means that the probability is equal to .9876. It could also be interpreted to mean that 98.76 percent of the observations will lie between 250 and 350. (The two preceding examples may also be interpreted in terms of percent of the product or the observations.)

Example 5–15 What is the probability that an element selected at random will be outside two specified points, one on each side of the mean; for example, if the upper and lower tolerance limits are 262 and 346, what percent of the product will be rejected?

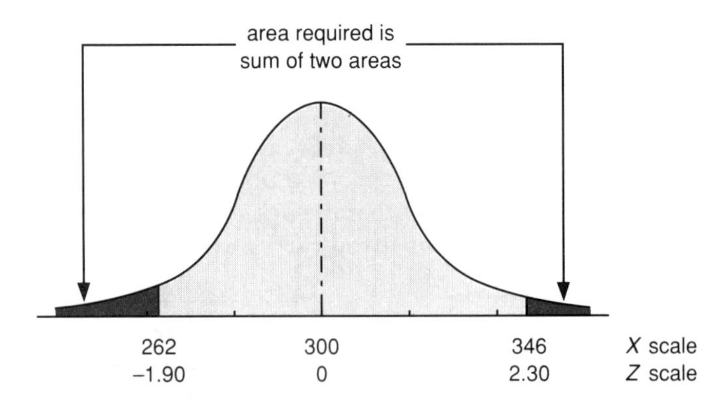

Solution

1. Compute the two Z scores:

$$Z_1 = \frac{262 - 300}{20} = -1.90 \qquad Z_2 = \frac{346 - 300}{20} = 2.30$$

2. Find the area between these two points:

The area between $Z_2 = -1.90$ and the mean is	.4713
The area between $Z_2 = 2.30$ and the mean is	.4893
The total area between the limits is	.9606

The area outside the tolerance limits is the complement of the total, $1 - .9606 = .0394$; alternatively, it is the sum of the areas in the two tails:

$$\begin{aligned} \text{Left tail} &= .5000 - .4713 = .0287 \\ \text{Right tail} &= .5000 - .4893 = .0107 \\ \text{Total} &= .0394 \end{aligned}$$

The percent of the product which will be rejected is the probability .0394 expressed as a percent = 3.94 percent.

Example 5–16 What is the area between two specified points, both lying above the mean; for example, what percent of the product will weigh between 310 and 345 pounds?

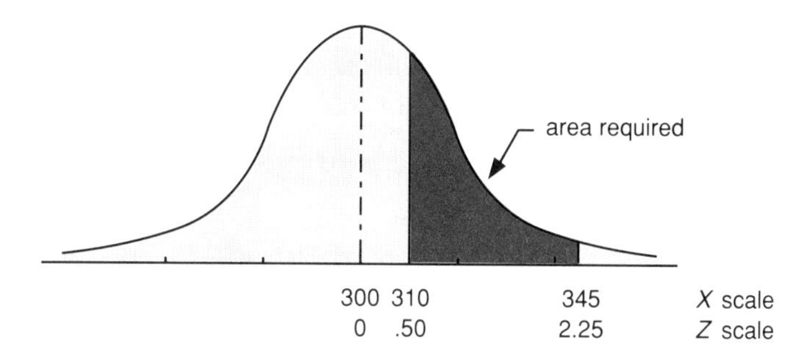

Solution

1. Compute the two Z scores:

$$Z_1 = \frac{310 - 300}{20} = .50 \qquad Z_2 = \frac{345 - 300}{20} = 2.25$$

2. Find the area within the segment 310 to 345:

The area from the mean to the upper limit, $Z_2 = 2.25$, is	.4878
The area from the mean to the lower limit, $Z_1 = .50$, is	.1915
The area between the specified limits is the difference	.2963

In the preceding five examples, the areas have been found from Z scores obtained from $Z = (X - \mu)/\sigma$. In the next examples, the areas are given, and the corresponding Z values must be determined by reading from the center of the table across to the Z column.

Example 5 – 17 A given percent of the elements will be equal to or less than what X value? For example, 90 percent of the boxes will be equal to or less than what value?

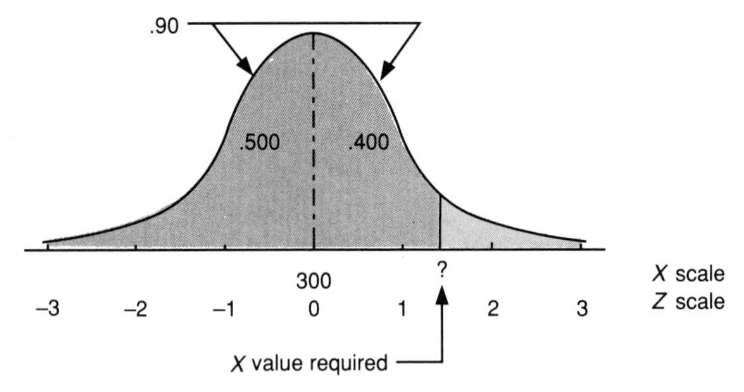

Solution

1. Decide whether the X value will be to the right or to the left of the mean. Fifty percent will be less than or equal to 300. Ninety percent will be less than or equal to a point located far enough to the right of the mean to include an additional 40 percent.
2. From the center of the table, find the number nearest .4000, and read the three-digit Z score. Since .3997 of the area lies between the mean and a point 1.28 standard deviations from the mean, the Z score for $.4000 \approx 1.28$.
3. The required point is 1.28 standard deviations to the right of the mean. Therefore, $X = \mu + 1.28(\sigma)$, or $X = 300 + 1.28(20) = 325.6$. Ninety percent of the scores will be less than or equal to 325.6 lbs.

Example 5–18 A given percent of the elements should be greater than or equal to some specified X value. For example, if the producer guarantees that 95 percent of his boxes will have weights greater than or equal to some specified value, what value can be specified?

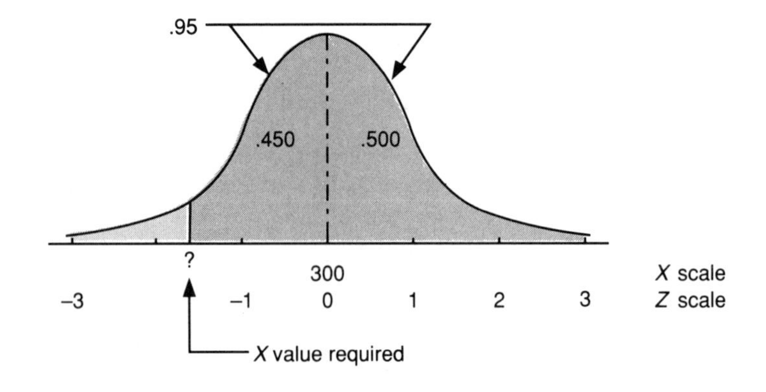

Solution

1. Determine whether the point is to the left or to the right of the mean. Fifty percent are larger than the mean. We must find a point far enough below the mean to contain an additional 45 percent of the scores.
2. Find the number nearest .4500 in the center of the table and read its corresponding Z score. Since .4495 is the area corresponding to 1.64, and .4505 is the area for 1.65, the Z score is halfway between, or 1.645. (The Z score for .4500 is so often used that its more exact value, 1.645, is used here and throughout the text, even though interpolation was not used in finding other Z scores.)
3. Locate a point 1.645 standard deviations below the mean. $X = \mu - 1.645(\sigma)$; $X = 300 - 1.645(20) = 267.10$. Ninety-five percent of the product should weigh 267.10 pounds or more.

Example 5–19 Between what symmetrical limits will a given percent of the area lie? For example, find symmetrical limits within which the middle 95 percent of the boxes will lie.

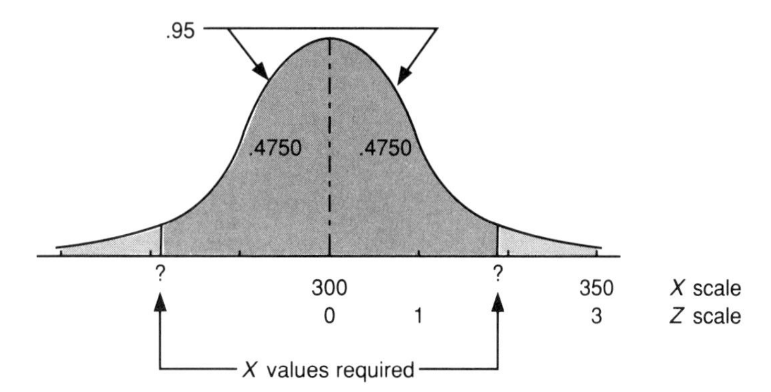

Solution

1. Determine what percent must lie on each side of the mean. Since $.95/2 = .4750$, 47.50 percent must lie on each side of the mean.
2. Find the figure nearest .4750 in the center of the table and its corresponding Z value. Since .4750 is at the intersection of 1.9 and column .06, the Z value is 1.96.
3. Compute the two X values which lie 1.96 standard deviations above and below the mean. $X = \mu \pm 1.96(\sigma) = 300 \pm 1.96(20)$. Therefore, the lower limit is $300 - 39.2 = 260.8$, and the upper limit is $300 + 39.2 = 339.2$. Ninety-five percent of the boxes should weigh between 260.8 and 339.2 pounds.

Example 5–20 The lowest given percent will lie below what X value? For example, if the firm were to discount the lowest 1 percent of the boxes, what will be the lowest acceptable weight?

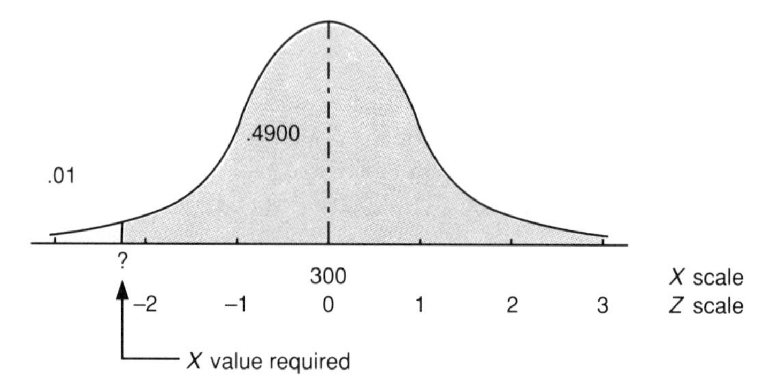

Solution

1. If the tail is to contain 1 percent, the area between the mean and the limit must be equal to 49 percent.
2. Find the figure nearest .4900 in the center of the table and read the Z value from the margins. Since .4901 is at the intersection of 2.3 and the column headed .03, we use the Z score 2.33.
3. The point lies 2.33 standard deviations *below* the mean.
   $X = \mu - 2.33(\sigma) = 300 - 2.33(20) = 253.4$. The score 253.4 is the lowest acceptable weight.

The preceding examples cover the application of the normal curve to populations which are assumed to be approximately normally distributed. (In chapter 9, we will see that the chi-square test could be used to test whether it is reasonable to assume that a variable really is normally distributed.)

THE NORMAL AS AN APPROXIMATION TO THE BINOMIAL

Although the binomial distribution is theoretically correct for solving many discrete probability problems, many binomial tables do not extend beyond a sample size of 20. For large numbers, the binomial equation can be tedious to calculate (without a computer). However, the solution to these problems can usually be approximated by either the Poisson or the normal distributions. If the proportion is very small (or very large), the Poisson approximation is best; otherwise the normal may be best. There is no set rule to follow, but a "rough guide" is to:

Use the **POISSON**: When $n \geqslant 20$ and $\pi \leqslant .05$ or $\pi \geqslant .95$

Use the **NORMAL**: When $n \geqslant 20$ and $\pi > .05$ and $< .95$ and $n\pi > 5$

Poisson limits of $\pi \leqslant .10$ and $\pi \geqslant .90$ may also suffice, depending upon the accuracy needed. Figure 5–11 shows how the sample size and π values affect the shape of the binomial distribution as n increases from 5 to 25. Even for small samples, the binomial approaches the normal as π nears .50. This is because the binomial is unimodal and is symmetrical at $\pi = .50$. And as the sample size gets larger, the approach to the normal distribution is very pronounced. But if the combination of n and π is quite small (< 5), the approximation may still not be satisfactory.

FIGURE 5–11 Binomial approach to normal distribution (as n increases and π approaches .50)

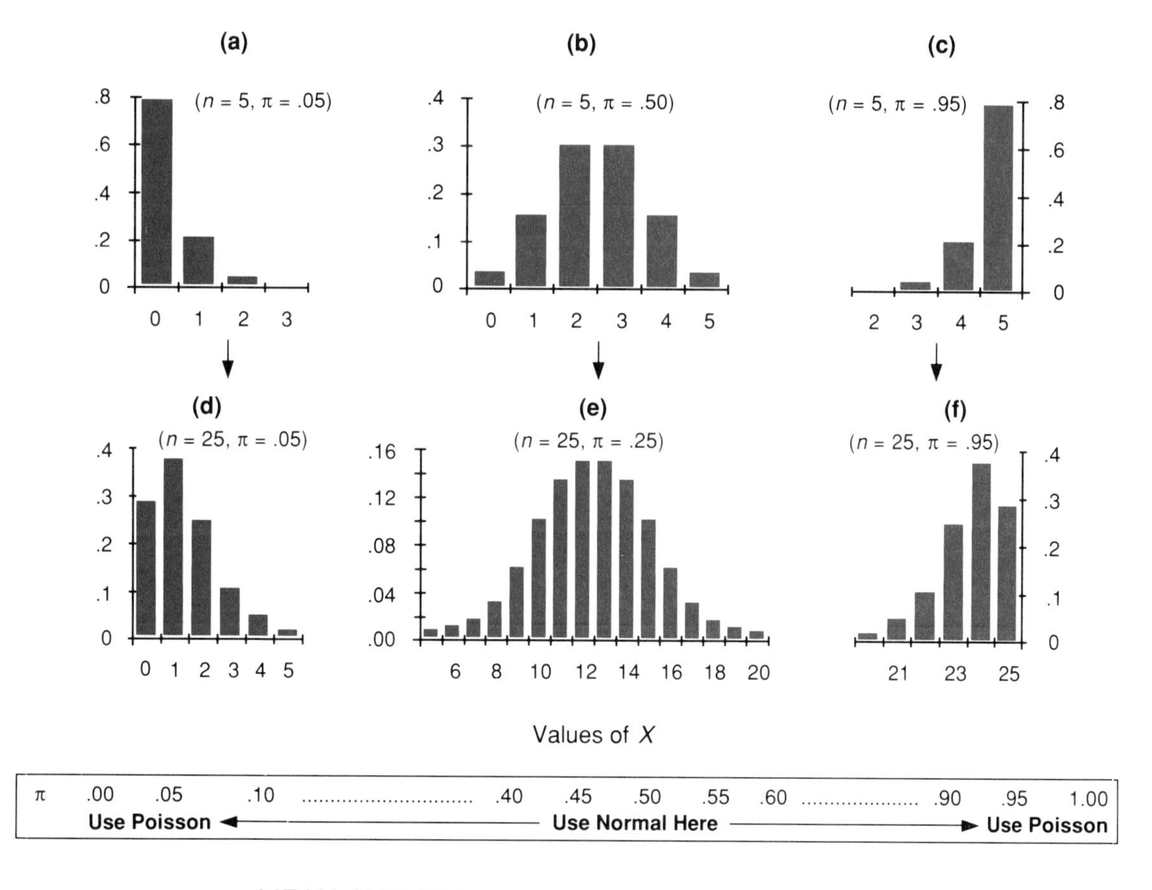

MEAN AND STANDARD DEVIATION We have already illustrated the use of the Poisson to approximate the solution to binomial problems. Using the normal distribution is also quite straightforward. We let the mean and standard deviation of the substitute normal distribution equal the corresponding statistics of the binomial:

$$\mu = n\pi \qquad \sigma = \sqrt{n\pi(1 - \pi)}$$

CONTINUITY CORRECTION In addition, because the normal is a continuous (rather than discrete) distribution, a *continuity correction* is sometimes needed. This is to account for the "gap" between discrete X values, e.g., between 8 and 9, and between 9 and 10. We do this by letting an integer X value include half the interval on each side. The correction becomes less important as the sample size increases.

Example 5–21 The binomial $P(X = 9 \mid n = 25, \pi = .40) = .1511$. What area under the normal distribution would be used to approximate this?

Solution

The conditions for normal approximation are satisfied because $n > 20$, $\pi > .05$ and $n\pi = (25)(.40) = 10.0$. The mean and standard deviation (in numbers) are thus:

Mean: $\mu = n\pi = (25)(.40) = 10.0$

Standard Deviation: $\sigma = \sqrt{n\pi(1 - \pi)} = \sqrt{(25)(.40)(.60)} = 2.45$

Thus, the normal approximation would have a mean of 10.0 and standard deviation of 2.45. The area corresponding to the $P(X = 9)$ can then be approximated as the area under the curve between 8.5 and 9.5. We can compute the approximating normal probability by finding the area between the mean and 8.5 and subtracting the area between the mean and 9.5.

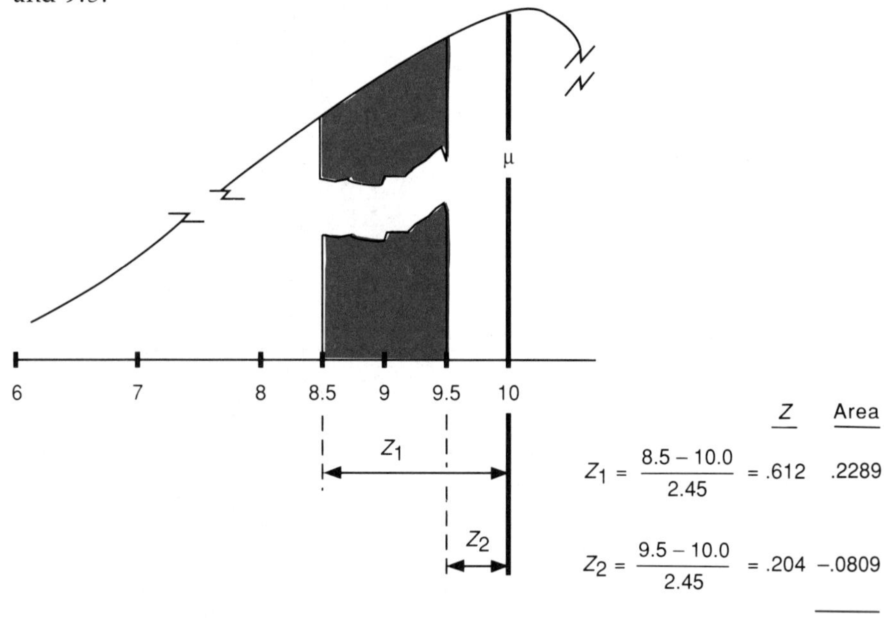

FIGURE 5–12 Approximation with continuity correction

The binomial result (.1511) and normal approximation result (.1489) are fairly close. Let us illustrate another approximation calculation with an example that has a sufficiently low sample size, so that we can look up the binomial probabilities and compare them with the normal approximation result.

Example 5–22 Most of the employees at Green Meadow Dairy Cooperative favor a profit sharing plan, but 30% of them oppose a current proposal. If a committee of twenty employees is selected at random, what is the probability that four or more of the committee members will oppose the proposal?

Solution

The *binomial probability* that four or more are opposed is:

$$P(X \geq 4 \mid n = 20, \pi = .30) = 1.00 - P(X \leq 3)$$
$$= 1.00 - [P(0) + P(1) + P(2) + P(3)]$$
$$= 1.00 - (.0008 + .0068 + .0278 + .0718)$$
$$= 1.00 - .1072 = .8928 \text{ (exactly correct)}$$

The normal approximation using the continuity correction results]in a very close value (.8888) as shown below.

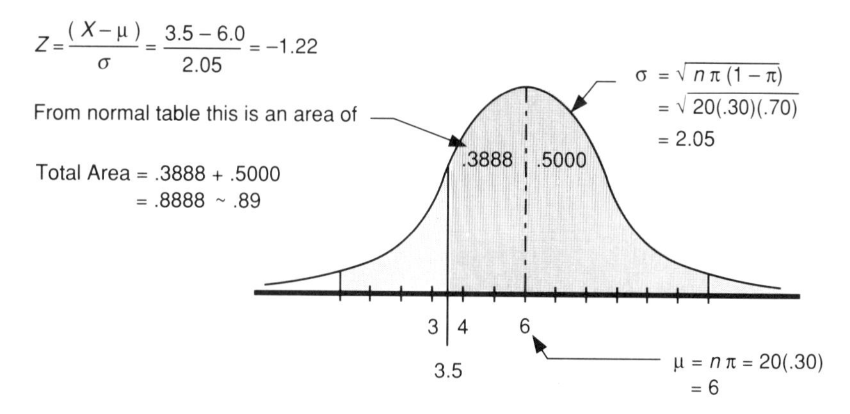

Both of the example approximations are close, partly because the continuity correction was made. This is unnecessary with large samples.

Example 5–23 Suppose a population is guaranteed to be no more than 20 percent defective. A sample of 400 taken from a population has 102 defectives. If the true proportion is really .20, what is the probability that a random sample of 400 will contain 102 or more defective items?

Solution

Using the binomial equation, it would be necessary to compute the probability of 0 defective, 1 defective, 2 defectives, etc., through 101 defectives, and subtract the sum of these probabilities from 1.00. This involves quite a few calculations (even for a computer) with values raised to exponents in the hundreds! With the normal approximation:

$$\mu = n\pi = 400(.20) = 80 \text{ and } \sigma = \sqrt{n\pi(1 - \pi)} = \sqrt{400(.20)(.80)} = 8.0$$

The probability of 102 or more defective items is then the area to the right of 102 (i.e., .0030) or 101.5 (i.e., .0036), if the continuity correction is made. (The correction is insignificant for such a large sample.)

EXERCISES IV

[Note: It is usually helpful to use a simple sketch to help solve normal distribution problems.]

32. What are the chief characteristics of a "normal" distribution curve?
33. In a normal distribution, what probability is associated with:
    (a) any particular point on the X scale
    (b) the area within $\pm$ 1 σ of the mean
    (c) the area within $\pm$ 2 σ of the mean
34. For a normal distribution, find the probability associated with:
    (a) the area between μ and $Z = 1.0$
    (b) the area to the left of $Z = -1.5$
    (c) the area between $Z = -1$ and $Z = 2$
    (d) the area to the right of $Z = 1.87$
35. For a normal distribution with $\mu = 400$ and $\sigma = 20$, find the area:
    (a) below 400, i.e., $P(X < 400)$
    (b) between 400 and 420
    (c) above 420, i.e., $P(X > 420)$
    (d) between 340 and 460?
36. Sketch a normal distribution with a mean of 80 and standard deviation of 10. Include an appropriate X scale that covers 99.7% of the range.

37. Use the table of areas under the normal curve to find the area between the mean and each of the following points:
    (a) a point one Z unit to the left of the mean
    (b) a point two σs to the right of the mean
    (c) a point 2.5 standard deviations above the mean
    (d) a point 1.96 standard deviations below the mean
38. A variable is normally distributed with $\mu = 250$ and $\sigma = 20$.
    (a) What percent of the area lies between the following pairs of points?
        (1) 250 and 290
        (2) 230 and 270
        (3) 270 and 280
        (4) 0 and 280
        (5) 100 and 400
        (6) 270 and 600
    (b) What percent of the area lies between:
        (1) $\mu \pm 1.00\ \sigma$
        (2) $\mu \pm 2.00\ \sigma$
        (3) $\mu \pm 3.00\ \sigma$
        (4) $\mu \pm 1.282\ \sigma$
        (5) $\mu \pm 1.645\ \sigma$
        (6) $\mu \pm 1.960\ \sigma$
        (7) $\mu \pm 2.326\ \sigma$
        (8) $\mu \pm 2.576\ \sigma$
    (c) Give the x-axis value (1) exceeded by 90 percent of the scores, (2) exceeded by 95 percent of the scores, (3) which exceeds 98 percent of the scores.
39. For a normally distributed variable with $\mu = 250$ and $\sigma = 20$,
    (a) the middle 95 percent of the scores lie between what two points?
    (b) the middle 99 percent of the scores lie between what two points?
40. The time required to service customers at a Rapid-Cut Hair Care center is normally distributed with a mean of 12.4 minutes and a standard deviation of 2 minutes. The firm usually makes $3 profit on each haircut, but if the service takes >16 minutes, it is free.
    (a) What percent of the customers would be finished in ≤ 10 minutes?
    (b) If 150 customers were serviced in one day, about how many would receive their haircut free?
    (c) Estimate the weekly cost (in lost profits) for 500 customers/week.
41. The normal curve is sometimes used to approximate the binomial distribution.
    (a) When is the binomial distribution totally symmetrical?
    (b) Explain when the normal distribution can safely be used, and when the Poisson distribution is better applied.
    (c) If $n = 20$ and π is only about .05, is the normal curve a good approximation? Why or why not?

42. What area under the normal distribution would be equivalent (in a normal approximation) to $P(X = 9 \mid n = 30, \pi = .35)$?
43. Use the normal curve to estimate the probability that a random sample of 400 will contain 375 or more good items if the population is 10 percent defective. (Hint: let $\pi = .90$.)
44. Forty percent of the employees of a construction firm have company medical insurance. If 90 employees required medical attention last year, what is the probability that at least 45 of them had company insurance?
45. The scores on a sales aptitude test are normally distributed with a mean of 80 points, and standard deviation of 8 points. If 50 applicants took the test, what is the probability that 2 (or more) received scores of higher than 90 points?

OTHER CHARACTERISTICS OF STATISTICAL DISTRIBUTIONS[2]

MEAN AND STANDARD DEVIATION OF DISCRETE DISTRIBUTIONS

ONE VARIABLE We have seen that the two most common descriptors of statistical distributions are the mean and standard deviation. And we know the mean is also the expected value, $E(X)$ of the distribution. For any discrete distribution with a probability distribution $P(X)$, the mean (μ) and standard deviation (σ) can be computed from the probability distribution as:

For discrete distributions

Mean: $\mu = E(X) = \Sigma[X \cdot P(X)]$ (5–11)

Standard deviation: $\sigma = \sqrt{\Sigma(X - \mu)^2 P(X)}$ (5–12)

[2] This section contains some related materials that extend beyond that covered in some introductory courses. Your instructor may choose to bypass it without any loss of continuity with other material. Coverage is intentionally brief.

TWO VARIABLES In some situations, we must work with a random variable that represents the sum of two other variables. For example, the volume of water in a reservoir may be a function of how fast water happens to flow in, as well as how much demand is experienced by pumping water out. If the two variables (X_1 and X_2) are both *random and completely independent*, the mean and standard deviation of the sum can be estimated as:

For two independent random variables

Mean: $\mu_{1+2} = \mu_1 + \mu_2$ (5–13)

Standard deviation: $\sigma_{1+2} = \sqrt{\sigma_1^2 + \sigma_2^2}$ (5–14)

Note that variances of independent random variables are additive (but standard deviations are not—so we must add in variance form).

CHEBYSHEV'S INEQUALITY

Sometimes we may wish to make statements about the probability of values lying within certain limits of the mean, without knowing whether the distribution is binomial, Poisson, normal, or whatever. The theoretical limits of the dispersion of values in *any* distribution are defined in a theorem known as Chebyshev's Inequality:

CHEBYSHEV'S INEQUALITY
For any distribution of a random variable (X), the proportion of items lying beyond k standard deviations from the mean will not exceed $1/k^2$, where $k \geq 1$.

Thus the proportion beyond 2 standard deviations will never exceed $\frac{1}{2^2}$ or 25%. Conversely the proportion within 2 σ's must be at least $1.00 - \frac{1}{2^2}$ or 75%. Similarly, within 3 σ's, the proportion must be $1.00 - \frac{1}{3^2} = 89\%$ of the observations, regardless of the shape of the distribution.

Example 5–24 A health maintenance organization is attempting to establish hospital care costs in order to help set insurance rates for customers in a metropolitan area. Analysts have researched data on all 840 people in hospitals in the area at a preselected time. Their study revealed that the mean length of stay was 9.0 days and $\sigma = 2.0$ days. However, they have not identified which form of distribution the length of stay data take.

Using Chebyshev's Inequality, (a) at least how many of the patients had a hospital stay between 5 and 13 days, and (b) at most, how many stayed less than 3 or more than 15 days?

Solution

(a) Given $\mu = 9.0$ days, and $\sigma = 2.0$ days, the 13 is:

$$Z = \frac{X - \mu}{\sigma} = \frac{13 - 9}{2} = 2 \text{ standard deviations away from } \mu$$

[Note: The lower limit of 5 days is also 2 standard deviations away.]

Per Chebyshev's Inequality, no more than $\frac{1}{k^2} = \frac{1}{2^2} = 25\%$ are outside the 2σ limits of 5 and 13. Therefore, at least (.75) (840) = 630 patients are within the 5- and 13-day limits.

(b) The 3- and 15-day limits are 3 σ's from the mean, and per Chebyshev's Inequality, no more than $\frac{1}{k^2} = \frac{1}{3^2} = 11\%$ are outside the 3-σ limits. Therefore, at most (.11) (840) = 93 patients stayed less than 3 or more than 15 days.

SUMMARY

Probability distributions show the values that a random variable can assume, along with the probability of each value. *Discrete distributions* are based upon counts (of discrete events) and *continuous distributions* upon measurements (along some interval). The long-run expected value of a distribution is its mean, μ.

Binomial probabilities arise from experiments that have (1) only two outcomes, (2) a constant probability of success, and (3) independent trials. The binomial equation can be used for computation, but tables are available for sample sizes up to about $n = 20$. The binomial has two parameters, its mean ($\mu = n\pi$), and standard deviation [$\sigma = \sqrt{n\pi(1 - \pi)}$].

Poisson probabilities describe situations that have (1) many possible outcomes over a continuum of time or space, (2) low and constant probabilities of success, and (3) independent trials. Either an equation or tables can be used. But again, tables are limited in size. The Poisson has only one parameter, its mean, λ, and describes "rare event" probabilities. It can, however, be used to approximate binomial distributions that have sample sizes of > 20 and π values of $< .05$ (or possibly .10). Then its mean is estimated as $\mu = n\pi$ (and its variance is also $n\pi$).

Normal probabilities arise from measurements on a continuous scale. The normal curve is continuous, unimodal, symmetrical, and bell-shaped, and is defined by two parameters (μ and σ). Although there is a different normal curve for every μ and σ combination, any curve can be converted to standardized (Z) form by expressing the X scale in numbers of standard deviations. To do this, we find the difference between the mean and any other value ($X - \mu$), and divide by the amount of one standard deviation. Then 68% of the area is within $\pm 1\,\sigma$, 95.5% of the area is within $\pm 2\,\sigma$, and 99.7% of the area is within $\pm 3\,\sigma$. These and other values are included in the Normal Distribution Table, where the total probability under half the curve is .5000. The normal can also be used to approximate the binomial and usually works well when $n \geqslant 20$ and $\pi > .05$ and $n\pi > 5$.

In addition to being useful for describing natural measurements, we will be using the normal distribution extensively for making statistical inference in later chapters.

Microcomputer software is readily available for computing binomial, Poisson, and normal probabilities. The programs typically ask you to enter the data, or the parameter values (e.g., μ, σ) and the desired X value. Then they calculate the probability (and possibly cumulative probabilities). You may wish to experiment with some of these programs on the end of chapter problems.

CASE: DETROIT LAKES POWER COMPANY

Tom Larkin knew it *wasn't fair* to throw a problem like this at Sandra, but what else was he to do? Tom was the Maintenance Superintendent for all of the Central Station power generation equipment. He had requested more staff assistance to try to "get a handle on the breakdown problems" the company was having at their main switching center. But he didn't expect to get an inexperienced greenhorn, fresh out of college—much less a girl who didn't know anything about electrical switches anyway.

Sandra Doyle listened intently as her boss explained the problem. The company had 420 switches at Central Station, with a sporadic pattern of breakdowns that was giving them fits—not to mention the complaints from some industrial customers. Tom was impressed with the questions Sandra asked.

SANDRA DOYLE: But what about the past breakdowns, Mr. Larkin. Exactly how many have occurred? And where have they been happening? How long do the switches normally operate before failure?

TOM LARKIN: You seem to want a life history on each switch. But we can oblige. It's all on computer tape.

Given the depth of her questions, Tom could see that Sandra was not going to do a superficial job. He returned to his office thinking the project would occupy her for about 2 months.

The next week, Sandra was in Mr. Larkin's office with a rough draft of her report. She had first mapped the switch breakdown data by location, but that had failed to reveal any discernible pattern.

TOM LARKIN: That's been the problem all along. These breakdowns are occurring all over the place—and at the most unpredictable times!

SANDRA DOYLE: Yes, that was my next avenue of inquiry. I was able to find complete data on 300 switches and grouped it by length of operating time before failure. Here's what it looks like.

Sandra showed Mr. Larkin a grouping of data for 300 switches arranged by length of operating time (i.e., $\leq$ yrs) until a failure occurred:

Time until Failure	6 mo	1 yr	2 yrs	3 yrs	4 yrs	5 yrs	6 yrs	7 yrs	8 yrs	9 yrs	10 yrs
Number of Relays	40	12	11	16	12	14	15	52	70	48	10

SANDRA DOYLE: This yields an interesting graph.

Sandra showed Mr. Larkin the graph, upon which she had marked approximately where the Poisson and normal distribution failures occurred.

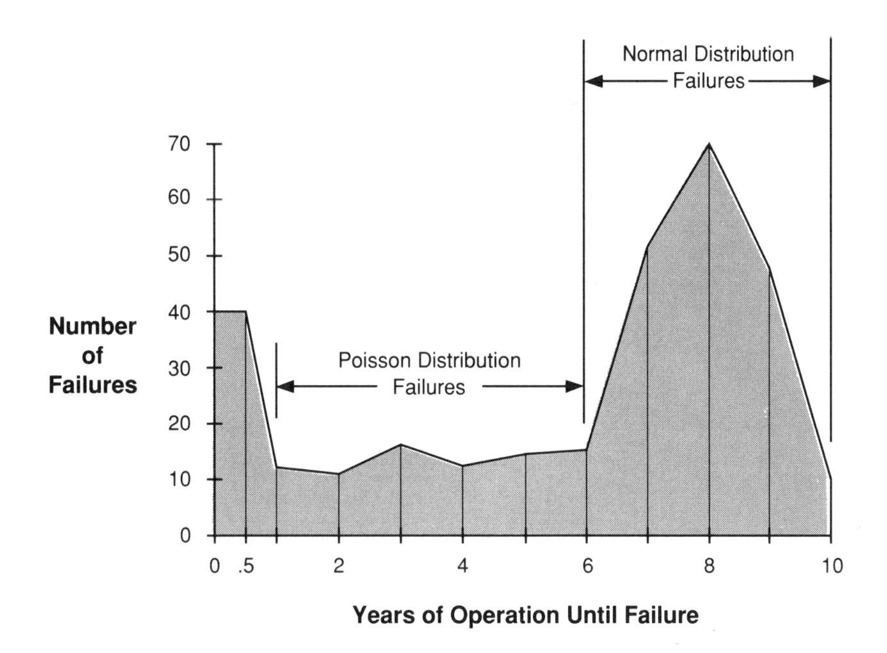

TOM LARKIN: It looks a little like an old time bathtub.

SANDRA DOYLE: It does? Anyway, there are several early failures in the first few months—then they level off for nearly 5 years. After that they go on up and peak in the eighth year. By the tenth year virtually all the switches have failed. I've compared the data to some statistical distributions, and this extended period can be described by a Poisson distribution, and the wear out period by a normal distribution.

TOM LARKIN: Well, isn't that something! I can explain those initial failures. They're usually due to some installation problem—a connection or faulty circuit protection. Does this mean you could predict the failure pattern for the whole installation, . . . all the relays in the shop?

SANDRA DOYLE: Oh yes, within reasonable limits of statistical accuracy. Do you think the report will be okay? Am I on the right track?

TOM LARKIN: Yes, yes . . . yes! This will save us a bundle of trouble, Sandra. The Production Manager is going to flip when he sees this! Say, we have another little problem in the generation plant. I was wondering if maybe you could . . . (Sandra interrupts him).

SANDRA DOYLE: But you *do* want me to get this finished up first, don't you?

TOM LARKIN: Oh yes. But I was thinking I might send my assistant down and you could have him put on the finishing touches. This other problem has been bugging us for some time, and I'd like to have you take a look at it.

SANDRA DOYLE: If you say so, but I don't know anything about generators.

TOM LARKIN: No problem. You'll pick up what you need.

Case Questions

(a) What might have prompted Sandra to consider using probability concepts to tackle this problem?

(b) Why was a distribution of failures over time more helpful than one of the number of failures at various locations?

(c) What do you suppose prompted Sandra to identify the 1–5 year failures as Poisson distributed, and the 6–10 year failures as normally distributed?

(d) Suppose Mr. Larkin used the above Poisson failure probability data to derive a proportion of failure for relays that had been in service for two to five years as $\approx$ 15 failures in 300 switches, or $\pi = .05$. If he wished to apply this to a section of the plant that has 60 switches in that category, what would be the probability that five or more of the switches would fail during that year?

(e) Suppose Mr. Larkin used the normal distribution wearout data above, assuming that the mean lifetime characteristics of those switches that survived the first five years were $\mu = 8$ yrs and $\sigma = .9$ yr. When should he plan to replace the switches (as a group) if he wants to replace them at a point when 30% have failed.

QUESTIONS AND PROBLEMS

[Note: The problems in this chapter are short enough to do by hand using a calculator. However, you may need to use a Poisson or normal approximation for selected binomial problems. Alternatively, most of the problems can be done on computer, if software programs are available.]

46. A shipment of 30 boxes of oranges is received from Cyprus under an agreement that the shipment can be rejected if more than 13% of the boxes contain spoiled fruit. Three boxes are selected at random and inspected.
    (a) What would be the random variable, X?
    (b) What kind of probability distribution would apply here? Would it be discrete or continuous?
    (c) Assuming 10% of the boxes have spoiled fruit, and the first two boxes are satisfactory, what is the probability of selecting a spoiled box on the third trial?
47. A tour bus service in Philadelphia charges $5 per person and has experienced a uniform demand varying from 13 to 20 tourists, inclusive.
    (a) Construct a probability distribution where the random variable is the amount of revenue from tourists.
    (b) What is $E(X)$?
    (c) Compute the mean and standard deviation of the distribution.
48. A precision casting process produces missile parts that are 20 percent defective. A sample of three parts is selected at random from this process.
    (a) How many ways are there to select 1 defective in a sample of 3?
    (b) Use the binomial equation to compute the probability distribution of defectives in a sample of 3.
    (c) What is the probability there will be at least one defective part in the sample?
    (d) Did you use $1 - P(G_1 \cap G_2 \cap G_3)$ to answer part (c)? If not, use this procedure and compare your answers.
49. A sample of 20 rayon automobile tires is drawn from a process which runs 20% defective.
    (a) Use the binomial equation to find the probability of exactly four defective tires in the sample.
    (b) Compare your answer with that in the binomial table for $n = 20$, $\pi = .20$, and $X = 4$.
    (c) What is the probability of 16 good tires in a sample of 20 if the process runs 80 percent good?

50. If a process produces parts which are 90 percent good, (a) what is the probability that a sample of 20 will contain exactly 18 good items? (b) what is the probability it will contain 18 or more good items?
51. The number of truck arrivals at a weighing station per hour follows a Poisson distribution, with a mean of 4.5 arrivals per hour. What is the probability that in an hour selected at random there will be (a) no arrivals, (b) exactly 5 arrivals, (c) 5 or more arrivals?
52. Beer bottles (that average 5% defective) are packed in boxes of 50.
    (a) Set up the equations to compute the probability of 48 or more good bottles in a box selected at random, using the binomial equation.
    (b) Approximate the answer which would be obtained in (a) using the Poisson approximation and values in the Appendix.
53. Find the probability of the following:
    (a) Binomial: $P(X = 7 \mid n = 9, \pi = .35)$
    (b) Poisson: $P(X = 3 \mid \lambda = 5)$
    (c) Normal: $P(X \geq 60 \mid \mu = 52, \sigma = 6)$
    (d) Normal: $P(X = 6 \mid \mu = 5, \sigma = 2)$
54. Use the appropriate probability distribution to compute the following:
    (a) $P(X < 3 \mid n = 8, \pi = .50)$
    (b) $P(X \geq 19 \mid n = 20, \pi = .85)$
    (c) $P(X = 14 \mid n = 30, \pi = .50)$
    (d) $P(20.0 < X < 25.0 \mid \mu = 18, \sigma = 6)$
55. A sample of $n = 50$ is drawn from a population where the probability of a defective is .04. Which probability distribution should be used to estimate the probability of 45 or more successes in the sample? Explain.
56. Sketch two normal distributions with a mean of 400 and show the mean value (400) and the value 450 on both. Let sketch (a) have a standard deviation of 20, and sketch (b) have a standard deviation of 50. (c) Show the Z value for the 450 score on both sketches.
57. A brand of 60-watt light bulbs has a normally distributed lifetime with a mean of 1,000 hrs and a standard deviation of 60 hrs.
    (a) What percent of the bulbs wear out between 940 and 1060 hrs?
    (b) What percent of the bulbs last more than 1100 hours?
    (c) 98% of the bulbs last longer than how many hours?
    (d) The middle 95.5% of the bulb life lies between what hours?
58. Boxes of fruit at a packing plant have a known $\mu = 60.0$ lbs and $\sigma = 3.0$ lbs. Assuming these parameters hold for a truckload of 1,000 boxes, use Chebyshev's Inequality to determine (a) the minimum number of boxes within 55.5 to 64.5 lbs and (b) the maximum number of boxes that lie outside the range 54.0 lbs to 66.0 lbs.

59. Assume 20 transportation firms are selected at random from a population that has the same proportion *not* paying dividends as the transportation firms in the database of Appendix A.

    (a) Use the binomial distribution to estimate the probability that 12 or more of the firms selected are not paying dividends.

    (b) Make a comparable estimate by using the Poisson approximation to the binomial.

    (c) How do you account for the discrepancy in the two estimates?

60. Assume the database of Price/Earnings (P/E) ratios in Appendix A is normally distributed with a standard deviation of 4.50. Assuming random selection, find the probability a company would have a P/E:

    (a) $< \$11.13$, (b) $> \$15.00$, (c) between \$10.00 and \$15.00,

    (d) $> \$10.00$, (e) Do 95.5 % of the P/E ratios lie within the $\pm 2\sigma$ range?

CHAPTER

Sampling and Sampling Distributions

INTRODUCTION

USING SAMPLES TO LEARN ABOUT POPULATIONS
- Why Sample?
- Sampling Design
- Errors in Sampling

USING A RANDOM NUMBER TABLE

JUDGMENT AND PROBABILITY SAMPLING
- Judgment Sampling
- Probability Sampling

SAMPLING DISTRIBUTIONS
- Infinite Versus Finite Populations
- Sampling Variation
- Sampling Distributions of Sample Means
- Sampling Distributions of Sample Proportions

SAMPLING DISTRIBUTION THEOREMS
- Theorem 1
- Theorem 2: The Central Limit Theorem

MEAN AND STANDARD ERROR OF SAMPLING DISTRIBUTION

FINITE POPULATION CORRECTION FACTOR

INFERENCE WHEN σ IS UNKNOWN: THE *t*-DISTRIBUTION

STANDARD ERRORS FOR MEANS AND PROPORTIONS
- Standard Error of Proportion
- Standard Error of Number of Occurrences

SUMMARY

CASE: DEBATE AT THE CITY COUNCIL

QUESTIONS AND PROBLEMS

Chapter 6 Objectives

1. Use a random number table to select a sample.
2. Distinguish between judgment vs. probability sampling.
3. Explain the meaning of a sampling distribution.
4. Compute standard errors of means and proportions.
5. Apply the finite population correction factor.

Consumer opinion is a vital source of information for nearly every organization. City officials need input from citizens to help decide whether taxes should be spent on streets, parks, or swimming pools. New firms need market information to help chart their strategy (as evidenced by the fact that more than half of all independent small businesses fail within five years). How will potential customers like their new products? Would a new "mood control" aerosol spray be as acceptable as hair spray or air freshener? What are the chances of success for a firm specializing in home health care? Will Japanese consumers patronize restaurants that feature the jukeboxes and bobby-soxed waitresses that were so popular in the U.S. in the 1950s? Do investors fear the stock market will drop again like it did in October, 1929, and October 1987?

These questions, like many others in business require information to guide decision making. Much of this information comes from sampling activities.

INTRODUCTION

IMPORTANCE OF SAMPLING Organizations need consumer information to help guide their financial, production, and marketing decisions. For example, a clothing manufacturer introducing a new line of women's sportswear would want to aim their advertising at the product features their consumers think are most important. Yet, they cannot survey all their future customers—so they must rely upon data from a sample of potential customers.

Suppose the clothing manufacturer's sample revealed that the most important determinants of merchandise value were:[1]

workmanship	23%	materials	14%
price	21%	looks	13%

This sample information suggests that the firm should focus its advertising mostly upon the workmanship and price aspects of their product.

Sample information such as that illustrated above is important to firms like Procter & Gamble, General Motors, and McDonald's, for each may spend over $150 million on television advertising alone in any given year! With that amount of money at stake, they must be quite confident that their sampling techniques are really helping them make the best advertising decisions.

CHAPTER CONTENT Statistical data comes largely from samples. But all sample information is not equally good, so we need to understand what constitutes a statistically valid sample. In the first section of this chapter we will learn how to select a kind of sample that will enable us to make statistical inferences about a population. Then we will draw some distinctions between types of sampling, and learn about sampling variation and sampling distributions. We also introduce the most important theorem of the book in this chapter—the *Central Limit Theorem*.

USING SAMPLES TO LEARN ABOUT POPULATIONS

Samples provide us with empirical data that can be statistically analyzed. In this section we consider (a) some reasons for sampling, (b) some types of sampling designs, and (c) some potential errors to be aware of in sampling.

[1] These percentages are actual survey results for (nonspecified) merchandise as obtained from *Chain Store Age* Magazine, and reported in *The Wall Street Journal*, 10/4/85.

WHY SAMPLE?

SOURCES OF SAMPLE DATA Statistical investigations that use empirical data usually obtain it from surveys, observations, or experiments. *Surveys* might include field inquiries to determine public opinion, competitive practices, business conditions, or any social or political conditions of interest. *Observations* are counts or measurements made to discover or estimate the specific properties of some population of interest. For example, one may make observations of the reliability of a product, its quality, or performance characteristics. *Experiments* are controlled tests for such purposes as determining the best material for a product, the preferred manufacturing process, or the most effective way to train employees. They often make use of a control group which is compared with an experimental group, or they may be designed to compare two or more competing products or procedures.

REASONS FOR SAMPLING Managers frequently use sample data to make inferences about the characteristics of populations. Three of the most common reasons for using sample, rather than population, data for making many business decisions are:

1. using the entire population data may be *impossible* (or impractical).
2. the *cost* of surveying the population may be prohibitive.
3. the *time* requirements may be prohibitive.

Firestone Tire Company would soon be out of business if they attempted to road test every tire they produced. Also, it would be impossible for the Kellogg Company to seek every potential consumer's opinion of a new breakfast cereal. In the same way, cost and time requirements often prompt firms to use sample data. For example, the local building materials firm could probably not afford to do a compression (strength) test on each concrete block they produce. But they may be able to test a few blocks out of each thousand, and use that information to draw some inference about the average compressive strength of the blocks produced during a given time period.

SAMPLING DESIGN

TARGET POPULATION The first step in planning any study that requires the collection of empirical data is to state exactly what questions the study is intended to answer. This requires a careful definition of the *target population,* which is the population that *should be* studied. The population that *actually is* sampled may or may not be the same as the target population. If it is impossible to study the target population, some other group may be sufficiently similar to the target population to provide reliable conclusions. Another possibility is to simulate the population characteristics on a computer, and "sample" from the simulated population distribution. Otherwise the study may have to be redesigned.

FRAME A list of all the members of the population to be studied is called a *frame*. The listing is frequently a written file, but it can be any well-defined group, such as items in a computer database, or in a container, or even locations on a map. If every element in the frame were observed, then we would have a census instead of a sample. However, we must normally work with only a sample of the population. So we will be computing and using *sample statistics* to make inferences about the *population parameters*.

TYPES OF SAMPLES Samples are frequently classified according to the manner in which the elements in the target population are selected. The two major categories are *judgment samples* and *probability samples*.

With judgment samples, the elements in the sample are deliberately chosen, so the samples are termed *nonrandom*. Judgment samples do not permit us to make statistical inferences about the population, so they are not our primary concern here. Probability samples involve some uncontrollable element of chance, and they do permit statistical inference. Business statistics is concerned mostly with probability samples.

ERRORS IN SAMPLING

Any sampling activity (or even a census-taking activity) is subject to error. We are all aware of the possibilities of recording the wrong information or keyboarding incorrect numbers into a computer. And there is also the (more remote) possibility of intentional errors—which can be difficult to detect and combat. But the two types of error of special concern in statistical analysis are *bias* and *sampling error*.

Bias is a "displacement" error due to a flaw or prejudice in the design, testing, or recording procedures. It is like having a rifle sight slightly off center, or a speedometer that registers 10% slow. Mail surveys can be biased—especially when the respondents are self-selected. For example, a mail survey designed to gather people's opinions about gun control legislation is likely to be biased if the only people that bother to respond are those that have strong feelings about gun control.

Sampling error is a random error due to the chance that any one sample may not accurately reflect the characteristics of the population. Whenever there is variability in the population, the potential exists for sampling error. For example, suppose a local chamber of commerce wished to use a sample of 50 households to estimate the median family income in their community. Median income levels range from about $43,000 for families with a college graduate, down to around $15,000 for some minority and single-person households. Any individual sample of 50 may not include a representative proportion of two-earner families, single men, minority families, single women, etc. So the sample value will most likely differ from the real median income of the population.[2] We refer to this difference as sampling error.

[2] The U.S. Census Bureau estimates the real median family income in the U.S. at $26,430.

If good sampling procedures are followed, the sampling error can usually be reduced by increasing the sample size. Thus we might expect a sample of 1,000 households to yield a closer estimate of median family income than a sample of only 50. However, increasing the sample size will not correct for bias (e.g., the bias of sampling only people who *own* their own homes). Correcting for bias usually requires an adjustment to the methodology or procedures used in the study.

EXERCISES I

1. Distinguish between surveys, observations, and experiments.
2. Explain why a sample would be used instead of a census to determine
   (a) the types of toys preferred by pre-schoolers in Canada.
   (b) the "crash resistance" of a new automotive bumper.
   (c) the best diet to follow in order to avoid having a heart attack.
3. What type (classification) of sampling design would be involved in
   (a) collecting opinions about the cause of a nuclear plant accident.
   (b) selecting a committee by drawing names out of a hat.
   (c) assessing public attitude toward a tax increase by sampling delegates to the Republican National Convention.
4. How does *bias* differ from *sampling error*?
5. A firm seeks to determine which is the most popular type of soft drink in the U.S. They place a large notice in *The Wall Street Journal*, inviting respondents to record their preference by calling a 1-800 (toll free) number and responding to an answering machine. Evaluate this sampling design in terms of its (a) target population, (b) type of sample, and (c) potential errors.

USING A RANDOM NUMBER TABLE

Probability samples require that items be selected by chance, or "at random," from the population. One of the most common methods of doing this is by using a table of random digits, where (a) each of the digits 0 through 9 has an equal chance of occurring, and (b) combinations of the digits also have the same probability of occurring.

Table 6–1 is a portion of a random number table. To use the table to select a sample, the population should first be serially numbered. Then start at any point in the table and select numbers by any uniform procedure, i.e., by reading across, down, or diagonally. The random numbers identify items from the target frame that are to be included in the sample. Continue until you have the sample size desired, omitting any numbers that do not correspond with the numbers in the population, or that happen to be repeated.

TABLE 6–1 Random numbers

1876	5498	4576	3982	1628	0337	6417
2917	3887	4620	1784	9365	2846	7420
1992	2387	6549	7804	1429	9840	7216
9081	9672	3002	6333	3795	6530	5787
0564	4138	1537	6652	8538	0599	3333
5618	0511	4089	6708	3252	4522	3195
8651	7723	7489	2870	8990	1450	9615

Example 6–1 A sales manager with 80 accounts would like to try a new promotional campaign on six accounts selected at random. Use the random number table to select the numbers of the six accounts.

Solution

The maximum number of customers (80) can be described by two digits, so we will need only two digits of the four digits shown in each column of the table. Assuming the customers are numbered from 01 through 80, let us arbitrarily start at the third column of the top line, use the last two digits of the set of four, and read across the table. Discard any duplicates or numbers greater than 80. The elements selected are the numbers

76 28 37 17 20 65

Thus, the sample would include the 76th account, the 28th account, and so forth.

In many cases, accounts, part numbers, etc., are included in the database of the organization's computer, so that listing may constitute a satisfactory "frame." If it is not possible to identify the elements of the population and assign numbers to them, a random sample might still be obtained if each element (and combination) has an equal chance of being selected; otherwise some other type of sampling plan must be used.

For small items such as bolts or electronic parts, a random selection can sometimes be made simply by stirring the elements themselves in some type of bowl or bin, and physically selecting the items for the sample. Thorough mixing is required after each element is selected, and care must be taken to ensure that lighter or smaller elements do not come to the top where they are more likely to be drawn. (This would be biasing the results.)

EXERCISES II

6. A stock brokerage firm maintains a list of 7,275 accounts, and would like to draw a random sample of 15 accounts to test-market a new bond issue. Beginning with the second column in Table 6–1, and working down, identify the 15 accounts to be used in the study.
7. The personnel manager of a firm with 21,200 employees would like to randomly sample 50 employees to assess their reaction to a recently implemented smoking policy.
   (a) How many digits must be used from the random number table?
   (b) Briefly describe how the sample might be selected.
8. Each of the 550 motors per week produced at an appliance factory in Cleveland is "bar-coded" with a serial number, and this week's numbers range from 4,200 to 4,750. Suppose a quality control inspector wished to select three motors for detailed testing. Working down the columns in Table 6–1, what three motors would be selected? (Start in the upper left corner.)
9. An IRS computer has suggested that six income tax returns be randomly selected for audit. Selection is to be from among social security numbers 428–50–0000 to 715–50–0000. Use groups of nine digits from Table 6–1 (working down) to select the sample.

JUDGMENT AND PROBABILITY SAMPLING

In this section we describe the most widely used subtypes of judgment and probability sampling mentioned earlier and depicted in Figure 6–1.

FIGURE 6–1 The two major types of samples

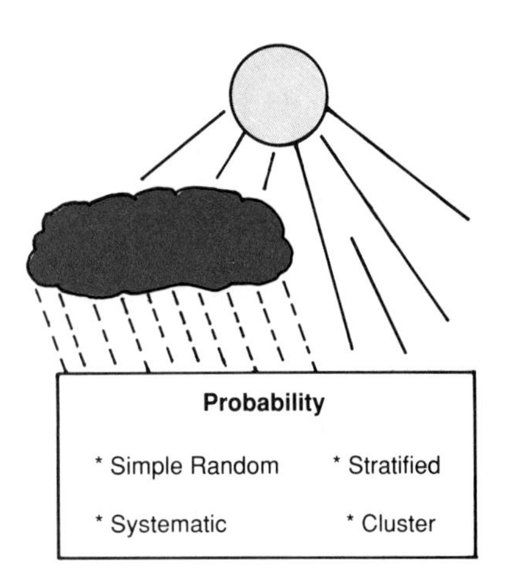

JUDGMENT SAMPLING

WHY USED We have noted that judgment samples involve some choice in deciding what elements will be included in the sample. For example, most exams by college instructors are "judgment samples" of the knowledge a student is supposed to have learned in a course. (Unfortunately, those exams may not always be truly representative samples—but we have to accept the results anyway.)

In some cases, judgment samples are the only practical way of obtaining information about a population—especially if it is very large or heterogeneous. In other cases, they are the most convenient or cost-effective method of sampling, and they are widely used.

LIMITATION The major limitation of judgment samples is, nevertheless, quite severe. They are subjective, and there is no satisfactory method of measuring their validity. Their analytical value is limited because we cannot validly use them for making statistical inference.

TYPES Three of the more commonly used subtypes of judgment samples are (1) *purpose samples*, (2) *quota samples*, and (3) *convenience samples*.

Purpose samples These samples are selected to satisfy a specific purpose, such as to estimate a firm's product acceptance or measure its productivity. The consumer price index (CPI) is one of the most widely used purpose samples in the U.S. It is used to measure the general level of prices in the economy. It is important because the wages of millions of workers are adjusted to this index annually. Note, however, that it does not include a random selection of items. Instead, it specifically includes items that significantly affect our cost of living, such as automobiles and rents. But it also excludes thousands of other items.

Quota samples These are samples of a specified number of respondents taken to study a given characteristic. For example, an auto dealer trying to decide how many cars to stock may ask for a sample of 40 households to determine potential buyer intentions for purchasing a new car. The "quota" may include 30 family households with two or more children, five apartment dwellers, and require that 60 percent of the respondents be female. The information may be useful, but quota samples can contain a selection bias because the subjects are chosen by interviewer and are not randomly selected.

Convenience samples Such samples are selected from subjects that are conveniently available, such as surveys at a supermarket or man-on-the-street TV interviews. Like quota samples, convenience samples are also open to selection bias. Some people are more "approachable" than others, or they may never go to that supermarket (or that street). These interviews may make entertaining news, but we cannot really attribute any statistical validity to them.

PROBABILITY SAMPLING

Probability samples are samples that result from some form of random sampling process that gives elements a measurable chance of being selected. Two key advantages over judgment samples are that probability samples:

(a) include a measure of the sampling *error* associated with the process.

(b) permit statements of statistical *inference* about the population.

ERROR As suggested earlier, sampling error accounts for the fact that sample statistics vary from one to another. Different samples might yield different results. However, the magnitude of the sampling error can be estimated (and to some extent controlled) in probability sampling. And given a knowledge of the sampling error, we can make valid statistical statements about the population being studied. This is, of course, the main purpose of most studies, so probability samples are widely used.

SUBTYPES There are four major subtypes of probability sampling:

(1) *simple random sampling*

(2) *systematic sampling*

(3) *stratified sampling*

(4) *cluster sampling*

Simple random sampling is the most common and the method we shall concentrate upon in this text. You may want to consult some advanced texts before using the other methods described below.

Simple random sampling **Simple random sampling** is the method of sample selection that gives each element (and each combination) in the population an equal chance of being included in the sample. As we suggested in Chapter 5, the term "random" really applies to the rule or sampling process itself—more than to the sample value obtained. If the population is discrete, this means the sampling process must give each element the same chance of being selected. If it is continuous, the probability of any interval being included in the sample must be equal to the proportion of the population that lies in that interval. Using random numbers is one of the best ways of ensuring that a sample is random. In Example 6–1, we used random numbers to help ensure randomness in selecting the six accounts.

Systematic sampling Systematic samples are often an acceptable substitute for simple random sampling, especially when one is sampling from a process. With systematic sampling, a random starting point is first selected (e.g., by using a random number table). Then additional elements are selected at fixed intervals of time or sequential locations. So the first element is randomly selected, and its position influences what other elements will also be selected.

Example 6–2 A large retailer wishes to obtain a sample of $n = 20$ charge accounts from a computer listing of $N = 3{,}000$ accounts. How might this be done using a systematic sampling procedure?

Solution

(1) To obtain the sample of 20 from 3,000, we must sample one account from each sequential listing of $N/n = 3{,}000/20 = 150$ accounts.

(2) Use a random number (RN) in the range of 1 to 150 to select the first account. Suppose the RN is 72. Then the first sample is account #72.

(3) Sample every 150th account after that until $n = 20$ accounts are sampled. The second sample will be account #222 (i.e., $72 + 150$).

Thus: #72, #222, #372, #522, #672, #822, . . . , #2922

Systematic sampling can be just as representative as simple random sampling and usually requires less time (and less cost). However, one must be alert to bias. Bias can occur if the sampling interval corresponds with some inherent periodicity in the data. For example, a sample of daily bank receipts would not be representative if the systematic interval turned out to be seven days, and the sample always reflected Monday's receipts (which tend to be higher because of weekend business).

Stratified sampling With stratified sampling, the population is divided into relatively *homogeneous* subgroups (or strata), and random samples are taken from each subgroup. The purpose of the groupings is to combine similar elements in a way that will result in less variability within the groups, and thus permit the use of smaller samples than might otherwise be required. (Thus, each element does not necessarily have the same chance of being selected.)

For example, suppose a state economist wished to use sample data to estimate the average length of time employees had worked for their respective firms. Suppose further that some industries, such as the electrical utility industry, had relatively stable employment over the years, whereas others, such as electronics, had experienced high turnover. The economist may benefit from first grouping the firms by industry, randomly sampling from each group, and then weighting the result in proportion to the percentage of employees each industry has of the total employees in the state.

The benefits of stratified sampling arise from the grouping of the population into strata that have similar characteristics, e.g., industry, geographical location, age, city size, etc. If the groups are homogeneous, then a relatively small sample may be sufficient to estimate the characteristic under study with the desired precision. (And if some groups have considerable variability, larger samples can be taken of those groups only.) Then the results from each group can be given their proper weight in the total.

Cluster sampling Cluster sampling is the counterpart of stratified sampling. With cluster sampling, the population is arranged into *heterogeneous* subgroups (or clusters) that have compositions similar to each other. The assumption here is that each "cluster" contains a representative mix, comparable to what exists in the population itself. A random sample is then taken from among the subgroups.

For example, suppose a textbook publisher wished to estimate the amount of money a typical university student spends on textbooks in a year. The sampling frame might be all of several million students at the numerous universities scattered across the country. A random sample of 50 students, for example, might mean contacting (or visiting) 50 different locations. However, the sampling frame (university students) might first be "clustered" into universities, recognizing that the costs to business, engineering, liberal arts, and other students are likely to be similar from one university to another. The publisher might randomly select just five universities as clusters. Then they could go into a second stage of the study by randomly selecting ten students at each of those five universities.

The book cost example would be referred to as a *multistage* random sampling approach. Each stage of the sampling process involves a random selection. However, clusters that are not included in the first stage do not have any chance of being selected in the second stage—so they do not retain "an equal chance of being selected."

The advantage of cluster sampling is that the travel and listing costs can be greatly reduced, e.g., by confining the survey to five universities and listing only those students in the clusters used. While it is often infeasible to group elements into heterogeneous subgroups, cluster sampling can sometimes provide a good basis for inference, and it can be done at a reasonable cost. It is especially advantageous in situations where simple random sampling is economically or physically impossible.

EXERCISES III

10. How does judgment sampling differ from probability sampling?
11. Which type of judgment sample would most likely be used:
    (a) by a newspaper reporter interviewing people fleeing from a flood
    (b) to determine by how much wholesale prices have increased over the past year
    (c) by a legislative assistant who needs opinions from 50 residents of a certain irrigation district in Iowa
12. Distinguish between simple random sampling and systematic sampling.
13. Distinguish between stratified sampling and cluster sampling.
14. A random telephone survey is to be made for a city parks department. Interviewers are instructed how to use a random number to obtain a name from the first page of the phonebook. Then they are to call the person listed in that same location on every fifth page in the book.
    (a) What type of sampling design is this?
    (b) Would this likely yield valid sample results, i.e., from both the sampling and bias error standpoints?

SAMPLING DISTRIBUTIONS

All four of the probability sampling methods described above used some type of random samples. This is essential because it is the random nature of the sampling process (and the associated error) that enables us to make quantifiable inferences about the populations from which samples are drawn. Without knowledge of the degree of sampling error, there would be no statistical inference. This section defines that error and describes how it is measured. It is most convenient to start with a distinction between *infinite* and *finite* populations.

INFINITE VERSUS FINITE POPULATIONS

Infinite populations theoretically contain an infinite number of items, such as the number of measurements that could be taken of radioactivity in the atmosphere. Although no discrete (countable) populations are really infinite, we act as if they are infinite if we cannot enumerate all the samples that could be taken. Thus, we would act as if the number of pine trees one could sample in Canada is infinite, even though there are a limited number of pine trees in the country. For sampling purposes, we also assume that many continuous processes produce infinite populations. For example, we might treat a sample of aspirin tablets (from a production line) as coming from an infinite population, even though the number of tablets produced is technically not infinite.

Finite populations contain a fixed, or limited, number of items, such as the number of savings accounts at a bank, or the number of students in a class. To begin our study of sampling distributions, we shall assume that we are dealing with finite populations that have a countable number of items.

SAMPLING VARIATION

Sampling error arises from the inherent variation in the population. If individual values in a population differ, then random samples from that population will also differ, depending upon what elements happen to be included in the sample. To illustrate sampling variability, consider a common example such as the age of students in a classroom.

Example 6–3 An economics class contains thirty students, and two random samples of $n = 3$ students each are selected to estimate the mean age of the population of students in the class. The sample means are $\bar{x}_1 = 20.0$ and $\bar{x}_2 = 27.0$, as shown below. (Although the population parameter would not normally be known, for purposes of this example assume that data from the registrar's office reveal that the population mean age for this class is $\mu = 21.0$ yrs.) Identify the variability due to sampling.

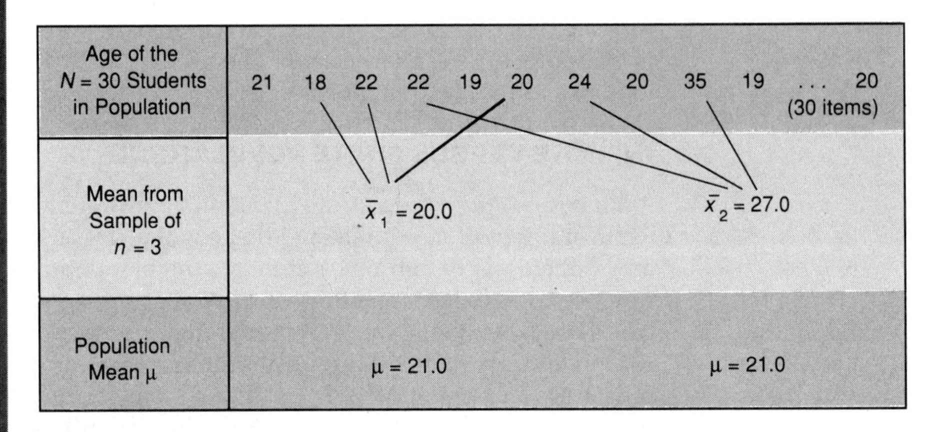

Solution

The variability in the individual samples due to sampling error is:

$$\text{For } \bar{x}_1\text{: variability} = \bar{x}_1 - \mu = 20.0 - 21.0 = -1.0 \text{ yr}$$

$$\text{For } \bar{x}_2\text{: variability} = \bar{x}_2 - \mu = 27.0 - 21.0 = 6.0 \text{ yr}$$

The differences between the sample means and μ are due simply to chance. And, although $\bar{x}_1$ provides a closer estimate of μ than $\bar{x}_2$, both sample means are in error. Furthermore, if we didn't know the value of μ (which would normally be the case), we wouldn't know which estimate was closer to μ.

Insofar as our purpose in using sample statistics, such as $\bar{x}$, is to estimate the value of (unknown) population parameters, we would like to be fairly sure that our statistic is close to the parameter value. Fortunately, *we can control the amount of error (at a given level of probability) by adjusting the size of the sample*. This brings us to consideration of the number and size of samples from a population, and the topic of sampling distributions. (Although we will explain sampling distributions in terms of sample means, the same theory applies to sample proportions. The extension to proportions will follow quite easily.)

SAMPLING DISTRIBUTIONS OF SAMPLE MEANS

NUMBER OF SAMPLE MEANS How many different samples of $n = 3$ students could really be drawn from a class of only 30 students? At first it may seem impossible, but over 4,000 possible sample combinations exist. Recall from Chapter 4 that combinations are the number of different ways that n items can be selected X units at a time, if no duplications are permitted, and the order of selection makes no difference. For the example above:

$$_nC_X = \frac{n!}{X!\,(n - X)!} = \frac{30}{3!(30 - 3)!} = 4{,}060 \text{ samples of } n = 3$$

Of course, not many classes could afford the time to even identify all the 4,060 different samples—much less calculate a mean for each one. (And with this small population, it would certainly make more sense to find the population mean by taking a census of the 30 students!) But a computer could provide such a listing if one had any doubts.

FORMING A SAMPLING DISTRIBUTION Nevertheless, let us suppose for a moment that we *did* take all 4,060 possible samples of $n = 3$ from the population of 30 students. And we proceed to calculate a mean for each sample, just as we calculated the $\bar{x}_1 = 20.0$ and $\bar{x}_2 = 27.0$ in the example above. After we calculated $\bar{x}_{4,060}$ we could form a *frequency distribution of the sample means*, just as we worked with frequency distributions of individual values earlier in the text. The difference here is, of course, that our distribution would be of *sample means* rather than of individual *sample values*. Figure 6–2 is an illustration of the distribution, wherein means are recorded to the nearest whole number.

FIGURE 6–2 Frequency distribution of sample means

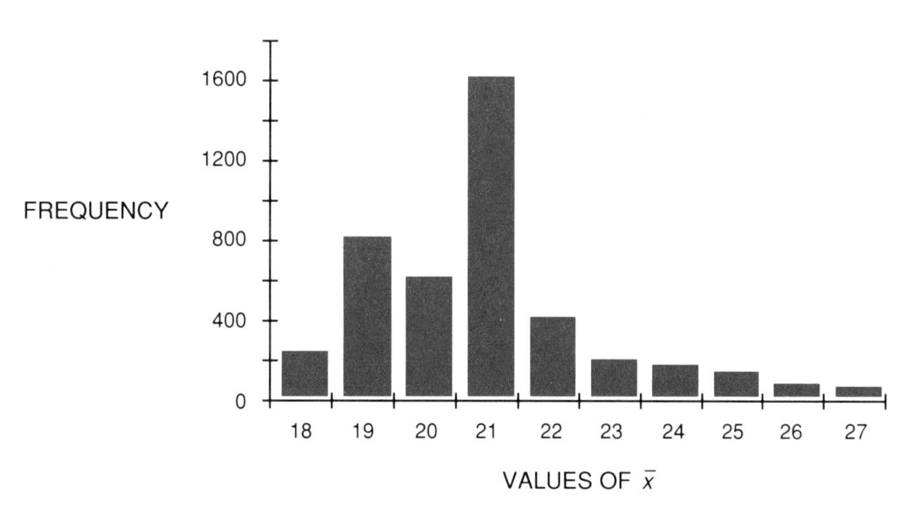

SAMPLING DISTRIBUTION If we expressed the frequencies with which various values of the sample means occurred as relative frequencies (or probabilities), we could refer to the probability distribution of the sample means as a *sampling distribution of the means* (with an emphasis on the *ing*).

> A **Sampling Distribution of the Sample Means** is a probability distribution of the sample mean (or sampl*ing* distribution of the mean). It tells
>
> (1) all possible values of $\bar{x}$ that can occur from samples of a given size, n
>
> (2) the probabilities of each $\bar{x}$

Table 6–2 illustrates the sampling distribution that would result from the sample means of Figure 6–2.

TABLE 6–2 Sampling distribution of ages of students (from samples of $n = 3$)

MEAN AGE $\bar{x}$	PROBABILITY $P(\bar{x})$
18	.05
19	.20
20	.14
21	.40
22	.10
23	.04
24	.03
25	.02
26	.01
27	.01
	1.00

Remember that a sampling distribution applies to one sample size only (in our case for a sample of $n = 3$). Moreover, it is "theoretical" in the sense that we would rarely (if ever) have occasion to bother to take all possible samples of a given size from a population, calculate their means, and compute the respective probabilities of the various means. But the concept of a sampling distribution of the mean will be extremely useful to us later in the text.

MEAN AND STANDARD DEVIATION OF THE SAMPLING DISTRIBUTION The statistics that will be most useful to us are, as before, estimates of the central tendency and dispersion of the distribution. For a distribution of sample means, the mean of all the $\bar{x}$s is simply the summation of all the possible sample means, divided by the number of sample means (4,060 in our illustration above). It is designated as $\mu_{\bar{x}}$ (or sometimes as $\bar{\bar{x}}$). The standard deviation of the sampling distribution is also designated with the subscript $_{\bar{x}}$ to distinguish it from the standard deviation of the population, σ.

Mean of a sampling distribution of $\bar{x}$	$\mu_{\bar{x}}$
Standard deviation of sampling distribution of $\bar{x}$	$\sigma_{\bar{x}}$

SAMPLING DISTRIBUTIONS OF SAMPLE PROPORTIONS

To take full advantage of statistical theory, we should pause at this point to note the similarity in working with proportions to working with means. For this and future discussions, we shall use the symbol π to designate a population proportion and p for the sample proportion.

Continuing with the age illustration, suppose we used the samples to estimate the proportion of the population (π) that was $\geqslant$ 22 yrs old, instead of the mean of the population. The number of possible samples of size of $n = 3$ would be the same (i.e., 4060). But this time we would compute the proportion (p) of each sample that was $\geqslant$ 22 years old. For example, only one of the three values in sample n_1 (of 18, 22, and 20) is $\geqslant$ 22. So the sample proportion from n_1 would be ⅓ or .333.[3]

If all the sample proportions were available, we could form a *sampling distribution of proportions*, where the average, $\bar{p}$, was the summation of all the sample p values divided by 4060. The standard deviation of this sampling distribution of sample proportions is designated σ_p.

Mean of a sampling distribution of p	μ_p or $\bar{p}$
Standard deviation of sampling distribution of p	σ_p

OTHER SAMPLING DISTRIBUTIONS We could also obtain sampling distributions of medians and other statistics, but they are not as widely used as those of means and proportions, so we shall not consider them here. Instead, we move on now to two very powerful theorems associated with the sampling distributions we have introduced.

[3] Insofar as the samples here consist of only 3 students, and either none, one, two, or all three of the students will be $\geqslant$ 22 years old, the only p values in this sampling distribution would be 0%, 33.3%, 66.7%, and 100%. However, if the sample size were increased to 10, the sampling distribution would have (11) classes of 0%, 10%, 20%, etc., up to 100%. The larger samples allow for the sample estimates to be closer to the true π value. This is one reason why small samples are seldom used to estimate population proportions.

SAMPLING DISTRIBUTION THEOREMS

THEOREM 1

Business analysts would rarely, if ever, bother to obtain all the samples necessary to formulate a sampling distribution. Nevertheless, two statistical theorems allow us to draw upon the characteristics of sampling distributions—even if we only have *one* sample. Both theorems assume the sample is obtained by simple random sampling.

> **THEOREM 1** The mean of the sampling distribution of $\bar{x}$ always equals the population mean, μ (i.e., $\mu_{\bar{x}} = \mu$).

Theorem 1 means that the estimate provided by individual sample means will, on average, tend to equal the population mean, μ. This is a consequence of the fact that a sample mean, $\bar{x}$, is an *unbiased estimator* of the population mean.[4] In addition, the sample mean yields values that are closer to μ than are the values from any other statistic—that is, $\bar{x}$ is also an *efficient estimator* of μ.

THEOREM 2: THE CENTRAL LIMIT THEOREM

As suggested in Figure 6–2, the sampling distribution for the very small samples (drawn from the economics class) had no symmetric pattern. Suppose the sample size used to estimate their average age was increased from $n = 3$ to $n = 5$ or $n = 10$. If you were to calculate the number of sample combinations, you would find a dramatic increase in the number of means in the (theoretical) sampling distribution:

SAMPLE SIZE	$n = 3$	$n = 5$	$n = 10$
NO. OF SAMPLES	4,060	142,506	30,045,015

As the sample size increases, the shape of the sampling distribution reveals two effects. *First*, the extreme (high and low) values have less representation (weight) in the average, so the sample $\bar{x}$s will be drawn closer to the population mean, μ. Carried to an extreme, this means that if the sample size were increased to the same size as the population ($N = 30$), then the $\bar{x}$ and μ values could not help but be the same value.

[4] The sample variance, s^2, is also an unbiased estimator of σ^2.

Secondly, as the sample size increases, the $\bar{x}$s tend to form a symmetrical distribution about their mean $\mu_{\bar{x}}$. And for large samples this approaches a normal distribution. Figure 6–3 illustrates this effect, which is formalized in the expression of the *Central Limit Theorem*.

FIGURE 6–3 Sampling distributions approach normality as sample size increases

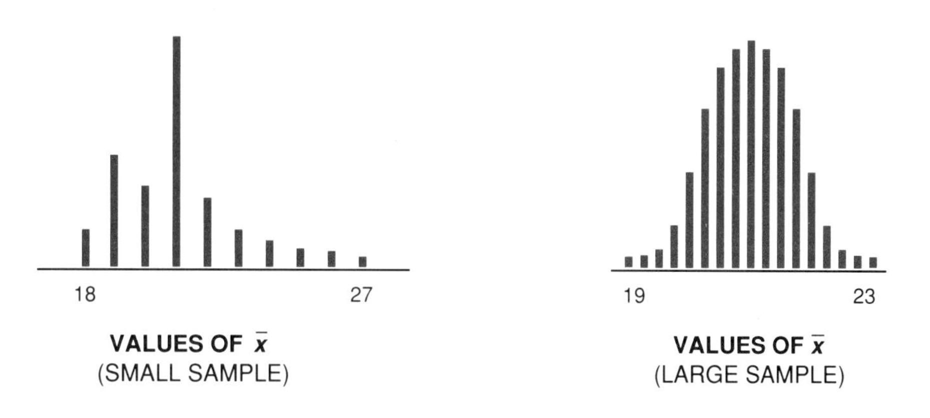

THEOREM 2: Central Limit Theorem The sampling distribution of $\bar{x}$ is approximately normal if the simple random sample size is sufficiently large.

Statisticians consider a sample size of 30 to be "sufficiently large." But even for samples as small as 20, the sampling distribution approaches normality, especially if the population is not highly skewed. In addition, if the population itself is normally distributed, the sampling distribution will always be normal, even for very small sample sizes.

The Central Limit Theorem is probably the most significant theorem in basic statistics because it enables us to use the normal distribution to make inference about population values, regardless of the way the population is distributed. For example, as depicted in Figure 6–4, large samples drawn from rectangular, triangular, skewed, or any other type of distribution, all yield sample means that tend to follow a normal distribution.

FIGURE 6–4 Sample means tend to be normally distributed

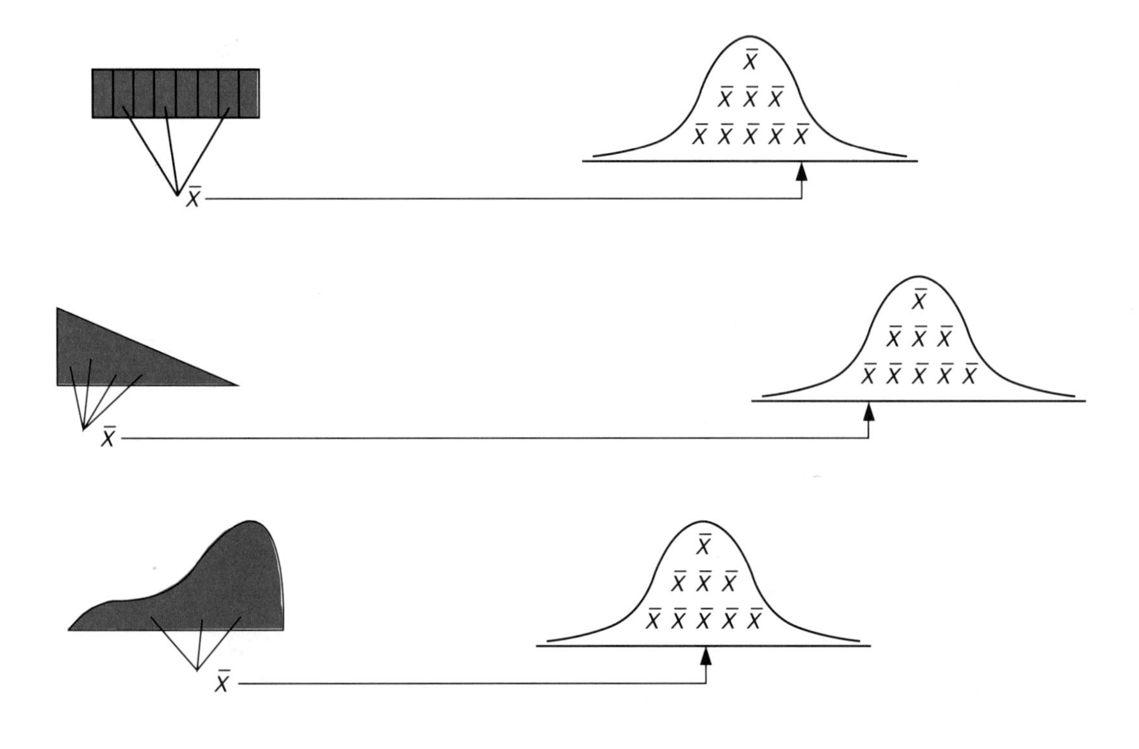

MEAN AND STANDARD ERROR OF SAMPLING DISTRIBUTION

MEAN From Theorem 1, the mean of a sampling distribution equals the *population mean*, μ. However, we will normally be working with *sample means*, $\bar{x}$s, which are unbiased estimators of μ. Moreover, as the sample size (n) increases, the accuracy of the sample means improves as they take on values closer to the population mean.

STANDARD ERROR OF THE MEAN The standard deviation of the distribution of sample means is similar to a population standard deviation, except instead of individual X values, we are working with $\bar{x}$ values. It is thus a measure of the error, or amount by which the sample means differ from the population mean, μ. This standard deviation of the $\bar{x}$s is called the *standard error of the mean*, and is designated $\sigma_{\bar{x}}$. It can be computed by summing the squared deviations of each $\bar{x}$ from μ, and dividing by the number of sample means, N', as illustrated in Equation 6–1.

$$\text{Standard Error of the mean} = \sigma_{\bar{x}} = \sqrt{\frac{\Sigma f(\bar{x} - \mu)^2}{N'}} = \sqrt{\frac{\Sigma f\bar{x}^2 - (\Sigma f\bar{x})^2/N'}{N'}} \qquad (6\text{–}1)$$

The form on the right in Equation 6–1 is easier to compute. However, we will almost always be working with one sample, rather than all the possible samples, N'. And Equation 6–2 is a more commonly used expression for computing $\sigma_{\bar{x}}$, where σ is known, and we have one sample of size n.

$$\sigma_{\bar{x}} = \frac{\sigma}{\sqrt{n}} \qquad (6\text{–}2)$$

The example that follows will demonstrate the equivalence of Equations 6–1 and 6–2. Because the probabilities of remaining sample combinations change as samples are removed from small finite populations, we shall assume we are sampling "with replacement" for now. That is, an element can be chosen more than once—so it can be used in other arrangements. This way we can illustrate the theory that applies to *infinite* populations while still working with a population of manageable size. Later, we will show what adjustment is needed for finite populations.

Example 6–4 (with replacement)

A population of 5 items (with values of 10, 12, 12, 16, and 20) has a mean of $\mu = 14$, and standard deviation of $\sigma = 3.58$. The distribution of sample means for samples of size $n = 2$ is as shown (assuming sampling *with replacement*).[5] Compute the standard error, $\sigma_{\bar{x}}$ by two methods: (a) using Equation 6–1, and (b) using Equation 6–2.

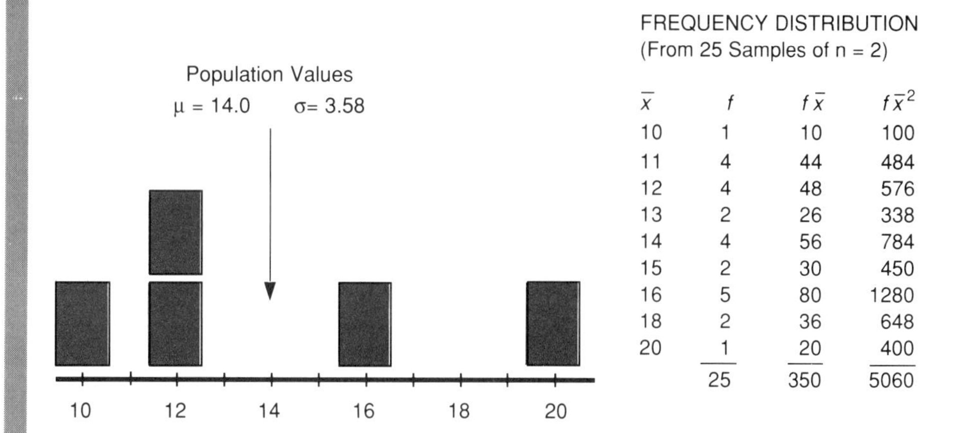

FREQUENCY DISTRIBUTION
(From 25 Samples of n = 2)

$\bar{x}$	f	$f\bar{x}$	$f\bar{x}^2$
10	1	10	100
11	4	44	484
12	4	48	576
13	2	26	338
14	4	56	784
15	2	30	450
16	5	80	1280
18	2	36	648
20	1	20	400
	25	350	5060

[5] Squaring the population standard deviation (of 3.58) yields a variance of 12.8, which could be computed from the equation $\sigma^2 = \Sigma(\bar{x} - \mu)^2/N$. If the sample variance is computed from each of the 25 samples of $n = 2$, using the equation $s^2 = \Sigma(x - \bar{x})^2/(n - 1)$, we would have 25 variance scores ranging from 0 to 50. Their sum is 320, and their average is 320/25 or 12.8, which is exactly equal to the true variance. This illustrates (though does not prove) that the $(n - 1)$ correction results in a **sample variance that is an unbiased estimator** of the population variance when sampling either with replacement, or from an infinite population.

However, if all 25 standard deviations are computed, summed (to 67.880), and divided by 25, their average (2.715) is not equal to the true standard deviation (3.58). In other words, the **sample standard deviation is not an unbiased estimator** of the true standard deviation.

Solution

For sampling with replacement, there are $n^x = 5^2 = 25$ different possible sample choices. We can envision these by assuming we start with a selection of the value 10, replace it, and draw a 10 again. This first sample of two 10s, when divided by 2, yields our first sample mean of 10 (the first row in the table above). Insofar as there are two distinct items with a value of 12, there are four sample choices that yield a mean of 11. These are the pairs (10, 12), (10, 12), (12, 10), and (12, 10). In total there are $N' = 25$ sample means, and we can compute the standard error as:

$$\text{(a)}\quad \sigma_{\bar{x}} = \sqrt{\frac{\Sigma f\bar{x}^2 - (\Sigma f\bar{x})^2/N'}{N'}} = \sqrt{\frac{5060 - (350)^2/25}{25}} = 2.53$$

$$\text{(b)}\quad \sigma_{\bar{x}} = \frac{\sigma}{\sqrt{n}} = \frac{3.58}{\sqrt{2}} = 2.53$$

Equation 6–2 is obviously the easier of the two. An examination of that equation shows that the variation in a distribution of sample means depends upon (1) the variation in the population itself and (2) the size of the sample. If every member of the population had the same value, then σ would equal zero, and the samples would all have the same mean, so $\sigma_{\bar{x}}$ would also equal zero. Moreover, the larger the sample, the closer $\bar{x}$ is to μ, so the smaller the value of $\sigma_{\bar{x}}$.

FINITE POPULATION CORRECTION FACTOR

The commonly used Equation (6–2) for computing the standard error of the mean applies to *populations that are infinite or when samples are taken with replacement*. But the equation does not yield a correct value for $\sigma_{\bar{x}}$ if samples are taken without replacement from finite populations. The error is more significant if the sample is a noticeable portion of the population. For example, to take an extreme case, assume a population of $N = 36$ accounts had a $\sigma = \$48$. If one took a sample of all 36 accounts, the sample mean would equal the population mean and the standard error of the mean should be zero. But Equation 6–2 would give the incorrect result that:

$$\sigma_{\bar{x}} = \frac{\sigma}{\sqrt{n}} = \frac{\$48}{\sqrt{36}} = \$8$$

Whenever the *sampling fraction* (n/N) is 5 percent or more of the population, the effect can be noticeable, and some correction is necessary. However, for sampling fractions of less than 5 percent, the factor is so close to 1.00 that it is usually omitted.

Correction for sampling (without replacement) from a finite population is accomplished by multiplying the standard error (Equation 6–2) by the *finite correction factor* of $\sqrt{(N - n)/(N - 1)}$.

$$\sigma_{\bar{x}} = \frac{\sigma}{\sqrt{n}} \sqrt{\frac{N - n}{N - 1}} \qquad (6\text{–}3)$$

The finite population correction factor is a fraction that reduces the size of the standard error to account for the fact that the computed error is too large when some samples are removed from the population (and not replaced).

Example 6–5 (without replacement)

Samples of $n = 2$ are taken, *without replacement*, from a finite population of $N = 5$ items, where $\sigma = 3.58$. Ten different sample means are possible as shown. Using the definitional form, the true standard error of the mean should be:

$$\sigma_{\bar{x}} = \sqrt{\frac{\Sigma (\bar{x} - \mu)^2}{N'}}$$

$$= \sqrt{\frac{48}{10}} = 2.19$$

Use Equation 6–3, with the finite population correction factor, to compute the value of $\sigma_{\bar{x}}$.

ARRAY OF TEN MEANS

x	(x − μ)	(x − μ)²
11	−3	9
11	−3	9
12	−2	4
13	−1	1
14	0	0
14	0	0
15	1	1
16	2	4
16	2	4
18	4	16
140		48

Solution

$$\sigma_{\bar{x}} = \frac{\sigma}{\sqrt{n}} \sqrt{\frac{N - n}{N - 1}} = \frac{3.58}{\sqrt{2}} \sqrt{\frac{5 - 2}{5 - 1}} = (2.53)(.87) = 2.19$$

Note that the finite population correction factor adequately corrects the standard error from 2.53 down to the correct value of 2.19.

To summarize, we have previously seen that the sample variance, s^2, is an unbiased estimator of σ^2, when sampling either with replacement or from an infinite population. So we can directly calculate the standard error of the mean for infinite populations. But when working with finite populations, the standard error has an oversize bias.[6] We can obtain the same (correct) result as we would get from using the (theoretical) sampling distribution if we multiply the standard error of the mean by the finite population correction factor.

[6] When the population is finite, the variance of the sample, s^2, is not an unbiased estimator of the population variance, σ^2, even when the $(n - 1)$ correction is made. This is apparent from Example 6–5. If the sample variance scores are computed from the 10 possible samples, their sum is 160, and their average is $^{160}/_{10} = 16$. But the true variance of the population is only 12.8. So the variance estimate of a finite population made from the sample is still oversize—even when the degrees of freedom correction is made. Still, s^2 is usually very close to σ^2 if the sample size is large, and is usually a satisfactory estimator of the population variance. But the finite correction factor presented here is advised for calculations of $\sigma_{\bar{x}}$ when the (n/N) ratio is $\geqslant 5\%$.

EXERCISES IV

15. Distinguish between infinite and finite populations. Can discrete populations be infinite? Explain.
16. A random sample of 10 castings revealed an average weight of 37 lbs although the population mean was later found to be 41 lbs. Explain how this could happen.
17. How many different sample *combinations* would be in the sampling distribution if samples of size n=6 are taken from a population of 20?
18. Suppose a valid random sample of $n = 49$ records showed a family of four, in a given income class, paid an average tax of $5,972/year, whereas a survey of IRS records of all such taxpayers revealed the population value was actually $6,172, and $\sigma = \$1400$.
    (a) Explain how this difference might happen.
    (b) What is the standard error of the mean?
    (c) Would you consider the sample a reasonably good estimate?
19. What are the essential ingredients of a sampling distribution of the means?
20. A population contains 3 items with values of 2, 8, and 4. Compute (a) the mean and (b) the standard deviation.
21. Assume that a small population containing only three items has the parameters $\mu = 4.667$ and $\sigma = 2.494$. The values are 4, 2, and 8.
    (a) Formulate the frequency distribution of the means of samples of $n = 2$ (with replacement) that comprise the sampling distribution.
    (b) Compute $\sigma_{\bar{x}}$ using equations (6–1) and (6–2).
22. Formulate a sampling distribution for samples of size $n = 2$ from a population of 3 items if the sample means of 2, 3, 4, 5, 6, and 8 occur with respective frequencies of 1, 2, 1, 2, 2, and 1.
23. Why do $\bar{x}$ values tend to move toward the value of μ as n increases?
24. A population has a mean of $385 and standard deviation of $39. Find the standard error of the mean for samples of (a) $n = 9$, (b) $n = 100$. (c) How would your answer differ if $\mu = \$420$?

INFERENCE WHEN σ IS UNKNOWN: THE *t*-DISTRIBUTION

The previous discussion was concerned with sampling from a population with mean μ and standard deviation σ. In addition to identifying the standard deviation of the sampling distribution as $\sigma_{\bar{x}} = \sigma/\sqrt{n}$, the discussion brought out two important conclusions:

1. If the population is normally distributed, the distribution of sample means is normal for any size sample (large and small).
2. If the population is not normally distributed, the distribution of sample means is still approximately normal if the sample size is sufficiently large (≥ 30).

WHEN σ IS UNKNOWN Although the earlier discussions assumed the population standard deviation was known, most business decisions are based upon sample values where σ is not known, so s is used as an estimator of σ. When s is used in place of σ, the standard deviation of the sampling distribution of $\bar{x}$ is somewhat larger, with a value of $s/\sqrt{n}$. The increased dispersion of the sample means is more apparent with small samples than with larger ones.

Suppose you were to take repeated samples of size n from a normal (μ, σ) distribution, compute s, and divide the difference ($\bar{x} - \mu$) by $s/\sqrt{n}$. The resultant statistic would follow what is referred to as Student's *t*-distribution[7] rather than a normal distribution.

> The ***t*-distribution** is a sampling distribution of continuous, symmetrical, bell-shaped curves which are slightly flatter (more spread out) than the normal curve, but approach the normal curve as the sample size increases.

The ***t*-distribution** thus describes a family of curves, each with a mean of zero and an area of 1.000. Two such curves are illustrated in Figure 6–5.

FIGURE 6–5 Normal distribution and *t*-distribution

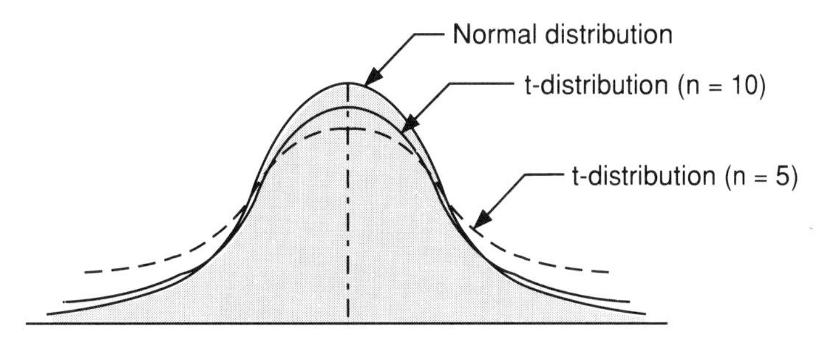

[7] Named for W. S. Gossett, an employee of the Guiness Brewery in Dublin, who signed his articles "Student" because employees were not allowed to publish this type of research under their own names.

LARGE SAMPLES σ UNKNOWN As noted above, the t-distribution is appropriate for describing the distribution of the means of any size sample from a normal population, when σ is unknown. Tables are available for determining the probability areas under the t-distribution for values up to sample sizes of 120 or more. However, for samples of $n \geqslant 30$, the areas under the t-distribution are so close to those of the normal distribution that most analysts simply use the normal (Z) rather than the t-distribution—even though the t is theoretically correct.

SMALL SAMPLES σ UNKNOWN For sample sizes of less than $n = 30$, the difference between the normal and the t-distribution is enough to require that the t be used. In theory, the t is appropriate only if the population being sampled is normal. In practice, this stipulation has more relevance for relatively small samples because the effect of a nonnormal population lessens as the sample size increases to $n = 30$ (and beyond).

t-TABLES Because the t-distribution is a family of curves whose shapes depend on n, it would take many pages to show the area figures for each sample size up to the point where the normal curve could be applied. This is not practical, so most t-tables show only the most commonly used values. Also, unlike the normal curve, most t-tables list t-values for the area in one tail of the distribution (or two tails combined), rather than the area between the mean and a point located some distance from the mean.

The t-table we shall use in this text is in Appendix F, and a portion of it is included here in Table 6–3. It shows the t-value for selected probabilities (or relative areas) in one tail and in both tails of the distribution. Use the table by going down the degrees of freedom (df) column to the appropriate row, and then proceed to the right until you get to the column that has the tail probability you are interested in. (More about df shortly.)

For example, assume degrees of freedom, $df = 4$. To find the t-value corresponding to *5% of the area in one tail* of the distribution, go down the df column to $df = 4$, and right to the .05 column (for one-tail probability) to find the number 2.132. Because the t-distribution is symmetrical, *this t-value for 5% of the area in one tail is also the t-value for 10% of the area divided into two tails* of the distribution.

The t-values are used in calculations in a similar manner to the Z-values of the normal distribution. The similarity is most evident from the bottom row of the t-table, which contains the same Z-values as we have been using for areas under the normal curve. Thus, the 1.960 in the bottom row of the (two-tail) .05 column corresponds to the Z-value that incorporates 95% of the area under the normal curve (i.e., with 5% outside the $\pm$ 1.960 limits).

For condensed notation, we frequently denote the area in the tail (or tails) of the distribution as α (or $\alpha/2$), and the degrees of freedom as df. These symbols, when applied as subscripts to the letter t, completely define the t statistic as $t_{\alpha,\, df}$. For example, assuming the context pertained to a one-tailed probability, the t-table value for $t_{.01,5}$ would be 3.365. To desig-

TABLE 6–3 Student's *t*-distribution

One Tail Probability	.10	.05	.025	.01	.005
Two Tail Probability	.20	.10	.05	.02	.01
d.f.	*Values of t*				
1	3.078	6.314	12.706	31.821	63.657
2	1.886	2.910	4.303	6.965	9.925
3	1.638	2.353	3.182	4.541	5.841
4	1.533	2.132	2.776	3.747	4.604
5	1.476	2.015	2.571	3.365	4.032
.					
.					
.					
30	1.310	1.697	2.042	2.457	2.750
120	1.289	1.658	1.980	2.358	2.617
Normal Distribution	1.282	1.645	1.960	2.326	2.576

nate the *t*-value for a distribution that contained 10% of the area divided equally into two areas outside the limits, we could write $t_{\frac{\alpha}{2}, df}$, or $t_{\frac{.10}{2}, df}$, which for $df = 5$ would have a value of 2.015.

CALCULATING THE *t*-VALUE In the same manner that a *Z*-value can represent the number of standard deviations or standard errors under the normal distribution curve, a *t*-value represents the number of standard errors under one of the *t*-distributions. A calculated *t*-value is thus a way of stating how far an $\bar{x}$ value is away from the mean of the distribution. With the *t*-distribution, however, we shall be using the *estimated* standard error ($s_{\bar{x}}$) as the unit of measure, instead of the population value $\sigma_{\bar{x}}$

t-value for area under sampling distribution
$$t = \frac{\bar{x} - \mu}{s_{\bar{x}}} \quad (6\text{–}4)$$

$$\text{where: } s_{\bar{x}} = \frac{s}{\sqrt{n}} \quad (6\text{–}5)$$

Example 6–6 A random sample of $n = 5$ customers of a large insurance company showed they carried an average of \$15,000 in life insurance, with a standard deviation of \$3,150. Assume the insurance amounts are normally distributed.

(a) Compute the estimated standard error of the mean, $s_{\bar{x}}$.

(b) Suppose the true (population) mean insurance amount carried by policy holders of the company is $\mu = \$18{,}000$. What calculated t-value would be used to determine the probability of getting a sample mean of $\bar{x} = \$15{,}000$ from a random sample of 5?

Solution

(a) $$s_{\bar{x}} = \frac{s}{\sqrt{n}} = \frac{\$3{,}150}{\sqrt{5}} = \$1{,}409.$$

(b) t-value for sampling distribution

$$t = \frac{\bar{x} - \mu}{s_{\bar{x}}} = \frac{\$15{,}000 - \$18{,}000}{\$1409} = -2.130$$

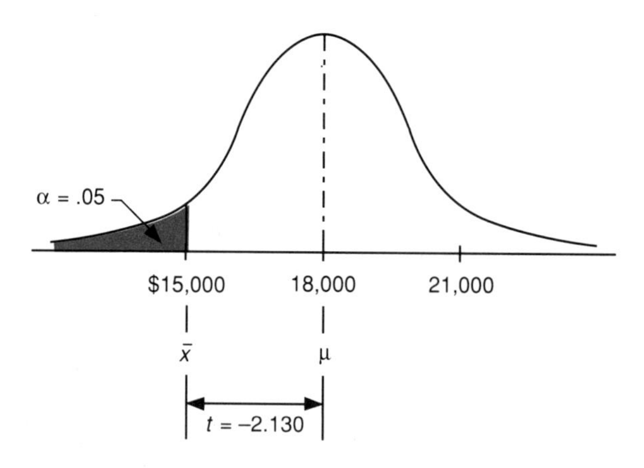

df AND INTERPRETATION Note that the calculated t-value of -2.130 is (negative, but) very close to 2.132, which is the t-value corresponding to 5% of the area in the left tail of the t-distribution for the row labeled 4 *df* The 4 *df* row is appropriate because one degree of freedom is lost when using the sample mean ($\bar{x}$) as an estimator of μ. We will have more to say about degrees of freedom later. For now, however, we can assume the *df* corresponds to the same correction we used in Chapter 3, which was sample size minus one, or $n - 1$.[8]

The calculated t-value of -2.130 can be interpreted as meaning that in a sampling distribution with a mean of \$18,000, and an estimated standard error of \$1,409, about 5% of the sample means would have a value of \$15,000 or less. And, insofar as the t-distribution is symmetrical, we could also conclude that about 5% of the sample means would have values greater than 2.130 standard errors above the mean. (This would correspond to the point \$18,000 + (2.130) (\$1,409), or about \$21,000.) In other words, the t-distribution table with 4 df, tells us that for samples of size $n = 5$, about 10% of the area lies outside the limits of $\pm$ 2.132 standard errors of the mean.

STANDARD ERRORS FOR MEANS AND PROPORTIONS

We have noted previously that s^2 is an unbiased estimator of σ^2 when sampling is from an infinite population, and when the degrees of freedom correction $(n - 1)$ is made in the denominator. However, s^2 is not an unbiased estimator of σ^2 when sampling from a finite population. And a correction factor should be applied if the sample size exceeds 5% of the population. Moreover, when s is used as an estimator of σ, the appropriate sampling distribution for small sample data is the t-distribution. In this last section, we summarize the equations for computing the standard errors of means and proportions when σ is unknown.

DISTRIBUTIONS OF SAMPLE MEANS Most of our theory thus far has been explained in terms of sample means. We have seen that for large samples, the $\bar{x}$s are normally distributed and that small sample means are distributed according to the t-distribution. When σ is unknown, the standard deviation of the sampling distribution of the means is estimated from the sample data as $s_{\bar{x}} = s / \sqrt{n}$. Now it is time to extend this same theory to sample proportions.

DISTRIBUTIONS OF SAMPLE PROPORTIONS Suppose you were asked to do some statistical analysis involving the proportion of customers favoring a new product, or the proportion of computer chips that are likely to be acceptable from a manufacturing operation. You would need to work

[8] The degrees of freedom concept is not intuitively apparent. However, for this example, you can envision it as the number of independent deviations (of $x - \bar{x}$) that are incorporated in the calculation of the sample standard deviation, s. We know the summation of the deviations $\Sigma(x - \bar{x})$ must equal zero. Thus, for $n = 5$ observations, only 4 of the deviations are really independent deviations, because the 5th must cause the sum to equal zero. We characterize this as the loss of one degree of freedom, and attribute it to the fact that in calculating s, the deviations are measured around the sample mean, $\bar{x}$, instead of the true mean, μ.

with a sample proportion rather than a sample mean. And your (one) sample proportion would be only one member of a (theoretical) sampling distribution consisting of all possible sample proportions that could be obtained from samples of the given size, n.

Earlier we designated π as the population proportion and p as any individual sample proportion. If we were to form a sampling distribution of all proportions from samples of size n, the mean of the sampling distribution would be Σp divided by the number of samples, or $\overline{p}$, and the standard deviation of the sampling distribution would be $\sigma_{\overline{p}}$. Paralleling the discussion on means, we can now apply the two theorems described earlier to sample proportions, and observe that:

1. The mean of the distribution of sample proportions, $\overline{p}$, will always equal the population proportion, π. That is, $\overline{p} = \pi$.
2. The sampling distribution of sample proportions, ps, can be approximated by the normal distribution if the sample size is fairly large and the proportion is not too close to either zero or one.

Regarding (1), sample proportions are unbiased estimators of the population proportion, π, so the mean of all the sample proportions (of a given size) will always equal π. Regarding (2), if the sample size is less than 20, the binomial tables (or equation) can be used to compute probabilities for each outcome. But for larger samples, the normal approximation is usually satisfactory. The criteria we used earlier (Chapter 5) would suggest that the sampling distribution of proportions can be approximated by the normal distribution whenever $n > 20$ and $n(p) > 5$ and $n(q) > 5$.

STANDARD ERROR OF PROPORTION

Assume that a sampling distribution of proportions has been formed from large samples of size n taken from a very large (or infinite) population. The standard deviation of that sampling distribution is designated the *standard error of proportion*, σ_p.

$$\sigma_p = \sqrt{\frac{\pi(1 - \pi)}{n}} \qquad (6\text{–}6)$$

The population proportion, π, can be expressed in either a decimal proportion or in a percentage.

Example 6–7 A computer chip manufacturing process has a known yield rate of 60% (i.e., 40% are defective). A random sample of 100 chips are collected for shipment to a customer overseas. Compute the standard error of proportion.

Solution

If we let π = the proportion defective, then $\pi = .40$.

$$\sigma_p = \sqrt{\frac{\pi(1-\pi)}{n}} = \sqrt{\frac{(.40)(.60)}{100}} = .049$$

Note that if the standard error were computed in percent, then $\pi = 40\%$ and $(1 - \pi) = 60\%$, so the standard error would be stated as 4.9% instead of .049.

WORKING WITH SAMPLE VALUES The population proportion, π, is frequently unknown, however, so we must normally work with sample statistics. In this situation, the sample values of p and q replace the population parameters π and $(1 - \pi)$. If we let s_p designate the estimated standard error of proportion as calculated from sample data, we have:

$$s_p = \sqrt{\frac{pq}{n}} \qquad (6\text{–}7)$$

STANDARD ERROR OF NUMBER OF OCCURRENCES

Whenever a proportion of a population (or sample) is multiplied by the number in the population (or in the sample), we have an expression of the number of occurrences, rather than the proportion of occurrences. Thus we can express the standard error of proportion in terms of numbers if we multiply the right side of the equations by the sample size, n. Designating the *standard error of the number* with the subscript np, we can rewrite equations 6–6 and 6–7 as follows.

STANDARD ERROR OF NUMBER

π known: $$\sigma_{np} = \sqrt{n(\pi)(1-\pi)} \qquad (6\text{–}8)$$

π unknown: $$s_{np} = \sqrt{npq} \qquad (6\text{–}9)$$

Example 6–8 In a sample of 400 restaurant customers, 57% said their favorite feature was a self-service salad bar. (a) What is the mean number who favor a salad bar? Compute the standard deviation of the sampling distribution in terms of (b) proportions, and (c) numbers.

Solution

(a) The *mean number* = np = (400)(.57) = 228 customers.

(b) $s_p = \sqrt{\frac{pq}{n}} = \sqrt{\frac{(.57)(.43)}{400}} = .0248$ (or 2.48% of customers)

(c) $s_{np} = \sqrt{npq} = \sqrt{(400)(.57)(.43)} = 9.90$ customers

Table 6–4 summarizes the most commonly used equations for computing the standard errors for means and proportions when working with sample data.

TABLE 6–4 Standard error equations for means and proportions

STANDARD ERROR OF	INFINITE POPULATION (AND WHEN $n/N \leqslant 5\%$)	FINITE POPULATION (WHEN $n/N > 5\%$)
means	$s_{\bar{x}} = \frac{s}{\sqrt{n}}$	$s_{\bar{x}} = \frac{s}{\sqrt{n}} \sqrt{\frac{N-n}{N-1}}$
proportions	$s_p = \sqrt{\frac{pq}{n}}$	$s_p = \sqrt{\frac{pq}{n}} \sqrt{\frac{N-n}{N-1}}$
numbers	$s_{np} = \sqrt{npq}$	$s_{np} = \sqrt{npq} \sqrt{\frac{N-n}{N-1}}$

EXERCISES V

25. When is t the appropriate distribution to use (a) if all theoretical restrictions are adhered to and (b) in most applied situations?
26. Find the following values associated with the t-distribution.
    (a) t-value for 3 degrees of freedom and .01 area in one tail
    (b) t-value for 24 *d.f.* with 20% of the area in the two tails of the distribution
    (c) $t_{.05,\ 14}$ where α is the area in one tail of the distribution
27. For a sample of $n = 16$ drawn from a normal distribution having a mean of 150 and a standard deviation of $\sigma = 20$,
    (a) find the mean $\mu_{\bar{x}}$ of the sampling distribution
    (b) find the standard deviation of the sampling distribution
    (c) What proportion of the sample means will have values of < 150?
28. Determine the appropriate standard error for each of the following:
    (a) a population where $\mu = 180$, $\sigma = 28$, and $n = 100$
    (b) $n = 20$, $s = 3.4$, and $\bar{x} = 42.6$
    (c) a sampling distribution with mean = 300 and standard deviation = 12
    (d) $\pi = .65$, $n = 60$
    (e) $p = 22\%$, $n = 45$
29. Confidential records in a state agency show that the true average retirement income of married persons who have worked for 40 years is $30,000 with a standard deviation of $3,000.
    (a) Compute the standard error of the mean for samples of 25 persons.
    (b) Suppose a random sample of 25 persons conducted by a real estate firm reveals an average retirement income of $28,762. What t-value would this sample mean correspond with?
    (c) If $30,000 is the correct population mean, what is the probability of obtaining a sample mean (of $28,762) this far away from the true mean (i.e., ± $1,238 or more away)?
30. A random sample of 24 customers taken by a city newspaper showed that 18 of the 24 households clipped coupons from their papers at least once a week. Estimate the standard error for the proportion of customers in this market area that clip coupons.
31. A sample of 500 customers at a First National Bank branch (that has 4,000 accounts) revealed that 105 customers had some form of a loan. Estimate the standard error in terms of (a) the proportion of customers and (b) the number of customers who have loans in samples of 500 customers.

SUMMARY

Managers use sample data for cost, time, and other practical reasons. But samples can be subject to *bias* (or displacement) and sampling error. Random selection (e.g., using a random number table) helps eliminate bias. Sampling error is the difference between the *sample statistic* and the *population parameter*; it occurs because of the inherent variability in populations.

Judgment samples (e.g., purpose, quota, and convenience samples) involve a choice in deciding what elements are included. *Probability samples* result from a random selection and (a) include a measure of sampling error, and (b) permit statements of statistical inference. The four major subtypes of probability sampling are (1) simple random, (2) systematic, (3) stratified, and (4) cluster sampling.

A *sampling distribution* is a probability distribution of a sample statistic such as sample means ($\bar{x}$s) or proportions (ps). It includes all possible values of the statistic, along with the probabilities of each. Two important theorems about sampling distributions state that:

1. The mean of a sampling distribution (of $\bar{x}$s or ps) always equals the population value (i.e., μ or π), and
2. The sampling distribution of the statistic ($\bar{x}$ *or* p) is approximately normal, if the simple random sample size is sufficiently large.

The standard deviation of the (theoretical) sampling distribution is called the *standard error of the mean* ($\sigma_{\bar{x}}$) or *of the proportion* (σ_p). However, these standard errors must usually be estimated from sample data as:

Estimated Standard error of mean $$s_{\bar{x}} = \frac{s}{\sqrt{n}}$$

Estimated Standard error of proportion $$s_p = \sqrt{\frac{pq}{n}}$$

If the sampling is done from a finite population where the sample constitutes $\geqslant 5\%$ of the population, then a finite population correction factor should be multiplied by the standard error to correct for a bias.

Theoretically, whenever σ is unknown, the sampling distribution is distributed as a t-distribution, which is like a normal distribution but flatter. There is a different t-distribution for each degree of freedom (which we have computed thus far as sample size minus 1). Unlike the normal distribution, however, most t-tables show the t-values associated with preselected areas in one tail (or two tails) of the distribution. The t-values are essentially the number of standard errors to a given probability area out away from the mean.

CASE: DEBATE AT THE CITY COUNCIL

City Manager Terry Malone was at his wits' end. It was already late February and the budget for the fiscal year (beginning July 1) was still far from balanced. Nearly all of the North Dakota city's departmental managers pleaded for more money for next year, and budget requests now totaled almost $1 million more than the city had available. The Parks and Recreation (P & R) Manager even claimed the citizenry were on his side. As evidence, he cited the random questioning he did of 20 people while on a walk through Riverside Park last summer. He concluded that about 60% of the people favored better park facilities over "anything else" in the budget. "Ask the people. It's well over 50%. Even a bureaucrat would rate that support as convincing," he concluded.

Terry Malone decided to follow the P & R manager's suggestion, and sample public opinion. He contacted the instructor of a marketing research class at a local university, and asked if his class would be interested in a "real world" project that was of vital interest to the city. The response was an enthusiastic yes, but the project would take at least a month to complete. The class would use a simple random sampling procedure to select names from city housing and mail delivery records. Then there would be mail and/or phone contacts.

When the results were in, members of the class presented the report to the City Council at their April 20 meeting. The students had hoped to sample 900 residents, but ended up with 730 usable results from the city of 15,000. Regarding the parks and recreation issue, 292 respondents said they would favor better park facilities over anything else. The P & R Manager acknowledged that the findings were not entirely consistent with his own information, but didn't think it was sufficient evidence to change his thinking. When questioned about whether he felt it was invalid because the sample was too small, he said,

> No, it's just that when you get a bunch of schoolkids doing a sample like this from behind their desks—well, the results could just as well go one way as the other. Another group of 725 people might just likely show 492 in favor instead of 292. You've got to get out and talk to the people that count—that's the only way to get decent results.

The City Manager asked if the P & R manager would be agreeable to combining the results, if he had the class do a follow-up sample with another 350 personal interviews. The P & R manager agreed, and the new results were reported in late May. They showed 154 of the 350 favored better park facilities over anything else in the budget.

At the next council meeting, the P & R Manager noted the improved result in the second sample, and said it confirmed that his estimate of over 50% was "well within the ballpark" of the combined figures.

Case Questions

(a) Comment on the validity of the P & R Manager's original sample.

(b) What is the standard error for the first sample of 730?

(c) What is the standard error, assuming the data can be combined?

(d) Comment on the P & R Manager's comment that his over 50% estimate was "in the ballpark" of the combined results. Given that the combined result is accepted, what is the chance of getting a sample result that shows 50% (or more) of the people in favor?

(e) The proportion favoring parks and recreation money in the second class sample was indeed higher than in the first class sample. Considering the two major sources of error (bias and sampling error), how might you explain this difference?

QUESTIONS AND PROBLEMS

32. What are the most common reasons for sampling?
33. What is (a) a target population, (b) a frame, (c) bias, (d) sampling error?
34. A chemical company has 6550 employees and needs to select five at random for a safety study. Starting with the fourth column of the four-digit random numbers in Table 6–1, select a random sample of five employees. (Assume records contain employee numbers beginning at number one.)
35. What type of sampling is (a) quota sampling, (b) stratified sampling, (c) random sampling? (d) Which of the three permit statistical inference?
36. A state forestry official proposes to obtain an estimate of the number of campers using a recreation site during the summer by randomly selecting a day, counting the campers, and then systematically counting the campers on that same day each week for the entire season. Discuss the validity of such a plan.
37. How does a sampling distribution of the means differ from a sampling distribution of the proportions?
38. Can sampling error be reduced by being more careful in taking the random sample? Explain.
39. Suppose you planned to formulate a sampling distribution for samples of size 6 from $n = 36$ units. How many different combinations would be included?
40. A market research firm took a sample of ten loan company customers to estimate the mean interest rate being paid on outstanding loans.

CUSTOMER NUMBER	1	2	3	4	5	6	7	8	9	10
INTEREST RATE (%)	23	13	17	19	20	21	16	18	17	20

(a) Find the sample mean, $\overline{x}$.

(b) Find the sample standard deviation.

(c) Estimate the standard error of the mean.

41. Assume the mean number of visitors during July at 600 state parks is 240 per day, with a standard deviation of 80. If samples are randomly taken from 100 parks, what is the standard error of the mean?

42. What is meant by the statement that the sample mean is an unbiased estimator of the true mean? What is an unbiased estimator of any statistic?

43. State the *Central Limit Theorem* and explain why it is so important for statistical reasoning.

44. Under what conditions is the sample variance an unbiased estimator of the population variance?

45. The standard deviation of a statistic is frequently called the standard error of that statistic.

    (a) What is the standard error of the mean for samples of 100 drawn from a process which has a mean of 500 and a standard deviation of 40?

    (b) What is the effect on the standard error of the mean if the sample size is increased to 400?

    (c) What are the two factors which affect the size of the standard error of the mean?

46. Regrading the finite population correction factor:

    (a) When should it be used in computing the standard error of the mean?

    (b) Apply it in computing $\sigma_{\overline{x}}$, assuming that samples of 100 are selected from a population of 1,000 elements when $\sigma = 20.0$.

    (c) A sample of 100 oranges is selected from a large box to estimate the mean weight of the oranges. What additional information is required to make the estimate?

47. Use the t-table to find each of the following:

    (a) t-value for 26 df with .05 area (total) in the two tails of the distribution.

    (b) t-value for 26 df with .05 area in the lower tail of the distribution.

    (c) number of df associated with a t-value of 2.76 if an area of .01 is included in each tail of the distribution.

48. A random sample of 1200 families revealed that 768 owned their own home. Compute the estimated standard error of proportion.

49. Explain the relationship between the equations:

$$\sigma_{np} = \sqrt{n\,(\pi)(1 - \pi)} \quad \text{and} \quad \sigma_p = \sqrt{\frac{\pi(1 - \pi)}{n}}$$

CHAPTER 7

Estimation

INTRODUCTION: INTERVAL ESTIMATION
- Point and Interval Estimates
- Assumptions and Conditions
- Logic of Estimation

CONFIDENCE INTERVAL ESTIMATES OF THE MEAN: LARGE SAMPLES, σ KNOWN

CONFIDENCE INTERVAL ESTIMATES OF THE MEAN: SMALL SAMPLES, σ UNKNOWN

FINDING THE SAMPLE SIZE FOR ESTIMATING THE MEAN
- Error
- Sample Size: σ Known
- Sample Size: σ Unknown
- Adjustment for Finite Population: $n/N \geq 5\%$

CONFIDENCE INTERVAL ESTIMATES OF PROPORTION
- Small Samples
- Large Samples
- Samples from Finite Populations

FINDING THE SAMPLE SIZE FOR ESTIMATING THE PROPORTION

SUMMARY

CASE: NORTHWESTERN GRAIN COOPERATIVE

QUESTIONS AND PROBLEMS

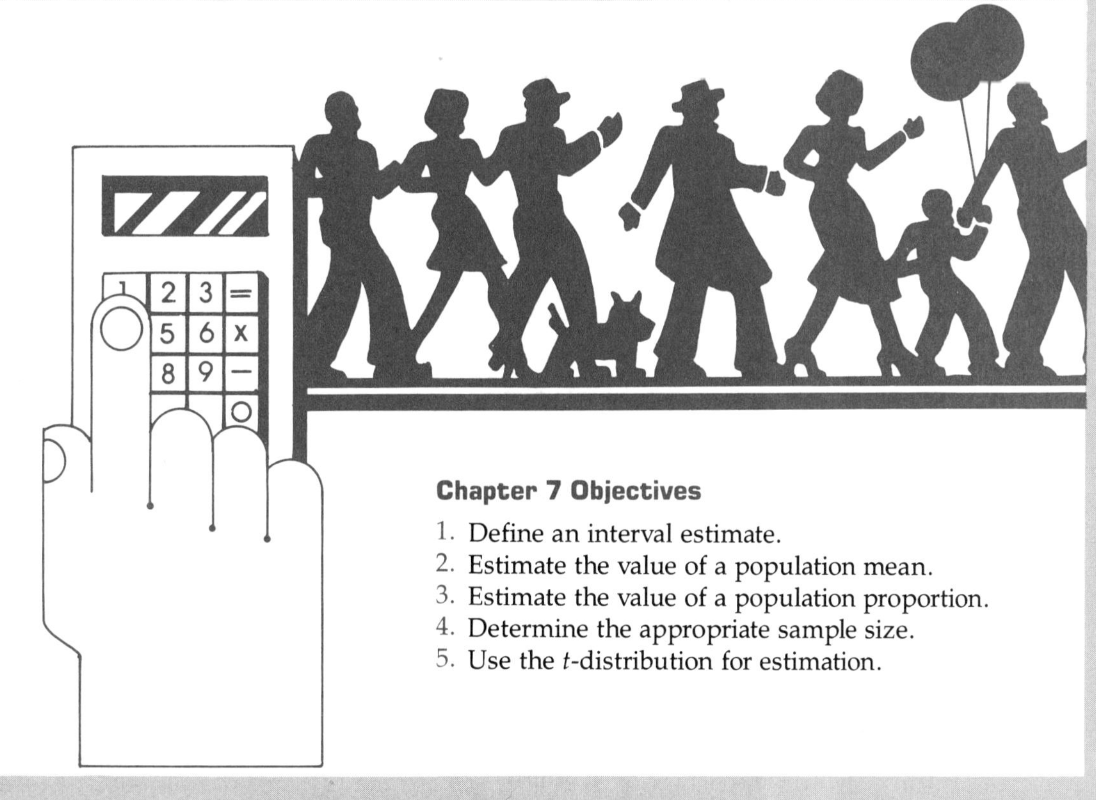

Chapter 7 Objectives

1. Define an interval estimate.
2. Estimate the value of a population mean.
3. Estimate the value of a population proportion.
4. Determine the appropriate sample size.
5. Use the t-distribution for estimation.

Physically counting the more than 250 million Americans is a tremendous task. And the census, which is taken every ten years, is an expensive headcount—the last census cost approximately a billion dollars, and was believed to be about 98% accurate. A statistical sample having an accuracy of 95% could probably have been done for about one thousandth of that cost (which is still $1 million!).

Most businesses can't afford the luxury of census data. For example, suppose you wished to open a bakery, and wanted to specialize in either doughnuts, cookies, or cakes and pastries. Which is likely to be the most successful today? You might take a survey of other bakeries to find what is successful for them. But there are 35,000 bakeries in the U.S.—so a survey would be infeasible. You would probably make your estimate from a sample.[1]

Samples give managers a wide range of information—from where consumers shop for their products, what wages are being paid by competitive firms, and how employees feel about their medical and dental plans, to whether safety regulations are being followed, the accuracy level of accounts receivable, and the quality level of the products being produced. Firms rely heavily on sample information. That is why statistical estimation is such a central feature to courses in business statistics.

[1] The Retail Bakers of America estimates that about 41% of bakeries are doughnut shops, followed by cookie shops at 18%, and cake and pastry shops at 11%. (*The Wall Street Journal*, April 30, 1986, p. 29).

INTRODUCTION: INTERVAL ESTIMATION

POINT AND INTERVAL ESTIMATES

Point estimates are single values of sample statistics used to estimate population parameters. Thus far, we have worked primarily with the sample statistics $\bar{x}$ and p, which are point estimators of the parameters μ and π:

	Population Parameter	Sample Statistic	Standard Error of Statistic
Mean:	μ	$\bar{x}$	$s_{\bar{x}}$
Proportion:	π	p	s_p

In addition, we have used the sample standard deviation, s, as a point estimator of the population value, σ.

We have also noted that point estimates are most useful if they are *consistent* (get closer to the value being estimated as the sample size increases), are *efficient* (have a smaller standard error than other estimators of the same parameter), and are *unbiased* (will, on average, equal the parameter value). Unfortunately, however, point estimates are rarely on target. Worse yet, a point estimate does not tell us anything about how close the estimate is to the population value. Decision makers really need to know (1) the limits within which the parameter is believed to lie, and (2) some measure of how likely those limits are to be correct.

Interval estimates are pairs of values (computed from sample statistics) within which the population parameter is expected to lie. Moreover, confidence interval estimates also convey a measure of likelihood, or probability, that the limits contain the unknown population parameter. And the interval estimates can apply to either population means or proportions. For example, confidence interval statements might read as follows:

Means: We are 95% confident that the true mean wage (μ) being paid by competitors in our state lies within the interval \$14.60 to 16.30 per hour.

Proportions: We are 98% confident that the true proportion (π) of employees favoring the new medical plan lies within the interval 58.5% to 62.8%.

Both confidence interval statements use the standard error (of the means or of the proportions) as a means of quantifying the uncertainty associated with the estimate. Let us review this underlying logic—which stems from the central limit theorem.

ASSUMPTIONS AND CONDITIONS

The theory underlying confidence interval estimation is not difficult, and applying it is really quite a straightforward process. For purposes of explanation, we shall assume that we are estimating a population mean, μ, from a large random sample of n observations, where the population standard deviation, σ, is known. Once the theory is understood, it is a short step to extend the reasoning to (a) population proportions, (b) smaller samples, and (c) situations where σ is unknown.

Before embarking on the explanation, we should note some points that may appear obvious, but are worth reiterating:

1. We are *estimating the population mean*—not the sample mean.
2. We are *using sample data*—an interval estimate is not needed if census data is available.
3. Each estimate should state the *limits* within which the parameter is expected to lie, along with a measure of the *amount of confidence* in the estimate.
4. The *width of the confidence limits* depends upon the amount of dispersion in the sampling distribution of the means (i.e., standard error of the means, $\sigma_{\bar{x}}$.)

LOGIC OF ESTIMATION

Recall that every sample (of a given size) taken from a population is really a member of the (theoretical) sampling distribution of that statistic. Also, from the central limit theorem, we know that the sampling distribution of sample means is approximately normal, if the sample size is sufficiently large. Suppose we let the normal curve shown below represent the true (but unknown) sampling distribution. Insofar as the sample means are normally distributed about the mean of the sample means $\mu_{\bar{x}}$, this means that:

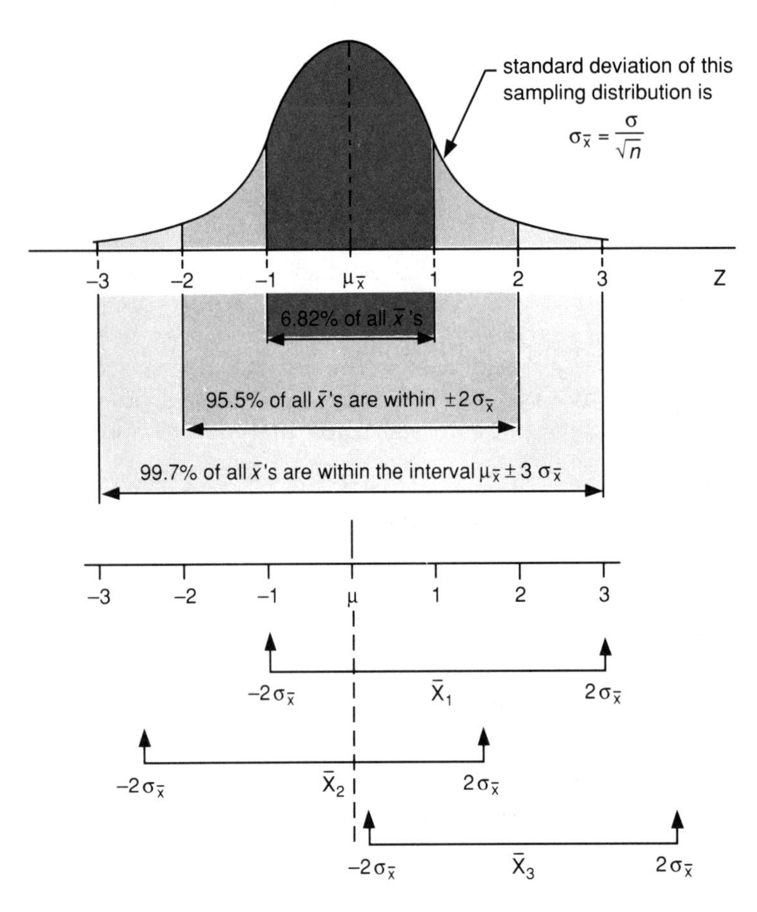

% of all $\bar{x}$s	**distance from $\mu_{\bar{x}}$**
68.2%	within $\pm$ 1 $\sigma_{\bar{x}}$
95.5%	within $\pm$ 2 $\sigma_{\bar{x}}$
99.7%	within $\pm$ 3 $\sigma_{\bar{x}}$

Conversely, 32.8% of all $\bar{x}$s are outside $\pm$ 1 $\sigma_{\bar{x}}$, 4.5% are outside $\pm$ 2 $\sigma_{\bar{x}}$, and .3% are outside $\pm$ 3 $\sigma_{\bar{x}}$.

Assuming we take only one sample, that sample mean $\bar{x}_1$, is one of many (perhaps thousands) of $\bar{x}$s that could be obtained. We would not know if our one sample mean is very close to μ or not, but we would know that 95.5% of all the sample means are within $\pm$ 2 $\sigma_{\bar{x}}$ of μ. So if we established an interval of $\pm$ 2 $\sigma_{\bar{x}}$ about any individual sample mean, and said that interval included μ, we would be correct 95.5% of the time. In a similar manner, if we establish an interval of $\pm$ 3 $\sigma_{\bar{x}}$, we would be correct in claiming that interval included the population mean 99.7% of the time.

Thus we can express the confidence interval for large samples from populations where σ is known as:

Confidence Limits: (means: σ known) $$CL = \bar{x} \pm Z\,\sigma_{\bar{x}} \qquad (7\text{–}1)$$

where Z represents the "confidence coefficient," or number of standard errors needed to encompass a specified proportion of area under the normal curve. The lower and upper confidence limits are frequently designated as *LCL* and *UCL* respectively.

INFERENCE BASED ON PROCEDURE Our expression of confidence that the stated interval includes the true mean is not based upon some special insight we have, or some unique knowledge we get from the one sample. Instead, it is *based upon the procedure we follow* in making the estimate. In repeated samplings, if one follows the correct procedure of setting confidence limits at $\pm$ 2 $\sigma_{\bar{x}}$ from the sample mean, those limits will indeed encompass the population parameter 95.5% of the time. The theory is as simple as that.

CONFIDENCE INTERVAL ESTIMATES OF THE MEAN: LARGE SAMPLES, σ KNOWN

Our first example will illustrate the most "ideal" situation, where we are using a large sample to estimate μ and σ is known. Then we will see what modifications are needed for smaller samples, and for an unknown σ.

Example 7–1
($n \geq 30$ and σ known)

A California winery wishes to estimate the true weight of over a thousand boxes of grapes which are being received from growers for processing. A sample of 49 boxes revealed a mean weight of 197.0 lbs, and from past seasons, the population standard deviation is known to be $\sigma = 14.0$ lbs. Compute a confidence interval estimate of the true mean weight per box (μ) that gives the winery 95% confidence of being correct.

Solution

$$CL = \bar{x} \pm Z\,\sigma_{\bar{x}}$$

where: Z value for .95 area in two tails (or .4750, in one tail) $= 1.96$

$$\sigma_{\bar{x}} = \frac{\sigma}{\sqrt{n}} = \frac{14.0}{\sqrt{49}} = 2.0$$

$$\therefore \quad LCL = \bar{x} - Z\,\sigma_{\bar{x}} = 197.0 - 1.96\,(2.0) = 193.1 \text{ lbs}$$

$$UCL = \bar{x} + Z\,\sigma_{\bar{x}} = 197.0 + 1.96\,(2.0) = 200.9 \text{ lbs}$$

Conclusion: The winery can be 95% confident that the true mean weight of the boxes of grapes lies within the interval 193.1 to 200.9 lbs per box.

CONFIDENCE VERSUS PROBABILITY In estimation, we use only one sample of the many combinations possible. Any one of numerous confidence intervals *could* emerge, depending upon what elements happen to be included in the sample. But only one combination is "lucky enough to be chosen"! In this sense, it is the interval itself that occurs by chance—and with a given probability. So the confidence interval is the random variable.

Once a particular confidence interval is selected, it either *does* or *does not* include the parameter μ. For example, the true mean weight of the boxes of grapes in the illustration above is either within the range of 193.1 to 200.9 lbs, or it isn't. It does not change or "fall" into some interval in a probabilistic sense. The population parameter, μ, *is not a random variable*. That is why statisticians refer to the chance that an interval includes the parameter as a *level of confidence*, rather than a measure of probability. Probability refers to a random variable (i.e., the interval), but parameters are not random variables.

CONFIDENCE INTERVAL ESTIMATES OF THE MEAN: SMALL SAMPLES, σ UNKNOWN

In the previous chapter, we saw that when σ is unknown, we use s as an estimator of σ, and the (sampling) distribution of sample means is properly described by the t-distribution. If the sample size is small, there may be quite a difference between the t-distribution and the normal curve, and the t-distribution is used for setting confidence limits. [The t-values should reflect areas in two tails of the distribution with $(n - 1)$ df]

$$CL = \bar{x} \pm ts_{\bar{x}} \tag{7–2}$$

Example 7–2
($n < 30$ and σ unknown)

Midwest Farm Cooperative has randomly sampled 25 farmers in a tri-state area to estimate the weight used (lbs/acre) of a certain chemical weed killer. The sample mean was 120 lbs and standard deviation was 10 lbs. Establish a 98% confidence interval estimate for the true mean, μ.

Solution

$$CL = \bar{x} \pm ts_{\bar{x}}$$

where: t-value is for $t_{\alpha/2,\ df} = t_{.02/2,\ 24} = 2.492$

$$s_{\bar{x}} = \frac{s}{\sqrt{n}} = \frac{10}{\sqrt{25}} = 2 \text{ lbs}$$

$$CL = 120 \pm (2.492)(2)$$

$$\therefore \quad LCL = 120 - 5 = 115 \text{ lbs}$$
$$UCL = 120 + 5 = 125 \text{ lbs}$$

Conclusion: The Farm Cooperative can be 98% confident that the true mean weight of weed killer applied in the area is within the interval of 115 lbs to 125 lbs.

In Example 7–2, we do not know for certain whether the interval 115 to 125 lbs includes the true mean, μ, or not. But we have followed a procedure that will give us an interval that encompasses μ approximately 98% of the time.

Recall from Chapter 6 that if the sample size is large, the t-distribution values are so close to the normal curve values that the normal curve may be used instead of the t-distribution. The next example has a sufficiently large sample to use the normal distribution ($n = 400$). In addition, the sample is more than 5% of the population size, so we must apply the finite population correction factor to the calculation of the standard error.

Example 7–3
(σ unknown, but $n > 30$, and $n/N > 5\%$)

A stockbroker has taken a random sample of 400 of her company's 4,000 accounts to find the average number of mutual fund shares held by her firm's clients. The sample mean was $\bar{x} = 900$ and $s = 380$ shares. What is the 95% confidence interval for μ?

Solution

$$CL = \bar{x} \pm t\, s_{\bar{x}}$$

where: t-value for $t_{.05/2,\ 399}$ is equivalent to $t_{.025,\ \infty}$

$\therefore\ t_{.025,\ \infty} = Z_{.95} = 1.96$ (use normal)

and: $n/N = 400/4{,}000 = .10$, so finite correction factor is required

$$s_{\bar{x}} = \frac{s}{\sqrt{n}}\sqrt{\frac{N-n}{N-1}} = \frac{380}{\sqrt{400}}\sqrt{\frac{4{,}000-400}{4{,}000-1}}$$

$$= 19.00(.95) = 18.03$$

$$CL = \bar{x} \pm Z\, s_{\bar{x}}$$
$$CL = 900 \pm (1.960)(18.03)$$

$\therefore\ LCL = 900 - 35.33 = 865$ shares (rounded)
$UCL = 900 + 35.33 = 935$ shares (rounded)

Conclusion: The broker can be 95% confident that the true mean number of mutual fund shares held by her firm's clients lies within the interval 865 to 935 shares.

By saying the true mean is within the interval of 865 to 935 shares, we mean that if many intervals were established in this way and statements of this type made many times, we could expect to be correct about 95% of the time.

In this example, we began with the t-distribution because σ was unknown. But the large sample size permitted us to move to use of the normal (Z). In subsequent problems, we will simply skip this first (illustrative) step and go directly to use of the normal when $n \geq 30$.

EXERCISES I

1. Why are point estimates insufficient for many business decisions?
2. What is an interval estimate?
3. What is wrong with the following interval estimates?
   (a) "I am quite sure the sample mean lies within the interval 54 to 60 hours."
   (b) "There is a 95% probability the population value will fall within the range 345 to 365 pounds."
4. Explain the statistical logic of estimation. That is, how can we make a valid estimate about a population mean when we are using only *one* of possibly thousands of different sample means (of size n)?
5. Identify all the differences involved in computing a confidence interval for μ when σ is known, versus computing one when σ is not known?
6. For a sample of 225, find the following confidence limits:
   (a) 95.5% limits when $\bar{x} = 160$ and $\sigma = 30$
   (b) 99.7% limits when $\bar{x} = 540$ and $\sigma = 45$
7. Use the t-distribution to establish the 90% confidence limits for a sample of 20 when $\bar{x} = 4.500$ inches and $s = .200$ inches.
8. Five dozen bags of chocolate candy are randomly selected from the output of a packaging machine to estimate the average package weight. The population standard deviation is known to be 15.0 grams. If the sample mean is 907.0 grams, compute a 98% confidence interval estimate of the population mean, μ.
9. A toy manufacturer has produced 150 battery operated robots. A random test of 18 of the robots revealed they ran (without recharge) an average of 14.60 hrs, with $s = 1.20$ hrs. Make a 95% confidence interval estimate of the mean number of hours the 150 robots will function without recharge.

FINDING THE SAMPLE SIZE FOR ESTIMATING THE MEAN

Up to this point, we have assumed that the sample size was given. But in a realistic business situation, one of the first questions we must face is how large a sample to take. If the population is fairly homogeneous, a small sample may suffice. (In fact, we would need only one element to determine the mean of a population if it were perfectly uniform.) However, for most business studies, larger samples are needed to give the desired precision. Unfortunately, larger samples increase the cost of the estimate. So the key to efficient estimation is to take a sample that is large enough to yield the desired precision, but not so large as to incur any unnecessary cost.

ERROR

The precision of an estimate refers to the closeness or amount of error in the estimate of the true value. More specifically, *estimation error* is the difference between the sample statistic and the population parameter. In the distribution of sample means, this difference is measured in units of standard error. Recalling that the standard error is $\sigma/\sqrt{n}$, it is perhaps obvious that as n increases, the standard error decreases (and the precision increases).

Figure 7–1 uses the confidence interval logic to depict the estimation error. It shows the error interval around one sample mean (which, for convenience, we will assume is located at μ). In terms of a confidence interval, the error, e, is equal to one half of the interval width. The width is the product of the number of standard errors needed for the specified level of confidence, Z, times the amount of each standard error ($\sigma_{\bar{x}}$).

Maximum Error $$e = Z\frac{\sigma}{\sqrt{n}} \tag{7–3}$$

FIGURE 7–1 Error in a confidence interval estimate

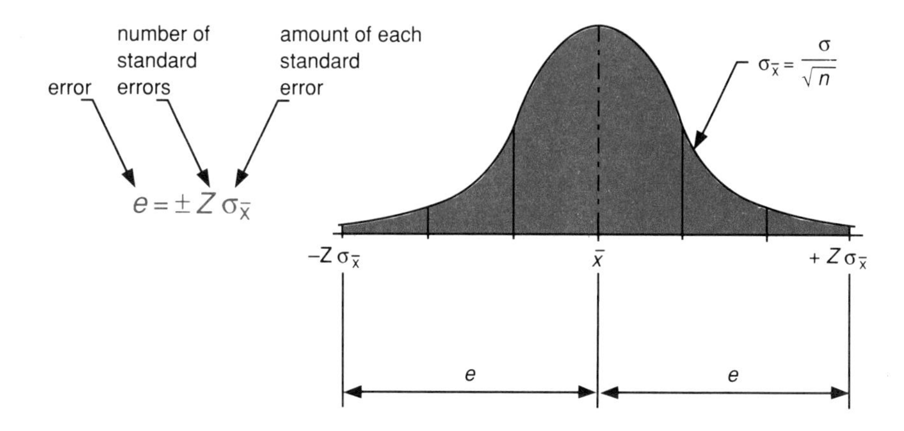

SAMPLE SIZE: σ KNOWN

The standard error in Figure 7–1 is, of course, dependent upon the population standard deviation and the sample size. If we restate $\sigma_{\bar{x}}$ in the estimating error equation as $\sigma/\sqrt{n}$, and solve for n, we have an expression for the sample size required to provide a specified level of confidence.

Sample size (means: σ known)

$$n = \left[\frac{Z\sigma}{e}\right]^2 \tag{7–4}$$

The sample size equation essentially dictates the "spread" of the sampling distribution. That is, if one specifies the number of error units for a level of confidence (Z-value), along with a measure of the variability in the population (σ) and the maximum allowable error (e), the sampling distribution will be *forced* into a sample size that will satisfy all the specified conditions.

Example 7–4 An economist at Atlantic National Bank needs to develop a 95% confidence interval estimate of the mean annual family expenditure on housing in Maryland, within an accuracy of plus or minus \$100. The population standard deviation is known to be \$2,000. How large a sample should be taken?

Solution

$$n = \left[\frac{Z\sigma}{e}\right]^2 = \left[\frac{(1.96)(\$2{,}000)}{\$100}\right]^2 = 1537$$

SAMPLE SIZE: σ UNKNOWN

The "puzzling" aspect of computing a sample size is that equation 7–4 suggests that one must know σ *before* the sample is taken. In practice, researchers have found that an estimate of σ is a satisfactory starting point. This estimate might come from dividing the estimated range of the data by six (a very rough approximation) or perhaps from a previous study (a better source). A preliminary sample may yield an even better estimate of σ, and the sample size can be recalculated (and adjusted) as data become available. When we must use sample values, and perhaps recalculate the sample size to be sure it is sufficient, the σ in equation 7–4 is replaced by s.

Sample size (means: σ unknown)

$$n = \left[\frac{Zs}{e}\right]^2 \tag{7–5}$$

As might be expected, the more confident one chooses to be, the larger the Z-value and the larger the sample must be. If the decision maker is satisfied by being correct 95 times in 100, a Z-value of 1.96 will suffice. But if the loss from an incorrect conclusion would be very severe, a 99% level ($Z = 2.58$) or 99.7% level ($Z = 3.00$) may be desirable. On the other hand, if one is too cautious, sampling costs may be excessively high, or the manager may fail to take action when it is justified.

ADJUSTMENT FOR FINITE POPULATION: $n/N \geq 5\%$

Insofar as sampling distributions of means tend toward normality, finite populations generally pose no special problem, so long as the sample size is larger than 30. However, if the population is finite and the sample is likely to constitute 5% or more of the population, the finite population correction factor may be applied. Its effect is to reduce the sample size from what might otherwise be required.

Letting the error, e, again equal half the interval width, we have:

$$e = Z\left[\frac{s}{\sqrt{n}}\sqrt{\frac{N-n}{N-1}}\right]$$

This expression can then be transformed into an equation for n. Recognizing that the -1 has a negligible effect on the solution, we can drop it to obtain an approximate value for n of:

$$n = \frac{s^2}{\dfrac{e^2}{Z^2} + \dfrac{s^2}{N}} \tag{7–6}$$

Example 7–5

A study is planned to determine the average expenditure for travel by 900 customers of a Boston travel agency. There will be no serious consequences if the estimate is incorrect, so a 90% confidence interval is decided upon. A preliminary sample of twenty customers indicated a mean of approximately \$800 and a standard deviation of \$250. If the sample mean is to be within about 5% of the true mean, how large a sample should be taken?

Solution

This problem introduces two ramifications to the basic sample size expression given in equation 7–4. *First*, the population is finite, and a 5% sample would only be 45 customers. So we may want to consider using the finite population adjustment provided in equation 7–6. *Second*, the precision is stated relative to the mean, instead of as a specific dollar amount. However, for purposes of estimating the sample size, we can use 5% of the sample mean value of \$800 as our maximum allowable error amount. Thus we will let $e = (.05)(\$800) = \$40$. Also, for 90% confidence, $Z = 1.645$. Then:

$$n = \frac{s^2}{\dfrac{e^2}{Z^2} + \dfrac{s^2}{N}} = \frac{(\$250)^2}{\dfrac{(40)^2}{(1.645)^2} + \dfrac{(\$250)^2}{900}} = 95$$

The question of sample size is taken up again, as needed, in relation to estimation of proportions and hypothesis testing in later sections of the text. However, the basic approach is similar to that presented in equation 7–4.

EXERCISES II

10. (a) With respect to estimation, explain what is meant by the "estimation error" associated with a sample mean? (b) In what units is estimation error measured?
11. What Z-value would be used to compute the sample size needed to have: (a) 98% confidence in the results, (b) 92% confidence in the results?
12. For an infinite population, how does the sample size equation differ when σ is known versus when σ is unknown?
13. Compute the sample size required for the following:
    (a) 95% confidence that the interval is accurate within $\pm$ \$40, $\sigma =$ \$280
    (b) 90% confidence, $e = 3$ minutes, $s = 18$ minutes
14. A fruit packing company manager wishes to take a sample of a packing line in order to be 95.5% confident in estimating the true mean weight per box, with an accuracy of plus or minus .15 lb. Previous data reveal that the standard deviation of weights is .60 lb. How large a sample should be taken?
15. The medical director of Mercy Hospital has asked a staff analyst to develop a 98% confidence interval estimate of the time required for a certain type of heart operation. She would like to be accurate within $\pm$ 6 min. Only 420 such operations have been performed nationwide. For the 8 operations performed in the region of Mercy Hospital, $\bar{x} =$ 3.20 hrs and s = .30 hr. Getting the data on all the different operations would require too much correspondence, so the analyst plans to take a random sample. How many operations should be included in the sample?

CONFIDENCE INTERVAL ESTIMATES OF PROPORTION

SMALL SAMPLES

Suppose you wished to estimate the proportion of defective electronic chips that are being produced by a constant, automatic process. In this case (a) there are only two categories—defective or good, (b) the probabilities are constant from trial to trial, and (c) each trial is independent of the previous trial. The binomial distribution would be (theoretically) correct for establishing the confidence interval estimate of the population proportion.

This is because the sample proportions are distributed according to the binomial distribution. And for samples of less than 20, we could use tables to look up the binomial probabilities.

Estimates of the true value of π are usually not made from small samples, however, because proportions derived from small samples are not very precise. For example, if a sample of five is used to estimate a true proportion of defectives, the "sample space" would be no defectives in five, one in five, etc., on up to all five defective. Thus the only values of sample proportions that could result are .00, .20, .40, .60, .80, or 1.00. If, for example, the true value of π were .32, the sample estimate could not be very close because the possible proportions from a sample of $n = 5$ are so limited. As a result, small sampling procedures (and the t-distribution) are not used when estimating sample proportions.

LARGE SAMPLES

Large samples are the more common situation. As we saw in Chapter 6, if the sample size is fairly large, and the proportion is not too close to either zero or one, we can use the normal curve approximation to the binomial. The sampling distribution of proportions is well approximated by a normal distribution whenever $n > 20$ and $n(p) > 5$ and $n(q) > 5$. We also noted earlier that for infinite populations such as a process, the standard error of proportion can be expressed in two ways: (1) as a proportion, σ_p, and (2) as a number, σ_{np}.

CONFIDENCE INTERVAL FOR PROPORTION The procedure for computing a confidence interval for the true proportion, π, is similar to what we used for constructing a confidence interval for the true mean. That is, the confidence limits are "established" around the sample statistic, in this case p. The letter Z identifies the confidence level, and σ_p is the true (and constant) value of the standard error of proportion.

Confidence Limits: (proportions: π known) $$CL = p \pm Z\,\sigma_p$$

Note, however, that in order to calculate the standard error (either σ_p or σ_{np}), we must first know the value of π, the same parameter we are trying to estimate! As a result, the equation above would not likely be used. Instead, the confidence limits must be calculated using the sample standard error, s_p.

Confidence Limits: (proportions: π unknown) $$CL = p \pm Z\,s_p \qquad (7\text{–}7)$$

WHY INTERVAL IS ONLY APPROXIMATELY CORRECT Unfortunately the confidence interval constructed about a sample proportion is not as theoretically correct as the confidence interval about a sample mean. The reason again relates to the calculation of the standard error.

$s_{\bar{x}}$ AND s_p The standard error of the mean, $s_{\bar{x}} = s/\sqrt{n}$, is not dependent upon the value of the true mean. That is, $s_{\bar{x}}$ does not vary as our sample means come in with different estimates of μ. But s_p must be calculated from the sample proportion, and it varies with each different sample proportion that is used to calculate it. For this reason, some advanced texts use quadratic equations or special charts that yield wider confidence limits to allow for this extra uncertainty. However, neither of these techniques is totally satisfactory. Even the charts can only contain confidence intervals for a limited number of sample sizes. Moreover, insofar as the amount of error is not usually large, the approximation obtained by using the sample standard error of proportions is adequate for most business decisions. But remember that confidence interval estimates of proportions are not technically as accurate as confidence interval estimates of means.

Example 7–6 A random sample of 250 employees of a large multinational firm showed that 50 were in favor of a proposed employee stock plan. Develop a 99% confidence interval estimate of the true proportion who are in favor of the plan.

Solution

The sample proportion in favor is $p = 50 / 250 = .20$

$$CL = p \pm Z\, s_p \quad \textit{where: } Z = 2.58$$

$$s_p = \sqrt{\frac{pq}{n}} = \sqrt{\frac{(.20)(.80)}{250}} = .025$$

$$\begin{aligned} \therefore \quad CL &= .20 \pm (2.58)(.025) \\ LCL &= .20 - .065 = .135 \\ UCL &= .20 + .065 = .265 \end{aligned}$$

Conclusion: The managers of the multinational firm can be 99% confident that the true proportion of employees of the firm who are in favor of the employee stock plan lies within the interval 13.5% to 26.5%.

SAMPLES FROM FINITE POPULATIONS

When the proportion in the sample constitutes 5% or more of the population, the finite population correction factor should be applied to correct the standard error, just as it is with sample means.

Example 7–7 Irish Exports Inc. has accumulated 900 hand knit sweaters for shipment overseas, but some of them must be sold as "seconds." A buyer is interested in making an offer for them if she can first make a sample inspection. Suppose her random sample of 150 sweaters reveals that 36 would be graded as seconds. (a) What is the 95% confidence interval for the proportion of seconds in the shipment? (b) How many seconds are in the lot? (c) How could the buyer get a "closer" estimate?

Solution

(a) The sample proportion of seconds is $p = 36/150 = .24$.

The sample size relative to the population is $n / N = 150/900 = .167$, so the finite correction factor should be applied:

$$CL = p \pm Z\, s_p$$

$$\textit{where: } Z = 1.96$$

$$s_p = \sqrt{\frac{pq}{n}}\sqrt{\frac{N-n}{N-1}} = \sqrt{\frac{(.24)(.76)}{150}}\sqrt{\frac{900-150}{900-1}}$$

$$= (.0349)\,(.9134) = .032$$

$$\therefore \quad CL = .24 \pm (1.96)(.032)$$
$$LCL = .24 - .062 = .178$$
$$UCL = .24 + .062 = .302$$

Conclusion: The buyer can be 95% confident that the proportion of seconds in the lot lies within the interval 17.8% to 30.2%.

(b) We can convert these proportions to numbers by multiplying by the number in the lot: (.178)(900) = 160 and (.302)(900) = 272.
Thus the buyer could have 95% confidence that the number of seconds in the lot is some value between 160 and 272 sweaters.

(c) The buyer could get a closer estimate (narrower interval) by changing either the Z value or the n value (since the p value is determined). So the confidence level could be dropped (e.g., to 90%) or the sample size increased.

FINDING THE SAMPLE SIZE FOR ESTIMATING THE POPULATION PROPORTION

The procedure for finding the sample size required to estimate a population proportion, π, is similar to that used for estimating a population mean. As with means, the maximum allowable error of a proportion, e, is half the confidence interval width. However, when working with proportions, the sample statistic, p, and allowable error, e, are stated as decimals (or percentages). Setting e equal to the confidence coefficient Z, times the standard error σ_p, and solving for n, we would have:

$$n = \frac{Z^2 (\pi)(1 - \pi)}{e^2}$$

We are again, however, faced with the same predicament of having to know the population value of π, in order to compute the size of the sample—which is to be used to estimate the value of π! In practice, we would either use a (conservative) estimate for π, or perhaps take a preliminary sample to estimate which p value to use. When sample data is used, the sample size equation for estimating a population proportion becomes:

$$n = \frac{Z^2 \, p \, q}{e^2} \tag{7–8}$$

Example 7–8 A random sample of county residents is to be taken to estimate the proportion in favor of a business and occupations (B & O) tax to raise funds for a new coliseum. The estimate is to be accurate to within ± two percentage points. How large a sample is needed to establish a 95% confidence interval? (The county commissioners estimate that 70% of the residents favor the tax.)

Solution

$$n = \frac{Z^2pq}{e^2} \quad \text{where: } Z^2 = (1.96)^2$$

$$p = .70 \therefore q = 1 - .70 = .30$$

$$e^2 = (.02)^2$$

$$n = \frac{(1.96)^2\,(.70)(.30)}{(.02)^2} = 2{,}017 \text{ residents}$$

NO PRELIMINARY ESTIMATE AVAILABLE If the researcher has no idea of the true value of π or no estimate of π from a sample, he or she can ensure that the sample will be sufficiently large by letting $p = .50$. This is because the $p\,q$ combination of $(.5)(.5) = .25$ is larger than any other $p\,q$ multiple. That is, $(.4)(.6) = .24$, and $(.3)(.7) = .21$, and so forth. By using

.5 for p (and thus .5 for q also) the sample will be large enough to yield the desired precision—or possibly more precision than needed. Of course, larger samples cost firms more time and money. To minimize the overall cost, some analysts intentionally begin with a small sample and recompute the sample size needed as p values become available from the initial sample data.

EXERCISES III

16. Why are small sampling procedures not used for estimating proportions?
17. Is a confidence interval estimate of a proportion likely to be as accurate as a confidence interval estimate of a mean? Explain.
18. For a sample of $n = 300$, find the 99.7% confidence intervals ($Z = 3.00$) if the sample proportions are: (a) .30 (b) .50.
19. Compute the 92% confidence intervals for the following:
    (a) $n = 220, p = .15$ (b) $n = 400, p = .15$
20. For a sample of 100, construct a 95.5% confidence interval estimate if the sample proportion is (a) .10, (b) .40. (c) What conclusion can you draw about the precision of an estimate of π by comparing the width of the confidence intervals in (a) and (b)?
21. The personnel manager of a Houston defense contractor with 1720 employees has taken a random sample of 125 employees in an attempt to determine the political affiliation of the firm's employees. If 35 of the those questioned claimed to be independents, what is the 90% confidence interval estimate for the proportion of independents in the firm?
22. Find the sample size required for 95% confidence that the population proportion is within $\pm$ .04, if (a) $p = .20$, (b) $p = .45$.
23. The marketing manager of a major grocery chain store wishes to obtain an estimate of the proportion of shoppers who shop at discount warehouses (as opposed to conventional stores and supermarkets). He thinks the proportion is about 12%. How large a sample should be taken if he wishes to have 95.5% confidence that the results are accurate within two percentage points?
24. An informal survey by enthusiasts of bicycling suggests that 70% of all the new participants in the sport are women. A bicycle manufacturer has asked you to take a random sample to make a more formal, 90% confidence interval estimate that is accurate to within $\pm$ 3% of the correct value. (a) Which sampling distribution applies (means or proportions)? (b) What Z value should be used? (c) How much error is allowed (that is, what e value applies)? (d) How large a sample of new participants is required?

SUMMARY

Point estimates (single values) of population parameters are useful, but they are typically wrong because of sampling error. *Interval estimates* (pairs of values) are more useful because they state a range within which the parameter is expected to lie. *Confidence interval estimates* also quantify the likelihood (level of confidence) that the parameter lies within the stated interval.

We are able to make a confidence statement not because we know the interval does or does not include the parameter. Our confidence lies in the *procedure* we follow, which will result in correct statements the proportion of time specified by the confidence coefficient. The theoretical basis for estimation rests upon the central limit theorem, which assures us that the sampling distributions of means (and proportions) are normal, if the sample size is sufficiently large.

The confidence intervals we studied in this chapter consist of the sample statistic ($\bar{x}$ or p) plus and minus a specified number (Z) of standard errors ($\sigma_{\bar{x}}$ or σ_p). Because population values are unknown, we normally use sample standard errors to compute confidence intervals. The t-distribution is theoretically correct whenever σ is unknown, and is used for confidence intervals for means when $n < 30$. However, the t-distribution is not used for proportions, and the normal distribution is satisfactory for $n \geq 30$. So the most widely used expressions for confidence intervals are:

For Means: $CL = \bar{x} \pm Z\, s_{\bar{x}}$

For Proportions: $CL = p \pm Z\, s_p$

For finite populations, if the sample is $\geq 5\%$ of the size of the population, the standard error of the mean should be corrected by multiplying it by the finite population correction factor:

$$\sqrt{\frac{N - n}{N - 1}}$$

The sample size required to make a confidence interval estimate is the n value that results from setting the maximum allowable error, e, equal to half the confidence interval width, which is equal to Z times the standard error. For the usual situations (σ and π unknown), the sample sizes are:

For Means

$$n = \left[\frac{Zs}{e}\right]^2$$

For Proportions

$$n = \frac{Z^2 p\, q}{e^2}$$

Both equations require some prior estimate (either of s or of p), which may come from an estimate, historical data, or from a preliminary sample. Using $p = .50$ will, of course, result in a sufficiently large sample for any estimate of proportions.

CASE: NORTHWESTERN GRAIN COOPERATIVE

Bill Springer had never seen any wheat raised within the city limits of Little Rock, Arkansas, where he grew up. But his job search had landed him in Colfax, Washington—the heart of the Northwest wheat country.

Bill's job title was Administrative Analyst for a farm cooperative, and one of his first assignments was to make a recommendation whether or not to contract for additional grain elevator storage space. The situation was like this:

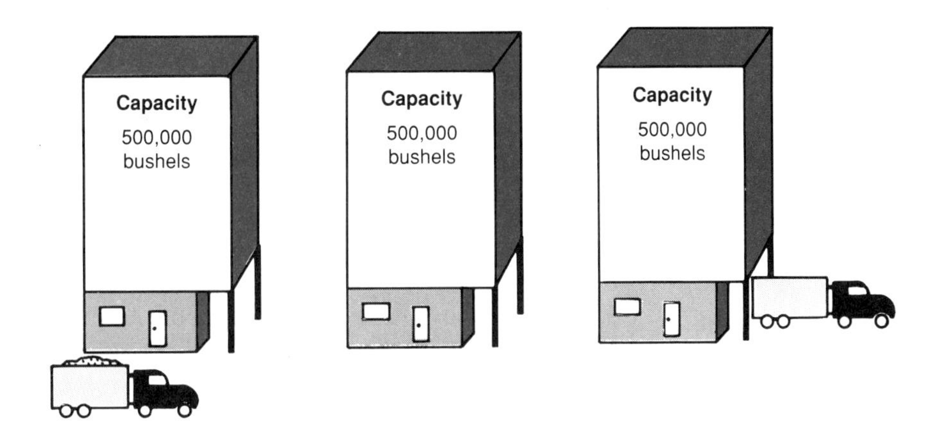

The cooperative owned three elevators, each capable of receiving 500,000 bushels of wheat. These elevators served as storage bins for a surrounding wheatland of 25,000 acres. But the yield from those acres was uncertain. Some years it was low, and the cooperative had excess capacity. If it was high, they could rent additional storage space—but there was a catch. The extra space cost $10 per ton of wheat, if arranged for within the first two days of the wheat harvest, but $40 per ton if done any time later.

That made the situation somewhat more complicated. Bill knew by now that a bushel of wheat weighed 60 lbs, and that there were 2,000 lbs in a ton. So he was able to figure out that the comparative costs/ton for storage were equivalent to $.30/bushel (if arranged early) and $1.20/bushel (if later). But the big question was—would the storage be needed, and if so, how much?

Inasmuch as harvest was about ready to start, the business manager of the cooperative suggested that Bill get out of the office and "do some sampling" so he could find out whether he would need additional space or not. The word *sampling* triggered Bill's thinking to realize that he had an estimation problem. A few more questions of the business manager and some review of data from past years revealed to Bill that:

(1) wheat had a value of about $4.50 per bushel;

(2) the average yield per acre varied from year to year;

(3) the standard deviation of yield was constant at 5 bushels per acre;

(4) with a difference of $1.20 − .30 = $.90/bushel, Bill needed to be quite confident in his results; 98% confidence would be required.

Bill used the available data to determine that a sample of 34 randomly selected acres would give him the confidence level he needed. By using the cooperative's data files, along with some county maps, a random number table, and some phone calls to obliging farmers, he was able to arrange for his samples.

The average yield of his sample was 57.8 bushels/acre. He felt that was a rather good yield. But upon working out his calculations, he concluded that no additional storage space would be needed. When he went in to give his report to the business manager, the manager's only comment was: "Well, all I know is we ran out of space last year. And it cost us $90,000 more than it should have, if we had been on our toes. I sure hope you're right!"

Case Questions

(a) Show calculations to support the $.30/bushel cost for early arrangement of storage space.

(b) What would be the maximum cost of an error of 5 bushels/acre (that is, assuming all the extra wheat had to be stored at the extra cost of $.90/bushel).

(c) What amount of error in yield per acre did Bill assume in his sample size calculation and what would be the maximum cost of this error, using the same $.90/bushel as in (b) above?

(d) Do you feel Bill was correct in recommending against contracting for more storage space?

QUESTIONS AND PROBLEMS

25. Explain the difference between a point estimate and an interval estimate of a population mean. Why do we never make an interval estimate of the sample mean?

26. A mining firm in Nevada delivers ore to a concentrating plant in containers that are guaranteed to contain an average of 820 lbs of ore.
    (a) If a sample of 400 containers revealed a mean of 808 lbs and a standard deviation of $s = 32$ lbs, what is the 95% confidence interval for the true mean?
    (b) Because the true standard deviation is not known, both the sample mean and sample standard deviation are random variables. Unless the population is normal, sample means are only approximately normally distributed. What distribution should be used to establish confidence intervals when σ is not known?
    (c) Establish a 95% confidence interval estimate for the true mean if a sample of 25 has an $\bar{x} = 808$ and $s = 32$.
    (d) Why is the interval for part (c) wider than that for part (a)?

27. A trainee in an advertising firm in Atlanta has been asked to find the average monthly expenditure by secretaries in a particular city for computer-related supplies.
    (a) What four things would the trainee have to decide or determine before she can decide the sample size required?
    (b) Assume that the answer required in (a) should be within $2 of the correct answer, the number of secretaries in the city is 15,000, the standard deviation based on a previous study is believed to be about $20, and the desired confidence level (that the interval will really bracket the true mean) is to be .95. What sample size should be taken?
    (c) Why can you ignore the fact that the population was finite in computing the sample size in part (b)?
    (d) What size sample is required if there are only 1,000 secretaries in the city? Why is n smaller than in part (b)?

28. A sample is to be taken for a 99% confidence interval estimate of the true mean of a process where the half-width of the confidence interval is to be within 1 percent of the mean—believed to be about 200.
    (a) How large should the sample be if the standard deviation is believed to be about 12.0?
    (b) If the sample of the size indicated in (a) is taken and the sample standard deviation is only 10.0, what is the effect on the solution obtained?
    (c) What is the result of overestimating the sample size required?

29. Sample proportions for a Bernoulli process are approximately normally distributed around the true proportion if the sample is large and neither π nor $(1 - \pi)$ is too small.
    (a) Make a point estimate of the true mean proportion defective if a random sample of 400 from a production process contains 24 defective items?
    (b) Use the data given in part (a) to make a 95 percent confidence interval estimate of the true proportion defective.
    (c) Explain why the procedure used in part (b) is only an approximation.
    (d) Why is the *t*-distribution seldom used in establishing confidence intervals for the true proportion?
30. Assume that a random sample of 400 items from a production process contains 24 defective items.
    (a) Make a point estimate of the number of defective items which will be found in lots of 400.
    (b) Compute the standard error of the number of defectives in a sample of 400.
    (c) Make a 95 percent confidence interval estimate of the number of defective units in lots of 400.
    (d) Explain the difference between the answers found in problems 29 and 30.
31. A student working on a take-home test for marketing comes to you with a problem. He wants to determine how large a sample he should take to make a 95 percent confidence interval estimate of the true proportion of the population who use a particular product. The answer is to be within ± 2 percentage points.
    (a) What is the maximum sample size required?
    (b) If a preliminary sample of 100 indicates that the true proportion is only about .20, revise the estimate of the sample size required.
    (c) Why can you set a maximum sample size in this problem when you cannot set a maximum sample size for estimates of μ unless you have an estimate of σ?
32. Suppose the retail firms listed in the Corporate Statistical Database (Appendix A) represent a preliminary sample taken for the purpose of preparing to estimate the proportion of retail firms (*R*) in the U.S. that have profit-sharing plans. If an accuracy of ± 3% is desired, how many retail firms should be sampled? Assume we wish to have 98% confidence in the results.

33. Using the Corporate Statistical Database (Appendix A), assume the first 36 firms (listed alphabetically) represent a random sample. Make a 90% confidence interval estimate of the current year earnings per share.
34. Assume the Corporate Statistical Database (Appendix A) is a random of 100 firms listed on recognized stock exchanges. Use the data to make a 99% confidence interval estimate of the average dividend paid by firms listed on recognized stock exchanges.

CHAPTER 8

Testing Hypotheses: One-Sample Tests

INTRODUCTION
- Purpose of Hypothesis Testing
- Method of Testing
- Types of Tests
- Logic of Testing

NULL AND ALTERNATE HYPOTHESES

TYPES OF ERRORS
- Type I and Type II Errors
- How Errors Can Occur

LEVEL OF SIGNIFICANCE

TWO-TAILED VERSUS ONE-TAILED TESTS

A STANDARDIZED TESTING PROCEDURE

TESTING THE POPULATION MEAN: σ KNOWN

TESTING THE POPULATION MEAN: σ UNKNOWN
- Large Samples
- Small Samples

TESTING THE POPULATION PROPORTION
- Large and Small Samples
- Determining the Significance *p* Value
- Computer Programs for Testing Hypotheses

COMPUTING THE TYPE II ERROR

POWER CURVES AND OPERATING CHARACTERISTIC CURVES
- Power Curve
- Operating Characteristic Curve

SUMMARY

CASE: INTERNATIONAL PRINTING AND PUBLISHING COMPANY

QUESTIONS AND PROBLEMS

Chapter 8 Objectives

1. Explain the logic of hypothesis testing.
2. Formulate null and alternative hypotheses.
3. Test hypotheses about a population mean.
4. Test hypotheses about a population proportion.
5. Compute the type II error of a test.

What is the prime responsibility of managers and administrators? They are decision makers. They are responsible for making the decisions that will best lead their employees toward the organization's goals. To make those decisions, managers frequently need to collect data, analyze them, and draw conclusions. Needless to say, their conclusions are not always correct. But by following the procedures discussed in this chapter and the next, they can ensure that they have a very good (and well-defined) chance of being correct.

INTRODUCTION

This first section of this chapter discusses the purpose, types, and logic of testing hypotheses. After that we will delineate the procedure for conducting a test for certain types of problems. When you finish the chapter, you should be able to formulate a hypothesis, select some data from a population, analyze them, and draw a conclusion about the statistical validity of the hypothesis.

PURPOSE OF HYPOTHESIS TESTING

A *hypothesis* is a statement, or claim, about a population parameter. For example, a manufacturer may advertise that its car batteries last an average of 40 months. Or a pharmaceutical producer may claim that lecithin significantly improves a person's memory. Hypothesis testing is a systematic technique of using random samples to test such claims.

Hypothesis testing is a statistical procedure for using sample data to support or reject a hypothesis about a population parameter.

The *purpose of hypothesis testing* is to help decide whether the hypothesis should be rejected or not. It does not eliminate the possibility of error, but it does make use of proven statistical theory that enables the decision maker to measure and control the amount of error associated with the decision. In other words, hypothesis testing replaces what might otherwise be an intuitive decision making process with an objective procedure that includes a probabilistic measure of the risk of making an incorrect decision.

METHOD OF TESTING

The procedure for testing a hypothesis is relatively standardized and consists of the following steps:

1. State the hypothesis and the criteria to be used to judge it.
2. Collect some sample data and compare them to the hypothesized value.
3. Determine whether the difference between the two is significant.
4. Draw a conclusion about the hypothesis (i.e., reject or do not reject it).

TYPES OF TESTS

At first glance, mastering the subject of hypothesis testing can seem like an overwhelming task; statistics texts are full of different types of tests of hypotheses. However, we can simplify the task considerably by recognizing some major distinctions that dictate what type of test is appropriate for a given situation.

First, it is most helpful to categorize hypothesis tests into *one-sample* tests (which we discuss in this chapter) and *two-sample* tests (discussed in the next chapter). Beyond that—and within each chapter—we shall distinguish clearly between (a) tests of population means, μ, and (b) tests of population proportions, π. (And within each of these categories we can have either one-tailed or two-tailed tests—but we will get to that in due course.) For now, the major types of tests of hypotheses we shall be concerned with are

One-Sample Tests (Chapter 8): Does the sample evidence support the (hypothesized) population value?
(a) Tests using one sample *mean*, $\bar{x}$.
(b) Tests using one sample *proportion*, p.

Two-Sample Tests (Chapter 9): Do the two samples reveal a significant difference between the (two hypothesized) population values?
(a) Tests using two sample means, $\bar{x}_1$ and $\bar{x}_2$.
(b) Tests using two sample proportions, p_1 and p_2.

Hypothesis tests of the differences among three or more population means and three or more population proportions are discussed in later chapters. The chapter on chi square analysis also discusses tests of differences between observed and theoretical *frequencies* (as distinct from means or proportions). And later chapters contain other tests of data relationships. Tests of hypotheses are indeed a major component of any study of statistics.

LOGIC OF TESTING

The logic underlying tests of hypotheses rests upon two concepts we have studied and used earlier: sampling distributions and the central limit theorem. Let us briefly summarize the theoretical basis for tests in terms of (a) means and (b) proportions. (Assume for now that we are working with samples of $n \geq 30$.)

MEANS Suppose a nylon rope maker claims his product has a mean breaking strength of 800 lbs. The logical way to test that claim (hypothesized value) would be to take some pieces of rope from his production output, measure their breaking strengths, and compare the sample result with the claimed mean of 800 lbs. This is exactly what takes place with a test of an hypothesis. But we go beyond a casual recognition of the result by taking advantage of our knowledge of the statistical distribution to which the sample statistic belongs.

Figure 8–1 illustrates how our knowledge of sampling distributions and the Central Limit Theorem enables us to extract more information about the population parameter than might otherwise be expected. First, recall that random samples of a given size form a sampling distribution with mean, μ, and standard deviation, $\sigma_{\bar{x}}$. Figure 8–1 uses $\bar{x}$s to depict the density function of the distribution. The Central Limit Theorem tells us that *if the hypothesis is true*, those sample means will be approximately normally distributed about the population mean. And the mean and standard deviation of that distribution of sample means will equal μ and $\sigma / \sqrt{n}$, respectively.

FIGURE 8–1 Distribution of sample means for hypothesis $\mu = 800$ lbs.

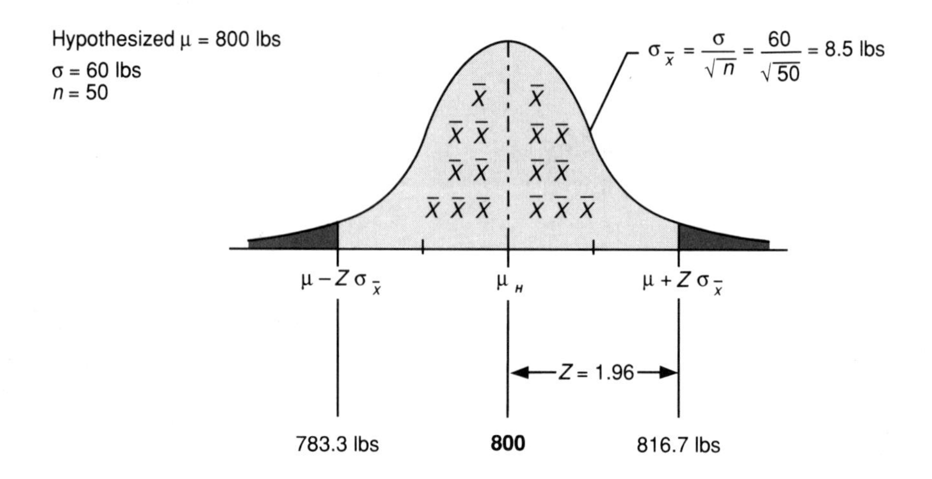

If the sample mean we happen to get is close to the hypothesized value (e.g., $\bar{x} = 795$ lbs), then it is reasonable to conclude that the difference between the hypothesized mean (800 lbs), and sample mean (795 lbs), is probably due to chance. However, if $\bar{x}$ is a great distance from the hypothetical value (for example, if it is 650 lbs), then we can feel quite safe in concluding that the true mean of the population is not 800 lbs. Our assurance rests upon the knowledge that if the hypothesis is true, then 95.5% of the sample means will be within $\pm\, 2\,\sigma_{\bar{x}}$, and 99.7 within $\pm 3\,\sigma_{\bar{x}}$. So if $\sigma_{\bar{x}} = 8.5$ lbs, as in Figure 8–1, we could expect 95% of all the sample means to be within $\pm\, 1.96\,\sigma_{\bar{x}}$, or within the range of 783.3 lbs to 816.9 lbs. A sample value of 650 lbs would be *highly* unlikely.

PROPORTIONS The same reasoning applies to tests of proportions, except the random variable is discrete, rather than continuous. And although the binomial distribution is correct for describing sample proportions, in order to have sufficient precision most tests of hypotheses necessitate that large samples be used. So for most tests of proportions, the distribution of sample proportions is based upon the normal approximation to the binomial distribution.

For an example involving proportions, suppose a computer software firm wished to know what percent of personal computer owners in its market area use database software (as distinguished from word processing, games, spreadsheets, and graphics). Assume management hypothesizes that 20% of computer owners use database software, and they plan to sample 400 computer owners to see if the sample data do or do not support the hypothesis. The sampling distribution of proportions for samples of size $n = 400$ is also normal (like the distribution of means). As illustrated in Figure 8–2, for a hypothesized proportion of $\pi = .20$, the standard error of proportion is .02. Thus we would expect 95% of all sample proportions to be within $\pm$ 1.96 σ_p, or .04 of the hypothesized value of .20, if the hypothesis is true. So a sample proportion of $p = .45$, for example, would suggest the hypothesis was false.

FIGURE 8–2 Distribution of sample proportions for hypothesis $\pi = .20$

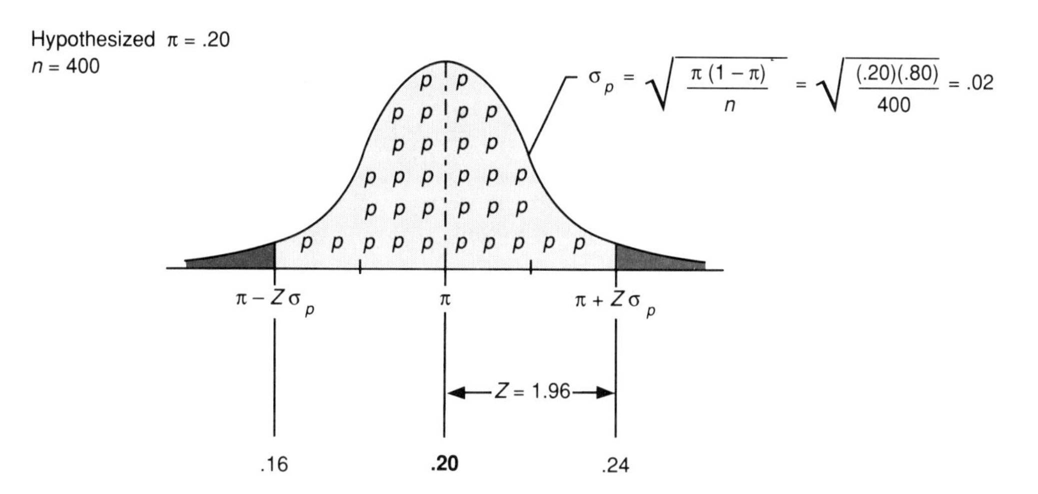

SUMMARY OF TEST LOGIC To recap the logic of a test, we first establish the hypothesis, or claim. Then we collect sample data and compare them to what might be expected in a sampling distribution centered upon the hypothesized value. This (distribution) is the criterion by which we judge the validity of the hypothesis. If the difference between the sample statistic and parameter of the hypothesized distribution is small, we conclude it is probably just due to sampling error, and *accept* the hypothesis. If large, the difference is probably real, so we *reject* the hypothesis.

We refer to a large difference as being *statistically significant*. This means that it is due to more than just sampling error, and that the sample is probably not from the hypothesized population. It does not mean that the difference is significant in other respects, however. Even though a difference is statistically significant, it may be too small to be of any economic or social significance.

NULL AND ALTERNATIVE HYPOTHESES

Tests of hypotheses begin with two statements. One is usually expressed in a way that supports what is thought to be the existing condition, or claimed status. It is often a statement that an assumed condition exists—for example, that "the mean strength of rope is 800 lbs," or that "20% of computer owners use database software." This quantifies a past condition or state of affairs. The second statement describes the other (alternative) status in the event the original claim does not hold.

NULL HYPOTHESIS The (first) hypothesis that there is no difference is known as the *null hypothesis*, and designated H_0. Null is frequently taken to mean *no* change in the status quo, being of *no* consequence, or equal to or amounting to *nothing*. It is a way of saying that conditions are normal. We hypothesize that they are as they have been in the past, or as they should be. And unless there is a substantial amount of evidence to the contrary, we will continue to assume nothing has changed.

The null hypothesis is typically "set up" as a test of one value (described by the hypothesized sampling distribution). It is formulated in a way that will take some *very solid evidence to disprove it*. The test is then to determine whether the sample data are sufficient to "shoot it down." Null hypotheses for the two illustrations we described earlier would be:

Type of Test	Null Hypothesis	Meaning
Means	H_0: μ = 800 lbs	Rope maker claims that true mean strength of rope is μ = 800 lbs
Proportions	H_0: π = .20	Software manager thinks true proportion of PC owners using database software is π = .20

ALTERNATIVE HYPOTHESIS Every null hypothesis should be accompanied by one or more alternative hypotheses. The alternative is a statement that conditions have changed. Alternative hypotheses for the illustrations above might be:

Type of Test	Alternative Hypothesis	Meaning
Means	H_1: $\mu \neq 800$ lbs	True mean strength of rope is not 800 lbs
Proportions	H_1: $\pi \neq .20$	True proportion of PC owners using a database is not .20

In many cases, the alternative identifies some new condition a company would like to support, such as a lower error rate, a longer lasting bearing, or a faster acting medicine. In this sense, the alternative is the research or test hypothesis that is under study. Sample data may then either act to discredit the null hypothesis or to confirm it.

REJECTING THE NULL Rejecting the null gives credibility to the alternative and is equivalent to accepting the alternative. However, *it does not prove the alternative is true*—proof comes only with a scientific investigation of all alternatives. Moreover, there is still some chance we may be making an error in rejecting the null. But for operational purposes, managers may choose to act as though the alternative is correct, even though it is not proven to be true.

NOT REJECTING (ACCEPTING) THE NULL If the null is not rejected, we conclude that the evidence is not sufficient to disprove it, so our conclusion is "Do not reject." Thus, we still retain faith in the null hypothesis, or at least consider the null tenable and reserve judgment on it.

To say "Do not reject" is not as strong as saying "Accept" the null outright (as confirmed), although many analysts do not distinguish between the two statements. The choices of *accept* and *reject* are sometimes more appropriate from a business decision making standpoint. You may encounter both types of statements in your statistical studies, although the intended meaning of *accept* is *do not reject.* We will use the term *accept* or *acceptance region* when it is pedagogically helpful to contrast acceptance and rejection. The understanding here is, of course, that acceptance of the null hypothesis does not constitute its proof.

Testing hypotheses is sometimes likened to a trial. The defendant is charged (hypothesis stated) but is presumed to be innocent (null hypothesis) until the evidence proves otherwise. If the data show the defendant is guilty beyond any reasonable doubt, the presumption of innocence (null hypothesis) is rejected. Otherwise, the defendant is acquitted. But being acquitted does not mean that we believe the defendant is innocent. It only means that there was not enough evidence (data) to prove the defendant guilty (reject the null hypothesis).

TYPES OF ERRORS

Statistical decisions generally entail some risk of error. In this section we define two types of error that analysts should be aware of and show how they can occur.

TYPE I AND TYPE II ERRORS

The errors associated with hypothesis testing are commonly designated as *type I* and *type II* error:

Error	Meaning	Risk Symbol
Type I	Rejecting the null hypothesis when it is true	α
Type II	Accepting the null hypothesis when it is false	β

Type I error is the mistake of rejecting the null hypothesis when it is really true and should be accepted. For example, it would be rejecting the claim that the mean strength of the nylon rope is 800 lbs, when the population mean, μ, really equals 800 lbs. The risk of type I error is a probability designated by the Greek letter alpha (α).

Type II error is the mistake of accepting the null hypothesis when it should be rejected. For example, it would be concluding that the mean strength of the nylon rope is 800 lbs when it really is less than 800 lbs, or greater than 800 lbs. The risk of type II error is a probability designated by the Greek letter beta (β).

Figure 8–3 summarizes the two types of error in a convenient table form.

FIGURE 8–3 Possible conclusions from a test of an hypothesis

		NULL HYPOTHESIS IS EITHER	
		True	**False**
DECISION IS EITHER TO	**Reject**	Type I Error	(correct)
	Accept	(correct)	Type II Error

HOW ERRORS CAN OCCUR

Why is it necessary to contend with errors in such a precise science as statistics? We saw earlier that without error there would be no need for statistical analysis. The errors are inherent in the use of samples and are due to the randomness or chance that given sample values will be selected.

Assume that we wished to test a hypothesis about the mean breaking strength of some nylon rope. If one could take every section of rope produced in the factory and measure its breaking strength—an unlikely but not impossible task—the strength values might appear as shown in Figure 8–4. Let us use these data to show how the two types of error could occur.

FIGURE 8–4 Sample of 100 sections of rope from production process

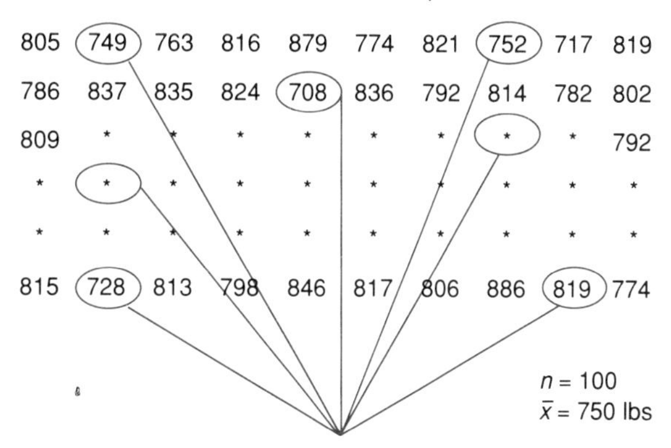

CASE I (TYPE I ERROR) Assume the (unknown) population mean μ is really 800 lbs. We take a sample of $n = 100$ sections of rope to test the hypothesis:

$$H_0: \mu = 800 \text{ lbs}$$

$$H_1: \mu \neq 800 \text{ lbs}$$

Suppose we obtain a sample mean of $\bar{x} = 750$ lbs as illustrated in Figure 8–4. The low value of 750 would prompt us to *reject the true hypothesis* H_0: $\mu = 800$ lbs, when it really should be accepted. We would be committing a type I error.

CASE II (TYPE II ERROR) Suppose the population mean is still really 800 lbs, but we don't know this, and we think it is 750 lbs. We take a sample of $n = 100$ to test the hypothesis

$$H_0: \mu = 750 \text{ lbs}$$

$$H_1: \mu \neq 750 \text{ lbs}$$

Again we obtain a sample mean of $\bar{x} = 750$ lbs as illustrated in Figure 8–4, but this time the sample agrees with our hypothesis (which is wrong), and we *accept the false hypothesis* that should be rejected. We would be committing a type II error.

In both cases, the errors were due simply to the sample that we happened to get when we made the random observations of the rope strength. In the first case, we happened to get some low values that caused us to reject a true hypothesis (type I error). In the second, we also happened to get some low values, but here they caused us to accept a false hypothesis (type II error). Both errors are to be expected, and our hypothesis testing procedure should be designed to recognize them. It can also help us "manage" them by assuring us that the chance of making one or the other of these errors is small.

LEVEL OF SIGNIFICANCE

The control over the risk of type I error is embodied in the initial specifications of a test of hypothesis. It is referred to as the level of significance of the test and is sometimes called the α level.

The **level of significance of a test** (α level) is the risk that one is willing to take of rejecting a hypothesis that is correct (i.e., the risk of type I error).

The level of significance is typically specified before sample data are collected (although we can also determine precisely "how significant" a result is after the data are collected). Commonly used levels of significance are 10%, 5%, 2%, and 1%.

Specifying the level of significance of a test defines the limits of how far a sample mean can be from the hypothesized mean μ_H and still be considered part of the sampling distribution centered around μ_H. As a means of distinguishing between the two areas, statisticians sometimes refer to the region around the hypothesized parameter as the *region of acceptance*, and the region outside the specified limits as the *region of rejection*.

Figure 8–5 depicts the acceptance and rejection regions for the nylon rope example we have been working with. Recall that we have hypothesized H_0: $\mu = 800$ lbs. And we know that if H_0 is true, samples will have $\bar{x}$'s that are normally distributed around the hypothesized mean, μ_H. In Figure 8–5, the limits around μ_H are $\pm 1.96\,\sigma_{\bar{x}}$ away from μ_H, so 95% of all

FIGURE 8–5 Regions of acceptance and rejection

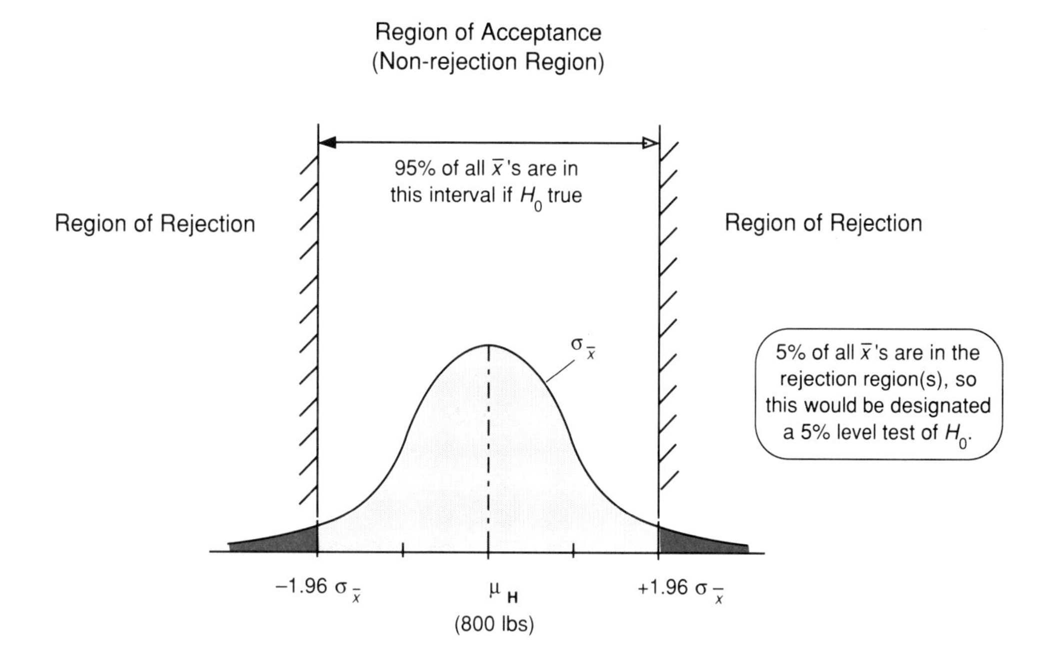

sample means will lie within those limits (if H_0 is true). This also means, of course, that 5% of the $\bar{x}$s will lie outside those limits (in the region of rejection).

If the mean of the random sample we take happens to be one of those $\bar{x}$'s in the rejection region, we would reject the H_0: $\mu = 800$ lbs, even though it was correct. This, of course, would be a type I error (rejecting an H_0 that is true). The risk of making this error is the α level, or level of significance of the test. So the test situation depicted in Figure 8–5 would be said to have an alpha level of 5%, or be "a 5% level test". The other point here is that the level of significance determines the region of rejection for a test of an hypothesis.

Another way to view the level of significance is the probability that we will obtain a sample mean that falls into the rejection region when the null hypothesis is true. When there is only a 1% chance of rejecting a true null hypothesis, analysts sometimes refer to a rejection as being "highly significant." This means the hypothesis would have been rejected by chance $\leq$ 1% of the time. If a hypothesis is rejected at the 5% level, analysts would probably term the results "significant." On the other hand, as the chance of rejecting a true null hypothesis moves to the 10% level, the results are deemed less significant. That is, rejecting a hypothesis at a 10% level of significance means that there is a 10% chance the decision may be in error.

EXERCISES I

1. Explain the purpose of hypothesis testing.
2. Is a test of an hypothesis a test of population parameters, or a test of sample statistics? Explain.
3. Distinguish between
   (a) one-sample and two-sample tests
   (b) tests of means and tests of proportions
   (c) null and alternative hypotheses
   (d) type I and type II error
4. Briefly summarize the logic underlying tests of hypotheses.
5. What are the (four) possible conclusions from a test of an hypothesis?
6. Since statistics is such a precise science, how is it that we can incur errors in the testing of hypotheses?
7. Explain what is meant by the "significance of a test." How is significance level related to type I error?
8. Suppose the hypothesis H_0: μ = \$15,000 is tested and rejected at the 5% level of significance.
   (a) What (exactly) would be meant by rejection at the 5% level?
   (b) Would the conclusion be "stronger" if it were rejected at the 10% level?
9. The text uses the nylon rope example to show how both type I and type II errors could occur. Use the other illustration (about the true proportion of PC owners using a database) to illustrate how (a) a type I error and (b) a type II error could occur.

TWO-TAILED VERSUS ONE-TAILED TESTS

Tests of hypotheses are designated as two-tailed or one-tailed tests depending upon the location of the rejection region. As illustrated in Figure 8–6, two-tailed tests have the rejection area divided into equal portions in both tails of the distribution, whereas one-tailed tests have the rejection area concentrated in one end of the distribution only.

FIGURE 8–6 Rejection regions for two-tailed and one-tailed tests

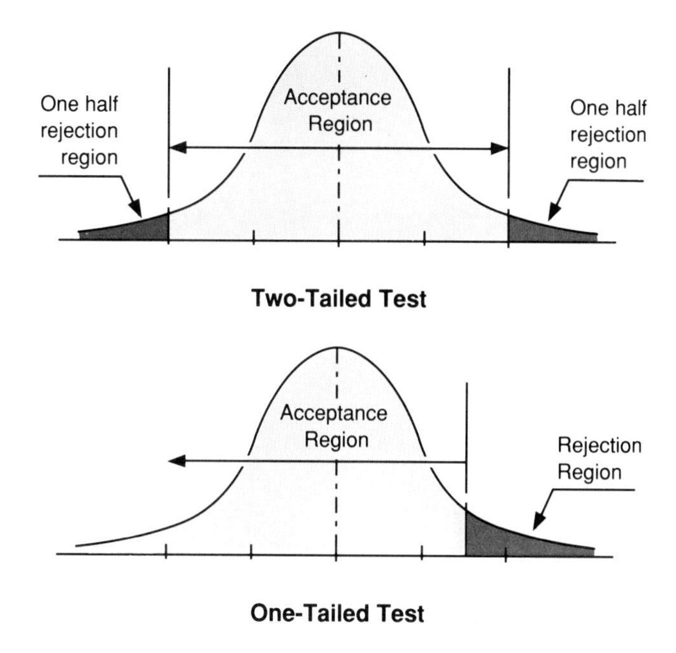

TWO-TAILED TESTS Two-tailed tests are appropriate for situations where either a higher or lower value of the sample statistic should cause rejection of the null hypothesis. For example, suppose a nylon rope is designed to break at 800 lbs of tensile force. A breaking strength of much less than 800 lbs is unacceptable because consumers are depending on (and paying for) 800 lbs of strength. A breaking strength of much greater than 800 lbs is also unacceptable because material costs will be too high and the standard lengths of rope may not fit on the winding machines.

CRITICAL LIMITS For a two-tailed test of the claimed strength of the nylon rope, we have seen that the null and alternative hypotheses are:

$$H_0\text{: } \mu = 800 \text{ lbs (null)}$$

$$H_1\text{: } \mu \neq 800 \text{ lbs (alternative)}$$

The ≠ sign in the alternate hypothesis signifies that the null hypothesis can be rejected by sample values either significantly greater or less than 800 lbs. The actual limits of the acceptance region (called the action or critical limits) are the criteria by which the acceptance or rejection of H_0 will be evaluated. They are designated C_1 and C_2 and can be stated either in terms of the distance away from H_0 (i.e., the number of standard errors), or in actual units of the variable being tested (lbs). We will use the number of standard errors because it is more consistent with the way other test criteria will be stated later in the text. But both ways are commonly used.

Example 8–1 A random sample of $n = 100$ strands of rope is taken to test the hypothesis H_0: $\mu = 800$ lbs against the alternative H_1: $\mu \neq 800$ lbs, at an $\alpha = 5\%$ level of significance. Assuming $\sigma = 40$ lbs, determine the critical limits for the test of the hypothesis, and express them in terms of (a) number of standard errors and (b) lbs of tensile strength.

Solution

(a) The limits are to be set at a distance that will include 100% − 5% = 95% of the area, with half on each side of the mean. Dividing .95 by 2, we therefore need the Z value for .475, which (from the normal distribution table) is 1.96. So the lower limit (C_1) is $Z = -1.96$ and the upper limit (C_2) is $Z = +1.96$ standard errors away from H_0.

(b) The limits may be converted to units of lbs by multiplying Z times the amount of the standard error, and then subtracting this amount from H_0 for C_1, and adding it for C_2. In this case, $\sigma_{\bar{x}} = \sigma/\sqrt{n} = 40 / \sqrt{100} = 4.0$ lbs.

Limit in Zs	Conversion to units		Limit in lbs
$C_1 = -1.96$	$C_1 = \mu - Z\sigma_{\bar{x}} = 800.0 - 1.96\,(4.0)$	=	792.2 lbs
$C_2 = +1.96$	$C_2 = \mu + Z\sigma_{\bar{x}} = 800.0 + 1.96\,(4.0)$	=	807.8 lbs

Thus, the critical limits are 792.2 lbs and 807.8 lbs. A sample mean outside these limits should cause one to reject the hypothesis.

ONE-TAILED TESTS One-tailed tests are appropriate where the null hypothesis should be rejected only if the sample statistic is significantly less (for lower-tailed test) or greater (upper-tailed test) than the hypothesized value. The sign in the alternative hypothesis indicates whether the test is lower-tailed or upper-tailed, as it points in the direction of the rejection region. The alternative hypothesis is normally stated in a way that will clearly discredit the null.

Example 8–2 Suppose the rope manufacturer in Example 8–1 claims the mean breaking strength of his nylon rope is 800 lbs *or more*. Using the same sample size ($n = 100$) and standard deviation $\sigma = 40$ lbs, (a) state the null and alternative hypotheses. (b) Determine the critical limit for a 5% level test.

Solution

(a) In this case, the null would be discredited by sample data that had a mean strength significantly lower than 800 lbs. So the hypotheses are

$$H_0: \mu = 800 \text{ lbs}^{1}$$

$$H_1: \mu < 800 \text{ lbs}$$

(b) Because all 5% of the rejection region is in the lower tail of the distribution, the critical limit corresponds to the Z value that encompasses 45% of the area, or $Z = -1.645$. Using the same format as above, we have:

Limit in Zs	Conversion to units	Limit in lbs
$C = -1.645$	$C = \mu - Z\sigma_{\bar{x}} = 800.0 - 1.645\ (4.0)$ =	793.4 lbs

Note that the presumption of credibility still rests with the null hypothesis. That is, in order to reject the H_0: $\mu = 800$ (or more), the sample mean must be 793.4 lbs or less—which is unlikely if the true mean is in close proximity to 800 lbs. So the "burden of proof" rests heavily upon the alternative hypothesis; otherwise we retain faith in the null (status quo).

[1] The meaning of the null hypothesis here is that $\mu \geq 800$ lbs (and some texts use that form). But the sampling distribution being used for the test really pertains to the one value $\mu = 800$, so we will use this form. Remember also that each sampling distribution pertains to all samples of one size (n) only.

FIGURE 8–7 Rejection limits for two-tailed and one-tailed tests (5% level, $n = 100$, $\sigma = 40$)

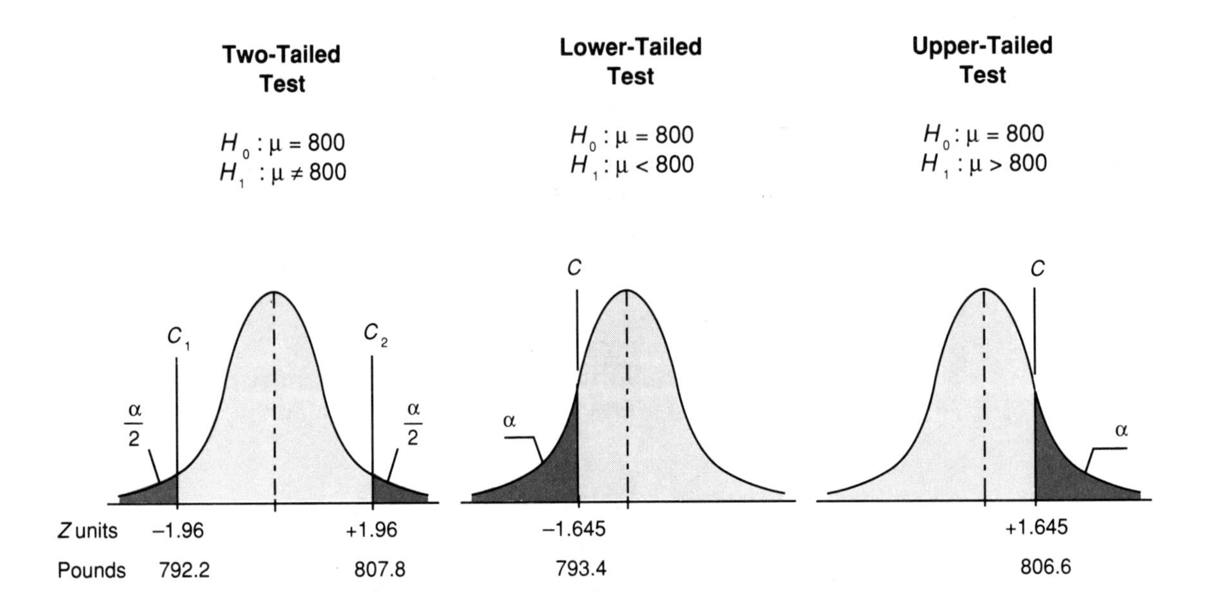

Figure 8–7 summarizes the rejection limits for the three possible types of tests that could be made of the null hypothesis H_0: $\mu = 800$ lbs. Two-tailed tests permit rejection of H_0 if the sample data vary from H_0 in *either* direction, whereas one-tailed tests permit rejection of H_0 if the sample data are significantly different in *one* direction only. Note that the alternative hypothesis clearly specifies the direction of the test, indicating whether the test is (a) two-tailed, (b) lower-tailed, or (c) upper-tailed.

USING THE ALTERNATIVE AS A RESEARCH HYPOTHESIS Consider a condition under which the upper-tailed test depicted in Figure 8–7 might be appropriate. Suppose the rope manufacturer felt that a new production process gave their rope an average strength of *over* 800 lbs. The firm may wish to make that claim on the package label.

To establish that the new process is producing a rope that is definitely stronger than 800 lbs would call for an upper-tailed test with the alternative hypothesis of H_1: $\mu > 800$ lbs. In this case, the manufacturer would be *looking to reject the null* that $\mu = 800$ lbs, which would imply acceptance of the alternative (that $\mu > 800$ lbs). By using the alternative as a research, or test, hypothesis like this, the manufacturer can gain evidence that the new rope strength is significantly greater than the hypothesized value. Moreover, the manufacturer would be making that claim with a well-defined risk of being wrong (e.g., 5%, 10%).

A STANDARDIZED TESTING PROCEDURE

TEST PROCEDURE Having discussed the theory underlying tests of hypotheses, let us now review a testing procedure that is widely followed in practice. We will explain and illustrate it with examples that follow in this chapter. In addition, you will find that this same procedure is applicable to numerous other tests in subsequent chapters of the text.

PROCEDURE FOR TESTING HYPOTHESES CONCERNING POPULATION MEANS

1. Establish the null and alternative hypotheses (H_0 and H_1)
2. Identify the level of significance, α
3. State the test criteria (distribution used & reject values) in either:
   (a) allowable standard errors (e.g., $-1.96\ \sigma_{\bar{x}}$ and $+1.96\ \sigma_{\bar{x}}$
   (b) units of the data (e.g., 792.2 lbs and 807.8 lbs)
4. Calculate the test statistic (e.g., the Z_c value or t_c value)
5. State the conclusion (reject, or do not reject the H_0)

TEST STATISTIC The test statistic (step 4) is simply a measure of the number of standard errors the sample statistic is from the hypothesized population parameter of μ_H or π_H.

$$\text{Test Statistic}_{(\text{calc.})} = \frac{\text{statistic} - \text{parameter}}{\text{standard error}} \tag{8–1}$$

In this chapter (and the next), where we will be working with distributions of means and proportions, we will be calculating either a normal distribution (Z_c) value or a t-distribution (t_c) test statistic. The calculation is quite straightforward.

Example 8–3 Suppose a test of the H_0: $\mu = 800.0$ lbs revealed a sample mean of $\bar{x} = 806.0$ lbs. Calculate the test statistic if the standard error is $\sigma_{\bar{x}} = 4.0$ lbs.

Solution

In this case we are working with a sampling distribution of sample $\bar{x}$'s, which is normal. So the normal distribution applies and the test statistic is measured in Z units:

$$Z_c = \frac{\text{statistic} - \text{parameter}}{\text{standard error}} = \frac{\bar{x} - \mu}{\sigma_{\bar{x}}} = \frac{806.0 - 800.0}{4.0} = 1.500$$

The Z_c value tells us that the sample mean is 1.5 standard errors (Z_c units) away from the hypothesized mean. If the number of Z_c units exceeds the distance to the rejection region, we should, of course, reject the hypothesis.

Table 8–1 summarizes the test statistics that apply to the tests of means and proportions which we will be working with in this chapter. Note that the major grouping is into (a) tests of means and (b) tests of proportions. Beyond that, we distinguish tests where σ is known from those where it is unknown, and small sample tests from large sample tests.

TABLE 8–1 Test statistics for one-sample tests of hypotheses

	MEANS	PROPORTIONS
TYPE OF TEST	**Does $\bar{x}$ support the value of μ_H?**	**Does p support the value of π_H?**
σ **KNOWN**	**Large Sample:** Use Z-distribution $Z_c = \frac{\bar{x} - \mu_H}{\sigma_{\bar{x}}}$ (8–2)	No applications (σ_p always unknown)
σ **UNKNOWN**	**Large Sample:** Use Z-distribution $Z_c = \frac{\bar{x} - \mu_H}{s_{\bar{x}}}$ (8–3)	**Large Sample:** Use normal approximation to binomial (Z-distribution) where $Z_c = \frac{p - \pi_H}{\sigma_p}$ (8–4) NOTE: Value of σ_p is computed using the hypothesized value of π in H_0
	Small Sample: Use t-distribution $t_c = \frac{\bar{x} - \mu_H}{s_{\bar{x}}}$ (8-5)	**Small Sample:** Infrequent applications but can use binomial $P(x \mid n,\pi)$

EXERCISES II

10. Explain (a) why you would use a two-tailed versus a one-tailed test of an hypothesis and (b) what is meant by a test statistic.
11. Regarding the critical limits for a test of an hypothesis:
    (a) How many limits are used for a two-tailed test?
    (b) On which side of the mean is the limit for a lower-tailed test?
    (c) In what two ways can limits be stated?
12. Why are critical limits stated in terms of number of standard errors (as opposed to number of standard deviations)?
13. State the appropriate null and alternative hypotheses for the following:
    (a) A test is to be conducted for a health insurance firm. Past records suggest that the average hospital costs per patient per day are $860, but the insurance firm would like to show that costs are now significantly greater than $860.
    (b) Quality control standards call for .50 inches of tread depth on a certain grade of tire. Random samples are to be taken to test whether the tires are being manufactured according to specifications.
14. Calculate the appropriate test statistic for the following:
    (a) H_0: $\mu = 16.0$ oz, $\sigma = .8$ oz, $n = 64$, and the sample value is $\bar{x} = 16.3$ oz.
    (b) H_0: $\pi = .35$, $\sigma_p = .02$, and the sample value is $p = .32$.

TESTING THE POPULATION MEAN: σ KNOWN

We turn now to some examples of tests of hypotheses, beginning with the case where the population standard deviation is known. When σ is known, the sampling distribution of sample means is normal (a) for *large* samples ($n \geq 30$) from any population and (b) for *all samples* (including small ones) from a normally distributed population. So a Z_c test statistic may be used here. But Z_c is not appropriate for small samples from populations that are not normal.

TWO-TAILED TEST We begin with a two-tailed test and use the example introduced earlier to illustrate the five steps of the testing procedure listed on page 275.

Example 8–4 A rope manufacturer plans to advertise that his EQ800 nylon rope has a mean breaking strength of 800 lbs, and wants to know if the strength is different (either higher or lower) from this. A sample of $n = 100$ strands revealed a mean of $\bar{x} = 790.0$ lbs, and σ is known to be 40 lbs. Test the H_0 at the $\alpha = 5\%$ level of significance.

Solution

1. The hypotheses are:
   $H_0: \mu = 800$
   $H_1: \mu \neq 800$
2. $\alpha = .05$
3. Use normal distribution:
   Reject if $Z_c < -1.96$ or $> +1.96$
4. $Z_c = \dfrac{\bar{x} - \mu_H}{\sigma_{\bar{x}}}$
   where: $\sigma_{\bar{x}} = \dfrac{40}{\sqrt{100}} = 4.0$

   $Z_c = \dfrac{790.0 - 800.0}{4.0} = -2.5$

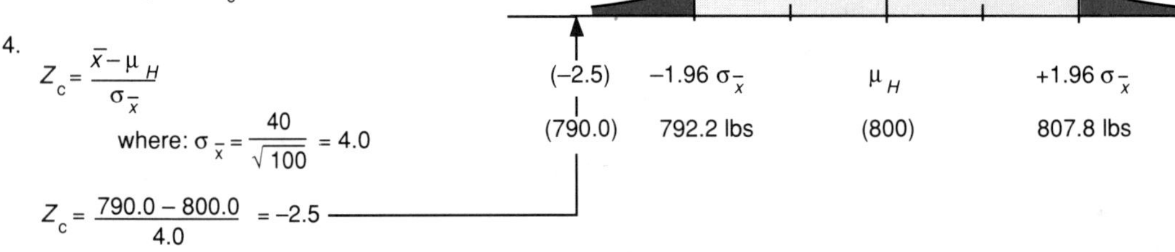

5. *Conclusion*: Reject $H_0: \mu = 800$ since $-2.5 < -1.96$
   Conclude that $\mu \neq 800$ lbs.

Computing the Z_c test statistic is a straightforward way of completing the test. However, expressing the action limits in units of pounds may have more intuitive value at this point, and they have also been included in the sketch. As computed previously, they are $C_1 = 792.2$ lbs and $C_2 = 807.8$ lbs. Using these action limits, it is perhaps more obvious that the sample mean of 790.0 lbs falls outside this range of acceptance.

AFTER REJECTION—WHAT THEN? Inasmuch as the rope manufacturer's hypothesis of a strength of 800 lbs is now rejected, he may well want to make a new estimate of the rope strength. This could easily be done now by using the procedures for estimation discussed in the previous chapter. The manufacturer need only specify the confidence level desired. Then, knowing $\bar{x}$, σ, and n, the limits could be readily established using the standard form $CL = \bar{x} \pm Z\,\sigma_{\bar{x}}$.

ONE-TAILED TEST One-tailed tests follow the same general format. Recall, however, that the alternative hypothesis points in the direction of the rejection region.

Example 8–5 The rope manufacturer of Example 8–4 has another rope, GT1200, made of nylon which has an advertised strength of 1200 lbs (or more). The manufacturer would be concerned if the rope strength turned out to be significantly less than 1200 lbs. A random sample of 36 strands yielded a mean strength of 1182 lbs; σ is known to be 60 lbs. Do a one-tailed test of the H_0: $\mu = 1200$ using a 2% level of significance.

Solution

The claim being tested is that the rope strength is 1200 lbs *or more*, so we will want to reject it if the sample evidence shows a strength of significantly less than 1200 lbs. This means the rejection region will be in the lower tail of the distribution, and the alternative hypothesis will be pointing to less than 1200 lbs.

1. The hypotheses are:
   $H_0: \mu = 1200$
   $H_1: \mu < 1200$
2. $\alpha = .02$
3. Use normal distribution:
   Reject if $Z_c < -2.05$
4. $Z_c = \dfrac{\bar{x} - \mu_H}{\sigma_{\bar{x}}}$

   where: $\sigma_{\bar{x}} = \dfrac{60}{\sqrt{36}} = 10$

   $Z_c = \dfrac{1182 - 1200}{10} = -1.80$

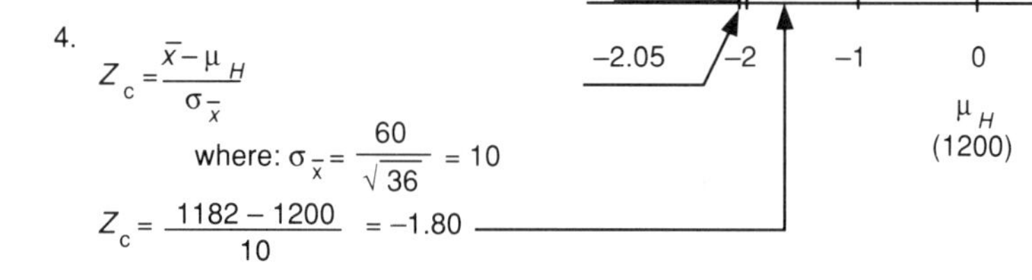

5. *Conclusion*: Do not reject $H_0: \mu = 1200$ because $-1.80 > -2.05$
   Data not sufficient to conclude that mean strength is < 1200 lbs if manufacturer is willing to be wrong only 2% of the time.

ANALYSIS If the action limit for Example 8–5 were expressed in pounds, it would be:

$$C = \mu_H - Z\sigma_{\bar{x}} = 1200.0 - 2.05\,(10.0) = 1179.5 \text{ lbs}$$

So the $\bar{x}$ value of 1182 lbs is not sufficient to reject the hypothesis that $\mu = 1200$ or more. This illustrates how the "burden of proof" is on the alternative hypothesis. Even though the sample mean was less than 1200 lbs, it was not enough less to discredit the null hypothesis—because the null hypothesis has the advantage of lying in the center of the sampling distribution. Recall, however, that the test does not "prove" the validity of the strength claim of H_0. It simply proclaims that the sample data are not sufficient to reject the claim, so the null is still tenable.

TESTING THE POPULATION MEAN: σ UNKNOWN

We have noted previously that for populations with an unknown standard deviation, the sample means are distributed according to the t-distribution. Moreover, the t-distribution assumes the samples are drawn from a normal population. For small samples, the t-distribution is theoretically correct and should be used for setting confidence levels (as we have seen in Chapter 7) and for testing hypotheses (as we shall illustrate in this section).

There is little change in our methodology for large samples—except that the Z (normal distribution) is a satisfactory approximation of the t sampling distribution. Moreover, the assumption of an approximately normal population is unnecessary if the sample size is sufficiently large. Therefore, for large samples, with σ unknown, we will be using the Z_c test statistic.

LARGE SAMPLES

For large samples ($n \geqslant 30$), the central limit theorem tells us that the sampling distribution of means (and proportions) is approximately normal. This allows us to follow the same procedure as that discussed above for testing hypotheses. The difference is that we must approximate the true value of the standard error of the mean, $\sigma_{\bar{x}}$, from sample data by using the expression $s_{\bar{x}} = s/\sqrt{n}$. This was the same substitution we followed for estimation in Chapter 7. Otherwise the procedure is identical to that used above when σ was known.

$$Z_c = \frac{\bar{x} - \mu_H}{s_{\bar{x}}} \qquad (8\text{–}3)$$

Let us illustrate with an upper-tailed test.

Example 8–6 The manager of a supermarket feels customers spend an average of no more than 6 min per week waiting in check-out lines at his company's store. A random sample of 100 customers revealed $\bar{x} = 6.4$ min and $s = 1.2$ min. Test the hypothesis that the mean waiting time is 6 min or less at a 1% level.

Solution

The sample must show significantly longer than 6 min of waiting time to reject the hypothesis, so this will be an upper-tail test, with only 1% of the area in the rejection region. Thus our Z_{table} value will correspond to .4900.

1. The hypotheses are:
   $H_0: \mu = 6.0$
   $H_1: \mu > 6.0$
2. $\alpha = .01$
3. Use normal distribution:
   Reject if $Z_c > 2.33$

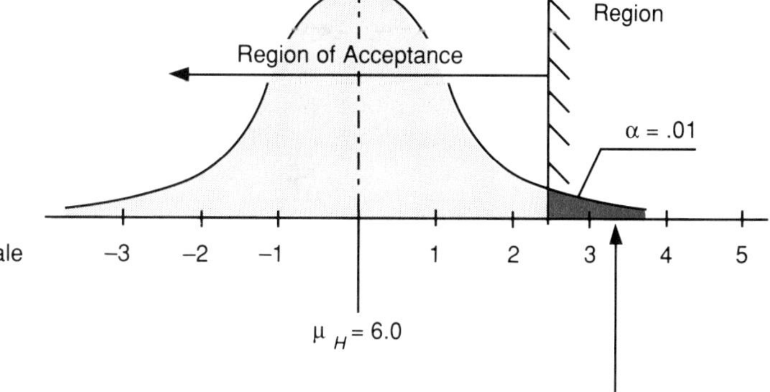

4. 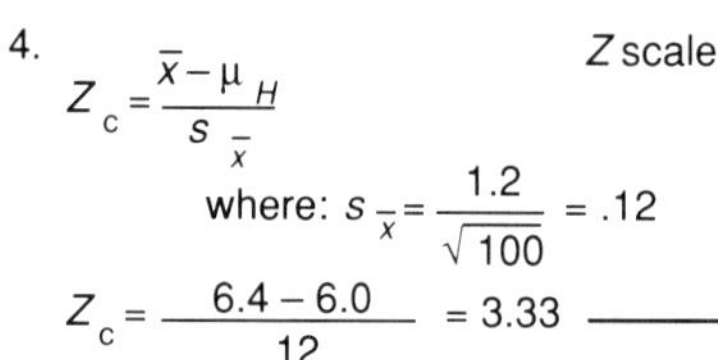

$$Z_c = \frac{\bar{x} - \mu_H}{s_{\bar{x}}}$$

where: $s_{\bar{x}} = \frac{1.2}{\sqrt{100}} = .12$

$$Z_c = \frac{6.4 - 6.0}{.12} = 3.33$$

5. *Conclusion*: Reject $H_0: \mu = 6.0$ because $3.33 > 2.33$
   Data are sufficient to reject claim at the 1% level of significance.
   Customers probably spend more than 6.0 min per week waiting in lines.

SMALL SAMPLES

With unknown σ, the test statistic, $(\bar{x} - \mu)/s_{\bar{x}}$, is distributed as a t-distribution with $n - 1$ degrees of freedom. So the test statistic for small sample tests where σ is unknown is:[2]

$$t_c = \frac{\bar{x} - \mu_H}{s_{\bar{x}}} \tag{8–5}$$

The t_c value is then compared with a t_{table} value of $t_{\alpha/2,\ df}$ or $t_{\alpha,\ df}$ depending upon whether we have a two-tailed or one-tailed test.

TWO-TAILED TEST We will illustrate the small sample hypothesis testing procedure first with a two-tailed test that calls for calculation of all values (including $\bar{x}$ and s). Then we will conclude this section on tests of population means with a one-tailed test of the same data, illustrating how the choice of significance level affects the conclusion.

---

[2] The $s_{\bar{x}}$ value must, of course, be corrected by the finite population correction factor if $n/N \geq 5\%$. (See Equation 6–3.)

Example 8–7 Powertech Chemical Co. has developed a replaceable fuel cell that converts stored chemical and nuclear energy directly into electrical current that can be used to power automobiles. They feel that the cells will power a car for an average of 60 hrs before being "recharged," and are now planning for the number of service centers that will be needed. A random test of 8 cells resulted in the following operating times.

CELL NO	1	2	3	4	5	6	7	8
Hrs Service	64	57	65	59	66	58	68	67

Using a 10% level of significance, test the hypothesis that the true mean service life is 60 hrs.

Solution

We must first estimate $s_{\bar{x}}$ by calculating $\bar{x}$ and s from the sample data.

$$\bar{x} = \frac{\Sigma X}{n} = \frac{504}{8} = 63.0$$

$$s = \sqrt{\frac{\Sigma (x - \bar{x})^2}{n - 1}} = \sqrt{\frac{132}{8 - 1}} = 4.34 \text{ hr}$$

$$s_{\bar{x}} = \frac{s}{\sqrt{n}} = \frac{4.34}{\sqrt{8}} = 1.54$$

X	$(X - \bar{X})$	$(X - \bar{X})^2$
64	1	1
57	−6	36
65	2	4
59	−4	16
66	3	9
58	−5	25
68	5	25
67	4	16
504		132

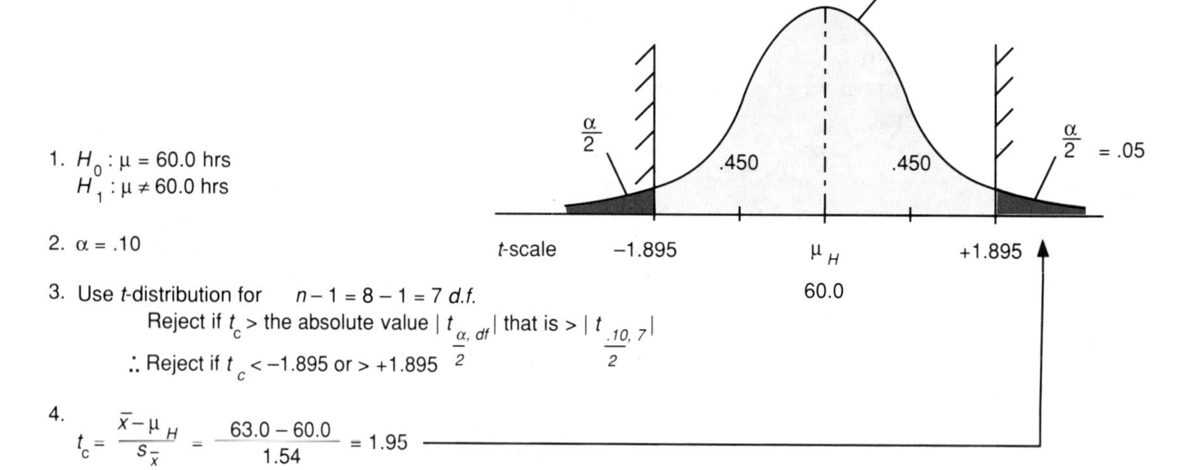

5. *Conclusion*: Reject $H_0: \mu = 60.0$ hrs because $1.95 > 1.895$
The sample evidence does not support the claim of 60 hrs.

Having rejected the H_0: $\mu = 60$ hrs with a sample mean greater than 60, we would be inclined to claim that the mean fuel cell operating time will be greater than 60 hrs. Note, however, that our $\alpha = 10\%$ level of significance means that we are willing to incur a 10% risk of being wrong in saying the true mean is not 60 hrs.

ONE-TAILED TEST To make a claim that the mean operating time of the fuel cell is greater than 60 hrs, we should use a one-tailed test. (When looking up t values for one-tailed tests, be sure to use the table heading from the one-tail probability row.)

Example 8–8 Use the fuel cell data from the previous example:

$$n = 8 \qquad \bar{x} = 63 \text{ hrs} \qquad s = 4.34 \text{ hrs} \qquad s_{\bar{x}} = 1.54$$

(a) Test the following: H_0: $\mu = 60.0$ hrs, H_1: $\mu > 60.0$ hrs. Use $\alpha = .05$ level.

(b) Suppose the test in (a) were conducted at the .01 level. Discuss any differences in the results.

Solution

(a) **1.** H_0: $\mu = 60.0$ hrs
H_1: $\mu > 60.0$ hrs

2. $\alpha = .05$

3. Use the $t_{\alpha,\,df}$ distribution for one-tail with $\alpha = .05$, $df = 8 - 1 = 7$
Reject if $t_c > t_{.05,\,7}$, or $t_c > +1.895$

4. $$t_c = \frac{\bar{x} - \mu_H}{s_{\bar{x}}} = \frac{63.0 - 60.0}{1.54} = 1.95$$

5. *Conclusion:* Reject H_0: $\mu = 60.0$ hrs because $1.95 > 1.895$.
Conclude that the mean operating time is > 60.0 hrs.

(b) For a .01 level test, the t_{table} reject value is farther from the mean:

Reject if $t_c > t_{.01,\,7}$, or $t_c > +2.998$.

The calculated test statistic is the same, $t_c = 1.95$, so the conclusion is to not reject the H_0: $\mu = 60$ hrs. In other words, we cannot conclude that the mean operating hrs are greater than 60 hrs if we are willing to be wrong only 1 % of the time.

ANALYSIS With the two-tailed test (Example 8–7) we were able to reject the claim of 60 hrs—but at a 10% risk of being wrong. With the one-tailed test at the 5% level, we also rejected it, but our risk of error was reduced to 5%. Thus, for the same data, the one-tailed test allowed us to make a "more powerful" statement than the two-tailed test. However, the data are not sufficiently strong to allow us to limit our risk of being wrong to 1%.

TESTING THE POPULATION PROPORTION

Tests of population proportions follow the same general format as tests of means. For example, suppose a business consultant wished to test whether 30% of companies offer their sales representatives a salary plus commission. The three null and alternative hypotheses for the test could be:

Two-tailed test	Lower-tailed test	Upper-tailed test
H_0: $\pi = .30$	H_0: $\pi = .30$	H_0: $\pi = .30$
H_1: $\pi \neq .30$	H_1: $\pi < .30$	H_1: $\pi > .30$

σ ALWAYS UNKNOWN When we must hypothesize about the value of the population proportion, π, we obviously do not know the value of π. This means we never know the true value of the standard error of proportion, σ_p—for if we did, we would also know π! But we *use the hypothesized value of* π_H *to compute* σ_p. This gives the benefit of any question to our hypothesized value of π, which is consistent with the philosophy of testing discussed earlier. [Note: This differs from the computation of the standard error of proportion used for estimation problems. In estimation, we use the sample value, p, to compute the standard error of proportion s_p.]

LARGE AND SMALL SAMPLES

Small sample tests of proportion can be made by comparing sample proportions to what might be expected from binomial distribution values. Advanced statistics texts contain procedures for this. However, most tests of proportions involve the use of large samples where the normal approximation to the binomial can be applied. Therefore, we shall limit our examples to large samples where the test statistic is:

$$Z_c = \frac{p - \pi_H}{\sigma_p} \quad (8\text{–}4)$$

Example 8–9 (two-tailed test, large sample)

The manager of a computer software firm feels that 20% of personal computer owners in its market area use database software. A random sample of 400 PC owners revealed that 68 used database software. Test the hypothesis at the $\alpha = .05$ level.

Solution

The problem statement does not specify a two-tailed test, but there is no evidence to suggest a claim of more or less than 20%, so a two-tailed test can be assumed. We will follow the same steps used for tests of means, with the major difference being in computation of the standard error σ_p.

1. $H_0: \pi = .20$
   $H_1: \pi \neq .20$

2. $\alpha = .05$

3. Use normal distribution:
   Reject if $Z_c < -1.96$ or $> +1.96$

4. $Z_c = \dfrac{p - \pi_H}{\sigma_p}$ where: $p = \dfrac{68}{400} = .17$

$$\sigma_p = \sqrt{\frac{\pi(1-\pi)}{n}} = \sqrt{\frac{(.20)(.80)}{400}} = .02$$

$$Z_c = \frac{.17 - .20}{.02} = -1.50$$

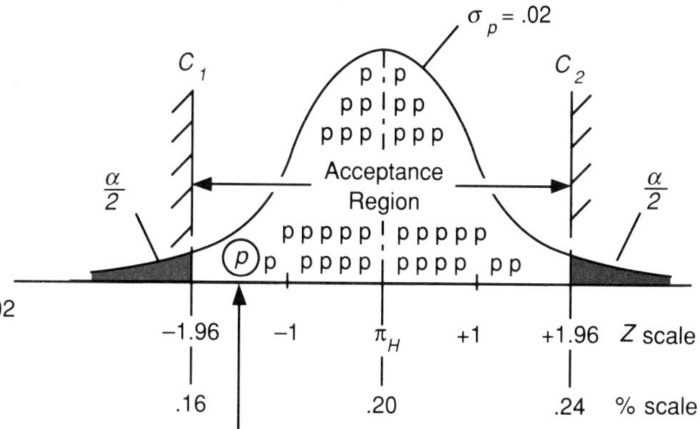

5. *Conclusion*: Do not reject $H_0: \pi = .20$ because $-1.50 > -1.96$.
   Sample data is not sufficient to conclude the proportion of database users is something other than 20%.

As with means, the action limits could also be stated in the units of the problem (in proportions) where we would have:

Limit in Zs	Conversion to units	Limit in %
$C_1 = -1.96$	$C_1 = \pi_H - Z\sigma_p = .20 - 1.96\,(.02) = .161$	16%
$C_2 = +1.96$	$C_2 = \pi_H + Z\sigma_p = .20 + 1.96\,(.02) = .239$	24%

The sample proportion of .17 lies within these limits so the null hypothesis retains its credibility.

Example 8–10 (one-tailed test, large sample)

A cosmetics firm has developed a new sunscreen product which they feel will be a real winner, but they do not want to market it if their share of the market is likely to be less than 30%. A sample of 900 prospective customers showed that 360 preferred and would buy the firm's product. Can the firm be sufficiently certain that π is at least 30%?

Solution

This example presents a difficult choice for establishing the alternative hypothesis. The null can be H_0: $\pi = .30$. If the firm managers wished to be very conservative and market the new product *only if* they were quite sure the market share was greater than 30%, the alternative hypothesis could be stated as H_1: $\pi > .30$. Then they would introduce the product if the null were rejected. That is, they would be insisting on substantial evidence that the sampling distribution was not centered upon $\pi = .30$ or values less than that.

The more traditional approach would be to trust the judgment of managers who already feel confident that the market will be at least 30%, and formulate an alternative hypothesis that will reject that assumption if the sample proportion is significantly less than .30. Following that tack we would have the rejection region on the lower tail of the distribution.

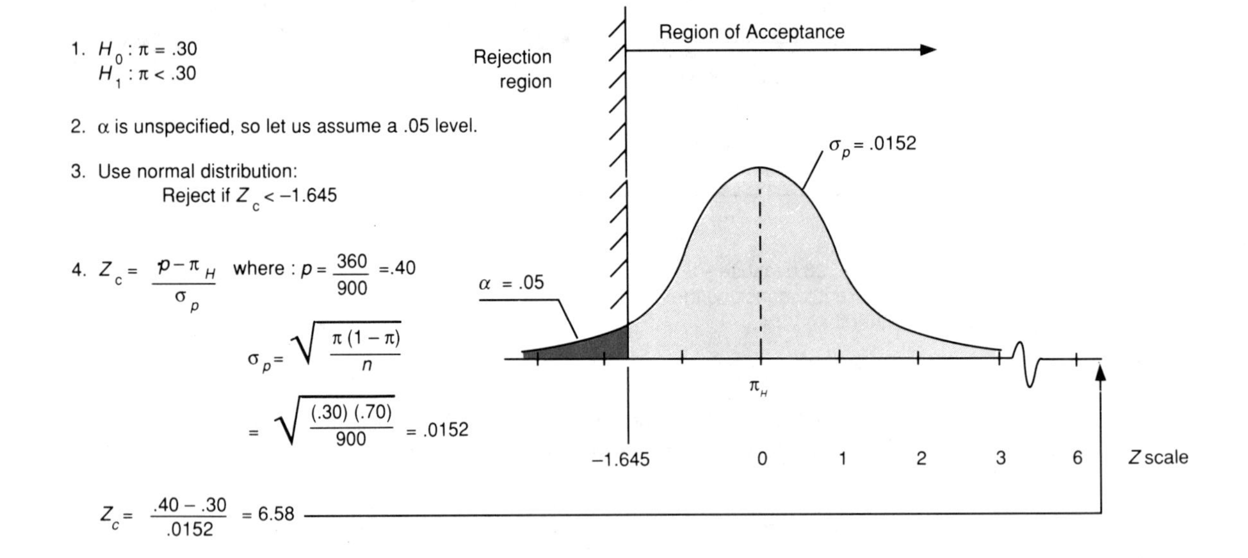

5. *Conclusion*: Do not reject $H_0: \pi = .30$ because $6.58 > -1.645$
The Z_c value of 6.58 lends strong support to the H_0 that the proportion preferring the product is 30% or more.

ANALYSIS In Example 8–10, the critical limit expressed in proportions would be $\pi_H - Z\sigma_p = .30 - 1.645\ (.0152) = .275$, or 27.5%. Note that the sample proportion of .40 was so much greater than the hypothesized value, that the population proportion is almost certainly > 30%. As a matter of fact, if H_1 had been formulated as H_1: $\mu > .30$, the null would have been rejected at the .01 level. (This realization should ease the manager's concern about having formulated the alternative in the traditional manner.)

DETERMINING THE SIGNIFICANCE *p* VALUE

In Example 8–9 where we tested the H_0: $\pi = .20$, we computed a test statistic of $Z_c = -1.50$ which was not below our action limit of -1.96. So we concluded the difference between the hypothesized and sample proportions was not significant at the .05 level. But might the hypothesis have been rejected at the .10 level? At what level of significance would we have rejected the hypothesis? This question is answered by the p value.

The ***p* value** is the smallest level of significance at which the null hypothesis can be rejected.

The p value is easy to determine, for one need only look up the area for the Z_c value in the appropriate distribution table and subtract it from .5000. In Example 8–9, the $Z_c = -1.50$ corresponds to an area of .4332 under the normal curve, so the p value would be $.5000 - .4332 = .0668$. Thus we could say the difference is significant at the 6.68% level for a one-tail test, or (doubling this) at the 13.3% level for a two-tailed test. (It would not be significant at the 10% level for a two-tailed test.)

COMPUTER PROGRAMS FOR TESTING HYPOTHESES

Numerous computer programs are available for nearly any type of hypothesis testing problem. The "unknown σ" programs typically accept raw data and make all the necessary calculations of $\bar{x}$ and s, in order to compute the test statistics. Insofar as it is easier to be more accurate with a computer, some of the programs use the (theoretically correct) t-distribution instead of the Z for samples larger than $n = 30$ when σ is unknown. A number of the programs also provide p value information.

EXERCISES III

15. Some tests of hypotheses are made using small samples. Under what conditions can the Z_c test statistic be used with small samples?
16. State the null and alternative hypotheses for the following tests.
    (a) The average hourly pay of factory workers is $11.62 per hour.
    (b) Bestway aluminum siding will last 30 years or more.
    (c) Less than 65% of families in this country own their homes.
    (d) British citizens spend an average of 18 hr/wk watching television.
    (e) More than three quarters of U.S. workers are satisfied with their jobs.
17. Rework Example 8–4 assuming that the rope is EQ1000 with an assumed mean breaking strength of 1,000 lbs and $\bar{x} = 990.0$ lbs. Let $\sigma = 60$ lbs and test at a 4% level of significance.
18. Could a hypothesis be accepted under a two-tailed test at a .10 level and rejected under a one-tailed test at the same level? Explain.
19. Administrators at Health Care Deluxe claim that, on average, patients are given attention within 14 minutes from the time they enter the offices. A state auditor seeking to test this claim sampled 400 customers and found their average time to obtain service was 15.0 min with a standard deviation of 10.0 min.
    (a) Use an $\alpha = .05$ level to test the claim. Show all (five) steps of the testing procedure, and state your conclusion.
    (b) State the reject limit(s) in units of minutes.
20. Calculate the test statistic for the following:
    (a) $\mu_H = 180$ $\bar{x} = 192$ $\sigma = 14$ $n = 49$
    (b) $\mu_\pi = .42$ $p = .37$ $n = 150$
    (c) $\mu_H = 640$ $\bar{x} = 625$ $s = 30$ $n = 25$
21. For the following test statistics (already calculated), state the reject criterion (Z value or t value) and conclusion (reject or do not reject).
    (a) $Z_c = 6.0$ (two-tailed test at the $\alpha = .05$ level)
    (b) $Z_c = -1.24$ (lower-tailed test at the $\alpha = .10$ level)
    (c) $t_c = -2.50$ (lower-tailed test at the $\alpha = .01$ level, $n = 25$)
    (d) $t_c = 1.80$ (two-tailed test at the $\alpha = .10$ level, $n = 10$)
22. Sketch the distribution showing the rejection area for the following:
    (a) H_0: $\mu = 180$, H_1: $\mu \neq 180$, $\alpha = .05$
    (b) H_0: $\pi = .42$, H_1: $\pi < .42$, $\alpha = .10$
    (c) H_0: $\mu = 640$, H_1: $\mu < 640$, $\alpha = .01$, $n = 25$

23. A social services agency is promoting the sale of lightbulbs in order to raise funds for handicapped persons. They propose to guarantee that the bulbs will last 3,000 hrs or more. A test of 12 boxes of bulbs (with 12 in each box) revealed an average lifetime of 2,960 hrs, with a sample standard deviation of 360 hrs.
    (a) Using a test of hypothesis at the 2% level of significance, are the sample data sufficient to cause the agency to question the proposed guarantee?
    (b) If you were in charge of this project, would you hesitate offering the guarantee?
24. A major retailing firm is considering using daily newspapers to offer coupons to potential customers, if the newspapers are a sufficiently large source of the coupon market (as opposed to magazines and direct mail). A random sample of 185 customers revealed that 44 had clipped coupons from their daily newspapers. Use these data to test the hypothesis that 25% of potential customers use coupons from daily newspapers against the alternative that the proportion is less than 25%. Test at the $\alpha = 10\%$ level.
25. During a city council meeting a council member said that 45% of her constituency was concerned about the amount of drug trafficking in the downtown city park.
    (a) State the null and alternative hypotheses if a two-tailed test is to be conducted to validate the claim.
    (b) What would be the reject value if the council wished to make an 8% level test of this claim?
    (c) Assume the council sampled 265 citizens and found that 146 were concerned about the drug traffic. Is this evidence sufficient to invalidate the council member's claim?
26. Production planners in a furniture factory are planning for the amount of time needed at each work center on a new assembly line. They feel 4.6 min is enough for an upholstery operation, and plan to allow that much time unless the sample time data are significantly less than that. Sample times on 14 pieces of furniture were as shown:

ITEM NO	1	2	3	4	5	6	7	8	9	10	11	12	13	14
Min Req'd	4.2	4.5	4.6	3.9	4.5	4.4	4.5	4.2	4.6	4.7	4.3	3.8	4.5	4.4

Test the hypothesis that the mean sample time is 4.6 min at the 5% level.

(a) Show all the steps of your procedure and state your conclusion.

(b) Comment upon the solution had you been asked to use the $\alpha = 1\%$ level.

27. A test of hypothesis was conducted and yielded the following test statistics when the difference $(\bar{x} - \mu_H)$ was divided by the standard error of the mean. What is the smallest level of significance (p value) at which the null hypotheses can be rejected if
    (a) $Z_c = 1.88$ as calculated from a one-tailed test where $n = 300$?
    (b) $Z_c = 2.24$ as calculated from a two-tailed test where $n = 85$?

COMPUTING THE TYPE II ERROR

Thus far we have concerned ourselves only with the α risk, or risk of type I error—which is the risk of rejecting the null hypothesis when it is true. For a two-tailed test, the hypothesis is true only when μ lies at the center of the sampling distribution. But for a one-tailed test, the hypothesis is also true for points below (for an upper-tailed test) or above (for a lower-tailed test) the mean.

TYPE II ERROR We saw earlier that a type II error is one of accepting a null hypothesis if it is false. We quantify that risk as the probability (or area of the sampling distribution) that lies within the acceptance limits when the null hypothesis is false. However, to compute the type II error, one must first specify what the (true) value of the parameter is. That is, each type II error calculation pertains to only one possible value of the true mean or proportion. Let us illustrate this with an example concerning the mean weight of luggage taken on an airline.

Luggage weight is a critical variable for air carriers because it affects both the load on the aircraft engines and the amount of fuel that is needed to transport that load. For this example, suppose we are interested in the chance of accepting a hypothesis that the mean weight of passenger luggage is 34 lbs, when it is really some value heavier than that, say 35 lbs.

Example 8–11 An airline baggage manager is testing the hypothesis that the true mean weight of baggage per passenger is 34.0 lbs, and has established rejection limits at $C_1 = 32.5$ lbs and $C_2 = 35.5$ lbs. (The standard error was found to be .75 lbs.) What is the probability of type II error if the true (but unknown) mean is actually 35.0 lbs?

Solution

A type II error can be committed because the true mean weight is really $\mu = 35.0$ lbs, instead of the hypothesized value of 34.0 lbs. The probability of type II error is equal to the area under the sampling distribution centered upon 35.0 lbs that lies within the acceptance region from 32.5 to 35.5 lbs.

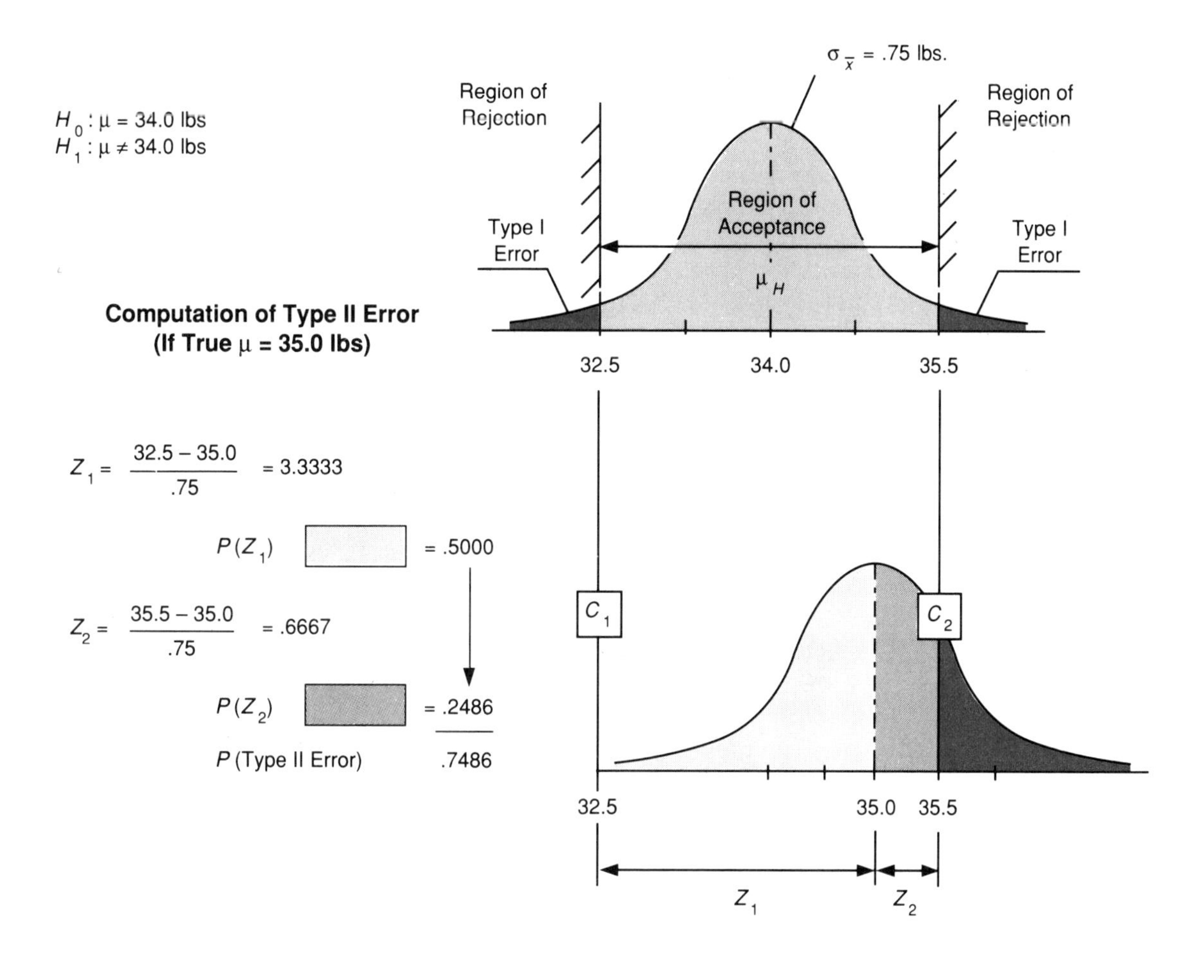

In Example 8–11, the distribution around the true mean is normal, so we can simply add up the relevant areas under the normal curve to obtain the probability of type II error. But there are also many other possible values of the true mean. It might be 25 lbs, 35 lbs, or perhaps even 65 lbs. So there are many possible values of type II error as well. Each is a response to a "what if . . . " question, such as "*What if* the true mean is really 25 lbs. Then what is the probability of still accepting the hypothesis that it is 34 lbs?" The choice of which possible means to evaluate depends upon the consequences of making the (type II) error, which may involve safety, legal, economic, or other considerations.

Example 8–12 carries forward with the calculation of several values of type II error to obtain probabilities that will then be used for graphing what we will later refer to as a "power curve."

Example 8–12 A metallurgical refining process is believed to produce an average of 800 gal/hr of a fluid that must be pumped to another location. Engineers propose to sample the output over $n = 100$ randomly selected hours to see if their hypothesis is correct. Assume $\sigma = 60$ gal/hr and $\sigma_{\bar{x}} = 6.0$. A two-tailed test is decided upon with the probability of type I error (α) set at .01. We shall let μ_H denote the hypothesized value of the mean (which may or may not be correct), and μ_T identify other possible values of the true mean. Compute the probability of type II error if the true mean, μ_T, is (1) 800, (2) 770, (3) 780, (4) 784.5, (5) 810, and (6) 820 gal/hr.

Solution

For a 1% level of significance, the action limits are set at $\mu_H \pm 2.58\ \sigma_{\bar{x}}$, or C_1 and $C_2 = \mu \pm 2.58(\sigma_{\bar{x}}) = 800 \pm 2.58\ (6.0)$. Rounding to one decimal, the limits are set at 784.5 and 815.5. The region between the limits C_1 and C_2 is the acceptance region because sample means falling between these limits lead to the acceptance of the null hypothesis, H_0: $\mu = 800$. Sample means outside these limits lead to rejection of H_0 and acceptance of the alternative hypothesis, H_1: $\mu \neq 800$. Each portion of Figure 8–8 is numbered to correspond to the numbered paragraphs which follow.

1. If the process mean remains at 800, the null hypothesis is true. The limits are set so that the probability of rejecting the hypothesis is .01. The probability of correctly accepting the null hypothesis is .99. There is no possibility of a type II error since a type II error can be made only if the null hypothesis is false.
2. If the process mean decreases to 770 due to faulty materials, operator error, and so on, then the null hypothesis is false. The probability of a type II error is represented by the area under the curve which lies between the limits C_1 and C_2. This area is equal to approximately .0078 since 770 lies about 2.42 standard errors of the mean below the C_1 limit; that is: $Z = (784.5 - 770)/6.0 = 2.42$, and the area between the mean and a point with a Z value of 2.42 is .4922. The probability of correctly concluding that the true mean is no longer 800 is $.5000 + .4922 = .9922$. Thus, if the true mean should decrease (or increase) by 30 points, H_0 will almost certainly be rejected and H_1: $\mu \neq 800$ accepted. There is no possibility of a type I error since, if the true mean is 770, H_0 is false and a type I error cannot be made if H_0 is false.
3. If the true mean, μ_T, decreases to 780, the probability of a type II error is computed as: $Z = (784.5 - 780)/6.0 = 0.75$. The area between 780 and the C_1 limit is .2734. The area above the C_1 limit is $.5000 - .2734 = .2266$, and almost all of this area is in the accept region between C_1 and C_2. The distance to the upper limit is: $Z = (815.5 - 780.0)/6.0 = 5.92$, so the probability of getting a sample mean above 815.5 is so small that it may be ignored. This

means that the probability that a sample mean will fall between 784.5 and 815.5, if the true mean is 780, is equal to .2266. If this occurs, the null hypothesis is accepted when it is false, and a type II error is made.

FIGURE 8–8 Probabilities of type I and type II errors for selected values of the population mean

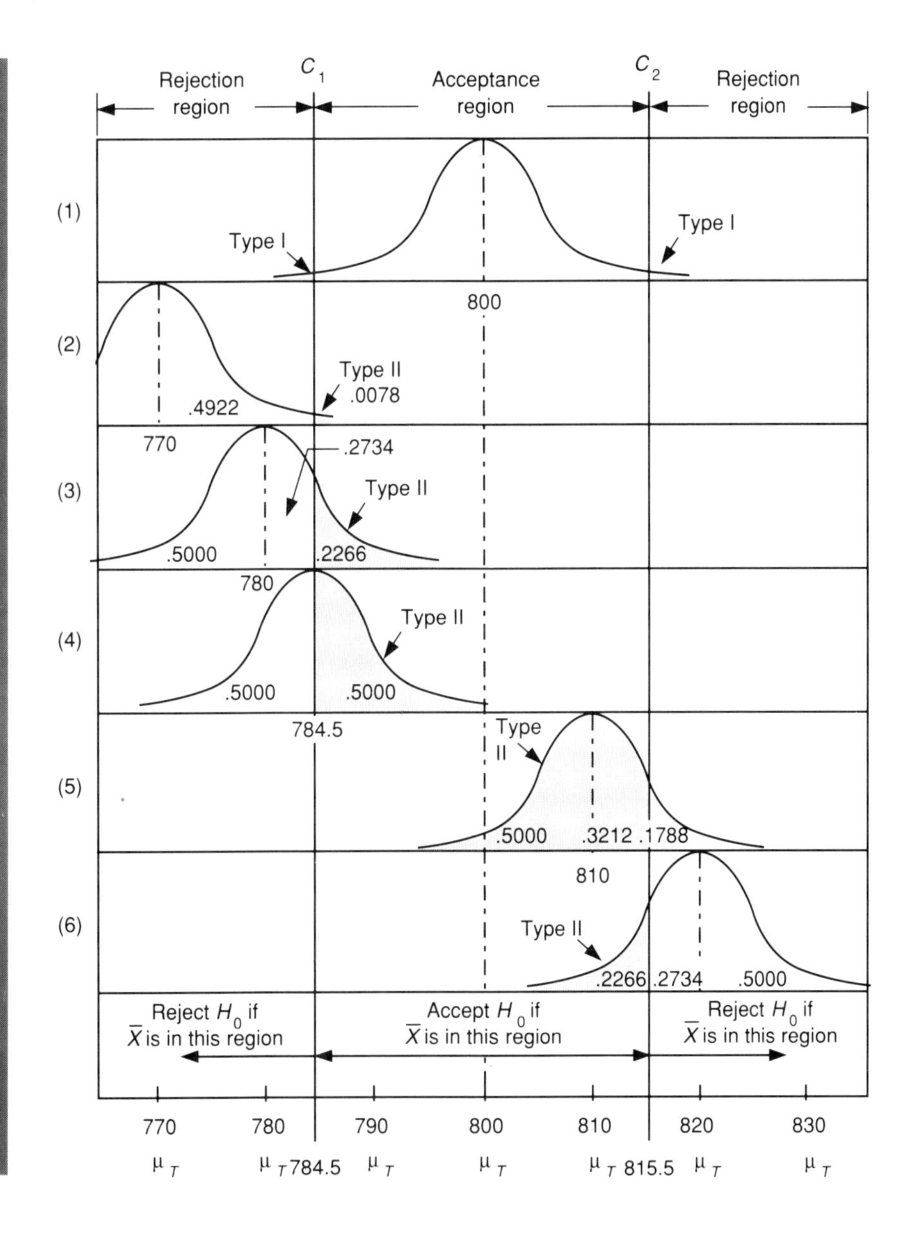

4. If, by chance, the process mean falls exactly on the C_1 limit, the probability that a sample mean will fall in the reject region below the limit is .5000, and the probability that it will fall in the accept region between the limits is almost .5000. There is a slight possibility that the sample mean will fall above C_2 since $Z = (815.5 - 784.5)/6.0 = 5.17$. However, this probability is almost .0000. Since a type II error results if the sample mean falls between C_1 and C_2, the probability of a type II error is .5000, and the probability of correctly rejecting H_0 is .5000. There is no possibility of a type I error because the null hypothesis is false.
5. If the process mean increases to 810, the probability of a type II error is quite high. It is represented by all the area between C_1 and C_2 and is computed as follows:

$$Z_1 = \frac{784.5 - 810}{6.0} = -4.25; \text{ Area} = .5000$$

$$Z_2 = \frac{815.5 - 810}{6.0} = .92; \text{ Area} = \underline{.3212}$$

$$\text{Total} = .8212$$

Thus, the probability of a type II error is .8212, and the probability of getting a sample mean outside the limits and correctly concluding that H_0 is false is only $1.000 - .8212 = .1788$.
6. If the process mean increases to 820, the probability of a type II error is still fairly high. It is computed as follows: $Z = (815.5 - 820)/6.0 = -.75$. The area between 820 and the C_2 limit is .2734, and the area between the two action limits is .2266, which is the probability of a type II error. The probability of correctly rejecting H_0 is $.5000 + .2734 = .7734$.

   If the mean increases to 830 (not illustrated), the probability of correctly rejecting H_0 is .9922, since 830 and 770 are equidistant with regard to the value of H_0 and the action limits, and the probabilities for $\mu_T = 770$ were computed in part 2.

POWER CURVES AND OPERATING CHARACTERISTIC CURVES[3]

POWER CURVE

If β is the type II error probability of accepting a false hypothesis, then $(1 - \beta)$ must be the probability of rejecting that false hypothesis and making a correct decision. Statisticians sometimes find it helpful to graph the probabilities of rejecting a hypothesis for various possible values of the

[3] This material extends beyond that covered in some introductory texts.

true mean so they can readily determine the probability of a correct conclusion, a type I error, or a type II error for any value of μ_T. Such a graph is called a power curve for the test of an hypothesis.

> A **power curve** is a graph showing the probability (or power) of rejecting a (false) null hypothesis. Its probabilities are computed as $1 - \beta$.

Like the computation of type II error values, the calculation of power curve values also necessitates individual calculations for each different value of the alternative hypothesis.

A power curve can be thought of as a graphic presentation of the power of a decision-making rule. The shape of the curve depends upon the hypothesis being tested, the sample size, the specified probability of type I error, the uniformity of the population, and whether a one- or two-tailed test is used.

The power curve for the two-tailed test described in the previous example (Example 8–12) is illustrated in Figure 8–9. Recall that the data for this example were: H_0: $\mu = 800$, $\sigma = 60$, $\sigma_{\bar{x}} = 6.0$, and $\alpha = .01$. The action limits are $C_1 = 784.5$ and $C_2 = 815.5$.

FIGURE 8–9 Power curve for a two-tailed test

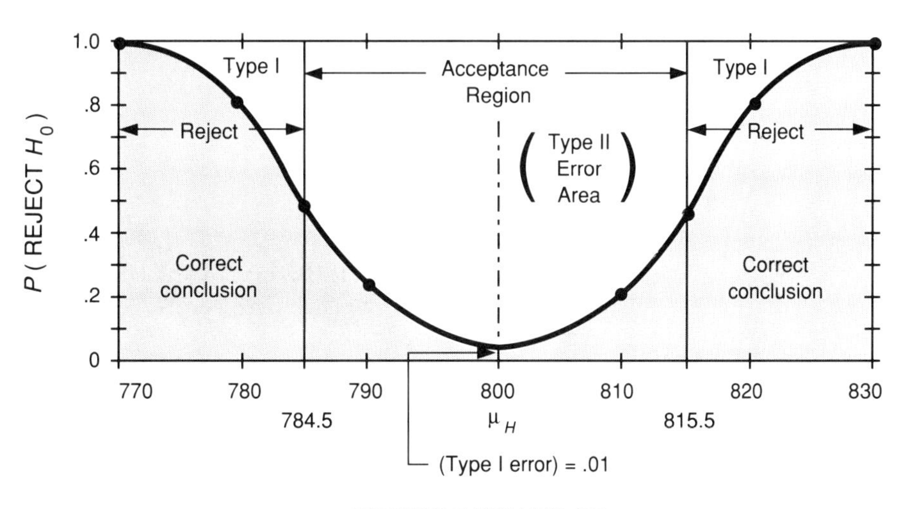

POWER CURVE CONSTRUCTION The vertical axis of a power curve shows the probability of rejecting H_0 and ranges between 0 and 1, whereas the horizontal axis shows possible values of the true population parameter, designated μ_T.

In the discussion of Example 8–11, the probability of rejecting H_0 was computed for the following possible values of the mean:

μ	Probability of Rejecting H_0	
770	.9922	
780	.7734	
784.5	.5000	
μ_H=800	.0100	(set when limits were set)
810	.1788	
820	.7734	
830	.9922	(same as 770)

The probability at 790 is also known to be .1788, the same as the probability at 810, since the limits are symmetrical around 800; and the probability at 815.5 is .5000, the same as the probability at 784.5, for the same reason. With these points and a little imagination, it is possible to draw the curve shown in Figure 8–9.

DECISION RULE The decision rule for deciding whether to accept or reject H_0: $\mu = 800$ could be stated as:

> Take a sample of 100 and compute the sample mean. If this sample mean is between 784.5 and 815.5, accept H_0 and conclude that the process mean is not significantly different from 800.0. If the sample mean is either less than 784.5 or greater than 815.5, reject H_0 and conclude that the true mean is no longer 800.0.

The probability of correctly concluding that H_0 is false for any selected μ_T other than $\mu_0 = 800$ is the height of an ordinate to the curve at that value of μ_T. The complement of each of these ordinates is the probability of a type II error at the same μ_T value. At exactly 800.0, the height of the curve is .01 and represents the probability of a type I error. The complement of the probability of a type I error at μ_H is the probability of correctly concluding that the mean has not changed. The graph was constructed to show the various probabilities; however, only the ordinates and their complement are of significance. The curve shows probabilities and does *not* depict the area under the curve.

Power curves for *one-tailed tests* have similar scales except that from the type I error point (at μ_H); they simply continue on down to the x-axis. Curves for lower-tailed tests (e.g., with an alternative H_1: $\mu < 800$) would taper on down to the right (and not rise back up), whereas upper-tailed test curves are the reverse. In both cases, the area under the curve between μ_H and the axis is representative of the type I error.

OPERATING CHARACTERISTIC CURVE

The complement of the power curve is a graph showing the probability of accepting H_0 that is called an operating characteristic (OC) curve.

> An **operating characteristic (OC) curve** is a graph showing the probability of accepting a hypothesis for (alternative) possible values of the true parameter.

The OC curve and the power curve are complements of each other. For a two-tailed test, the OC curve begins on the left on the x-axis where the probability of acceptance is zero. It then rises until it reaches a maximum where H_0 and μ_T are equal (and a type I error can occur). Then it declines back down to zero. So it is just the inverse of the power curve. The next example illustrates an OC curve for an upper-tailed test.

Example 8–13 For the following data, graph the operating characteristic curve.

H_0: $\mu = 30.0$, H_1: $\mu > 30.0$, $\alpha = .05$, $\bar{x} = 31$, $s_{\bar{x}} = .6$, $n = 100$
(Graph values for: $\mu = 29.0, 30.0, 30.5, 31.0, 31.5, 32.0, 32.5$)

Solution

This is a test with the rejection region in the upper tail, and the OC curve shows the probability of accepting the hypothesis. The critical limit is shown on the next page.

$$C = \mu_H + Z\, s_{\bar{x}} = 30.0 + 1.645(.6) = 30.0 + 1.0 = 31.0$$

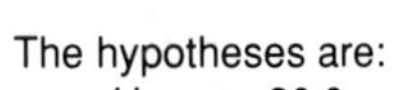

The hypotheses are:
$H_0 : \mu = 30.0$
$H_1 : \mu > 30.0$

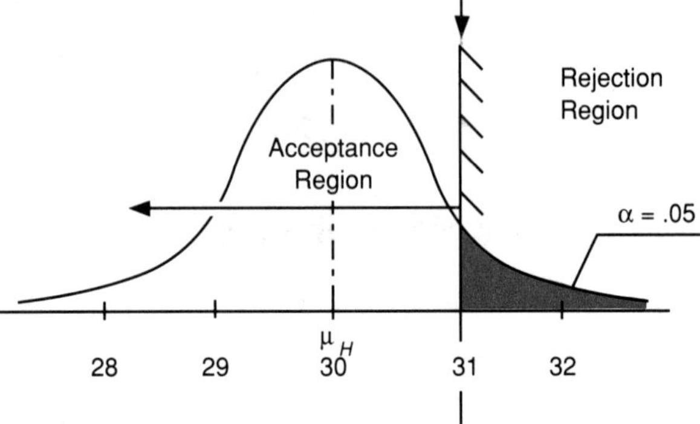

O.C. Curve probabilities are calculated as areas within the acceptance region (to the left of the critical limit 31.0)

For example, if true mean is $\mu = 30.5$

$$Z_c = \frac{\bar{x} - \mu}{s_{\bar{x}}} = \frac{31.0 - 30.5}{.6} = .833$$

$P(Z) = .2967$

$P(\text{Accept } H_0) = .5000 + .2967 = .7967$

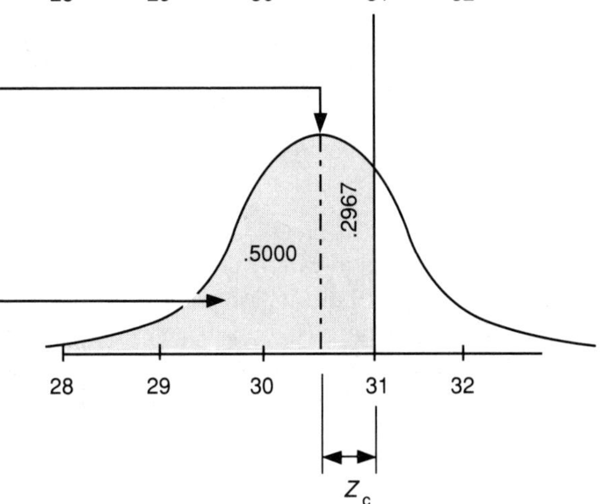

Other values of P (Accept) are as listed below and result in the O.C. Curve shown.

μ value	$P(\text{Accept } H_0)$
29.0	.5000 +.5000 = 1.0000
30.0	.5000 +.4500 = .9500
30.5	.5000 +.2967 = .7967
31.0	.5000 + 0 = .5000
31.5	.5000 −.2967 = .2033
32.0	.5000 −.4525 = .0475
32.5	.5000 −.4938 = .0062

OPERATING CHARACTERISTIC CURVE

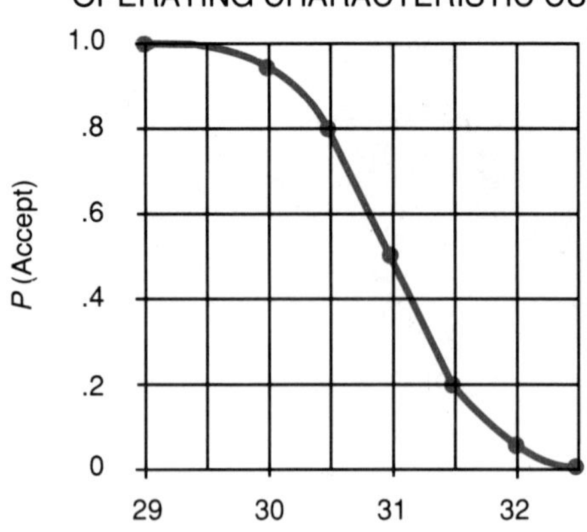

POSSIBLE VALUES OF TRUE MEAN μ

Operating characteristic curves are used extensively in the field of quality control. For example, they help managers evaluate acceptance sampling plans by revealing the probability that shipments of raw materials will be accepted from a supplier, given the different possible values of the true percent of defectives those shipments might contain. Both government and private industry make extensive use of OC curves for purchasing decisions.

SUMMARY

A *hypothesis* is a claim about a population parameter. *Hypothesis testing* is a statistical procedure for using sample data to accept or reject that claim. If a hypothesis is true, the distribution of sample means (or proportions) will form a normal distribution (if large samples) or t-distribution (small samples) about the hypothesized value. And the sample statistic will be part of that *sampling distribution*. The limits within which the sample statistic should lie (if H_0 is true) are the *action*, or *reject, limits*. They are influenced by the Z or t value corresponding to the specified level of significance:

$$\text{Reject limits} = H_0 \pm (Z \text{ or } t \text{ value})(\text{standard error})$$

The *null hypothesis* H_0 is a statement of "no change" from the expected, and the action limits are usually set up to "favor" the null. That is, the calculated *test statistic* (Z_c or t_c) must show that the sample statistic is significantly different from H_0 in order to reject the H_0. Otherwise we continue to assume the null has credibility—even though we do not consider acceptance of the null as any form of "proof." Rejecting the null hypothesis is typically a "stronger" conclusion than accepting the null hypothesis because the chance of making an error here is generally limited to the specified α risk (i.e., 1%, 5%, 10%). It implies acceptance of the alternative hypothesis.

The *alternative hypothesis* H_1 is frequently the research or test hypothesis that is under examination. The inequality sign in H_1 clearly indicates whether the test is a *two-tailed* ($\neq$), *lower-tailed* ($\leq$), or *upper-tailed* ($\geq$) test.

Two errors can occur. *Type I error* is the error of rejecting a null hypothesis that is true. The risk of doing this is α, which is the specified level of significance of the test. *Type II error* is the error of accepting the null hypothesis when it is false. This is the β risk. It corresponds to the area inside the action limits, but under the distribution centered upon a specific value of the true parameter. Some value of the parameter must be specified before one can calculate β.

Once some probabilities of accepting the null hypothesis have been calculated, they can be graphed to provide two useful curves. The *operating characteristic* (OC) *curve* is essentially a direct graph of the β values. It shows the probability of accepting the null hypothesis for alternative values of the true parameter. We know that only one value of the parameter can exist, but the curve is useful for decision making in situations where the true value is unknown, such as in quality control.

If the β values are subtracted from 1.0 and then graphed, the result is a *power curve* that describes the power of the test to reject a false null hypothesis. Single-tailed tests have more power than two-tailed tests because the rejection area is concentrated in one tail of the distribution—meaning that the reject limit is closer to the hypothesized parameter. This results in a higher probability (more power) of rejecting null hypotheses that are false. The operating characteristic (OC) curve is the complement of the power curve.

CASE: INTERNATIONAL PRINTING AND PUBLISHING COMPANY

A wrong decision was not likely to bankrupt the International Printing and Publishing Company. But $4.5 million were at stake. Two issues faced the company's board of directors at their monthly meeting: (1) Should the News Division be permitted to abandon its traditional (and profitable) typesetting methods? and (2) Is the proposed new *Photo-News Journal* magazine worth the investment the divisional manager, Charolette Martin, wanted them to approve?[4]

Regarding (1), the case for computer-controlled laser printers was well made. An investment of $28,000 for three small computers, a laser printer, and some desktop publishing software would give the division more editing flexibility than the $130,000 worth of traditional typesetting equipment the directors had approved at their earlier meeting. Given the division manager's assurance that the quality of printing would not suffer, it didn't take the board long to dispense with this item. In fact they congratulated themselves on saving $102,000 as the meeting agenda moved on to item (2).

The case for the new *Photo-News Journal* was also well documented. A preliminary study had suggested the magazine would capture just 15% of the market, which made it an unacceptable venture. Now the division had better market research data from a total of 800 potential customers in nine randomly selected cities stretching from Seattle to Miami. Yes, the competition was keen—from *Time, Newsweek, U.S. News,* and even *Life* and *Sports Illustrated*. But the potential was there. Ms. Martin emphasized that society is becoming increasingly information oriented. *Photo-News* would sell for about $1.25 and the division's promotional staff had already lined up 245 advertisers who were ready to put their money on the line.

Then came the market research data. The division argued that the target photo-news market was different from both the photo magazines market and from the straight news magazines market. Of those interviewed, 180 had said they would definitely purchase *Photo-News Journal*. During the discussion, one of the directors reminded the division manager of the company's policy of discounting buyer intentions by 20%

[4] The board had a policy that any new publication must show the promise of capturing more than 15.0 percent of the market before it could be approved.

because of their previous findings that intentions were not always converted into sales. The manager agreed that the number could be reduced by 20% but she felt it would still meet the company's criteria of more than 15% market share.

BOARD MEMBER: I'm afraid that when you take 20% off the 180, you're getting into a very questionable percentage, Charolette. In other words, it may be slightly over 15%, but it looks to me like it could be a sample from a market share of 15%—there's always going to be a little variation, you know. If it were definitely over 15% I'd go along with you.

CHAROLETTE MARTIN: George, would you accept a chance of less than 1 in 100 that it is not 15%?

BOARD MEMBER: Oh for Pete's sake, yes! I'm not asking for your right arm. But we've got a lot at stake here.

CHAROLETTE MARTIN: I agree we have—and I appreciate your probing—otherwise, I'd feel all my staff's research data went for nothing. We have prepared some calculations which I think will substantiate our position though, George. Might I leave them with you?

BOARD MEMBER: Certainly, Charolette. Knowing you, I had a feeling you might have more ammunition than we saw on the first volley. I'd like to review them and I propose we take action on the *Photo-News Journal* plan at the next meeting—after everyone has had a chance to see your calculations.

Case Questions

(a) Would the division's analysis be most likely based on a test of means or a test of proportions? Why?

(b) What kind of test would be most appropriate for this situation: a two-tailed test or a one-tailed test? Explain your reasons.

(c) Suppose you were a member of the board of directors and wanted to be quite sure the market was > 15% before proceeding. If the null were set up as H_0: $\pi = .15$, would you want the alternative set up as $\pi < .15$ or as $\pi > .15$? Why?

(d) Taking the 20% discount into consideration, and using the H_0: $\pi = .15$, H_1: $\pi > .15$, does the test data satisfy the criteria brought out in the discussion?

(e) Suppose the true market share to be realized by the firm were 16.2%.

  (i) What (in words), would constitute a type II error?

  (ii) How might it come about?

  (iii) What would be the probability of this type II error?

  (iv) What would be the economic implications of a type II error?

QUESTIONS AND PROBLEMS

28. Which of the following statements about tests of hypotheses are false?
    (a) The risk of committing a type I error is designated as β.
    (b) The null hypothesis value is assumed to lie at the mean of a sampling distribution.
    (c) The major cause of type I and II errors is from mistakes made in recording data and in calculating results.
    (d) If a hypothesis is tested at the $\alpha = .05$ level and is not rejected, then either the hypothesis is true or we are committing a type II error.
29. Both estimation (Chapter 7) and testing hypotheses (Chapter 8) involve the use of limits consisting of a number of standard errors. Distinguish between estimation and testing in terms of (a) the value in the center of each interval and (b) the basic purpose of these two statistical procedures.
30. Assume a null hypothesis for a two-tailed test is H_0: $\mu = 600$.
    (a) If the true mean is really 600, give the two conclusions one might reach on the basis of the sample data. What type of error is made if H_0 is rejected?
    (b) If the true mean is really 620, give the two conclusions one might reach on the basis of incomplete (sample) evidence. What type of error is made if H_0 is accepted when it is really false?
31. You are testing a hypothesis that a true mean is 400, have found the standard error to be 2.0, and have established rejection limits at 394 and 406. If the true mean is actually 404, what is the probability of type I error?
32. You are conducting a large sample upper-tailed test of H_0: $\mu = 500$ and have calculated a standard error of 3.0.
    (a) What would be the action (reject) limit for $\alpha = .12$?
    (b) Suppose $\bar{x} = 504.2$. Would H_0 be rejected?
    (c) What p value corresponds to a sample result of 504.2?
33. Suppose a sample of 25 is drawn from a production process to do a two-tailed test at the 5% level of whether the mean fabrication time of an item is 42.0 min. The results are $\bar{x} = 39.6$ and $s = 5$ min.
    (a) What sampling distribution would apply?
    (b) What reject values should be used?
    (c) Would a test at the .10 level have wider acceptance limits?
    (d) Calculate the value of the test statistic.
    (e) Should the hypothesis be rejected?
    (f) What is the closest p value from the table values of the appropriate distribution?

34. Test the H_0: $\pi = .45$ against H_1: $\pi \neq .45$ at the 10% level, assuming a sample of 300 resulted in $p = .41$.
35. Test the H_0: $\mu = 750$ hrs against H_1: $\mu < 750$ at the $\alpha = 2.5\%$ level:
    (a) assuming a sample of $n = 49$ resulted in an $\bar{x} = 720$ hrs and $s = 70$ hrs.
    (b) assuming a sample of $n = 9$ resulted in an $\bar{x} = 720$ hrs and $s = 70$ hrs.
    (c) Explain any differences in the conclusion from (a) and (b). Do the results support the credibility of the null hypothesis unless there is strong evidence to disprove it?
36. Set action or decision limits in each of the following cases, assuming that $\mu_H = 360.0$, $\sigma = 35.0$, and random samples are to be taken from an infinite population.
    (a) $n = 100$, probability of type I error specified at .05, both upper and lower limits
    (b) $n = 100$, probability of type I error specified at .01, with only an upper action limit
    (c) $n = 400$, probability of type I error specified at .05, and assuming that we are interested only in decreases in the true mean
    (d) $n = 100$, probability of type II error to be .05, if the true mean decreases to 352.0
37. Assume that a random sample was drawn from a population and the following data obtained: $\bar{x} = 224.2$, $s = 28.0$.
    (a) Make an estimate of the standard error of the mean if $n = 400$.
    (b) Test the null hypothesis H_0: $\mu = 225.0$ against the alternative H_1: $\mu \neq 225.0$ using the sample data and assuming a sample size of 400.
    (c) Make an estimate of the standard error of the mean for samples of 25.
    (d) Test the hypothesis of part (b), but assume the sample size was only 25.
    (e) Why did you use the t-distribution in part (d)?
    (f) Which line of the t-distribution is used if $n = 400$?
    (g) Which line is used if $n = 25$?
    (h) Why can the standard normal distribution (area under the normal curve) be used when n is large?
38. A sample of 400 was drawn from a population with $\sigma = 16.0$. If the sample mean was 286, test the hypothesis H_0: $\mu = 290$; H_1 $\mu \neq 290$:
    (a) using the .01 level
    (b) using the .05 level

39. Given the hypothesis testing data:

H_0: $\mu = 300$ $\quad n = 64$ $\quad \sigma_{\bar{x}} = 4$

H_1: $\mu \neq 300$ $\quad \sigma = 32$ $\quad \alpha = .08$

Find the P(type II error) if the true mean μ_T is:
(a) 290, (b) 293, (c) 299, (d) 300, (e) 305

40. A power curve is often used to provide a pictorial representation of the power of a decision-making rule.
    (a) What is on the vertical axis and what is the invariable range of the vertical scale?
    (b) What is on the horizontal axis if the H_0 concerns the population mean?
    (c) What is on the horizontal axis if the H_0 concerns the population proportion?
    (d) What is the difference between a power curve and an operating characteristic curve?

41. Make a rough sketch of the shape and the scales of a power curve, assuming:
    (a) $\mu_H = 600$, $\sigma = 30$, $n = 100$, both upper and lower action limits, and the probability of a type I error set at .05
    (b) $\mu_H = 600$, $\sigma = 30$, $n = 100$, and only one action limit located 1.645 $\sigma_{\bar{x}}$ below the value of μ_H
    (c) $\mu_H = 600$, $\sigma = 30$, $n = 400$, and only one action limit located far enough above μ_H to reduce the probability of a type I error to .01
    (d) For each of the curves sketched in parts (a), (b), and (c), indicate the x-axis points where ordinates to the curve represent type I error or correct conclusions and where their complements represent type II error or correct conclusions.

42. A manufacturer claims that his product never runs more than 10 percent in grades below A. A random sample of 400, selected from a shipment of several hundred thousand, contained 52 units below grade A. Assume that we want to test the hypothesis H_0: $\pi = .10$ using the .05 level of significance.
    (a) What is a logical alternative hypothesis for an upper-tailed test?
    (b) What is σ_p?
    (c) Find:

$$Z = \frac{p - \pi_H}{\sigma_p}$$

    (d) What is your conclusion?
    (e) What is σ_{np}?

(f) Find the Z score using the number of defective units in the sample minus the hypothetical number of defective units over σ_{np}.

(g) What is your conclusion using this procedure? Are the procedures equivalent?

(h) Solve parts (a), (b), and (c), assuming there were only 2,000 in the shipment.

43. A management analyst for a national health care firm is reviewing the time standard for their staff attendants to examine patients and prescribe some form of treatment. The firm has been allowing an average of 20 minutes. But if the time required is significantly less than this, there is too much idle time so the standard will be reduced. A random sample of 53 patient times revealed a mean of 19.4 min and a standard deviation of 4.0 min. Let $\alpha = .10$.

(a) State the null and alternative hypotheses.

(b) What is the critical limit?

(c) Would this sample data be sufficient to conclude that the mean time is less than 20 minutes?

(d) What, in words relevant to this problem, is the type II error?

(e) What is the value of the type II error if the true mean is 19.0 min?

(f) What is the probability of accepting the hypothesis if the true mean is each of the following:
18.0 min, 18.5 min, 19.0 min, 19.3 min, 19.6 min, 20.6 min?

(g) Graph the operating characteristic curve and explain its meaning.

44. Assume that the price earnings ratios listed in the Corporate Statistical Database of Appendix A represent a random sample of $n = 100$. Let H_0 be that the mean price earnings ratio of companies is \$10.50 per share, and test: (a) H_1: $\mu \neq$ \$ 10.50 at $\alpha = .05$ level, (b) H_1: $\mu >$ \$ 10.50 at $\alpha = .01$ level.

45. Assuming that the $n = 100$ firms listed in the Corporate Statistical Database of Appendix A represent a random sample, test the hypothesis that the proportion of firms with a profit sharing plan is 40%, against the alternative that it is less than 40%. Use a 10% level of significance.

46. Assume that the banking and insurance firms listed in the Corporate Statistical Database of Appendix *A* constitute a random sample of the industry. Test the following hypothesis at the $\alpha = .05$ level:

(a) H_0: the mean age of employees = 35 yrs against H_1: $\mu \neq 35$ yrs

(b) H_0: stock price per share = \$ 37.00 against H_1: $\mu >$ \$37.00

(c) Structure your test so as to conclude that the annual dividend is *significantly more than \$1.50 per share* if the sample data support this conclusion.

CHAPTER 9

Testing Hypotheses: Two-Sample Tests

INTRODUCTION

- Logic of Two-Sample Tests
- Large Samples: σ Known
- Large Samples: σ Unknown
- Small Samples: σ Unknown

TESTING THE DIFFERENCE BETWEEN MEANS FROM TWO INDEPENDENT SAMPLES

- Large Sample Procedure
- Confidence Interval Estimate for the Difference Between Means of Two Independent Samples
- Small Sample Procedure

TESTING THE DIFFERENCE BETWEEN MEANS OF TWO PAIRED SAMPLES

- Testing for Before and After Differences
- Confidence Interval for the Difference Between Means of Two Paired Samples

TESTING THE DIFFERENCE BETWEEN PROPORTIONS FROM TWO INDEPENDENT SAMPLES

- Difference Between Two-Sample Proportions: Large Independent Samples
- Confidence Interval for the Difference Between Two Proportions

DECISION MAKING RULES AND SAMPLE SIZE FOR QUALITY CONTROL

SUMMARY

CASE: TRANSCONTINENTAL COMMUNICATIONS COMPANY

QUESTIONS AND PROBLEMS

Chapter 9 Objectives

1. Test the difference between means of two independent samples.
2. Test the difference between means of two paired samples.
3. Test the difference between two proportions.
4. Make confidence interval estimates of differences.
5. Compute the sample size required for hypothesis testing situations.

Is the cost of credit equal for businessmen and businesswomen—or must women pay more for credit than men? The Equal Credit Opportunity Act is supposed to prohibit credit discrimination. But a survey of 230 members by the National Association of Women Business Owners found that of 172 who had applied, about 65 had been denied. And more than half of those who had been turned down felt they had been discriminated against.

What about interest rates paid for credit? Are the rates equal in the eastern and the western part of the country? A survey of 1,015 Americans found that 6 in 10 did not realize that credit card interest rates varied from bank to bank. Arkansas, for example, had state-regulated interest rates as low as 11.5%—compared to a national average of about 18%.[1]

How about the typical U.S. family's net worth? Is the net worth for householders in the 55 to 64 age group significantly different from those householders under 35? Data certainly confirm this—to the tune of $73,660 for the 55-64 age group compared to $5,180 for those under 35. In addition, for all categories, those with a college education have about twice the net worth of those who only graduated from high school ($60,420 versus $31,890).[2]

[1] *The Spokesman-Review Spokane Chronicle* (Spokane, WA) July 19, 1986, p. B9.
[2] *The Spokesman-Review Spokane Chronicle* (Spokane, WA) July 22, 1986, p. A11.

INTRODUCTION[3]

WHERE WE'RE GOING The questions illustrated above, along with countless others, call for an analysis of the difference between two groups. That's what this chapter is about. Is there a statistically significant difference between the means (or proportions) of two groups? As in the previous chapter, we generally begin with a null hypothesis of no difference, which is another way of saying that the two samples are both from a population having the same mean, μ (or proportion π). Then we let the data speak for itself at a specified level of significance. This allows us to draw statistically valid conclusions that take the risks of error into account.

In this introductory section (1) we review the theory underlying the test statistics used for the various tests. Then we (2) introduce tests of the *differences between means* for large and small samples, and (3) go on to tests of *differences between proportions*. The last section (4) describes how to determine the required *sample size* for a test of an hypothesis. After completing this chapter (plus the previous chapter on one-sample tests), you should have a comprehensive understanding of both the theory and the use of hypothesis tests.

LOGIC OF TWO-SAMPLE TESTS

The means of successive samples from a statistical universe are almost never equal. So it may be difficult to determine if the difference in two sample means ($\bar{x}_1$ and $\bar{x}_2$) is due to sampling error, or to a real difference in the populations from which the samples are drawn. Intuitively, we know that if there is a very large difference between two sample means, the means of the two populations (μ_1 and μ_2) are probably not equal. And conversely, if the two sample means are very nearly equal, there is probably no real difference between the means of the two populations.

THEORY OF DIFFERENCE TESTS As with one-sample tests, the procedures for working with two samples are slightly different for (a) large samples (of $n \geqslant 30$), and (b) small samples (of $n < 30$ when σ is unknown). Also, we assume the sampling procedures yield sample observations that are completely independent from each other within the groups, as well as independent between the groups.[4] If the difference between the sample means is too great to be attributed to chance, we conclude that it is probably due to a real difference between the populations, and say that the difference between the sample means is *statistically significant*.

---

[3] Portions of this chapter extend beyond that covered in some introductory texts, so some instructors may choose not to assign some of this material, or to consider parts of it optional.

[4] There are special tests for dependent samples which we shall deal with later.

The theory underlying tests of differences is similar to that used for one-sample tests, which was based upon our knowledge of the sampling distribution of means (x's) and of proportions (p's). This is because just as large sample means and proportions are normally distributed, so too the sampling distributions of the difference in sample means ($\bar{x}_1 - \bar{x}_2$) and the difference in proportions ($p_1 - p_2$) are also normally distributed. And for small samples the distribution of differences also follows the t-distribution—just as we used for one-sample tests. Figure 9–1 illustrates this.

FIGURE 9–1 Sampling distributions of differences

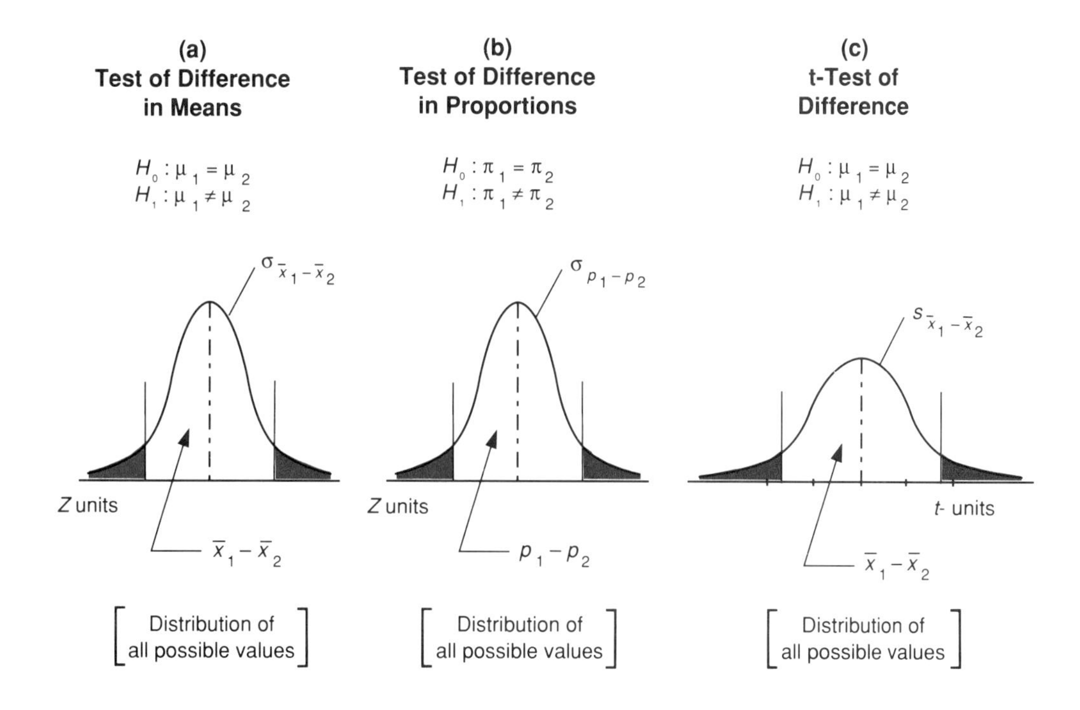

The same theoretical constraints apply to tests of differences. To compute the population standard error requires a knowledge of σ, and we assume that small samples are from a normally distributed population. But again we can use s as an estimator of σ, and $s_{\bar{x}}$ as an estimator of $\sigma_{\bar{x}}$ when σ is unknown.

TEST STATISTIC The test statistic used for testing differences in means and proportions from two independent samples is of similar form to the one-sample statistic. We saw in equation 8–1 that

$$\text{Test Statistic}_{(\text{calc.})} = \frac{\text{statistic} - \text{parameter}}{\text{standard error}}$$

Now, instead of $\bar{x}$ or p, our statistics are $(\bar{x}_1 - \bar{x}_2)$ and $(p_1 - p_2)$, which come from the two groups of sample data. The parameter is typically the hypothesized difference of $\mu_1 - \mu_2 = 0$, or $\pi_1 - \pi_2 = 0$. However, tests can be made to see whether the difference is a specific numerical amount, such as +10 lbs or −35 percent. In these situations we would usually be performing a one-tailed test, in contrast to the two-tailed tests depicted in Figure 9–1.

Stating the test statistic in terms of differences, we have

$$Z_c = \frac{\text{(observed difference)} - \text{(theoretical difference)}}{\text{standard error of the difference}} \qquad (9\text{–}1)$$

where the standard error of the difference is computed by combining the variances of the two samples. As before, the designation of $\sigma_{\bar{x}_1-\bar{x}_2}$ or $s_{\bar{x}_1-\bar{x}_2}$ depends upon whether σ is known. Table 9–1 shows the standard error equations for large sample tests.

TABLE 9–1 Standard error equations for large sample tests of differences

	σ KNOWN	σ UNKNOWN
TESTS OF MEANS:	$\sigma_{\bar{x}_1-\bar{x}_2} = \sqrt{\frac{\sigma_1^2}{n_1} + \frac{\sigma_2^2}{n_2}}$ (9–2)	$s_{\bar{x}_1-\bar{x}_2} = \sqrt{\frac{s_1^2}{n_1} + \frac{s_2^2}{n_2}}$ (9–3)
TEST OF PROPORTIONS:	(no test)	$s_{\hat{p}_1-\hat{p}_2} = \sqrt{\frac{\hat{p}_1\hat{q}_1}{n_1} + \frac{\hat{p}_2\hat{q}_2}{n_2}}$ (9–4)

LARGE SAMPLES: σ KNOWN

The (theoretical) test statistic for large samples where σ is known is

$$\text{For two means:} \quad Z_c = \frac{(\bar{x}_1 - \bar{x}_2) - (\mu_1 - \mu_2)}{\sigma_{\bar{x}_1-\bar{x}_2}} \qquad (9\text{–}5)$$

where $\sigma_{\bar{x}_1-\bar{x}_2}$ is the true standard error of the difference in means.

We do not include an equation for testing the difference between two proportions when σ is known, because if $\sigma_{p_1-p_2}$ can be calculated, this implies π_1 and π_2 must also be known, so there is no need for the test.

LARGE SAMPLES: σ UNKNOWN

For the usual situations where σ is unknown, the t-distribution is again the theoretically correct distribution to describe the differences in means or proportions. But as n becomes large, t approaches Z, and we can again use the normal to approximate the t-distribution.

For difference in two means:

$$t_c \longrightarrow Z_c = \frac{(\bar{x}_1 - \bar{x}_2) - (\mu_1 - \mu_2)}{s_{\bar{x}_1 - \bar{x}_2}} \qquad (9\text{–}6)$$

For difference in two proportions:

$$t_c \longrightarrow Z_c = \frac{(p_1 - p_2) - (\pi_1 - \pi_2)}{s_{\hat{p}_1 - \hat{p}_2}} \qquad (9\text{–}7)$$

In the equations above, $s_{\bar{x}_1 - \bar{x}_2}$ and $s_{\hat{p}_1 - \hat{p}_2}$ are expressions for the estimated standard error of the difference in means and proportions respectively, as calculated from the available sample data.

SMALL SAMPLES: σ UNKNOWN

Small sample tests of the difference between means generally assume that both samples are from normally distributed populations, and that the populations have equal variances.[5] Then the t-distribution is appropriate for describing the standardized differences. However, insofar as t-distribution values are affected by the sample's size, and the samples from the two groups may be of different sizes, the standard error of difference must weight each sample variance by its degrees of freedom. This is accomplished by computing the standard error from a "pooled" or combined estimate of the sample variances, designated s^2_{pl}.

Pooled variance estimate for small samples

$$s^2_{p1} = \frac{s^2_1(n_1 - 1) + s^2_2(n_2 - 1)}{n_1 + n_2 - 2} \qquad (9\text{–}8)$$

Once the pooled estimate has been computed, the test statistic for small sample tests of differences between means is

For difference in means of small samples:

$$t_c = \frac{(\bar{x}_1 - \bar{x}_2) - (\mu_1 - \mu_2)}{\sqrt{\dfrac{s^2_{pl}}{n_1} + \dfrac{s^2_{pl}}{n_2}}} \qquad (9\text{–}9)$$

---

[5] When σ^2_1 and σ^2_2 are not equal, the $(\bar{x}_1 - \bar{x}_2)$ difference is approximately t distributed, and equation (9–3) can be used directly to calculate $s_{\bar{x}_1 - \bar{x}_2}$. However, for the test, the df requires a special adjustment. See S. Christian Albright, *Statistics for Business and Economics* (New York: Macmillan Co., 1987) p. 362.

TESTING THE DIFFERENCE BETWEEN MEANS FROM TWO INDEPENDENT SAMPLES

LARGE SAMPLE PROCEDURE

This section illustrates the procedure for determining the significance of the difference between the means of two independent samples. We begin with a large sample example.

Example 9–1
(Large sample, σ unknown)

A research institute study of 100 smokers in the U.S. showed that they smoke an average of 23 cigarettes per day, with a standard deviation of 3.0 per day. Is this significantly different from a European country where 225 smokers had a mean consumption of 21 per day, with a standard deviation of 5.0 per day?

Solution

We can follow the same testing sequence as one-sample tests. Let the U.S. sample data be designated by the subscript 1 and the European data by subscript 2. Then:

$$n_1 = 100, \quad \bar{x}_1 = 23.0, \quad s_1 = 3.0$$

$$n_2 = 225, \quad \bar{x}_2 = 21.0, \quad s_2 = 5.0$$

The hypotheses can be stated either as:

$$\text{(I) } H_0\text{: } \mu_1 = \mu_2 \qquad \text{and } H_1\text{: } \mu_1 \neq \mu_2$$

$$\text{or (II) } H_0\text{: } \mu_1 - \mu_2 = 0 \quad \text{and } H_1\text{: } \mu_1 - \mu_2 \neq 0$$

Both statements express the same equality for two-tailed tests. However, for one-tailed tests (where a numerical difference is hypothesized), form (II) must be used.

The level of significance α is not specified, but the question is posed as to whether there is a *significant* difference. So we will assume a value of $\alpha = .05$. Also, with large n_1 and n_2 we can use the Z_c test statistic to approximate t_c.

1. H_0: $\mu_1 = \mu_2$
   H_1: $\mu_1 \neq \mu_2$
2. $\alpha = .05$
3. Use normal distribution:
   Reject if $Z_c < -1.96$ or $> +1.96$

4. $Z_c = \frac{(\bar{x}_1 - \bar{x}_2) - (\mu_1 - \mu_2)}{s_{\bar{x}_1-\bar{x}_2}}$ *where*: $\bar{x}_1 - \bar{x}_2 = 23.0 - 21.0 = 2.0$
$\mu_1 - \mu_2 = 0$

$$s_{\bar{x}_1-\bar{x}_2} = \sqrt{\frac{s_1^2}{n_1} + \frac{s_2^2}{n_2}} = \sqrt{\frac{3^2}{100} + \frac{5^2}{225}} = .45$$

$$Z_c = \frac{2.0 - 0}{.45} = 4.44$$

5. *Conclusion*: Reject H_0 of equal means at the 5% level because 4.44 > 1.96. The difference in cigarette consumption is significant. [Note: The H_0 could also be rejected at the 1% level (where $Z = 2.58$) so the difference is actually *highly significant*.]

CONFIDENCE INTERVAL ESTIMATE FOR THE DIFFERENCE BETWEEN MEANS OF TWO INDEPENDENT SAMPLES

When a hypothesis is rejected (and we conclude the population means are different), we may wish to establish a confidence interval estimate for the true difference between the two population means. The procedure is similar to that for setting a confidence interval about one sample mean, with the modification that we establish an interval about the *difference* in means, and we use the standard error of difference to help set the width of this interval.

$$CL \text{ for } (\mu_1 - \mu_2) = (\bar{x}_1 - \bar{x}_2) \pm Z\,(s_{\bar{x}_1-\bar{x}_2}) \qquad (9\text{–}10)$$

Example 9–2 Establish a 95% confidence interval estimate for the true mean difference in cigarette consumption from the previous example where $\bar{x}_1 - \bar{x}_2 = 2.0$ cigarettes per day, and $s_{\bar{x}_1-\bar{x}_2} = .45$.

Solution

$$\begin{aligned} CL_{(\mu_1-\mu_2)} &= (\bar{x}_1 - \bar{x}_2) \pm Z\,(s_{\bar{x}_1-\bar{x}_2}) \\ &= 2.0 \pm 1.96\,(.45) \\ &= 1.1 \text{ to } 2.9 \text{ cigarettes/day} \end{aligned}$$

From the above example, we can be 95% confident that the true difference in consumption in the two regions lies within the interval of 1.1 to 2.9 cigarettes per day. The fact that the 95% confidence interval for the difference between the two means does not include zero attests to the validity of our earlier test that the difference is significant.

SMALL SAMPLE PROCEDURE

Product line managers comparing the reliability of their product with that of a competitive product are sometimes forced to use small samples because testing may destroy the products. And marketing managers seeking to determine if there are significant differences in customer perceptions of their products and competitive products are often faced with high market research costs. So, small sample tests have wide application in business.

The small sample test procedure is similar to the large sample procedure, except that the *t*-distribution is used and the standard error of difference must be calculated from a pooled estimate. At this point, it may be helpful to illustrate the procedure by beginning with some original observations and carrying them through the testing procedure in a step-by-step fashion to the final conclusion.

Example 9–3 An automobile manufacturer in Yugoslavia is experiencing production delays because their motor supplier is unreliable. The firm can improve their delivery schedule and reduce their costs if they are permitted to have two motor suppliers instead of one. They have permission to use either a Japanese or a West German manufactured motor in their new model cars, "if the gas mileage of both motors is the same." The problem is to demonstrate this similarity.

To help make the decision, the manufacturer acquires and installs Japanese motors in 9 cars and German motors in 7 cars and collects data on the gas mileage under normal driving conditions. The data are shown in the following table, along with some deviations used to compute the variance of each sample. Test the hypothesis that the mileages are equal at the 10% level.

JAPANESE MOTOR TEST				WEST GERMAN MOTOR TEST			
Motor No.	mpg x_1	$(x_1 - \bar{x}_1)$	$(x_1 - \bar{x}_1)^2$	Motor No.	mpg x_2	$(x_2 - \bar{x}_2)$	$(x_2 - \bar{x}_2)^2$
1	28	−3	9	1	26	−3	9
2	29	−2	4	2	27	−2	4
3	29	−2	4	3	28	−1	1
4	30	−1	1	4	28	−1	1
5	31	0	0	5	30	1	1
6	32	1	1	6	30	1	1
7	33	2	4	7	34	5	25
8	33	2	4				
9	34	3	9				
Totals	279		36	Totals	203		42

Solution

$$\bar{x}_1 = \frac{\Sigma x_1}{n_1} = \frac{279}{9} = 31.10 \qquad \bar{x}_2 = \frac{\Sigma x_2}{n_2} = \frac{203}{7} = 29.0$$

$$s_1 = \sqrt{\frac{\Sigma(x_1 - \bar{x}_1)^2}{n_1 - 1}} \qquad s_2 = \sqrt{\frac{\Sigma(x_2 - \bar{x}_2)^2}{n_2 - 1}}$$

$$= \sqrt{\frac{36}{9 - 1}} = 2.12 \qquad = \sqrt{\frac{42}{7 - 1}} = 2.65$$

1. H_0: $\mu_1 = \mu_2$
   H_1: $\mu_1 \neq \mu_2$
2. $\alpha = .10$ (This constitutes a 10% chance of rejecting H_0, even if it is true.)
3. Use t-distribution with $n_1 + n_2 - 2 = 9 + 7 - 2 = 14\ df$.
   Reject if $|t_c| > t_{\frac{.10}{2},\ 14df} = 1.761$
4. $$t_c = \frac{(\bar{x}_1 - \bar{x}_2) - (\mu_1 - \mu_2)}{\sqrt{\frac{s_{pl}^2}{n_1} + \frac{s_{pl}^2}{n_2}}}$$

   where, given the assumption of equal population variances, the sample variances can be pooled as:

   $$s_{pl}^2 = \frac{s_1^2(n_1 - 1) + s_2^2(n_2 - 1)}{n_1 + n_2 - 2} = \frac{(2.12)^2(8) + (2.65)^2(6)}{9 + 7 - 2} = 5.57$$

   $$t_c = \frac{(31.00 - 29.00) - 0}{\sqrt{\frac{5.57}{9} + \frac{5.57}{7}}} = \frac{2.00}{1.19} = 1.68$$

5. *Conclusion*: Do not reject H_0: $\mu_1 = \mu_2$. The difference of 2.00 mpg would occur by chance more than 10% of the time, even if there were no real difference. So we conclude the difference is probably just due to sampling error, and not to any statistically significant difference in the motors.

If the evidence in the above example is not strong enough, sample sizes would have to be increased. If the (2 mpg) difference is real—and not just due to sampling error—larger samples would substantiate the difference, while reducing the standard error of the difference. This would make the t_c value larger, perhaps causing the hypothesis to be rejected.

EXERCISES I

1. Suppose you wished to test whether there was a difference in the planned increase in employment over the next six months in the mining industry versus the lumber industry. What null and alternative hypotheses would be most appropriate?
2. In what way is the theory applied to tests of differences similar to the theory used for one-sample tests?
3. Why does the chapter *not* provide a formula for computing the standard error $\sigma_{p_1 - p_2}$ for a large sample test of proportions when σ is known?
4. Why is it necessary to use a "pooled" estimate of the sample variances when computing the standard error of the difference for small sample tests?
5. An electronics firm considering two locations for a new plant has obtained the following data from random samples:

	Location A	Location B
Number of firms sampled	150	95
Average hourly pay	\$8.60	\$9.10
Standard deviation	1.00	2.40

   Test whether there is a statistically significant difference in the hourly pay at the two locations. Let $\alpha = .10$.

6. Compute the appropriate standard error for a test of differences given the following data:

   (a) $s_1 = 8,\ n_1 = 80,\ s_2 = 12,\ n_2 = 120$

   (b) $s_1 = 40,\ n_1 = 15,\ s_2 = 50,\ n_2 = 25$

7. Crescent Machine Shop, experimenting with two methods of welding, found that it took 245 min to complete 75 parts using method #1, and 135 min to complete 45 parts using method #2. The standard deviations were $s_1 = .2$ min/part and $s_2 = .4$ min/part.

   (a) Test whether there is a real difference in the mean time to weld a part at the $\alpha = .01$ level.

   (b) If your difference in (a) is significant, establish a 95% confidence interval for the amount of difference in the two methods.

8. A purchasing agent has tested (and rejected) a hypothesis that the prices for selected building materials are the same in Denver and Atlanta. If the difference between mean prices per unit at the two locations is \$3.65 per ton and the standard error of difference is \$.42 per ton, construct a 96% confidence interval of the difference.

9. Statistics students at two state universities in Michigan were randomly selected and given similar exams. At University #1, where 50 students were tested, the average score was 82 with a standard deviation of 8, whereas at University #2, where 40 students were tested, the average score was 88 with a standard deviation of 6. Test whether the difference in performance between students in the two universities is significant at the 2% level.
10. Test H_0: $\mu_1 = \mu_2$ against H_1:$\mu_1 \neq \mu_2$ at the $\alpha = .20$ level for the following data. Note: $s_1 = 5.08$ and $s_2 = 3.32$. State your steps and conclusion.

Sample #1 scores	43	48	54	34	44	45	39	42	46	48	40	45				Total = 528
Sample #2 scores	46	44	50	48	47	39	49	48	43	53	49	47	45	49	48	Total = 705

11. Industrial engineers at a defense plant have developed two methods of assembling a submarine component. Using the data shown below, test whether the difference between the mean times is significant at the 5% level.

METHOD A (min)	METHOD B (min)
28	25
25	29
27	28
25	27
25	30
	29
	28

TESTING THE DIFFERENCE BETWEEN MEANS OF TWO PAIRED SAMPLES

The t test for differences between two means discussed in the preceding section was based on the assumptions that (1) each sample is *independent* of the other, (2) both samples are *from normally distributed populations*, and (3) *the variances of both samples are equal*. Good approximations may be made when the normality and equal variance assumptions are only approximately correct. But a different procedure is required if the samples are not independent.

This section describes tests involving dependent (or paired) samples. For example, we may wish to evaluate the change in performance of employees attending a communications seminar. Dependent samples allow us to test for any differences in the same individuals (or elements) before and after some treatment, or to compare paired elements receiving different treatments.

The computations for paired samples are simpler than for independent samples, because both samples must be the same size. But instead of the standard error of the difference in means, $s_{\bar{x}_1 - \bar{x}_2}$, we use the standard error of the mean differences $s_{\bar{d}}$, which is equal to the standard deviation of sample differences, s_d divided by $\sqrt{n}$.

$$s_{\bar{d}} = \frac{s_d}{\sqrt{n}} \tag{9–11}$$

And the value of s_d is computed in a similar manner to the standard deviation of X values—except the *values used are differences* instead of original values.

$$s_d = \sqrt{\frac{\Sigma(d - \bar{d})^2}{n - 1}} \quad \text{or} = \sqrt{\frac{\Sigma d^2}{n - 1} - \frac{n\bar{d}^2}{n - 1}} \tag{9–12}$$

The test statistic for paired data ($_{pd}$) is again distributed as a t-distribution in the familiar form of the statistic minus the hypothesized parameter, divided by the standard error of difference:

$$t_{pd} = \frac{\bar{d} - \mu_d}{s_{\bar{d}}} \tag{9–13}$$

Although the new equations may seem a bit confusing at this point, an example will show that for the most part, we simply substitute differences for X values; otherwise the testing procedure remains largely the same.

TESTING FOR BEFORE AND AFTER DIFFERENCES

TWO-TAILED TEST We illustrate the paired data applications first as a two-tailed test, and then follow with a one-tailed test.

Example 9–4 Ten bank tellers are asked to help test whether a corrective drill significantly improves the speed and accuracy with which they enter customer transactions into the bank computer. They are first given a five-minute "Before Treatment" test, then the drill, and finally an "After Treatment" test. The statistical analysis is intended to test the hypothesis that the drill has not significantly affected their speed and accuracy scores.

Data below show the before and after scores, along with the difference values needed to calculate the standard deviation of the differences, s_d. (Remember that these are not random samples from two independent populations. The scores on the second test are related to scores on the first test because they were made by the same individuals.)

Differences between the paired observations for the two tests are provided below in order to help obtain the single sample of differences. Test whether there is any difference in scores at the 5% level of significance.

SCORES BEFORE AND AFTER CORRECTIVE DRILL FOR TEN BANK TELLERS

Teller Number	Before Drill Score	After Drill Score	Difference d	Deviation $(d - \bar{d})$	Deviation Squared $(d - \bar{d})^2$
A	45	49	4	1	1
B	52	56	4	1	1
C	34	31	−3	−6	36
D	38	46	8	5	25
E	47	54	7	4	16
F	42	39	−3	−6	36
G	61	68	7	4	16
H	53	55	2	−1	1
I	52	50	−2	−5	25
J	49	55	6	3	9
TOTAL			30		166

Solution

The mean difference value is needed to complete the table above:

$$\text{Mean difference:} \quad \bar{d} = \frac{\Sigma d}{n} = \frac{30}{10} = 3$$

Standard deviation of differences:

$$s_d = \sqrt{\frac{\Sigma(d - \bar{d})^2}{n - 1}} = \sqrt{\frac{166}{10 - 1}} = \sqrt{18.44} = 4.295$$

Standard error of the mean difference:

$$s_{\bar{d}} = \frac{s_d}{\sqrt{n}} = \frac{4.295}{\sqrt{10}} = 1.358$$

Our hypothesized difference is zero, and we can follow the standard steps for testing a hypothesis:

1. H_0: $\mu_1 - \mu_2 = 0$ (which is equivalent to H_0: $\mu_1 = \mu_2$)
   H_1: $\mu_1 - \mu_2 \neq 0$
2. $\alpha = .05$
3. Use t-distribution with $n - 1 = 10 - 1 = 9\ df$.
   Reject if $|t_c| > t_{\frac{.05}{2},\ 9\ df} = 2.262$

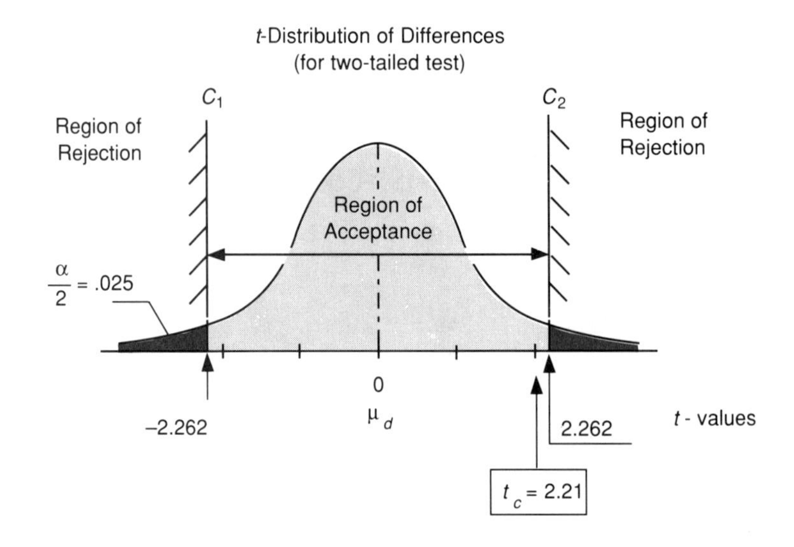

4.

$$t_c = t_{pd} = \frac{\bar{d} - \mu_d}{s_{\bar{d}}}$$

where: $\bar{d}$ = mean difference = 3

μ_d = hypothesized mean difference = 0

$s_{\bar{d}}$ = standard error of mean difference = 1.358

$$t_{pd} = \frac{3 - 0}{1.358} = 2.21$$

5. *Conclusion*: Do not reject the hypothesis of no difference because the difference is not significant at the 5% level for this two-tailed test.

ONE-TAILED TEST The example above asked to test for "any difference." However, insofar as the drill was expected to have a positive effect on the speed and accuracy scores, it may be more logical to use a (more powerful) one-tailed test for this situation.

Example 9–5

Using the data of Example 9–4, test the hypothesis of no difference against the alternative that the drill scores after the training are higher.

1. H_0: $\mu_1 - \mu_2 = 0$
   H_1: $\mu_2 > \mu_1$
2. $\alpha = .05$
3. Use t-distribution with $n - 1 = 10 - 1 = 9$ df.
   Reject if $|t_c| > t_{\frac{.05}{2},\, 9\, df} = 1.833$

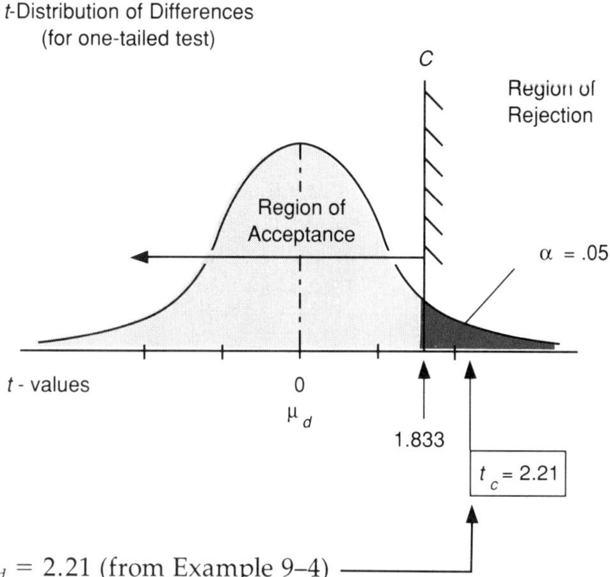

4. $t_c = t_{pd} = 2.21$ (from Example 9–4)
5. *Conclusion*: Reject hypothesis of equal means because $2.21 > 1.833$. Conclude that there is an improvement in the speed and accuracy score.

ANALYSIS We can reject the hypothesis of no difference in scores with a one-tailed test because all the error probability (α) is in the upper tail of the distribution, giving us the smaller reject value (1.833), versus 2.262 for the two-tailed test. As noted in Chapter 8, the *one-tailed test thus has more power* (to reject a false null hypothesis).

PAIRED ELEMENTS RECEIVING DIFFERENT TREATMENTS The paired difference test can also be used to test the effect of a specified treatment on pairs of observations from a population. For example, several pairs of equally qualified workers may be given different types of training:

- Worker #1 of the pair is given on-the-job training
- Worker #2 of the pair is given training in a company school.

After the training, the difference in performance, d, is measured to determine whether it is large enough to conclude that one training procedure is superior to the other. Like the before-after testing, this is also a dependent situation because the difference in performance is presumed to be explained by (or *dependent upon*) the difference in treatment (e.g., training).

To assure the validity of paired difference tests involving different treatments, it is important that both elements of the pair have equal capabilities to begin with. When applied to people, this may require careful matching in terms of age, experience, I.Q. score, and other background data.

CONFIDENCE INTERVAL FOR DIFFERENCE BETWEEN MEANS OF TWO PAIRED SAMPLES

When a test for the equality of differences is rejected, one may wish to establish a confidence interval for the true difference between the means of paired samples. Insofar as the samples are not independent, we need not assume equal variances—as was required for the confidence interval for the difference between means of two independent samples. However, the procedure for constructing the confidence interval is similar to that used before; the difference is we use the symbols for paired samples instead.

$$CL \text{ for } (\mu_1 - \mu_2) = \bar{d} \pm t_{\frac{\alpha}{2}, n-1}\, s_{\bar{d}} \tag{9–14}$$

Example 9–6 A sample of 16 telemarketing representatives were tested on their ability to hold customers on the phone before and after a sales training session. If $\bar{d}$ is 10 min and $s_d = 5$, what is the 95% confidence interval for the true difference?

Solution

$$CL \text{ for } (\mu_1 - \mu_2) = \bar{d} \pm t_{\frac{\alpha}{2}, n-1}\, s_{\bar{d}}$$

$$\text{where: } t_{\frac{.05}{2}, 15\, df} = 2.131$$

$$s_{\bar{d}} = \frac{s_{\bar{d}}}{\sqrt{n}} = \frac{5}{\sqrt{16}} = 1.25$$

$$= 10 \pm 2.131\,(1.25)$$
$$= 7.34 \text{ min to } 12.66 \text{ min}$$

Confidence intervals for differences between means of two paired samples are less widely used than for single means or proportions. It is usually sufficient to determine whether the difference between samples is large enough to reject the hypothesis that there is no difference.

EXERCISES II

12. Data collected for a test of H_0: $\mu_1 - \mu_2 = 0$ using the difference between means of paired samples revealed the following:

$$n = 12 \qquad \Sigma\, d = 32 \qquad \Sigma\,(d - \bar{d})^2 = 418.6$$

Find: (a) $\bar{d}$, (b) s_d, (c) $s_{\bar{d}}$, and (d) t_c

13. A production supervisor in a clothing factory is attempting to determine whether background music affects the productivity of her employees. While nine employees were working on a large order of the same items (blouses), she recorded the output during one week with no music, and another week with music. Results are shown below:

EMPLOYEE NAME	UNITS PRODUCED: NO MUSIC	UNITS PRODUCED: WITH MUSIC
Louise	92	99
Kathleen	87	83
Bridget	104	101
Merridy	76	83
Steven	88	91
Catherine	94	85
Jo	79	66
Bryant	96	83
Evelyn	94	92

(a) State H_0 and H_1.

(b) Test your hypothesis at the 5% level, being sure to state all steps.

14. Sixty marketing representatives for a cosmetics company were rated (on a zero to ten scale) according to their sales ability before and after a sales training seminar in Dallas. The results were: $\Sigma\, d = 78$ and $\Sigma\,(d - \bar{d})^2 = 408$. Assuming the seminar was expected to improve one's sales ability, perform a one-tailed test of H_0: $\mu_1 - \mu_2 = 0$ against H_1: $\mu_2 > \mu_1$. Use a 10% level of significance.

15. Ten pairs of students are selected for an experiment to determine if there is a difference in learning statistics by a class-paced versus a self-paced method. Each pair is carefully matched in terms of gpa, total course load, quantitative aptitude, and other variables. Then one student of the pair is assigned to each method. The final exam scores are as follows:

Pair No.	1	2	3	4	5	6	7	8	9	10
Class-Paced	78	85	69	92	87	88	75	91	83	84
Self-Paced	72	86	64	88	78	82	78	90	80	74

15. Compare the two different methods of instruction by testing the mean difference in the paired samples at the 2% level of significance.
16. (*One-tailed test of specified difference*) The manufacturer of a gasoline additive has, in the past, shown that adding a 6 oz bottle of their product (*Mileage-Maker*) to the gasoline tank will increase one's mileage. Now they have asked a consumer products research organization to test the claim that mileage is improved by 3 mpg. The research organization has agreed to give credence to the claim by testing the H_0: $\mu_1 - \mu_2 = 3$ mpg against the alternative H_1: $\mu_1 - \mu_2 < 3$ mpg at the $\alpha = .05$ level. Then they carefully measured the mileage obtained from a tank of gasoline on six different makes of cars both before and after adding the chemical.

AUTO NO.	1	2	3	4	5	6
MPG Before	28.6	35.4	32.2	41.8	27.2	39.4
MPG After	29.8	38.2	37.4	44.3	31.7	42.4

Use the data provided to conduct the test and state your conclusion.

17. In a test of the difference between two instructional techniques using ten pairs of equally qualified students, the students using method B scored an average of 4 points better than those using method A. If the standard error of the mean difference is 1.31 points, what is the 95% confidence interval for the true difference?
18. Given the following data resulting from a test of the difference between means of two paired samples:

$$n = 20 \text{ pairs} \qquad \Sigma d = 62 \qquad \Sigma(d - \bar{d})^2 = 220$$

(a) Test H_0: $\mu_1 - \mu_2 = 0$ against H_1: $\mu_1 - \mu_2 \neq 0$ at the $\alpha = .10$ level.

(b) Establish a 95% confidence interval for the difference.

TESTING THE DIFFERENCE BETWEEN PROPORTIONS FROM TWO INDEPENDENT SAMPLES

Like some other tests, the test for a difference between two proportions is a statistical procedure for deciding whether the difference observed in two independent samples is large enough to be reasonably certain that it stems from a real difference between the populations. Small sample procedures are not commonly used here because a precise estimate of a proportion requires a reasonably large sample. Moreover, if n is small, the normal curve (or t-distribution) is not a satisfactory approximation of the binomial. So our discussion shall be limited to large sample tests, where σ is unknown.

DIFFERENCE BETWEEN TWO-SAMPLE PROPORTIONS: LARGE INDEPENDENT SAMPLES

The statistic for testing the difference in proportions from large independent samples where σ is unknown was introduced earlier as equation 9–7:

$$Z_c = \frac{(p_1 - p_2) - (\pi_1 - \pi_2)}{s_{\hat{p}_1 - \hat{p}_2}}$$

In this equation, $s_{\hat{p}_1 - \hat{p}_2}$ is the estimated standard error of the difference between the two proportions. Note, however, that in computing the value of $s_{\hat{p}_1 - \hat{p}_2}$, we have only the sample values of p_1 and p_2 to work with—no hypothesized values of the population proportions are given. Therefore, before computing $s_{\hat{p}_1 - \hat{p}_2}$, we first combine the data from the two samples to obtain a better single estimate of the true value of the population proportion than either of the individual sample estimates.[6] The combined estimate (in the forms for both proportions and for numbers) is:

USING PROPORTIONS	USING NUMBERS	
$\hat{p} = \dfrac{n_1(p_1) + n_2(p_2)}{n_1 + n_2}$	$\hat{p} = \dfrac{x_1 + x_2}{n_1 + n_2}$	(9–15)

where x_1 and x_2 are the number of successes in samples 1 and 2, respectively. Given this weighted estimate $\hat{p}$, we can then combine the samples in the calculation of the estimated standard error of the difference (see equation 9–4):

$$s_{\hat{p}_1 - \hat{p}_2} = \sqrt{\frac{\hat{p}\,\hat{q}}{n_1} + \frac{\hat{p}\,\hat{q}}{n_2}}$$

An example will illustrate. Although it uses equal sample sizes, n_1 and n_2 need not be equal (but they should both be > 30).

Example 9–7 A manufacturer of computers has two suppliers for a keyboard component. Supplier A has a very compact unit and a lower price, whereas supplier B is located in the same region, and can respond faster to changing delivery requirements. A test for defectives in random samples of 400 units from each supplier revealed the following:

	SUPPLIER A	SUPPLIER B
Number of units	$n_1 = 400$	$n_2 = 400$
Number of defectives	$x_1 = 82$	$x_2 = 94$
Percent defective	$p_1 = 82/400 = .205$	$p_2 = 94/400 = .235$

[6] The procedure discussed herein is for tests of H_0: $\pi_1 - \pi_2 = 0$. In problems where the hypothesized difference between two proportions is some value greater than zero, the two estimates are not weighted into a common proportion. The standard error of the difference is then estimated from equation (9–17) in the next section.

The difference in percent defective is rather small (3%), and the manufacturer wishes to test whether the proportion defective from each supplier is equal, or (alternatively) whether it is statistically significant at the 10% level.

Solution

1. H_0: $\pi_1 = \pi_2$
   H_1: $\pi_1 \neq \pi_2$
2. $\alpha = .10$
3. Use normal distribution; reject if $|Z_c| > 1.645$
4. $$Z_c = \frac{(p_1 - p_2) - (\pi_1 - \pi_2)}{s_{\hat{p}_1 - \hat{p}_2}}$$

   where: $(p_1 - p_2) = .205 - .235 = -.03$
   $(\pi_1 - \pi_2) = 0$

   $$s_{\hat{p}_1 - \hat{p}_2} = \sqrt{\frac{\hat{p}\,\hat{q}}{n_1} + \frac{\hat{p}\,\hat{q}}{n_2}}$$

   $$\text{and } \hat{p} = \frac{x_1 + x_2}{n_1 + n_2} + \frac{82 + 94}{800} = .220$$

   $$s_{\hat{p}_1 - \hat{p}_2} = \sqrt{\frac{(.22)(.78)}{400} + \frac{(.22)(.78)}{400}} = .0293$$

   $$Z_c = \frac{(-.3) - 0}{.0293} = -1.024$$
5. *Conclusion*: Do not reject hypothesis of equal proportions because $-1.645 < Z_c$ (of -1.024) < 1.645.

ANALYSIS The difference is not sufficient to conclude that there is a real difference in the proportion defective of the two producers at the 10% level of significance. But how significant is the difference? What is the p value?

From the table of areas under the normal curve, the area corresponding to Z_c of 1.024 is approximately .3461. Adjusting this to the two-tailed test, the probability of a difference of three or more percentage points in favor of either A or B is (2)(.5000 − .3461) = .3087. This means that we should expect a difference this large or larger about 31% of the time—as a result of sampling error, even if there is really no difference.

ONE-TAILED TESTS One-tailed tests of the difference in proportions are computed in a similar manner to the example. However, the alternative hypothesis is stated as H_1: $\pi_1 > \pi_2$, or as H_1: $\pi_1 < \pi_2$, and the rejection region is all in one tail of the distribution, rather than being split into two tails. Otherwise, the computations are comparable.

CONFIDENCE INTERVAL FOR THE DIFFERENCE BETWEEN TWO PROPORTIONS

If desired, a confidence interval for the difference between two proportions can be established by following the standard confidence interval procedure.

$$CL \text{ for } (\pi_1 - \pi_2) = (p_1 - p_2) \pm Zs_{p_1 - p_2} \qquad (9\text{–}16)$$

There is one difference to take note of, however. Because confidence limits are based wholly upon sample data (with no hypothesized parameter involved), there is no reason to combine the data to get a pooled estimate of π. So we need not first calculate p. The standard error of the difference is estimated more directly from equation 9–17:

$$s_{p_1 - p_2} = \sqrt{\frac{p_1 q_1}{n_1} + \frac{p_2 q_2}{n_2}} \qquad (9\text{–}17)$$

Intervals with various confidence coefficients can thus be established by using different Z values.

EXERCISES III

19. Given the data shown, test H_0: $\pi_1 = \pi_2$ against H_1: $\pi_1 \neq \pi_2$ at $\alpha = .10$

SAMPLE #1: $n_1 = 50$, $p_1 = .22$ SAMPLE #2: $n_2 = 100$, $p_2 = .18$

20. A market research firm must determine whether the proportion of households with VCRs is the same in two marketing areas. Sample data are:

	REGION 1 (Northeast)	**REGION 2 (Southeast)**
Number households	150	200
Households with a VCR	57	90

Test the hypothesis that the proportion of households with VCRs is equal at the 20% level of significance.

21. A governmental agency wishes to determine whether the nation's wealth is becoming more concentrated. The group used for the study is a sample from the top one-half of one percent of households. A study before an extended economic expansion (1963) suggested these households controlled 25% of the nation's wealth ($p_1 = .25$), whereas a later study (1983) suggested a value of 35% ($p_2 = .35$). If the standard error of the difference, $s_{\hat{p}_1 - \hat{p}_2}$, has been computed to be .022 , test whether the apparent increase is statistically significant at the 1% level. (Hint: Use the H_0: $\pi_1 = \pi_2$, and H_1: $\pi_2 > \pi_1$)

22. A major automobile manufacturer plans to evaluate the effectiveness of an expensive advertising campaign by statistical analysis of consumer acceptance of the firm's image before and after the campaign. Samples of prospective customers revealed the following:

BEFORE ADVERTISING	AFTER ADVERTISING
$n_1 = 500$	$n_2 = 1{,}000$
Number favoring = 205	Number favoring = 500
Percent favoring = 205/500 = 41%	Percent favoring = 500/1,000 = 50%

Determine whether there has been a real change in consumer acceptance of the firm. [Note: This is different from determining whether advertising caused a change. If all other factors were held constant—which may be impossible—advertising may be a likely cause. But we *cannot prove causation* by this procedure.] Use a significance level of 1%.

23. Use the data from the previous problem to compute the 95% confidence interval estimate for the true difference between the two proportions of 41% and 50%.

24. A multinational firm has hired a consultant to determine if workers in their domestic plants are as satisfied with their jobs as employees in their foreign plants. Results of the study are as follows:

	DOMESTIC PLANTS	FOREIGN PLANTS
Number workers sampled	350	225
Number satisfied with job	245	189

(a) Test whether the proportions satisfied in the two locations are equal at the $\alpha = .05$ level. (Show all steps.)

(b) Establish a 90% confidence interval estimate of the difference between the two proportions. (Show all calculations.)

(c) Did you use the same standard error of difference in both calculations? Explain.

DECISION MAKING RULES AND SAMPLE SIZE FOR QUALITY CONTROL[7]

Managers use tests of hypotheses to analyze data so they can make good decisions. But some statistical decision making can be reduced to a simple decision making rule that, once established, requires no special knowledge or skill on the part of the user. One area of extremely wide usage for this is in quality control, where workers may need to apply statistical sampling and testing procedures, even though they may lack an understanding of the statistics. Other applications exist in marketing, personnel, and finance.

The theory that applies to decision making rules has already been explained in this chapter and the previous one. So we will conclude the chapter now with four examples that illustrate how simple instructions may be given to apply some of these statistical techniques to typical situations—especially in the quality control area. The first three examples relate to setting action, or C, limits, as we referred to them in the discussion on tests of hypotheses. Then the final example shows how the specification of the risk of type I and type II error essentially dictates what the sample size must be in order to meet the producer and consumer requirements.

Example 9–8 Assume that a company is buying a product where strength is important. The standard deviation is 25.0 and the hypothetical mean strength is 420. The probability of rejecting a shipment, using samples of 100, is not to exceed .05 if the mean is $\geqslant 420$. Provide a decision making rule which will enable anyone to determine whether a shipment should be accepted.

Solution

The standard error of the mean for samples of 100 is:

$$\sigma_{\bar{x}} = \frac{\sigma}{\sqrt{n}} = \frac{25.0}{\sqrt{100}} = 2.5$$

Only one limit is required since shipments will never be rejected because they are too strong. The action limit, C_1, should be set at $420 - 1.645\,(2.5) = 415.89$. The rule is: "Take a sample of 100 and compute the average (mean) strength. If the mean is less than 415.89, reject the shipment. If the mean is equal to or greater than 415.89, accept the shipment."

[7] This section extends beyond the material covered in some introductory texts to more advanced applications in the area of quality control. So your instructor may choose to designate it as optional; otherwise proceed on as usual.

Example 9–9 Assume that we are manufacturing a product where strength is a critical factor. The strength is typically equal to about 400. If the average decreases to 390, we want the probability of stopping the process to be .99. The population standard deviation is 20.0, and samples of 100 will be taken every hour. What rule should be followed in the plant?

Solution

The standard error of the mean is:

$$\sigma_{\bar{x}} = \frac{\sigma}{\sqrt{n}} = \frac{20.0}{\sqrt{100}} = 2.0$$

The probability of a type II error is specified at .01 if μ is equal to 390. The C_1 limit should be set far enough above 390 so that if $\mu = 390$, the probability will be .99 that the process will be stopped. $C_1 = 390 + 2.33\,(2.0) = 394.66$. The rule is: "Take a sample of 100 every hour and compute the mean for each sample. If the sample mean is less than 394.66, stop the process and look for the cause. If the sample mean is 394.66 or above, let the process run."

Example 9–10 Assume that a supplier guarantees that his product will run no more than 20 percent defective. He agrees that a shipment may be returned if a sample of 400 indicates an excessive number of defective units in the shipment. We want the probability of rejection to be .95 if the true proportion is as great as .30.

Solution

There need be only one limit, since we will never reject a shipment because it is too good. If the true proportion is .30, then $\sigma_p = \sqrt{(.30 \times .70)/400} = .0229$. The limit should be set far enough below .30 to include .4500 between .30 and the C_2 limit. $C_2 = .30 - 1.645(.0229) = .2623$ or 26.23%, and 26.23% of 400 is 104.9. The rule, in its simplest form, is: "Take a sample of 400 and determine the number of defective units. If the number defective is 104 or less, accept the shipment. If the number defective is 105 or more, reject the shipment."

Example 9–11 A buyer and a seller agree that the hypothetical mean is 500 and the standard deviation is 25.0. A decrease in the mean is detrimental to the buyer, and he wants his risk of acceptance of a shipment which has a mean of 495 to be no more than .01. The seller wants the risk of having a shipment rejected when its mean is ≥ 500 to be no more than .05. How large should the sample be, and what rule should be followed?

Solution

We can solve for n by setting up two equations for the (same) limit C:

FROM RIGHT ($\mu = 500$) SIDE:

$$C = \mu_H - Z_\alpha \frac{s}{\sqrt{n}} = 500 - 1.645 \frac{25}{\sqrt{n}}$$

FROM LEFT ($\mu = 495$) SIDE:

$$C = \mu + Z_\beta \frac{s}{\sqrt{n}} = 495 + 2.33 \frac{25}{\sqrt{n}}$$

SETTING $C = C$ AND SOLVING FOR n WE HAVE:

$$500 - 1.645 \frac{25}{\sqrt{n}} = 495 + 2.33 \frac{25}{\sqrt{n}}$$

$$5 = \frac{58.25}{\sqrt{n}} + \frac{41.13}{\sqrt{n}}$$

$$n = \left(\frac{99.38}{5}\right)^2 = 394$$

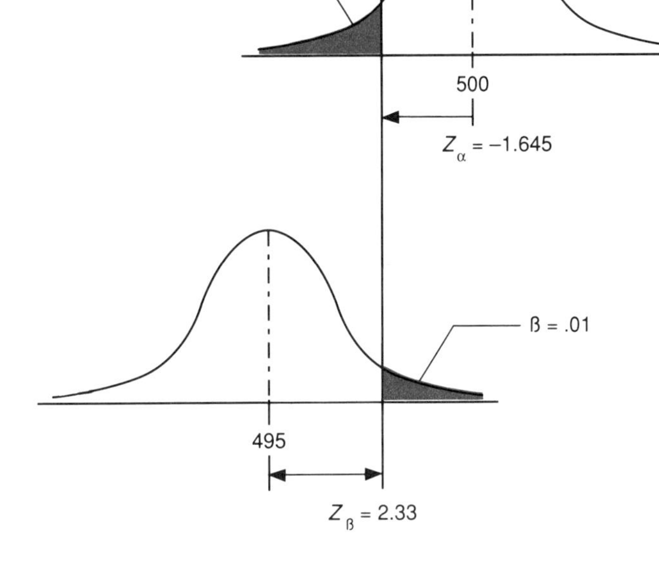

The sample size should be about 394. The limit should be set at $500 - 1.645(25)/\sqrt{394} = 497.93$ or at $495 + 2.33\,(25)/\sqrt{394} = 497.93$.

The rule is: "Take a sample of 394 and compute the sample mean. If the mean is less than 497.93, reject the shipment. If the mean is greater than or equal to 497.93, accept the shipment."

These four examples show how to predetermine the probability of type I error or type II error with a given sample size, and how the probability of both type I and type II errors may be specified if the sample size is not predetermined. Rules for both means and proportions have been illustrated. In every case, the rule is stated in such a way that no knowledge of statistics is required of the decision maker.

SUMMARY

Two-sample tests rely upon the normal distribution of the difference in sample means ($\bar{x}_1 - \bar{x}_2$), and sample proportions ($p_1 - p_2$) for large random samples, and upon the t-distribution for small samples. If the difference between the sample statistics is larger than what might be expected in a sampling distribution centered upon the hypothesized difference, then we conclude that it is *statistically significant*.

For tests of differences, we typically hypothesize that the two parameters are equal (and alternatively not equal, or one greater than another). The α level and reject values (Z or t) are then set in the same manner as one-tailed tests. But the test statistic (Z_c or t_c) is computed as:

$$\begin{bmatrix}\text{test}\\ \text{statistic}\end{bmatrix} = \frac{(\text{sample difference}) - (\text{hypothesized difference})}{\text{standard error of the difference}}$$

If the Z_c or t_c values exceed the Z or t-distribution limits, respectively, we reject the hypothesis of equality of the population parameters and conclude the samples are probably from different populations.

LARGE SAMPLES A major computational difference in conducting two-sample tests versus one-sample tests is in the computation of the standard error. Assuming σ is unknown (which is usually the case), the standard errors of difference for means and proportions of large samples are:

FOR DIFFERENCE IN MEANS	**FOR DIFFERENCE IN PROPORTIONS**
$s_{\bar{x}_1-\bar{x}_2} = \sqrt{\dfrac{s_1^2}{n_1} + \dfrac{s_2^2}{n_2}}$	$s_{\hat{p}_1-\hat{p}_2} = \sqrt{\dfrac{\hat{p}\hat{q}}{n_1} + \dfrac{\hat{p}\hat{q}}{n_2}}$

The $\hat{p}$ and $\hat{q}$ values for the proportion equation must, in turn, be estimated by combining the data from the two samples to obtain a single estimate of the population proportion. [For confidence interval estimates of the difference, however, the sample values (p_1, q_1, n_1 and p_2, q_2, n_2) are used to compute $s_{p_1-p_2}$ directly, because no hypothesized value is assumed.]

SMALL SAMPLES Small samples are not used to estimate the difference in proportions. The t-distribution is appropriate for small sample estimates of the difference in means, and the test statistic is:

$$t_c = \frac{(\bar{x}_1 - \bar{x}_2) - (\mu_1 - \mu_2)}{\sqrt{\dfrac{s_{pl}^2}{n_1} + \dfrac{s_{pl}^2}{n_2}}}$$

Because the samples are small (and possibly of different size), the test statistic equation for small samples uses a "pooled" estimate of the two sample variances, s_{pl}^2. This weights each sample variance by its respective degrees of freedom, and results in a more representative value than if either s_1^2 or s_2^2 were used individually.

$$s_{pl}^2 = \frac{s_1^2(n_1 - 1) + s_2^2(n_2 - 1)}{n_1 + n_2 - 2}$$

TESTS OF PAIRED DIFFERENCES The other major topic discussed in this chapter was tests of the difference between the means of two paired samples. These were distinguished as *dependent* tests because they were either "before" and "after" measures on the same person (or element), or they were comparisons of paired elements that received different treatments.

The test procedure for paired differences was similar to the other tests, except that the difference values, d, replaced the individual X values, and $\bar{d}$ replaced $\bar{x}$. Then the standard error of the mean difference was

$$s_{\bar{d}} = \frac{s_d}{\sqrt{n}}$$

And the standard deviation of differences, s_d, was computed as

$$s_d = \sqrt{\frac{\Sigma(d - \bar{d})^2}{n - 1}}$$

Finally, the test statistic for paired data was the familiar form of statistic minus hypothesized parameter, divided by the standard error of difference:

$$t_{pd} = \frac{\bar{d} - \mu_d}{s_{\bar{d}}}$$

The last section of the chapter showed how hypothesis testing theory is used in some firms by presenting some applications in the area of quality control. It also demonstrated how to determine the sample size required to control the risks of type I and type II error.

ESTIMATION AND TESTING SUMMARY EQUATIONS Table 9–2 summarizes the estimation and hypothesis testing equations encountered in the last few chapters. The confidence interval equations are in the column on the right; otherwise the equations are arranged for use in testing hypotheses. Begin on the left side and work to the right. For example, your inquiry might take the following pattern, or sequence:

1. *Sample size*: Is the sample large ($n > 30$) or small?
2. *Means or proportions*: Is the problem one of means or of proportions?
3. *Number of samples*: Do you have one sample or two?
4. *One-tail/two-tail*: Is it a one-tailed test, a two-tailed test, or an estimation problem?
5. *Sample statistic*: What is the sample statistic and what is its value?
6. σ: Is σ known or unknown? (This establishes which equation—and row—is used to compute the standard error.)
7. *Standard error*: Compute the standard error.
8. *Do the calculation*: Calculate the test statistic, or confidence interval.

TABLE 9–2 Summary of equations for testing hypotheses and for estimation

	Test of Hypothesis	Hypothesized Parameter	Sample Statistic	Standard Error	Calculated Test statistic	σ is	Confidence Interval for Estimation
Large Sample $n \geq 30$	**MEANS** **One sample**	$H_0: \mu = 80$ $H_1: \mu \neq 80$ *or* $\mu < 80$ > 80	$\bar{x}$	$\sigma_{\bar{x}} = \frac{\sigma}{\sqrt{n}}$	$z_c = \frac{\bar{x} - \mu_H}{\sigma_{\bar{x}}}$	Known	$\bar{x} \pm z\, \sigma_{\bar{x}}$
				$s_{\bar{x}} = \frac{s}{\sqrt{n}}$	$z_c = \frac{\bar{x} - \mu_H}{s_{\bar{x}}}$	Unknown	$\bar{x} \pm z\, s_{\bar{x}}$
	Two samples	$H_0: \mu_1 - \mu_2 = 0$ $H_1: \mu_1 - \mu_2 \neq 0$ *or* $H_0: \mu_1 = \mu_2$ $H_1: \mu_1 \neq \mu_2$	$\bar{x}_1 - \bar{x}_2$	$\sigma_{\bar{x}_1 - \bar{x}_2} = \sqrt{\frac{\sigma_1^2}{n_1} + \frac{\sigma_2^2}{n_2}}$	$z_c = \frac{(\bar{x}_1 - \bar{x}_2) - (\mu_1 - \mu_2)}{\sigma_{\bar{x}_1 - \bar{x}_2}}$	Known	$(\bar{x}_1 - \bar{x}_2) \pm z\, \sigma_{\bar{x}_1 - \bar{x}_2}$
				$s_{\bar{x}_1 - \bar{x}_2} = \sqrt{\frac{s_1^2}{n_1} + \frac{s_2^2}{n_2}}$	$z_c = \frac{(\bar{x}_1 - \bar{x}_2) - (\mu_1 - \mu_2)}{s_{\bar{x}_1 - \bar{x}_2}}$	Unknown	$(\bar{x}_1 - \bar{x}_2) \pm z\, s_{\bar{x}_1 - \bar{x}_2}$
	PROPORTIONS **One sample**	$H_0: \pi = .20$ $H_1: \pi \neq .20$ *or* $\pi < .20$ $\pi > .20$	p	$\sigma_p = \sqrt{\frac{\pi_H(1 - \pi_H)}{n}}$	(no test)	Known or (assumed)	(no estimate)
				$s_p = \sqrt{\frac{p\,q}{n}}$	$z_c = \frac{p - \pi_H}{\sigma_p}$	Unknown	$p \pm z\, s_p$
	Two samples	$H_0: \pi_1 - \pi_2 = 0$ $H_1: \pi_1 - \pi_2 \neq 0$	$p_1 - p_2$		(no test)	Known	(no estimate)
				$s_{\hat{p}_1 - \hat{p}_2} = \sqrt{\frac{\hat{p}\,\hat{q}}{n_1} + \frac{\hat{p}\,\hat{q}}{n_2}}$ *where* $\hat{p} = \frac{x_1 + x_2}{n_1 + n_2}$	$z_c = \frac{(p_1 - p_2) - (\pi_1 - \pi_2)}{s_{\hat{p}_1 - \hat{p}_2}}$	Unknown	$(p_1 - p_2) \pm z\, s_{p_1 - p_2}$ *where* $s_{p_1 - p_2} = \sqrt{\frac{p_1\, q_1}{n_1} + \frac{p_2\, q_2}{n_2}}$
Small Sample $n < 30$	**MEAN** **One sample**	$H_0: \mu = 70$ $H_1: \mu < 70$	$\bar{x}$	$s_{\bar{x}} = \frac{s}{\sqrt{n}}$	$t_c = \frac{\bar{x} - \mu}{s_{\bar{x}}}$	Unknown	$\bar{x} \pm t\, s_{\bar{x}}$
	Two samples	$H_0: \mu_1 - \mu_2 = 0$ $H_1: \mu_1 - \mu_2 \neq 0$	$\bar{x}_1 - \bar{x}_2$	$s_{\overline{pl}} = \sqrt{\frac{s_{pl}^2}{n_1} + \frac{s_{pl}^2}{n_2}}$ *where* $s_{pl}^2 = \frac{s_1^2(n_1 - 1) + s_2^2(n_2 - 1)}{n_1 + n_2 - 2}$	$t_c = \frac{(\bar{x}_1 - \bar{x}_2) - (\mu_1 - \mu_2)}{s_{\overline{pl}}}$	Unknown	$(\bar{x}_1 - \bar{x}_2) \pm t\, s_{\overline{pl}}$
	PAIRED DIFFERENCE	$H_0: \mu_1 - \mu_2 = 0$ $H_1: \mu_1 - \mu_2 \neq 0$	$\bar{d}$	$s_{\bar{d}} = \frac{s_d}{\sqrt{n}}$	$t_c = \frac{\bar{d} - \mu_d}{s_{\bar{d}}}$	Unknown	$\bar{d} \pm t\, s_{\bar{d}}$

CASE: TRANSCONTINENTAL COMMUNICATIONS COMPANY

Gene Duffy had his fingers crossed that everything would go well at the 3 PM meeting today in the president's conference room. That was when the top brass of the company were to preview his proposed television commercial. Transcontinental Communications Co. (TCC) was counting on this commercial to help secure enough long distance customers to justify a multimillion dollar investment in fiber optic transmission equipment. No doubt about it—it just had to be supercolossal.

After six months of planning, filming, and editing, plus an expenditure of nearly $325,000, Gene felt he was ready. Everyone who had worked with the 45-second commercial seemed to think it was great—*fantastic* was the word the marketing manager had used. Now it was up to the top management to give their approval to proceed with the final touches.

At the meeting, the presentation went well, but the questions were more penetrating than Gene Duffy had anticipated. Thomas Bryant, the legal counsel, immediately homed in on the claim in the film that the cost of TCC's service was the same as National Telephone's, but that TCC's quality was better.

BRYANT: You can conjure up all kinds of quality advantages—and we've got some real ones for sure. But what about some proof of that cost claim? Are the costs really equal? If we say so in the advertisement, we'd better be ready to back that up!

DUFFY: We don't have any hard facts, but we've called around to get prices and compare costs. And we seem to come out about the same.

BRYANT: I sympathize with you, but I'm afraid that kind of data wouldn't stand up under a competitor's lawsuit. Could you possibly cite some specific comparative costs—even a small sample, but something concrete?

The meeting ended with the recommendation that the commercial be approved if costs on a random sample of calls showed TCC to be the same as National. Gene Duffy was to collect some data, and Thomas Bryant was to make the final decision.

The data had to be gathered quickly. Gene proposed that it take the form of a sample of a dozen phone calls placed over both systems from the same locations at the same hours of the day (and night). The costs would be billed to an existing customer under each system, and then be compared.

As the billing notices came in, Gene was surprised at the difference in costs. He finally worked up his courage and walked into Tom Bryant's office ready to concede that TCC should not advertise equal costs.

DUFFY: Well, it sure surprised me! I guess it's a good thing we did the study. I'll get right onto a revision in the commercial this afternoon.

BRYANT: I can understand your disappointment. But let's not be too hasty about this, Gene. I talked to Sue Reicher down in Consumer Statistics about these costs. She seems to think we have nothing to worry about. In fact she says the difference between the two bills is insignificant. That's good enough for me. I say let's go with it.

COST COMPARISON FOR TELEPHONE CALLS TO SAME NUMBERS

CALL NUMBER	CLASS OF CALL	NET CHARGE TCC ($)	NET CHARGE NATIONAL ($)
1	ZB	.65	.68
2	D	12.07	13.20
3	ZB	.30	.28
4	ZD	29.02	20.50
5	B	15.43	10.44
6	B	19.82	20.10
7	A	7.45	5.15
8	B	.54	.55
9	ZD	2.52	1.20
10	D	16.20	22.10
11	D	23.70	24.00
12	ZB	4.30	1.80
Totals		132.00	120.00

Case Questions

(a) What would be a statement of the hypothesis Sue Reicher tested?

(b) How does the t_c value compare with the reject value for $t_{\frac{.10}{2},\ 22df}$?

(c) How could Sue consider the difference "insignificant" when the National Telephone bill was really less than the TCC bill?

(d) In this situation, what is accomplished by "pooling" the variances?

(e) Suppose the same results (with same means, $\bar{x}_{TCC} = \$11.00$ and $\bar{x}_{NT} = \$10.00$, and same variances) had resulted from a sample of 600 calls instead of only 12. Would the same conclusion hold?

(f) Do you think Mr. Bryant was wise in encouraging Gene Duffy to proceed by leaving the cost claim in the commercial? Explain.

QUESTIONS AND PROBLEMS

25. For testing the difference between means, what sampling distribution is theoretically correct for:
    (a) large samples when σ is known?
    (b) large samples when σ is unknown?
    (c) small samples when σ is unknown?
26. Suppose you are testing the difference between means from two small independent samples. What two assumptions apply?
27. Differentiate between
    (a) a test of the difference between means of two independent samples, and the difference between means of two paired samples
    (b) the standard error of difference in proportions used for (1) testing hypotheses and (2) setting a confidence interval for the true difference
28. An insurance company is planning to offer a new insurance/investment product to retired persons. They have sampled some retirees in two major market areas to see if there is a difference in their retirement income (from pension and social security) with the following results:

	Region 1	Region 2
Number persons sampled	300	225
Mean retirement income/yr	$23,000	$27,500
Standard deviation	$ 3,000	$ 6,000

    Is there a significant difference (i.e., at the 5% level)? Show all the steps of your test.
29. Two samples of $n = 16$ are drawn to test for the equality of means, and the difference turns out to be $\bar{x}_1 - \bar{x}_2 = 4.2$, with $s_1 = 4.0$ and $s_2 = 2.0$. What would be the t_c value for the test?
30. The following data were collected to test whether there was a difference in the salaries of secretaries who lived in the core area versus the suburban area of a major city:

CORE RESIDENTS	SUBURBAN RESIDENTS
$n_1 = 10$	$n_2 = 17$
$\bar{x}_1 = \$ 24,450$	$\bar{x}_2 = \$ 26,900$
$s_1 = \$ 860$	$s_2 = \$1,240$

Show all steps to test the hypothesis that both samples were drawn from the same population at the .01 level of significance.

31. In an attempt to determine whether there was a difference in performance between employees receiving their training on the job versus at a training school, fifteen employees were carefully matched in terms of ability, age, and other factors. Then one person from each pair was randomly assigned to each group. Scores reflecting their subsequent performance are given below. The firm wishes to determine whether the difference in scores is large enough to reasonably conclude that one training procedure is superior to the other. Test the H_0: $\mu_1 = \mu_2$ at the .02 level of significance.

Pair No.	1	2	3	4	5	6	7	8	9	10	11	12	13	14	15
On-Job Score	76	80	94	88	90	82	76	81	83	86	78	76	79	84	80
School Score	82	81	87	79	95	85	68	78	91	87	83	79	85	87	84

32. Sixteen students were tested on their ability to remember items presented to them prior to and following a month-long period when they were given lecithin every day. The mean difference in their memory score was $\bar{d} = +10$ (improvement), and the standard error of the mean difference was 1.25 points. Compute the 95% confidence interval for the true difference.

33. An automobile company is doing a preliminary study to determine which of two assembly plants to phase out over the next three years. Each plant employs several thousand workers, and they have sampled some to determine average age. A random sample of 100 workers from the California plant showed a mean age of 43 years and standard deviation of 7 years. A similar sample of 50 workers from the plant in Ohio showed a mean age of 41 years with a standard deviation of 6 years. Under the H_0: $\mu_{CAL} = \mu_{OHIO}$ (versus H_1: $\mu_{CAL} \neq \mu_{OHIO}$), at what specific level of significance is the difference in average age significant?

34. Nationwide Real Estate Company wishes to determine if the proportion of U.S. households that own their homes is any different in the $40 < 50$ age group versus the $50 < 60$ age group. A random sample of 1,200 households in the $40 < 50$ age group showed that 768 owned their homes, whereas a sample of 900 in the $50 < 60$ age group showed that only 504 owned their homes. Is the difference statistically significant (at the .05 level)?

35. A soft drink producer in California is evaluating an advertising campaign to promote the sale of its diet soft drink. To determine the effectiveness of the program, random samples of 2,000 persons were questioned before and after the advertising was done. The following results were obtained:

Before: $n_1 = 2{,}000$ Number favoring company's product = 600

After: $n_2 = 2{,}000$ Number favoring company's product = 800

These data are to be used to test the hypothesis that there has been no increase in the percent of the population favoring the company's product. Use the .01 level of significance.

(a) What is the null hypothesis and what is the alternative hypothesis?

(b) What is the observed difference between the two sample proportions?

(c) What is the combined estimate of π from the two samples?

(d) Is the increase in customer preference statistically significant?

(e) If it is, is it necessarily due to the advertising campaign?

36. The following data were collected in a survey to determine whether a new detergent (A) was really superior to a proven product (B). The X-scores are ratings on a 500-point scale.

DETERGENT A	**DETERGENT B**
$\bar{X}_1 = 426.0$	$\bar{X}_2 = 415.0$
$s_1 = 30.0$	$s_2 = 31.0$

(a) Suppose one were to test the hypothesis H_0: $\mu_1 = \mu_2$ against the alternative H_1: $\mu_1 \neq \mu_2$ using a sample of 100. Could the H_0 be rejected at the .05 level?

(b) Assume that only 16 boxes of detergent were sampled from each brand. Could the H_0 be rejected at the .05 level?

(c) Summarize your results from part (a) and part (b).

(d) Why were you less likely to be able to reject H_0 in part (b)?

(e) Why did you have to use the t-distribution to solve part (b)?

37. Make a decision rule for accepting shipments from a supplier assuming that the supplier guarantees that his product contains no more than 10% defective. Assume that we will take samples of 400 and we want the probability of rejecting a shipment with 20% defective to be .99.

(a) State the rule in terms of number of defectives.

(b) State the rule in terms of percent defective.

38. A building materials manufacturer produces an insulation section that is supposed to weigh 60.0 lbs; the standard deviation is known to be 2.0 lbs. The Quality Control Supervisor wishes to test the hypothesis that the true mean weight is 60 lbs against H_1: $\mu < 60.0$ lbs. He wants to design the test so that the probability of type I error is .05. If the true mean weight is as low as 59.0 lbs, the probability of acceptance of the hypothesis (type II error) is to be .10. How large a sample is needed?

39. Assume the firms listed in the Corporate Statistical Database (Appendix A) constitute random samples from normally distributed populations (with the same variance). Test the hypothesis that the price/earnings ratio of the retail firms equals that of the banking and insurance firms. Use a significance level of 10%.

40. Using the Corporate Statistical Database (Appendix A), test the hypothesis that the advertising expenditures are equal in the banking and insurance versus the transportation firms. Let $\alpha = .05$, and assume the data constitute random samples of the industry drawn from normally distributed populations.

41. Using the electrical and electronic firms of the Corporate Statistical Database (Appendix A) as a sample, conduct a paired difference test to determine if the mean earnings per share increased from two years ago, EPS2 (before), to one year ago, EPS1 (after). Let $\alpha = .05$.

42. A management analyst doing a plant location study has narrowed the choice of site down to a western location (in Washington, Oregon, or California), or an eastern one (in Massachusetts, New York, or New Jersey). Now the analyst wishes to compare income and salary levels in the two locations. Use the data from Appendix A.

    (a) Is the mean state per capita income different in the two locations? [Note: The per capita income figures are from census data.]

    (b) Assuming the salary data for the firms listed are representative of all salaries in the areas, use the HQ data to test whether there is a difference in the average annual salary of employees in the two locations. [Note: You may wish to sort the data according to states first.] Use $\alpha = .10$.

CHAPTER

10

Chi-Square and Analysis of Variance

INTRODUCTION

THE CHI-SQUARE DISTRIBUTION

TESTS OF GOODNESS OF FIT

Fit of Data to a Uniform Distribution
Fit of Data to a Poisson Distribution
Fit of Data to Other Distributions

TESTS FOR INDEPENDENCE OF CLASSIFICATION

Contingency Tables
Test Procedure

ADDITIONAL APPLICATIONS OF CHI-SQUARE

THE *F*-DISTRIBUTION

ONE-WAY ANALYSIS OF VARIANCE

USING THE ANOVA TABLE

TWO-WAY ANOVA TESTS

SUMMARY

CASE: AEROSPACE INDUSTRIES

QUESTIONS AND PROBLEMS

DO SALARIES FOR GRADUATES DIFFER ACROSS THE COUNTRY?

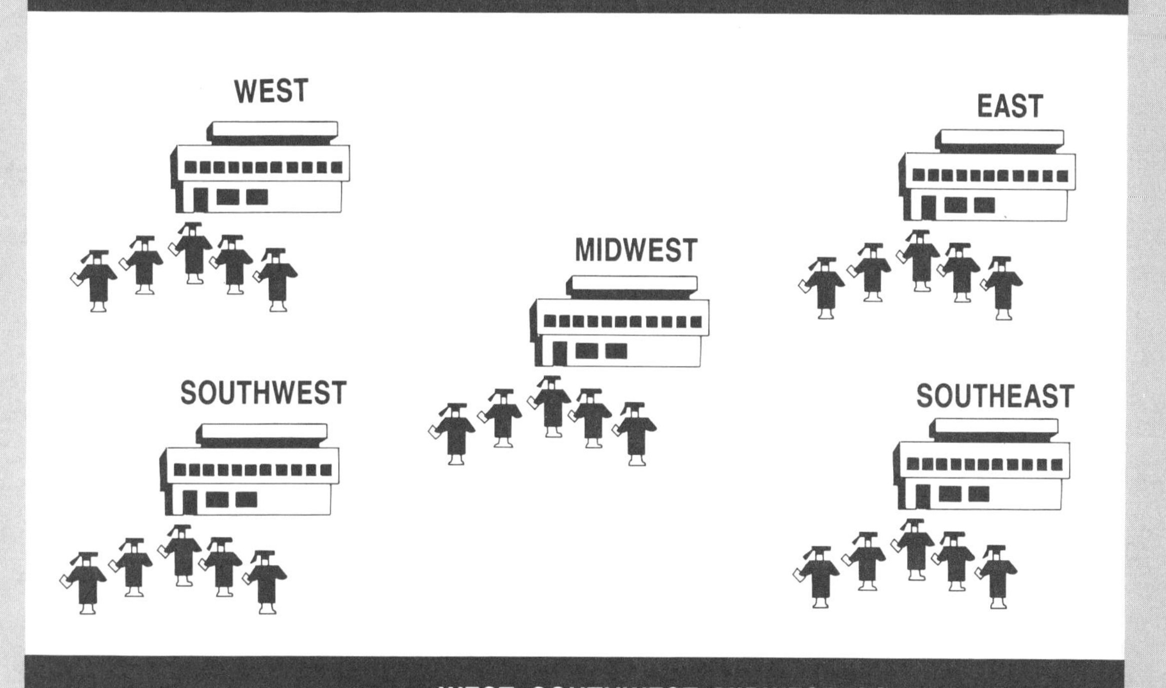

	WEST	SOUTHWEST	MIDWEST	SOUTHEAST	EAST
Number earning < $24,000	26	37	19	62	109
Number earning ≥ $24,000	16	33	19	50	89

Chapter 10 Objectives

1. Use chi-square to test whether a population conforms to (fits) an assumed pattern.
2. Use chi-square to test whether tabular data are independent of the principles of classification.
3. Recognize suitable applications, and apply the *F*-distribution.
4. Use ANOVA to test for the difference among means of more than two groups of samples.

Does the salary you might expect upon graduation depend upon where you happen to live? Is there really a difference in salaries offered to marketing, finance, personnel, and economics majors? And do your employment prospects differ depending upon whether you graduate from a national university, or a state college, or a private college?

INTRODUCTION

MULTINOMIAL DATA Suppose a random survey of 460 recent marketing graduates in different regions of the country revealed the annual salary offers shown above. Are the data sufficient to suggest a real difference in salaries in the geographic classifications? Or is any such difference more likely due to chance? The methods discussed in this chapter enable us to make statistically valid comparisons among several classifications of data. In this sense, they are said to apply to *multinomial* (many categories of) data—in contrast to binomial, or two-category, data, for example.

CHI-SQUARE AND *F*-DISTRIBUTIONS This chapter introduces us to two new distributions—the chi-square, χ^2 (like Ki and rhymes with "pie" square), and the *F*-distributions. Both are used to analyze data that are classified into two or more categories. However, chi-square is appropriate for testing multiple sets of data expressed in *frequencies* or *counts,* as above. Moreover, the attributes under study need not be numerically measurable. For example, they can be nominal (name) classifications, such as geographic areas, colors, or opinions (agree, neutral, disagree). The *F*-distribution is used primarily to test the equality of *several means.* But it has other uses as well.

We will first study two major types of chi-square tests. Then we will go on to show how the *F*-distribution is used for testing means in a procedure that statisticians refer to as the Analysis of Variance (ANOVA).

THE CHI-SQUARE DISTRIBUTION

Suppose you have a normally distributed variable with a mean of zero and a standard deviation of one, as illustrated in Figure 10–1. Now draw a random sample of two values from that distribution—assume you get -1.6 and $+.8$.

FIGURE 10–1 A random sample of two from a standard normal ($\mu = 0, \sigma = 1$) distribution

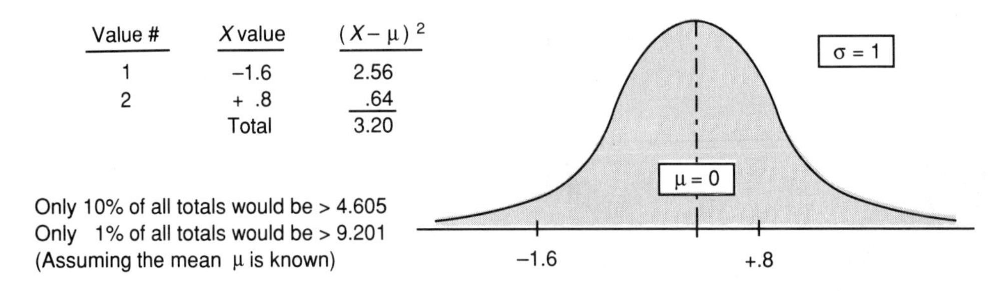

Because the mean of the distribution is already zero, these two values are essentially deviations from the mean. If you square the deviations and add

them, the total (3.20) would be the sum of squared deviations (*SSD*) from a normal ($\mu = 0, \sigma = 1$) distribution for a sample of size $n = 2$. If you were to do this thousands of times, you would find that only 10% of the totals would be greater than 4.605, and only 1% greater than 9.201. Of course, if you took a sample of three, the comparable limits would be higher (6.251 and 11.345).

These *SSD* values for various sample sizes constitute elements of what we refer to as the *chi-square distribution.*

Chi-square, χ^2, is a probability distribution of the sum of squared deviations that result from sampling from a normal distribution (where $\mu = 0, \sigma = 1$). There is a different chi-square distribution for each sample size; the mean of each equals its number of *dfs*, and the standard deviation equals $\sqrt{2df}$.

CHI-SQUARE TABLES As illustrated in Appendix F, chi-square distribution tables show the sums of squared deviations that one might expect from samples of the different sizes. Moreover, they show what proportion of the time those sums are likely to be exceeded. Note, however, that the sums are tabulated according to the (only) parameter of the chi-square distribution—which is the number of degrees of freedom (*df*). This corresponds to the number of squared deviations that are summed up if we know μ, or to $n - 1$ if we use $\bar{x}$ as an estimator of μ. Thus, in Appendix F, the 4.605 and 9.201 values above correspond to the limits for the right tail area (or level of significance) of .10 and .01 respectively, for tests involving 2*df*. (The number of *df* differs depending upon the test being conducted.)

CHI-SQUARE CURVES Being a sum of squared deviations, chi-square can theoretically take on any positive values from zero to infinity. And there is a different curve for each number of degrees of freedom. So chi-square is really a *family of continuous distributions.* They are skewed to the right for small numbers of *df*, but approach the normal curve as the *df* gets larger. As illustrated in the sketch in Appendix F, the peak of the curve (the mode) is at zero for curves with one or two degrees of freedom, and at degrees of freedom minus 2 for distributions with three or more degrees of freedom.

CHI-SQUARE TEST STATISTIC We learned earlier that sample means and proportions tend to be normally distributed. Chi-square tests rely upon the similarity of a distribution of deviations. If the observed frequencies (f_o) and expected frequencies (f_e) are reasonably close, the squared deviations can be expected to follow the same pattern as one would get by comparable sampling from a normal (0,1) distribution—which are the values listed in the chi-square distribution table. If not, the squared values will be larger than the values in the chi-square table. This tells us that there is a significant difference among the groups of data.

The test statistic to be compared with the tabled chi-square values in Appendix F is computed as

$$\chi^2 = \Sigma \frac{(f_o - f_e)^2}{f_e} \quad (10\text{–}1)$$

where f_o denotes the observed frequencies and f_e denotes the expected, or theoretical, frequencies. We will use this same expression in both types of tests that follow.

TESTS OF GOODNESS OF FIT

Goodness of fit tests are used to determine how closely observed frequencies of occurrence agree with the frequencies that would occur if the distribution followed some specified theoretical pattern. For example, chi-square tests can be used to determine whether empirical data actually conform to a statistical distribution such as the uniform, Poisson, normal, or binomial. We illustrate the procedure with examples using the uniform and Poisson distributions. Other procedures follow the same general idea.

FIT OF DATA TO A UNIFORM DISTRIBUTION

A uniform or equally likely distribution has an equal proportion of observations in each class. If actual data conform to the theoretical distribution, then the distribution of differences between observed and theoretical values should be normally distributed, and the calculated chi-square values should not exceed the χ^2 table (test criterion) values.

STEPS IN THE TEST A chi-square test of the hypothesis that data are distributed according to the uniform distribution follows the same (five) steps we used earlier for other tests of hypotheses. We begin with the null hypothesis that the data are distributed according to the uniform distribution—or that the proportions are equal—either H_0 would be satisfactory. The alternate hypothesis is that the data are not uniformly distributed or that the proportions are not equal:

(1) H_0: $\pi_1 = \pi_2 = \pi_3 = \pi_4 = \pi_5$

H_1: at least one pair not equal

Then we designate (2) the level of significance, α, and (3) the test criterion, or reject value, from the chi-square table. This, of course, depends upon the degrees of freedom. (For uniform distribution tests, *df* is equal to the number of categories minus one.) Next (4) a test statistic is calculated using equation 10-1, and finally (5) a conclusion is reached to either reject or not reject the null hypothesis.

Example 10–1 Is it reasonable to believe that the frequency of accidents in a chemical plant is uniformly distributed among the five workdays of a week? Or might one expect more accidents on Mondays (after a busy weekend) or on Fridays? If so, company managers may wish to search for some causal factor. Assume that a review of safety records shows a plant had fifty accidents in the past two years. The days on which they occurred are shown in the f_o column of the table below. Test the hypothesis that the accident data is uniformly distributed at the $\alpha = .05$ level of significance.

Solution

Given our hypothesis that the distribution is uniform, there should be an equal number of accidents on each day of the week. The total, 50, divided by the number of working days per week, 5, gives us the expected number of accidents per working day, of $f_e = 10$.

DAY	f_o	f_e	$f_o - f_e$	$(f_o - f_e)^2$	$(f_o - f_e)^2/f_e$
Monday	8	10	−2	4	.4
Tuesday	10	10	0	0	.0
Wednesday	12	10	2	4	.4
Thursday	14	10	4	16	1.6
Friday	6	10	−4	16	1.6
Totals	50	50	0		4.0

The $(f_o - f_e)$ column shows how far the observed frequencies deviate from the expected frequencies. Then the deviations are squared and "averaged" by dividing by the number of expected frequencies. The total for the five categories is 4.0, which will be compared with the reject value from the chi-square table. Completing the steps of the hypothesis testing procedure, we have:

1. H_0: $\pi_M = \pi_{TU} = \pi_W = \pi_{TH} = \pi_F$

   H_1: Proportions not all equal
2. $\alpha = .05$
3. Use χ^2: Reject if $\chi^2_{calc} > \chi^2_{.05,\,4df} = 9.488$
4. Calculate $\chi^2 = \Sigma \dfrac{(f_o - f_e)^2}{f_e} = .4 + 0 + .4 + 1.6 + 1.6 = 4.0$
5. *Conclusion*: Do not reject H_0, because $4.0 < 9.488$

As is evident from Figure 10–2, the computed chi-square value of 4.00 lies at a point between the .50 and .10 levels of significance, but is much closer to .50. Thus the probability is almost .50 that we would find the frequency of accidents to be distributed in a pattern that differs from the uniform distribution as much as the chemical company data shown above. It would probably be a waste of time to try to find a cause for the high frequencies of Wednesday and Thursday, because that is very likely just a coincidence.

FIGURE 10–2 Chi-square distribution with 4 degrees of freedom

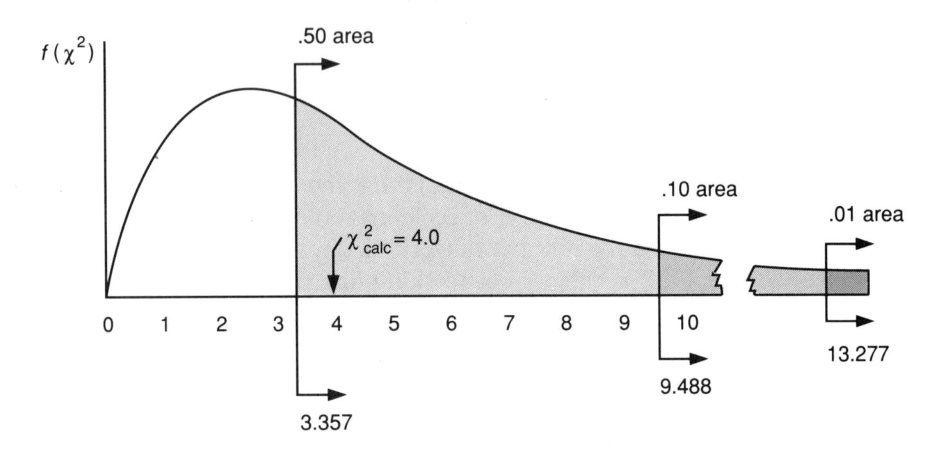

FIT OF DATA TO A POISSON DISTRIBUTION

There are many occasions for testing whether observations follow a Poisson distribution. A common application is in the analysis of waiting lines, where the researcher wants to determine whether the number of customers per unit of time follows a Poisson distribution. If it does, the waiting line can be simulated on a computer to determine the number of service facilities (service attendants, tellers' stations, dock facilities, and so on) which minimize costs while providing the desired level of service.

DEGREES OF FREEDOM For the uniform distribution where no population parameters had to be estimated, the *df* was equal to the number of classes or categories of data minus one. The one *df* is automatically lost because the individual values must match a fixed total. As a more general rule, the *df* for chi-square tests of goodness of fit are equal to the number of classes (k) minus one, minus the number of parameters (m) that must be estimated from sample data.

$$df = (k - 1) - m \qquad (10\text{–}2)$$

For distributions such as the uniform, where no parameters need to be estimated, the *df* is thus equal to $k - 1$. For Poisson distributions where the sample mean is used as an estimate of λ, then $m = 1$, so the $df = (k - 1) - 1$. (However, If the true mean is known beforehand, then only the one *df* is lost.)

To illustrate the fit of data to a Poisson distribution, we use data which we may assume represent the number of ship arrivals per day at a major seaport, as tabulated over a 365-day period.

Example 10–2 Data have been collected on the number of ship arrivals per day for the purpose of developing a computer simulation model that will enable planners to project docking needs for the next several years. In 365 days of data, the number of arrivals per day ranged from zero to ten, as shown in column 2 of the table below. Test whether the frequencies of ship arrivals are Poisson distributed, using a significance level of .10.

Solution

Table 10–1 has been extended to include some of the needed calculations, which are explained below.

TABLE 10–1 Number of ships arriving at a port per day for 365 days

(1)	(2)	(3)	(4)	(5)	(6)	(7)	(8)	(9)	(10)
No. of Arrivals x	Observed (No. Days) f_o	f_oX	Revised Arrivals x	Revised (No. Days) f_o	Poisson Prob.	$(6)\times\Sigma(2)$ Expected f_e	$(f_o - f_e)$	$(f_o - f_e)^2$	$\frac{(f_o - f_e)^2}{f_e}$
0	20	0	0	20	.0498	18.2	1.8	3.24	.178
1	53	53	1	53	.1494	54.5	−1.5	2.25	.041
2	80	160	2	80	.2240	81.8	−1.8	3.24	.040
3	83	249	3	83	.2240	81.8	1.2	1.44	.018
4	64	256	4	64	.1680	61.3	2.7	7.29	.119
5	34	170	5	34	.1008	36.1	−2.8	7.84	.213
6	18	108	6	18	.0504	18.4	−.4	.16	.009
7	8	56	≥7	13	.0335	12.2	.8	.64	.052
8	3	24							
9	1	9							
10	1	10							
Totals	365	1,095		365	1.0000	365	0		.670

This example presents two new challenges: (1) the need to calculate the frequencies expected under the assumption of a Poisson distribution, and (2) avoiding the error that results when the f_e in any class is less than 5.

POISSON PROBABILITIES Poisson probabilities can be taken directly from the Poisson distribution (Appendix D) once the mean, λ, of the distribution is estimated from the sample. And the mean can easily be estimated from columns 1 and 2, which constitute a frequency distribution with a class interval equal to 1.0.

$$\lambda\bar{x} = \frac{\Sigma f_o x}{n} = \frac{1{,}095 \text{ ship arrivals}}{365 \text{ days}} = 3.0 \text{ ship arrivals/day}$$

The first Poisson probability in column 6 is thus the $P(X = 0 \mid \lambda = 3.0)$, which, from Appendix D, is .0498. Other values follow in the same manner. Note that the "frequencies" we are concerned with are the number of days that fall into each category. The expected (or theoretical) frequency values are then obtained by multiplying these Poisson probabilities by the sample size of $n = 365$ days. So the first expected frequency is $f_e = (.0498)(365) = 18.2$ days of zero arrivals. (Expected frequencies need not be whole numbers.)

SMALL FREQUENCY ADJUSTMENT Cells with expected frequencies of less than 5 observations do not provide a sufficiently accurate estimate of values that might be *expected* from the (continuous) chi-square distribution. This limitation presents a problem when we compute the Poisson probability for the classes representing 8, and 9, and 10 arrivals. For example, the $P(X = 8 \mid \lambda = 3.0)$ is .0081, so the expected frequency would be $f_e = (.0081)(365) = 2.96$, or only 3 days.

The small-frequency-error problem is usually avoided by combining the classes having very low expected frequencies with other classes, so that no cell in the f_e column contains less than 5 observations. (*Actual* f_o values may, however, be less than 5.) In this example, we combine the frequencies (from column 2) for the classes representing 7, 8, 9, and 10 arrivals into a single group of $\geqslant 7$ arrivals, with a frequency of 13 days (in the revised column 5). The corresponding Poisson probability (Column 6) is then the sum of the probabilities of 7, 8, 9, and 10 arrivals (or more).

$$P(x \geqslant 7 \mid \lambda = 3.0) = [P(x = 7) + P(x = 8) + P(x = 9) + P(x = 10) + P(x > 10)]$$
$$= .0216 + .0081 + .0027 + .0008 + .0003 = .0335$$

For the $\geqslant 7$ class, the theoretical f_e value is then this sum (.0335), multiplied by the sample size (365), which yields the value shown (12.2). (Note that we cannot combine these cells until the mean has been calculated because the mean is needed to compute the Poisson probabilities. But it cannot be calculated from an open-ended frequency distribution with a class of $\geqslant 7$.)

The remaining columns of the table follow the standard format. Column 8 shows the difference between the observed and expected values, column 9 squares these differences to eliminate signs, and column 10 divides the squared differences by the theoretical frequencies and sums to yield the computed chi-square value of .670. The completed test is thus:

1. H_0: Population data are Poisson distributed
   H_1: Population data are not Poisson distributed
2. $\alpha = .10$
3. Use χ^2 : Reject if $\chi^2_{calc} > \chi^2_{.10,\ 6df} = 10.645$
4. Calculate $\chi^2 = \Sigma \dfrac{(f_o - f_e)^2}{f_e} = .670$
5. *Conclusion:* Do not reject H_0, because $.670 < 10.645$

The *df* for the test in the example above is the number of classes minus two. With arrivals shown in eight classes (0 through ≥ 7), the $df = 8 - 2 = 6$. From Appendix F, the chi-square reject value is 10.645, so the hypothesis of a fit to the Poisson distribution cannot be rejected. In fact, a closer review of the $df = 6$ row of the chi-square table shows that the computed value of .670 is even smaller than the .99 level of significance value. This means that over 99% of samples of this size taken from a Poisson distribution would vary from the expected frequencies by more than the amount shown in our example. In other words, the fit is almost perfect. While such a perfect fit would never cause us to reject the hypothesis that the data are Poisson distributed, such close results might well cause us to check our procedure and computations, or question the authenticity of the data.

FIT OF DATA TO OTHER DISTRIBUTIONS

NORMAL A common assumption in statistics is that a sample is from a population that is normally distributed. To check this assumption, one might plot the data in the form of a frequency polygon and see if it conforms to a normal curve drawn on the same graph—using a table of heights of ordinates of the normal curve. A second procedure is to plot the cumulative relative frequencies on preprinted probability paper. If the data are normally distributed, the graph of the cumulative relative frequencies forms a straight line diagonally across the probability paper. Both procedures are somewhat subjective.

A third procedure is to use the chi-square goodness of fit test for the null hypothesis that the population is normally distributed. While we will not illustrate that procedure here, it is similar in concept to the Poisson distribution approach. The mean and standard deviation must first be estimated from the sample data. Then the areas (under the normal curve) are computed for each of the class intervals, and these are used to develop expected frequencies for each class. The expected frequencies are then compared with the actual frequencies in each class using the standard chi-square calculation format.

For the fit of data to normal distributions, the number of degrees of freedom is equal to the number of classes minus three degrees of freedom lost. One *df* is lost by virtue of fixing the final observed total, and one each for the parameters μ and σ, which have to be estimated from the sample data.

BINOMIAL AND OTHER DISTRIBUTIONS For other distributions, the expected frequencies for each class are calculated using the appropriate distribution (e.g., the binomial), and the chi-square statistic is calculated in the standard manner. If no population parameters need to be estimated for the calculation, the degrees of freedom are equal to the number of classes minus one (i.e., $k - 1$). However, an additional *df* should be subtracted for each population parameter that must be estimated from the sample data. Otherwise, the tests are similar to what we have done above.

EXERCISES I

1. Suppose you draw random samples of $n = 9$ from a normal (0,1) distribution, and square the values obtained. What percent of the time will the sum of the squared values exceed (a) 8.343, (b) 14.684, (c) 21.666?
2. How many parameters are needed to define the following distributions?
   (a) binomial, (b) Poisson, (c) normal, (d) chi-square
3. With respect to chi-square:
   (a) Is the distribution discrete or continuous?
   (b) What would be the mode of a distribution with 25 *df*?
   (c) For any kind of empirical data, what type of pattern could one expect from squared deviations of observed and expected values that are reasonably close?
4. What are the appropriate number of degrees of freedom for the following goodness of fit tests?
   (a) Uniform distribution
   (b) Poisson distribution where μ is known
   (c) Normal distribution where neither μ nor σ are known
   (d) Other distributions in general
5. A chi-square test of the goodness of fit of data to a uniform distribution with 20 classes revealed a $\Sigma[(f_o - f_e)^2/f_e]$ value of 29.375. Show the steps for testing the hypothesis that the proportions are equal at the .10 level.
6. Random samples of some stockroom items were counted to see if the actual (observed) counts agreed with the inventory records (expected). Compute the chi-square value and compare it with the .05 level to determine if the observed and expected counts are in significant agreement.

ITEM NO.	1	2	3	4	5	6	7	8	9	10	Total
Observed f_o	15	18	65	80	80	55	20	35	18	74	460
Expected f_e	14	20	65	76	92	50	22	33	20	68	460

7. When a manufacturing process is under control, 90 percent of the units produced are of grade A and 10 percent are of grade B. A sample of 400 units contained 56 units of grade B. Use the chi-square test of goodness of fit at the $\alpha = .02$ level to determine if the process is in control (or producing up to expectations). *Hint:* The categories are grades A and B, and the hypothesized proportions are

$\pi_A = .90$ and $\pi_B = .10$. These result in expected values of .90 times the sample total and .10 times the sample total, respectively.

8. A new study is made of accident occurrences in an industrial plant to determine if accidents are equally likely on any day of the week. Given the results shown below, test whether the data are uniformly distributed. Use a .10 level of significance, and show all the steps of your test.

Day	No. Accidents
Monday	15
Tuesday	20
Wednesday	25
Thursday	28
Friday	12
Total	100

9. A state lottery official claims that the winning digits are randomly selected and occur with an equal likelihood. However, the frequency of the last digit in the winning numbers has had the pattern shown below. Are the data sufficient to question the randomness of the winning numbers? Use a chi-square goodness of fit test at the .10 level of significance.

Winning Digit	0	1	2	3	4	5	6	7	8	9	**Total**
No. Occurrences	4	8	12	9	17	8	12	6	14	10	100

10. Test whether the following data conform to the Poisson distribution. Use an $\alpha = .02$ level, and compute the mean, λ, from the data given.

No. of insulation defects	0	1	2	3	4	5	**Total**
Observed frequency	20	28	36	40	22	14	160

11. A staff planner at a Miami hospital wishes to test whether the emergency patients arrive according to a Poisson distribution with a mean of $\lambda = 3.4$ per hour. A random sample of 80 hours of operation revealed the following actual number of patient arrivals:

No. Arrivals	0	1	2	3	4	5	6	7	8	**Total**
Frequency f_o	13	7	12	16	15	10	4	3	0	80

Test the fit of the data to the Poisson distribution using a significance level of .05.

TESTS FOR INDEPENDENCE OF CLASSIFICATION

Up to this point, we have used chi-square to test the fit of data to preestablished distributions such as the uniform or Poisson. However, the chi-square procedure can be extended to compare other observed and expected frequencies that are not related to specific statistical distributions (i.e., that are *distribution free,* or "nonparametric"). This second major use of chi-square, which also relies upon the use of frequencies, enables us to determine whether two attributes of a variable are independent of each other, or whether there is a relationship between the two principles of classification.

The question posed at the beginning of the chapter illustrates the general tenor of chi-square tests of independence of classification. As repeated below in Table 10–2, the data in that illustration described the number of jobs secured by marketing graduates. Chi-square tests of independence could help to determine whether the beginning salary was in any way influenced by the geographical area—or vice versa.

TABLE 10–2 Data on jobs secured by marketing graduates

	WEST	SOUTHWEST	MIDWEST	SOUTHEAST	EAST	TOTALS
Number earning $<$ \$24,000	26	37	19	62	109	253
Number earning $\geqslant$ \$24,000	16	33	19	50	89	207
TOTALS	42	70	38	112	198	460

CONTINGENCY TABLES

DIMENSIONS Table 10–2, which presents data grouped according to two major classifications, is referred to as a *contingency table.* The row classification relates to salary and the column classification to the geographic area. Inasmuch as this particular contingency table has two rows (r) and five columns (c), it would be referred to as a two-row-by-five-column contingency table, or a "2×5 table." (The total row and total column are not counted in the designation.)

DEGREES OF FREEDOM The number of degrees of freedom for problems of independence of classification is the product of $(r-1)(c-1)$ where r is again the number of rows, and c the number of columns.

$$\textit{For contingency tables:} \quad df = (r-1)(c-1) \qquad (10\text{–}3)$$

Intuitively, the *df* reflects the number of cells of a contingency table that are free, F, to accept values (as opposed to not free, N), once the row and column totals are specified. Figure 10–3 illustrates the degrees of freedom for a 3×8 contingency table, which has 14 *df*. Row and column totals are designated RT and CT respectively, and GT represents the grand total.

FIGURE 10–3 Degrees of freedom for 3 × 8 contingency table

$$df = (r-1)(c-1) = (3-1)(8-1) = (2)(7) = 14$$

								TOTALS
F	F	F	F	F	F	F	N	RT_1
F	F	F	F	F	F	F	N	RT_2
N	N	N	N	N	N	N	N	RT_3
TOTALS CT_1	CT_2	CT_3	CT_4	CT_5	CT_6	CT_7	CT_8	GT

TEST PROCEDURE

The independence of classification test procedure is analogous to that already described for chi-square goodness of fit tests. (1) The null hypothesis is typically formulated as a statement that one classification is independent of the other. This is equivalent to saying that the *proportions* in each row (salary) class are equal across each of the column (geographic) categories. For example, we might hypothesize that the proportion of marketing graduates with jobs paying less than $24,000 is equal in the West, Southwest, Midwest, Southeast, and East. Then (2) the α level of the test is specified, and (3) the reject level of chi-square determined. Following this, (4), the test statistic is calculated using the same chi-square equation we previously used. Observed frequencies are usually available, or collected, whereas the expected frequencies must be computed. Finally, (5) the conclusion is drawn to either reject or not reject the null hypothesis.

EXPECTED FREQUENCIES The expected frequencies for each cell are computed by multiplying the ratio of each row total (RT) to the grand total (GT), by each column total (CT):

$$\text{Cell } f_e = \frac{RT}{GT}(CT) \tag{10–4}$$

For example, in the marketing illustration, 253 (which is RT_1) of the 460 (GT) marketing graduates took jobs earning < $24,000. That is a ratio of 55%. If the salaries are independent of geographic area, we would expect that 55% ratio to hold across all geographic areas. So, in the West, our expected frequency for graduates earning < $24,000 would be 55% of the CT of 42 graduates in the West, or $(.55)(42) = 23.1$. Similarly, for the < $24,000 category in the Southwest, our expected frequency would be $(.55)(70) = 38.5$. This same approach is carried out to obtain the other expected frequency values. However, the final row values can also be obtained by subtraction, because the totals in the expected frequency table will always equal those in the observed frequency tables.

Example 10–3 Using the marketing job data from Table 10–2, test whether the salaries obtained are independent of geographic region. Let $\alpha = .10$.

Solution

1. H_0: Salary is *independent* of location.
   (i.e., H_0: $\pi_W = \pi_{SW} = \pi_{MW} = \pi_{SE} = \pi_E$)

   H_1: Salary is *dependent* upon location. (i.e., H_1: class proportions $\neq$).
2. $\alpha = .10$
3. Use χ^2: Reject if $\chi^2_{calc} > \chi^2_{.10,\ 4df} = 7.779$

   where $df = (2 - 1)(5 - 1) = (1)(4) = 4$
4. Calculation of χ^2 is best illustrated by formulating separate tables for observed and expected frequencies. The observed frequencies are given.

OBSERVED FREQUENCIES	W	SW	MW	SE	E	TOTALS	
< $24,000	26	37	19	62	109	253	Row % = 253/460 = 55%
⩾ $24,000	16	33	19	50	89	207	Row % = 207/460 = 45%
TOTALS	42	70	38	112	198	460	

The expected frequencies are computed by multiplying each row % by each column total (or subtracting for the last row values)

	W	SW	MW	. . .
< $24,000	(.55)(42) = 23.1	(.55)(70) = 38.5	(.55)(38) = 20.9	. . .
⩾ $24,000	(.45)(42) = 18.9	70 − 38.5 = 31.5	. . .	. . .

Entering these values into an expected frequency table we have

EXPECTED FREQUENCIES	W	SW	MW	SE	E	TOTALS
< $24,000	23.1	38.5	20.9	61.6	108.9	253
⩾ $24,000	18.9	31.5	17.1	50.4	89.1	207
TOTALS	42	70	38	112	198	460

The calculated χ^2 value is the summation of all (10) computations from both rows:

$$\chi^2 = \Sigma \frac{(f_o - f_e)^2}{f_e} = \frac{(26 - 23.1)^2}{23.1} + \frac{(37 - 38.5)^2}{38.5} + \ldots + \frac{(89 - 89.1)^2}{89.1} = 1.329$$

5. *Conclusion:* Do not reject hypothesis of independence (and equality of salaries) because $1.329 < 7.779$. The data support the independence of salary and geographic attributes.

ADDITIONAL APPLICATIONS OF CHI-SQUARE

Chi-square is one of the more widely used statistics, and it has a number of other applications that may be of interest, depending upon the business situation. These applications will not be discussed in detail, but additional descriptions of them are available in advanced level texts on statistics.

TESTS OF PROPORTIONS When only two categories are involved, the chi-square goodness of fit test is equivalent to a one-sample test of proportions—and it will produce identical results. Indeed, when there is only one degree of freedom, the chi-square table values are equal to the square of the familiar normal distribution table values. For example, at 1 *df* and the .05 level, the chi-square value of 3.841 is equal to the Z^2 value $(1.96)^2$. So chi-square can also be used for some of the more traditional so-called "parametric" tests. In the same manner, it can be applied to two-sample tests of the difference between sample proportions.

YATES CONTINUITY CORRECTION While reviewing the goodness of fit test, we saw that when expected frequencies were less than 5, the low-frequency cells had to be combined in order to minimize the error associated with working with discrete observations. The same requirement holds for chi-square contingency tests—low-frequency categories should be combined with the next higher category when the category frequency is < 5.

For frequencies between about 5 and 10, however, there is still some bias toward rejecting hypotheses that should not be rejected. *Yates continuity correction* is often used to reduce this and enhance the validity of the conclusions. It adjusts the computed values of chi-square to the continuous chi-square curve by reducing each of the $(f_o - f_e)$ values by .5. However, if the expected frequencies are reasonably large (10 or more), the correction is typically insignificant.

LARGE CONTINGENCY TABLES Appendix F shows chi-square values for *df* up to and including 30. For situations where the $df > 30$, the chi-square procedure can be modified to allow use of the table of areas under the normal curve. The computed Z value is

$$Z = \sqrt{2\chi^2} - \sqrt{2(df) - 1} \qquad (10\text{–}5)$$

The test conclusion is then based upon the comparison of this computed value of Z with the Z value from the normal distribution table.

INFERENCE ABOUT POPULATION VARIANCES A final application of chi-square is its use in establishing confidence interval estimates of the population variance, or to test a hypothesis regarding σ. For example, a tire manufacturer may wish to establish a 95% confidence interval estimate for *both* the *mean* tire mileage, μ, *and* the *variance* in tire mileage (σ^2) for a new tire they plan to market. Using a procedure somewhat like that used for computing a confidence interval for the true mean, an analyst can use the chi-square distribution to establish a confidence interval estimate for the true variance, σ^2. Computations are not difficult, but these procedures are not as commonly used as the chi-square applications discussed earlier in this chapter, and they are outside the scope of our presentation here. Nevertheless, it is good to be aware of them, for such procedures are available and can be used if the occasion arises.

COMPUTER PROGRAMS Numerous computer programs are available for chi-square (and ANOVA, which follows). Most programs allow you to enter your observed values. Then they calculate the expected values, apply the appropriate *df*, and present you with tabulated results. For chi-square calculations involving more than two or three rows and columns, the computer software can save considerable time.

EXERCISES II

12. What is the appropriate reject level for the following chi-square contingency tests?
    (a) A .02 level test that has 8 column classes and 4 row classes
    (b) A 3×6 table at the $\alpha = .10$ level
13. Data were collected to determine whether people's attitude toward tax legislation was influenced by their educational level. Using the data below, state all the steps to test whether the classifications are independent at the .025 level.

	OBSERVED FREQUENCY			
Attitude	High School	College	Advanded	Total
Favor	30	10	20	60
Oppose	20	10	10	40
Total	50	20	30	100

	EXPECTED FREQUENCY			
Attitude	High School	College	Advanced	Total
Favor	30	12	18	60
Oppose	20	8	12	40
Total	50	20	30	100

14. An automobile manufacturer has conducted some field tests to determine whether the age of passengers has an influence on the use of a new rear seat safety feature. The results are as given below. Test the hypothesis that the use of the safety device is independent of age at the 2% level of significance.

	AGE CLASSIFICATION (YRS)			
	11–25 (A)	26–40 (B)	41 & over (C)	Total
Use safety device	36	24	60	120
Do not use device	14	26	40	80
Total	50	50	100	200

15. An insurance company is studying the accident records of some of its clients in an attempt to determine whether the number of accidents a driver has is dependent upon the driver's smoking habits. Data are as shown below. Test whether the two attributes are independent at the .01 level.

	NUMBER OF ACCIDENTS		
	None	One	$\geqslant$ One
Smoke	5	15	20
Do not smoke	30	20	10

16. A survey was taken to learn whether employees' understanding of the pay and benefits system was related to their perception of the fairness of the system. The number of employees responding in each category is as shown below.

Fairness Perception	RATING OF EMPLOYEES' UNDERSTANDING OF SYSTEM				
	Poor	Not Well	Fairly Good	Very Good	Total
Unfair	23	27	4	6	60
Not Very Fair	41	123	50	28	242
Fair	31	141	124	61	357
Very Fair	5	9	22	25	61
TOTAL	100	300	200	120	720

Test whether the employee's perception of the fairness of the pay and benefits system is independent of his or her understanding of the system. Use the $\alpha = .05$ level and comment on your result. [Note: Even though one cell has an observed frequency of < 5, the expected frequency is $\geqslant 5$, so no cells need be combined.]

THE *F*-DISTRIBUTION

In recent years, industry and government managers have become concerned about counterfeit components that have unwittingly been used in critical defense and space applications. For example, investigations have shown that some of the bolts used to assemble tanks and space exploration equipment have failed to meet specifications. These components constitute serious safety hazards, for even if the average strength of a counterfeit bolt meets a tensile strength standard (e.g., of 60,000 psi), failures can occur if there is too much variability. It can be risky to use bolts that vary in strength from 20,000 psi to 100,000 psi, even though their average strength meets a 60,000 psi standard.

This section introduces the *F* statistical distribution which is used to compare variances. For example, the *F*-distribution could be used to compare the variance in tensile strength of similar looking bolts from two suppliers. The discussion appropriately follows our study of χ^2 because the ratio of two independent χ^2 variables (divided by their respective degrees of freedom) is distributed as an *F*-distribution.

> Suppose samples of size n_1 and n_2 are randomly selected from two populations that are both normal and have equal variances. The ratio of $\frac{s_1^2}{\sigma_2^2}$ to $\frac{s_2^2}{\sigma_2^2}$ is distributed as an *F*-distribution with $n_1 - 1$ and $n_2 - 1$ degrees of freedom.

The $(n_1 - 1)$ refers to the df_1 in the numerator, and the $(n_2 - 1)$ to the df_2 in the denominator of the variance ratio. Because the two population variances are equal, the ratio of σ_1^2 to σ_2^2 should equal 1.00 (and the mean ratio of any sample variances from these populations should also equal 1.00).

For tests of hypotheses, the test statistic $\frac{s_1^2}{s_2^2}$ has an *F*-distribution.[1]

$$F_{calc} = \frac{s_1^2}{s_2^2} \qquad (10\text{-}6)$$

This F_{calc} can then be compared with the table value for $F_{(n_1 - 1),(n_2 - 1)}$.

[1] Insofar as each sample variance is $\Sigma(x-\bar{x})^2/(n-1)$, which is *SSD*/*df*, each sample variance is essentially a χ^2 / df distribution. And the ratio of χ^2_1 / df_1 divided by χ^2_2 / df_2 is distributed as *F*. So the *F*-, chi-square, and normal distributions are all related.

All F values are positive, and the F-distribution is unimodal and skewed to the right. However, the distributions tend to become more symmetrical as df_1 and df_2 increase.

USING THE *F* TABLE F is a continuous distribution ranging from 0 to infinity, and there is a different F-distribution for each sample size combination. So it would be impractical to table all the F values. However, Appendix G is a table of F values for selected variance ratios. The numerator df runs across the top of the table and denominator df down the side. The table shows the values that are likely to be exceeded 5% or 1% of the time, if indeed the two population variances are equal. The lightfaced values at each intersection of df_1 and df_2 are the ratios that will be exceeded only 5% of the time, and boldfaced values are ratios that will be exceeded only 1% of the time, if the two variances are equal.

Table 10–3 illustrates the use of the F-distribution table. Suppose, for example, that you wanted to find the $\alpha = .01$ reject value of F for variances calculated from samples of size $n = 15$ in the numerator and $n = 11$ in the denominator. Letting $df = n - 1$, you would then read down the $df = 14$ column and across the $df = 10$ row to obtain the intersection value of $F_{reject} = 4.60$.

TABLE 10–3 Finding values in the *F*-distribution table

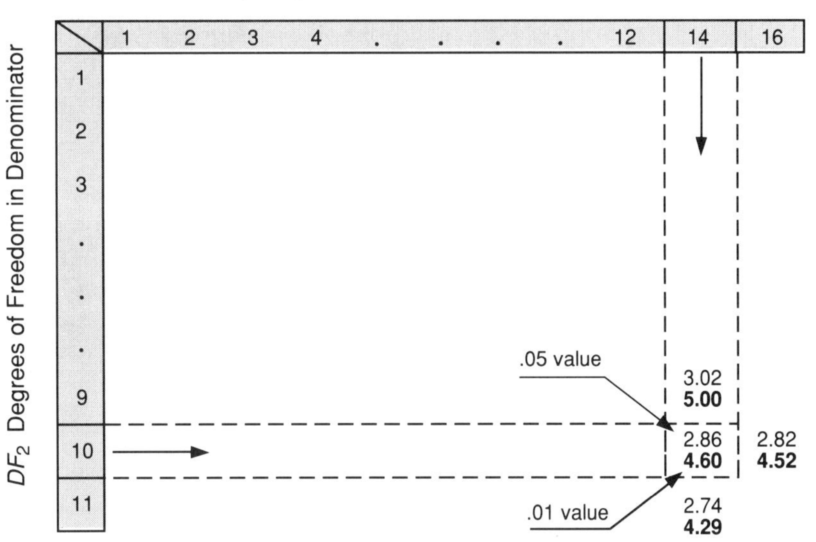

When using the F table for tests about two population variances, the calculated F value will be ≥ 1. We ensure this by always assigning the larger variance to the numerator. For one-tailed tests, this simply amounts to letting the upper-tail area equal α. With two-tailed tests, however, this automatically excludes one tail of the F-distribution. For this reason, we use F-table values corresponding to one-half α for two-tailed tests. In summary:

Type of Test	Null H_0:	Alternate H_1:	F-table Value
Two-tailed	H_0: $\sigma_1^2 = \sigma_2^2$	H_1: $\sigma_1^2 \neq \sigma_2^2$	$\alpha/2$
One-tailed	H_0: $\sigma_1^2 = \sigma_2^2$	H_1: $\sigma_1^2 \geq \sigma_2^2$	α

If the two variances are from the same population, the two estimates of variance should approach 1.0, or be close enough to each other so that the ratio is within the limit given in the F table. Otherwise, the ratio becomes increasingly larger as the difference between the sample variances increases.

Example 10–4 A defense contractor in Arizona has received similar looking steel bolts from two suppliers. Strength testing of random samples has revealed the following estimates of variance:

From supplier A: sample of $n_1 = 15$ $s_1^2 = 38{,}400$ psi

From supplier B: sample of $n_2 = 11$ $s_2^2 = 17{,}800$ psi

Use the F-distribution to test whether these variance estimates are significantly different at the .10 level.

Solution

We can follow the standard steps of a hypothesis test, beginning with the null hypothesis that the two variances are equal. Because it is a two-tailed test at $\alpha = .10$, and we use only the upper-tail area, we shall use $\alpha/2 = .05$.

1. H_0: $\sigma_1^2 = \sigma_2^2$

   H_1: $\sigma_1^2 \neq \sigma_2^2$
2. $\alpha/2 = .05$
3. Use F-distribution.

   Reject if $F_{calc} > F_{\alpha/2,\ (n_1-1)(n_2-1)}$
   $> F_{.05(15-1)(11-1)} = 2.86$ (Appendix G)
4. $$F = \frac{s_1^2}{s_2^2} = \frac{38.4}{17.8} = 2.16$$

FIGURE 10–4 *F* probability distribution for 14 and 10 *df*

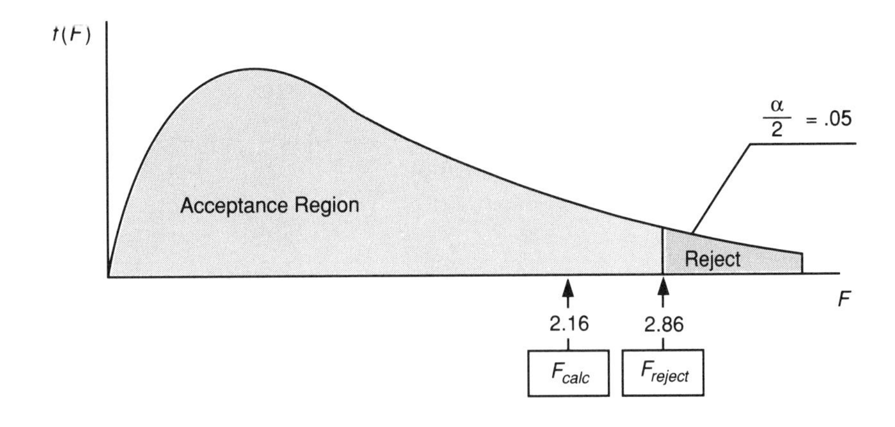

5. *Conclusion:* Do not reject the null hypothesis that the two variances are equal. That is, the ratio of the two variances is not significantly different (at the .10 level) to conclude that the samples are from populations with different variances.

The example above again illustrates the "favor" given to the null hypothesis. As set up, the null hypothesis requires that a significant difference exist to discredit it. A larger sample may provide sufficient evidence to reject the hypothesis.

ONE-WAY ANALYSIS OF VARIANCE

In Chapter 9 we used the normal and *t*-distributions to test for the significance of the difference in two sample proportions and two sample means. Then in the first part of this chapter, we learned that chi-square can be used to test the equality of *more than two sample proportions*. Now, in the latter part of the chapter, we focus attention on a technique that enables us to compare *more than two means*.

Analysis of variance (ANOVA) is a method of testing the equality of two or more means by using variances from the sample data.

HYPOTHESIS The null hypothesis we test using ANOVA is simply a statement that the means of populations (from *k* different groups) are equal.

$$H_0: \quad \mu_1 = \mu_2 = \mu_3 = \ldots = \mu_k$$

$$H_1: \quad \text{Means are not all equal}$$

We realize that there will almost always be some difference in the sample means—even if they come from the same population. But if the difference among any of the group means is more than can be expected from normal sampling variation, we would like to consider it "significant," reject the hypothesis, and conclude the population means are not equal.

ASSUMPTIONS OF ANOVA Like other statistical tests, ANOVA entails some assumptions. As with the *t*-distribution, we assume (1) the samples come from normally distributed populations. Moreover, ANOVA procedures are based upon (2) independent samples that (3) have a common variance. However, the procedure is not significantly affected by minor deviations from normality, and the equal variance assumption is not critical if the sample sizes are comparable. So we find fairly widespread use of ANOVA. Computer programs for ANOVA are readily available for both micros and main frames.

LOGIC OF ANOVA At first thought, one might well question the logic of testing for the equality of means by "comparing variances," but this is really what is done. The ANOVA procedure is basically a comparison of two measures of the variance associated with the data sets being tested.

A **variation** is a sum of squared deviations, *SSD.* As illustrated in Figure 10–5, the first estimate of the variation (*Variation between groups*) stems from the deviation of each of the group means ($\bar{x}_i$) from the grand mean ($\bar{\bar{x}}$) of all the data points. When each of the group mean deviations is weighted by the number in the group, we have one estimate of variation. The second estimate (*Variation within groups*) is derived from the deviation of each of the individual values from their own group mean. When all of these individual deviations are summed, we have our second estimate of variation. Both estimates are sums of squared deviations (*SSDs*), and when divided by their appropriate degrees of freedom, they yield the two measures of variance.

FIGURE 10–5 Structure of an ANOVA test

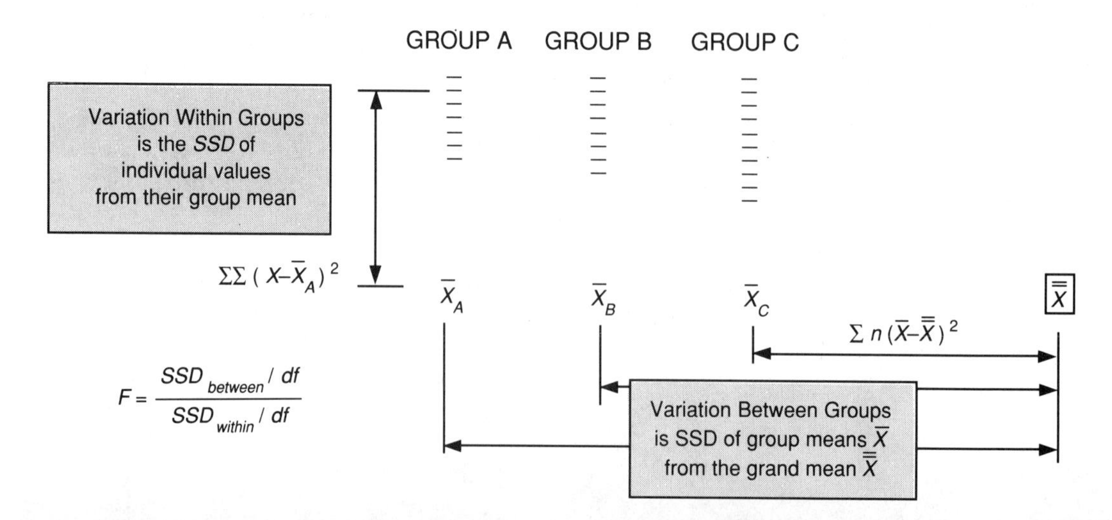

ANOVA USES AN *F*-TEST It is probably no surprise that the test statistic used in ANOVA is from the *F*-distribution. The *F*-distribution is appropriate for comparing two variances, and our comparison is made by dividing the variance **between** (or among) groups by the variance **within** groups.

$$\text{For ANOVA:} \qquad F = \frac{\text{Variance Between Groups}}{\text{Variance Within Groups}}$$

The *within-group* estimate is a measure based on the inherent or average variance that is characteristic of the data being analyzed. It does not reflect any difference in the groups because the deviations are all of the values from their own group mean. The *between-group* variance does reflect these differences, however. It is computed by using the differences between the individual group means and the grand mean. If some of the individual group means do not cluster closely around the grand mean, the between-group variance will reveal this by yielding a relatively large estimate of variance. This larger estimate then goes in the numerator of the F ratio which, in turn, can cause rejection of the null hypothesis of equal means. So the analysis of variance really does rest upon the difference in the group means from the grand mean—but we square that difference and measure it as a variance instead of simply as a subtracted (difference) amount.

If we let T represent the total number of observations, and k the number of groups, we can express the respective variances more precisely as sums of squared deviations between (SSD_B) and within (SSD_W), divided by their respective degrees of freedom.

$$\text{For ANOVA:} \qquad F = \frac{SSD_B \,/\, (k-1)}{SSD_W \,/\, (T-k)} \qquad (10\text{–}7)$$

In equation (10-7), the numerator *df* is simply the number of groups minus one. The denominator *df* is the total number of observations minus the number of groups, k. The SSD_B is equal to the squared deviation of each group mean from the grand mean, weighted by the number in the specific group, n. Then we sum (Σ) across all groups:

$$\text{Variation Between:} \qquad SSD_B = \Sigma\, n\, (\bar{x} - \bar{\bar{x}})^2$$

sum across all groups (10–8)

The SSD_W is equal to the squared deviation of each individual value from its own group mean. These deviations are summed within each group and then across all groups, so we need a double summation sign here.

$$\text{Variation Within:} \qquad SSD_W = \Sigma\, \Sigma\, (x - \bar{x})^2$$

sum within the group
sum across all groups (10–9)

Example 10–5 A mine safety testing service has tested a total of ten batteries from three manufacturers of mining equipment to learn if there is a statistically significant difference in the hours of emergency service they provide. Results were as follows (with times until failure in hrs):

Manufacturer A	Manufacturer B	Manufacturer C	
(n = 4)	(n = 3)	(n = 3)	
10	4	9	
4	6	7	
9	5	11	
5			
28	15	27	grand total = 70 hrs

Test whether the mean lifetimes of the three groups of batteries are equal. Use the .05 level of significance. [Note: $k = 3$ groups and $T = 10$ observations].

Solution

1. Our hypotheses are: H_0: $\mu_A = \mu_B = \mu_C$
   H_1: Means are not all equal
2. $\alpha = .05$
3. Use F-distribution. Reject if $F_{calc} > F_{.05,(k-1)(T-k)} = F_{.05,2,7} = 4.74$
4. For F_{calc} we need the grand mean, group means, and SSD_B and SSD_W.

Grand Mean $\bar{\bar{x}} = \frac{\Sigma x}{\Sigma n} = \frac{28 + 15 + 27}{4 + 3 + 3} = \frac{70 \text{ hrs}}{10 \text{ batteries}} = 7$ hr/battery

	MANUFACTURER A	**MANUFACTURER B**	**MANUFACTURER C**
Group Means:	$\bar{x}_A = \frac{28}{4} = 7$	$\bar{x}_B = \frac{15}{3} = 5$	$\bar{x}_C = \frac{27}{3} = 9$

Variation Between:

$$SSD_B = \Sigma n(\bar{x} - \bar{\bar{x}})^2 = 4(7-7)^2 + 3(5-7)^2 + 3(9-7)^2$$
$$= 0 + 12 + 12$$
$$= 24.00$$

Variation Within:

$$SSD_W = \Sigma\Sigma(x - \bar{x})^2 = \begin{bmatrix}(10-7)^2 + (4-7)^2 \\ +(9-7)^2 + (5-7)^2\end{bmatrix}\begin{bmatrix}(4-5)^2 + (6-5)^2 \\ +(5-5)^2\end{bmatrix}\begin{bmatrix}(9-9)^2 + (7-9)^2 \\ +(11-9)^2\end{bmatrix}$$

$$= 26 + 2 + 8$$

$$= 36.00$$

$$F_{calc} = \frac{SSD_B\ /\ (k-1)}{SSD_W\ /\ (T-k)} = \frac{24.00\ /\ (3-1)}{36.00\ /\ (10-3)} = \frac{12}{5.14} = 2.33$$

5. *Conclusion:* Do not reject, because 2.33 < 4.74. The data are not sufficient to conclude the mean hours of service by the batteries are different if we are willing to be wrong only 5% of the time.

ANALYSIS As with other tests of hypotheses, we tend to favor the null hypothesis unless the evidence is very strong against it. The sample means of 7 hrs, 5 hrs, and 9 hrs do suggest a difference, but we are working with small samples, and these results could happen on a random basis more than 5% of the time. A larger sample might enable us to reject the hypothesis, but we cannot do so on the basis of the data given here. Nevertheless, accepting the hypothesis on the basis of these small samples may be risky.

USING THE ANOVA TABLE

ANOVA test results are commonly presented in what is known as an ANOVA table. It shows the source and amount of variation, the *df* and variances, the calculated and table *F* values, and the conclusion. Table 10–4 shows the results of Example 10–5, arranged in an ANOVA table. In that table, as in most ANOVA tables, the variances are referred to as mean squared deviations, *MSDs*, which are simply the *SSDs* divided by the *dfs*.

TABLE 10–4 ANOVA table for test of three groups of batteries

SOURCE	*SSD*	*df*	*MSD*	F_{calc}	F_{table}
Between	24.00	2	12.00	$F = 2.33$	$F_{.05,2,7} = 4.74$
Within	36.00	7	5.14		
Total	60.00	9			

Conclusion: Do not reject equality of means, because 2.33 < 4.74

TREATMENT AND ERROR VARIANCE Early applications of ANOVA were in experimental (e.g., agricultural) work, and some of the initial terminology remains in use today. For example, the between groups variation is still sometimes referred to as the "treatment" variation, because different treatments (e.g., of fertilizer) might be the reason for different means. The within variation is frequently referred to as the "error," or residual, variation, because it is the inherent variation that cannot be explained by the treatments.

SAMPLE SIZES AND EQUATIONS The sample sizes we used in the example above were really quite small for drawing a useful inference about the difference in batteries from the three manufacturers. We used small samples to simplify the calculations, but in most business applications the sample sizes will be larger than we used.

In order to convey the meaning of ANOVA as clearly as possible, we chose to explain it with "definitional" type equations that apply to either equal or different sample sizes. However, you may find other more rigorous (or efficient) forms of ANOVA equations in other references. Many earlier texts have sets of ANOVA equations that apply to equal sample sizes and different (more formidable looking) sets that apply to different sample sizes. So when you use other references, be alert to the fact that some sets of equations require that the sample sizes be equal. Chances are that you will use a "friendly computer program" for any problem of substantial size, however, so it should handle the mechanics—leaving you to explain the meaning.

TWO-WAY ANOVA TESTS

The previous analysis is commonly referred to as one-way ANOVA, because the data are grouped according to one classification only (by manufacturer in our example). But suppose the data are classified two ways—say by manufacturer and by battery type. We can test whether there is a difference in the means of either (or both) classification by using two-way ANOVA.

FIGURE 10–6 Data classified for two-way ANOVA

HYPOTHESES TESTED UNDER TWO-WAY ANOVA

Column Effects:

$H_0: \mu_A = \mu_B = \mu_C$

H_1: Column means not all equal

Row Effects:

$H_0: \mu_{SD} = \mu_{HD} = \mu_{WP}$

H_1: Row means not all equal

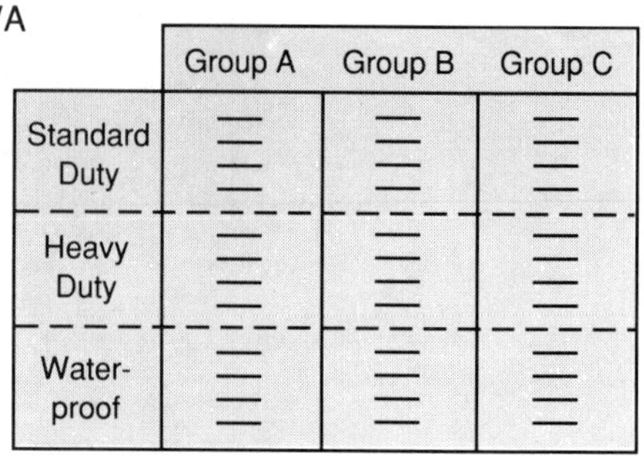

	Group A	Group B	Group C
Standard Duty			
Heavy Duty			
Waterproof			

Two-way ANOVA is similar to one-way, except two sets of hypotheses are tested, as illustrated in Figure 10–6. In addition to computing the variance between columns, the variance between rows must also be computed. The within variance is replaced by an error, or residual mean square, which again reflects the unaccounted-for variance in the matrix. Then two F ratios are calculated, and conclusions drawn for both the column and row variables.

The two-way ANOVA described briefly above is limited to models that have equal cell sizes. In reality, variance analysis is a powerful technique that extends much beyond one-way analysis, and adequate coverage would require treatment beyond the space available here. Computer programs are used extensively for the two-way and more advanced models.

EXERCISES III

17. Find the F_{table} value for
    (a) a 5% level test where $s_1^2 = 40$, $df_1 = 12$ and $s_2^2 = 28$, $df_2 = 16$
    (b) a 1% level test where $s_1^2 = 65$, $df_1 = 30$ and $s_2^2 = 28$, $df_2 = 20$
18. Samples of $n_1 = 10$ and $n_2 = 15$ revealed variances of $s_1^2 = 80$ and $s_2^2 = 18$, respectively. Are the variances significantly different at $\alpha = .01$?
19. What is the rationale for using the F-distribution and measures of variance to test for the equality of means? That is, why not use means to test for differences in means—instead of variances?
20. Suppose the testing agency in Example 10–5 collected some additional data, so the battery failure times were now as shown below.
    (a) Using all the available data, retest for the equality of means at the .05 level.
    (b) Does your conclusion agree with that of the example? Why or why not?

	Manufacturer A	Manufacturer B	Manufacturer C	
	(n = 7)	(n = 5)	(n = 6)	
	10	4	9	
	4	6	7	
	9	5	11	
	5	—	—	
	7	5	10	
	8	5	8	
	6	—	9	
Totals	49	25	54	grand total = 128 hrs

21. An analysis of variance was conducted by an oil company to determine whether there is really any difference in the mileage obtained by six new imported cars. The data and some results of the calculations are shown below. Use an ANOVA table to test the hypothesis of equal mileage at the .01 level.

MILES PER GALLON OF GASOLINE OBTAINED FROM IMPORT AUTOMOBILES

Model #1	Model #2	Model #3	Model #4	Model #5	Model #6
36.7	39.4	33.6	44.2	42.9	36.3
40.5	44.4	37.2	41.6	45.3	39.8
39.7	40.5	38.2	42.0	48.1	41.2
41.5	40.8	37.4	39.6	40.3	38.4
42.2	38.0	40.1	45.4	44.7	40.3
38.6	41.1	38.9	41.7	44.2	37.3

ADDITIONAL RESULTS: $SSD_B = 176.13$, $SSD_W = 136.31$

22. National Mortgage Company has sampled new customers from each of three lending institutions to learn if the length in years of their home mortgages differs. At the 5% level of significance, test whether there is a difference in the mortgages offered by the three financial institutions.

LENGTH OF MORTGAGE (YRS) OFFERED TO CUSTOMERS OF

(A) Commercial Bank	(B) Savings & Loan	(C) Credit Union
30	31	33
36	33	30
33	32	30

23. An airline operating a shuttle service between Boston and Washington is studying its customer base so as to provide the "right class of service." A sample of fliers revealed the following income (in $ thousands/yr) for four age groupings. Is there a significant difference in the average income of the four groups? Use the .05 level and present your results in an ANOVA table.

INCOMES OF SHUTTLE CUSTOMERS: $000/yr			
Under 30 yrs	**30 < 45 yrs**	**45 < 55 yrs**	**55 yrs & Over**
38	54	27	65
33	38	46	22
18	71	33	50
27	45	61	58
22		56	15
30		35	
168	208	258	210

24. A pharmaceutical company has developed five types of treatments for a new hormone designed to enhance growth in livestock. The five types were tested on 25 animals, and the following weight gains were reported (lbs per week). Test whether the mean gain is equal across all treatments. Let $\alpha = .01$, and report your results in an ANOVA table.

WEIGHT GAIN (LBS) FROM TREATMENT					
	#1	**#2**	**#3**	**#4**	**#5**
	4.2	3.7	4.8	6.4	5.3
	4.9	4.2	5.4	5.9	6.0
	4.0	3.5	5.2	7.4	4.2
	5.1	4.4	5.0	6.2	4.5
	4.8	4.2	5.6	6.6	5.0
Totals	**23.0**	**20.0**	**26.0**	**32.5**	**25.0**

SUMMARY

This chapter introduced two new techniques for testing hypotheses. Both extend our statistical capabilities by enabling us to work with more than two categories or groups of data. *Chi-square* allows us to test for the equality of two or more *proportions*, and *analysis of variance (ANOVA)* allows us to test for the equality of two or more *means*. And both use the sum of squared deviations (SSD) in their procedures:

TEST	HYPOTHESES	TYPE OF DATA	TEST STATISTIC
Chi-square	$H_0: \pi_1 = \pi_2 = \ldots = \pi_k$ H_1: some proportions $\neq$	frequencies or counts	
ANOVA (one way)	$H_0: \mu_1 = \mu_2 = \ldots = \mu_k$ H_1: some means $\neq$	measurements	

Chi-square, χ^2, is a distribution of the random *SSD* that results from sampling from a normal (0,1) distribution. There is really a family of chi-square curves, one for each degree of freedom. The two major types of chi-square tests discussed in the chapter were:

(a) *Goodness of fit tests* These compare the observed frequencies with those expected, if the data are distributed as a uniform, Poisson, normal, or other distribution. The calculated chi-square value is compared to the reject value from the chi-square table. For the table value, the $df = (k - 1) - m$, where k is the number of classes, and m is the number of parameters that must be estimated from sample data. If χ^2_{calc} is > the χ^2_{table} value, the hypothesis of fit to the given distribution is rejected.

(b) *Tests for independence* These are "nonparametric" tests, which enable us to test whether two attributes (or classifications) of a variable are independent of each other. Data are displayed in a two-way, or contingency, table that shows both row and column categories. Observed frequencies are typically given, or collected, and expected frequencies for each cell are computed by multiplying the ratio of each row total to the grand total (*RT*/*GT*) by the column total, *CT*. The chi-square value is calculated and compared to the reject value from the chi-square table. For contingency tables, the $df = (\text{row} - 1)(\text{col} - 1)$. If χ^2_{calc} is > the χ^2_{table} value, a relationship probably exists between the classifications, so the hypothesis of independence is rejected.

The *f*-distribution is a distribution of the ratio of two variances, and has two *df* associated with each *F* value—one applies to the numerator variance, and the other to the denominator variance. The *F* table in Appendix G gives the variance ratios, s_1^2/s_2^2, of random samples from two normal (0,1) distributions. Table values will be exceeded 5% of the time (lightfaced type), or 1% of the time (boldfaced), if the two variances are equal. So if an F_{calc} value > F_{table} value, the variances (and populations) are most likely different.

The analysis of variance makes use of the *F*-distribution in testing for the equality of two or more means. In one-way ANOVA, the numerator variance is the *between-group* variance, calculated as the SSD of group means from the grand mean, and divided by the *df*. The denominator variance is the residual, or error, variance called the *within-group* variance. It is calculated as the *SSD* of individual values from their group means, summed over all groups, and divided by its *df*. The logic of ANOVA holds that if some of the individual group means do not cluster closely around the grand mean, the between-group variance will be larger than the within-group variance, causing rejection of the null hypothesis.

ANOVA results are typically displayed in an ANOVA table. For data that are classified two ways, a two-way ANOVA procedure is also available. Most of this analysis would, however, most likely be done on computer, so the equations for two-way ANOVA were not included in the chapter.

CASE: AEROSPACE INDUSTRIES

Ted Stevens hoped it wasn't a connector from *his* firm that caused the loss of power that scrubbed the launch of the space capsule. The capsule had literally thousands of connectors, and Aerospace Industries was the supplier of only four types of them, the AZ series. They required close quality control, for it was critical that they all have equal (electrical) resistance, and he thought they did.

As part of the investigation, each supplier was asked to take a random sample of 30 connectors, test them, and send their test conclusions (and the connectors) to the Space Commission in Houston. Suppliers that revealed different resistances in their connectors *and* were unable to detect the difference were automatically precluded from bidding on future contracts for one year.

Aerospace Industries derived the following data from their random sample of 30 connectors:

RESULTS OF RANDOM TESTING OF AZ SERIES CONNECTORS: AEROSPACE INDUSTRIES

(Resistance readings in micro-ohms)

	SERIES AZ-4	SERIES AZ-5	SERIES AZ-6	SERIES AZ-7
	4.7	4.8	4.7	4.6
	4.9	4.7	4.8	4.7
	4.8	4.9	4.6	4.6
	4.7	4.8	4.7	4.5
	4.8	4.9	4.9	4.6
	4.5	4.7	4.7	4.6
	4.9	4.8	4.6	4.7
	4.8	4.7		
Mean	4.76	4.79	4.71	4.61

After examining the data, Mr. Stevens was less convinced of the equality of his firm's products because the Series AZ-7 appeared to be lower than the other three, though not by much. However, if the investigating commission proclaimed them different, Aerospace would no longer be an acceptable vendor—and this was their major market! So a correct decision was important.

Ted Stevens asked his two assistants, Sherry Tate and Matt Duggan, for independent opinions. After doing some analysis, they reported back to his office.

SHERRY TATE: I don't think we have anything to worry about, Mr. Stevens.

TED STEVENS: I'd like to think so too, Sherry. But those readings from AZ-7 have me worried.

SHERRY TATE: It may look like they're different, but I think it's just coincidence. All the readings are really pretty much the same.

TED STEVENS: Would you stake your job on that? We may have to! What's your conclusion, Matt?

MATT DUGGAN: Much as I hate to say so, Mr. Stevens, I'd say those readings come from different groups. In my opinion there is a difference. And I'm for acknowledging it, taking our licks, and correcting the problem.

SHERRY TATE: Tell you what, Mr. Stevens: If the commission finds they're different, I'll forfeit one week of my vacation. If not, I get a 10% raise. Okay?

TED STEVENS: I'm tempted to take you up on that, Sherry. But maybe I better read both of your reports first. Let's talk about it again tomorrow.

Case Questions

(a) Without doing any analysis of your own, do you tend to agree with Sherry's conclusion or with Matt's?

(b) What type of statistical analysis should Mr. Stevens' assistants have used to arrive at their conclusions?

(c) How is it possible for Sherry and Matt to come up with different conclusions when they both used the same data? Could you envision this happening in other business settings? What might help avoid this?

(d) Suppose Aerospace Industries takes the position that the difference is just due to chance. What percent of the time is such a conclusion likely to be wrong?

(e) What recommendation would *you* make to Mr. Stevens?

QUESTIONS AND PROBLEMS

25. For the following, indicate (1) the type, or name, of the statistical test to use, and (2) the type of statistical distribution used in the test:
    (a) A test to determine whether five sample proportions come from populations that have equal proportions
    (b) A test to compare three population means to learn if they are equal
    (c) A test to find out whether some data are uniformly distributed
    (d) A test to learn whether two sample variances have come from populations with equal variances
    (e) A test to determine whether there is a relationship between frequency data that is classified two ways (i.e., by both rows and columns)
26. Distinguish between a goodness of fit test and a test for independence of classification.
27. The chi-square distribution is formed from samples of a normal (0,1) distribution. How is it that the fit of data to Poisson, binomial, and other distributions can be tested using the chi-square distribution? (i.e., Shouldn't we use other distributions to test nonnormal data?)
28. In the chi-square test, what type of adjustment is required if
    (a) the f_e is less than 5 in any cell
    (b) the f_e is between 5 and 10 observations
    (c) the f_e is greater than 10, but f_o is less than 5 observations
29. What is the rule for determining the *df* for
    (a) Chi-square tests of goodness of fit
    (b) Chi-square tests of independence of classification
    (c) between groups variance in ANOVA
    (d) within groups variance in ANOVA
30. (a) Distinguish between one-way and two-way ANOVA tests. (b) How do the null hypotheses differ for one-way versus two-way tests?
31. Explain how the following terms differ:
    (a) sum of squared deviations (*SSD*) and variance
    (b) between groups *SSD* and within groups *SSD*
    (c) treatment variation and SSD_B
    (d) variation and variance
    (e) variance and mean squared deviation

32. The Tristate Appliance Repair Center is open five days a week, and the manager schedules the number of maintenance technicians according to the expected number of service calls. Records of the number of requests for the repair of electric dryers on each day of the week for one year showed a total of 800 calls distributed as follows: Monday 190; Tuesday 180; Wednesday 130; Thursday 160; Friday 140.
   (a) Test the hypothesis that calls for dryer service to follow a uniform distribution, using the .05 level.
   (b) What is the implication to the manager of the conclusion reached in part (a)?
33. Sketch a graph of the chi-square distribution with two degrees of freedom. Use a horizontal scale from 0 to 10, and mark the points above which 10%, 5%, 2.5%, 2%, and 1% of the area of the curve will lie.
34. A production process for computer chip manufacturing is expected to produce only 20% defective. However, a random sample of 100 chips showed that 28 were defective.
   (a) Using the chi-square procedure, test whether the observed frequency agrees with the expected frequency at the .05 level.
   (b) Suppose this hypothesis was tested using the values of Z from the normal curve. What would be the reject Z value, and how would it relate to the chi-square reject value?
35. A public accounting firm in Cleveland specializes in auditing of bank records. The auditing manager feels that the number of errors detected per day follows a Poisson distribution and would like to verify that. If it is so, she can be better prepared for the amount of computer time that will be required for reruns of the bank's statements. Records for 180 days showing the number of errors detected in bank customer accounts are as shown:

X Number Errors	*f* Number Days	*fX*
0	12	0
1	39	39
2	47	94
3	40	120
4	20	80
5	17	85
6	3	18
7	2	14
Totals	180	450

   (a) Compute the mean number of errors found by the auditors per day.
   (b) Compute the expected frequencies if the true mean is 2.5.

(c) Test the hypothesis that the data are Poisson distributed, using the .05 level.

(d) Suppose μ were known in advance of the test. How would the test differ from the above where no prior knowledge of μ is assumed?

36. A telephone company has prepared three versions of a set of instructions for placing long distance calls and has asked 1,200 persons which of the three forms was easiest to understand. Assume the following results:

Form	Number Preferring
A	450
B	390
C	360

Use chi-square to test the significance of the differences in the numbers of persons preferring the three forms at the .10 level.

37. In an effort to forecast economic conditions for the upcoming year, a survey was conducted of representative academic, business, and government leaders, with the results as shown below. Do the data suggest that one's forecast of the economy is independent of the source of his or her employment? Use the .01 level of significance.

	NUMBER OF PERSONS FORECASTING ECONOMIC				
	Depression	Recession	Stability	Expansion	Totals
Academic Economists	24	28	17	11	80
Business Managers	15	21	43	21	100
Government Economists	11	11	20	28	70
Totals	50	60	80	60	250

38. The data shown below constitute a two-way classification of 5,000 insurance company clients grouped by age level and income level. Use the test of independence of principles of classification at the .05 level to determine whether the age of clients is independent of their income level.

	NUMBER OF CLIENTS IN AGE CATEGORY			
	$<$ 20 yrs	20 $<$ 50 yrs	$\geqslant$ 50 yrs	Totals
High income	1,000	400	600	2,000
Medium income	900	600	300	1,800
Low income	800	300	100	1,200
Totals	2,700	1,300	1,000	5,000

39. An employment survey of twenty-one business women in each of two Midwestern cities revealed they worked approximately 46 hr/wk. However, the standard deviation in City A was 1.2 hr/wk, whereas that in City B was 1.8 hr/wk. Using the *F*-ratio, is the difference in variance more than could be expected to occur 5% of the time?

40. Are airport facilities being used effectively? The operations manager of a large airport wishes to determine if the air traffic is being allocated equally among four airport concourses (A, B, C, and D). Samples of the number of passengers in each concourse at random times during the week are shown below (with values in hundreds of passengers). Use the $\alpha = .05$ level to test whether the concourses receive equal use, or whether some have a significantly higher average use than others.

NUMBER OF PASSENGERS IN AIRPORT CONCOURSE

	A (red)	B (blue)	C (yellow)	D (green)
	6.2	4.3	1.8	3.3
	3.5	2.2	7.5	5.4
	1.2	5.8	2.0	5.8
	5.7	2.1	3.9	2.6
Totals	16.6	14.4	15.2	17.1

41. The Benefits Manager of a health maintenance organization (HMO) in St. Louis must decide whether her firm should provide different amounts of coverage for the same medical procedure in different cities. If the costs are "really higher" in some cities, she would like to propose different rates. But with the varying charges of individual doctors, it is difficult to tell. She has obtained sample costs of the procedure in the six cities as shown below. Using the $\alpha = .05$ level, test whether there is a real difference in average costs among the six cities. Show your results in an ANOVA table.

COST (IN $ HUNDREDS) OF MEDICAL PROCEDURE IN CITY

	City A	City B	City C	City D	City E	City F
	96	110	122	85	125	95
	108	110	110	140	90	110
	89	106	90	90	90	85
	129	108	92	110	100	130
	89	90	110	110	90	120
Totals	511	524	524	535	495	540
$\bar{x}$	102.2	104.8	104.8	107	99	108

42. Is the increase in earnings per share from the previous year (EPS1) to the current year (EPSC) independent across the utility (U), retail (R), transportation (T), and banking (B) industries? Using the Corporate Statistical Database (Appendix A), count the number of these firms in each group that showed an increase from EPS1 to EPSC and the number that show a decrease. Because there are so few observations in the transportation industry, combine it with the banking firms. Then your observed frequencies should fit into the following format.

	CHANGE FROM EPS1 (one year ago) TO EPSC (current)		
	EPSC is UP	**EPSC is DOWN**	**Total**
Utilities (U)			20
Retail (R)			15
Transportation (T) and Banking (B)			15
			70

Use the frequencies to test whether the proportion of increases (and decreases) in EPS is equal for the three industry groups (versus a better performance in one or more industries). Let $\alpha = .10$.

43. Using all the firms of the Corporate Statistical Database (Appendix A), sort the data so as to obtain the number of firms (frequencies) falling into the sales and employee classifications shown below. (It is easier to categorize the firms if you apply a record selection or sorting routine to your database—if one is available. However, the frequencies can also be tallied by hand. Subtotals are provided below as a check on your figures.)

	Sales of Firms in the Corporate Data Base ($ M)		
	$<$ $ 500 M	**$\geq$ $ 500 M**	
$<$ 4,000 employees			31
$\geq$ 4,000 employees			69
	28	72	100

Use chi-square to test whether the sales are independent of the number of employees. Test at the $\alpha = .05$ level, and state your conclusions.

44. Using the Corporate Statistical Database (Appendix A), use ANOVA to test whether the mean price earnings ratios in the chemical (C), electrical (E), and retail (R) industries are equal. Use a significance level of .05, and report your results in the ANOVA table provided by your computer program.

CHAPTER 11

Regression and Correlation Analysis

INTRODUCTION: WHAT IS REGRESSION AND CORRELATION?
- Univariate and Bivariate Data
- Simple Linear Regression

FINDING THE LINEAR REGRESSION EQUATION
- Using the Normal Equations
- Making a Point Estimate

USING THE REGRESSION MODEL FOR INFERENCE
- Assumptions Underlying Regression
- The Standard Deviation of Regression
- Approximate Prediction Interval for Individual Values
- More Accurate Prediction Interval for Individual Values
- Confidence Interval for Mean Values
- Inference about the Slope of the Regression Line

SIMPLE LINEAR CORRELATION
- Correlation Coefficient
- Coefficient of Determination
- Inference about the Correlation Coefficient
- Correlation Limitations

SUMMARY

CASE: KATHY DUBOIS, LEGISLATIVE ASSISTANT

QUESTIONS AND PROBLEMS

Chapter 11 Objectives

1. Define the terms regression and correlation.
2. Given two sets of data, derive a regression equation.
3. Use a regression equation to make inferences.
4. Compute and interpret the correlation coefficient.

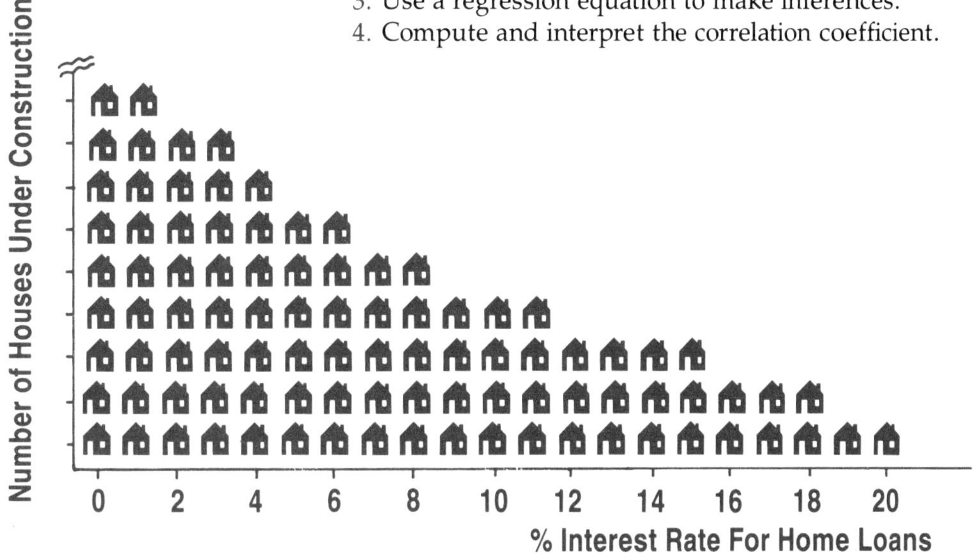

To what extent will housing construction "pick up" if the Federal Reserve Board causes interest rates to drop? Will lower tax rates prompt a higher level of capital investment by manufacturers? How do expenditures on advertising affect sales? Do the hours spent on quality control training really result in higher productivity? These questions, and countless others, concern the use of one or more variables to infer something about another variable of special interest to managers—such as sales or employee productivity.

Like millions of other business people, you probably do (or eventually will) keep pace with business activity by reading The Wall Street Journal. *In it, you may come across statements such as "The correlation between capital spending and the income tax rates has been found to be . . . ." If your boss (or maybe a subordinate!) wanted to discuss this further with you, would you be able to contribute to the discussion in a meaningful way? By the time you finish this chapter, you should have a good working knowledge of regression and correlation. So you should not only understand the precise meaning of the terms that others may be using, but you should also be able to call upon these very powerful tools of analysis whenever the opportunity arises.*

INTRODUCTION: WHAT IS REGRESSION AND CORRELATION?

UNIVARIATE AND BIVARIATE DATA

Suppose you wanted to learn how long it took cars to stop on a freeway. If you were able to go out on a freeway with a stop sign and collect such information (heaven forbid), you would no doubt find quite a difference in the stopping distances, as depicted in Figure 11–1 (a). And if you had to predict the distance, Y, required to stop a car selected at random, your best (point) estimate would probably be about 100 ft. This is because you would be working with *univariate* (one variable) data,—where the best estimator is the sample mean. We have been working mostly with univariate data until now.

FIGURE 11–1 Stopping distance for cars on freeway

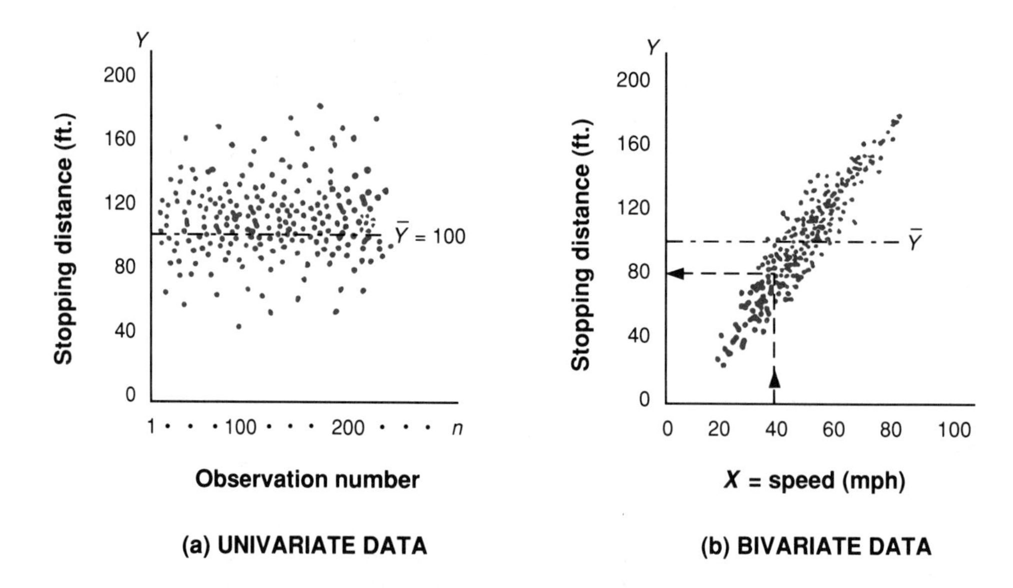

But everyone knows that stopping distance is a function of other variables, especially of speed. You might do a better job of predicting the stopping distance by recognizing that it was dependent upon the speed of the car, X. If you collected data on both the speed and stopping distance, you would be working with *bivariate* (two variable) data. Then, as illustrated in Figure 11–1 (b), you might predict a stopping distance of 80 ft if the car's speed is 40 mph, but one of 160 ft if the car is going 80 mph. Moreover, the closer the sample points lie to a single line, the better your estimate is likely to be.

In the bivariate case, the variable of interest, Y, is referred to as the *dependent* variable, and the variable used as a predictor, X, is called the *independent* variable. Using these terms, we can gain an intuitive meaning of the terms regression and correlation:

Regression is the process of estimating the value of a dependent variable, Y, from one or more independent variables, X(s).

Correlation is a measure of the degree to which two or more variables are related.

SIMPLE LINEAR REGRESSION

The most basic (and straightforward) type of regression is *simple linear regression*. In this section we define this term more fully, and see how a regression model is formulated.

SIMPLE AND MULTIPLE REGRESSION *Simple regression* analysis uses only one independent variable, X, to estimate the value of the dependent variable, Y. If, in addition to speed, we used data on the road condition, tire wear, and other variables to help predict stopping distance, we would be doing a multiple regression analysis. *Multiple regression* uses two or more independent variables, X(s), to better predict the (one) dependent variable, Y. (This chapter will be limited to simple regression and correlation models.)

LINEAR AND NONLINEAR REGRESSION The first step in a regression analysis is to correctly identify the form of the relationship between the independent and dependent variables. This relationship can frequently be perceived by constructing a scatter diagram.

A **scatter diagram** is a plot of observed data points on X-Y coordinates for the purpose of revealing a possible relationship between the variables.

Figure 11–1(b) was a scatter diagram of the bivariate data relating speed to stopping distance. It happens to be a positive relationship (i.e., Y generally increases as X increases), and appears to be fairly linear (i.e., the points lie around a diagonal, straight line).

Figure 11–2 shows some scatter diagram relationships that might be obtained from different data. (Dotted lines have been added to help describe the pattern of the points.) Note that (a) is a "stronger" relationship than (b) because the data points lie closer to the line that typifies the relationship. Figure 11–2 (c) illustrates a high positive relationship between X and Y, but the relationship is *curvilinear* rather than linear. It cannot be adequately described by a straight line, so a more complex regression model is needed. The variables in (d) form a "cloud" and do not appear to be related in a useful manner. For situations like this, the X values are of no help in predicting Y, and further analysis is not usually justified.

FIGURE 11–2 Scatter diagram of some X and Y relationships

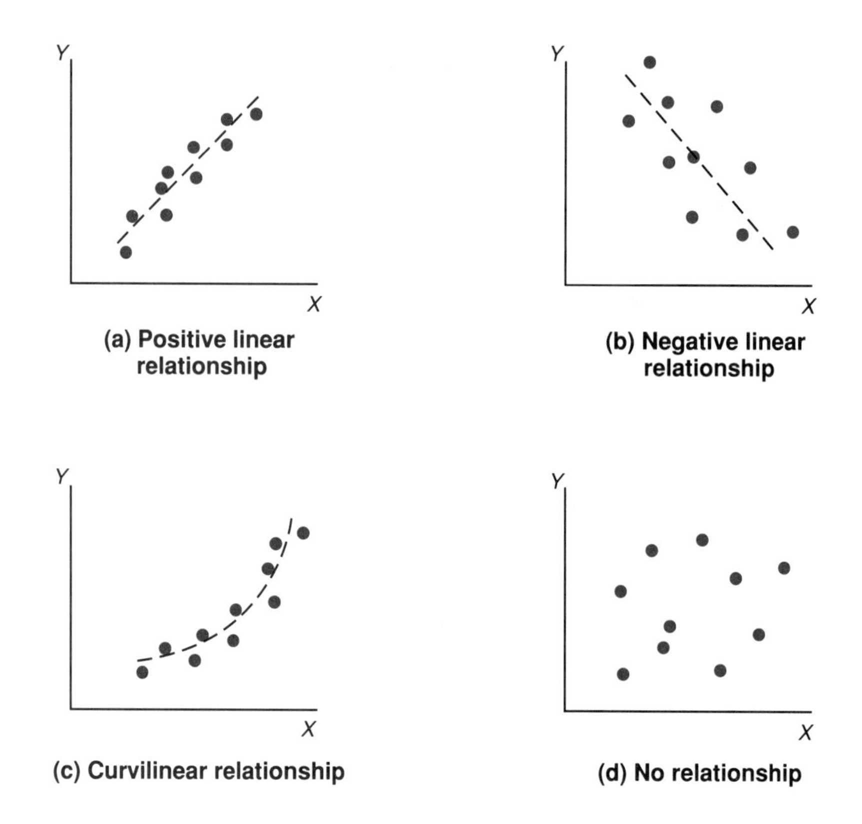

GENERAL REGRESSION MODEL In a regression analysis, if all the pairs of data points fell in a straight line (or on a smooth curve), X would be a perfect predictor of Y. Unfortunately that is rarely the case. As with the stopping distance example, there is usually some difference (or error) in making an estimate. If we let the Greek letter epsilon, ε, depict the error of an estimate, we can observe that the value of the dependent variable value, Y, is a function, f, of the independent variable, X, and the amount of error, ε. So the generalized form of a regression model is:

$$Y = f(X, \varepsilon)$$

SIMPLE LINEAR REGRESSION MODEL In this chapter, however, we will be constructing regression models for a specific (linear) type of relationship. When the data points fall roughly in a straight line pattern (either positive or negative), then a linear equation can be used to describe the relationship. At any fixed X point on the line, the mean of all the Y values lies on the line and can be designated $\mu_{y \cdot x}$. If we let α represent the Y-intercept, and β the true slope of the line, then the linear equation describing a general population relationship can be written as:[1]

$$\text{Population Regression Line:} \qquad \mu_{y \cdot x} = \alpha + \beta X$$

Population values are rarely (if ever) known, however, so our regression equations will be computed from sample data. Letting Y_c represent the calculated value of the dependent variable Y, we can express the sample linear regression model in an *operational* form as:

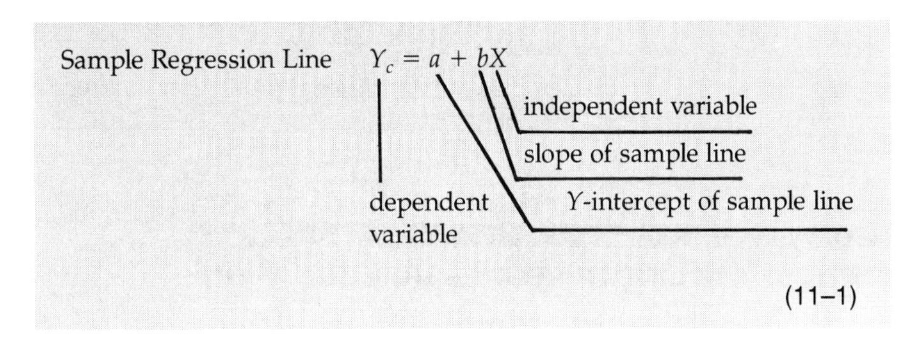

The intercept "a" and slope "b" values are the constants we must derive from the X and Y data points. The a value is an estimate of where the regression line crosses the Y-axis, and b estimates how much the dependent variable Y_c changes for each unit of change in the independent variable, X (i.e., the slope $\Delta y / \Delta x$).

TAKING ACCOUNT OF THE ERROR Note that the (operational) model above does not make specific mention of any error. Nevertheless, as we shall see later, the error of the points about the regression line is usually calculated as part of the analysis. It is measured in the same (deviation) units we have been using previously. The change here is that the error represents the deviation of points from the regression line, Y_c, instead of deviations of points from their sample mean. So our measure of error will rely upon the difference between the actual Y and calculated Y_c values, or $(Y - Y_c)$.

Once the sample regression equation is derived, it can be used to predict Y_c values. These will be points that lie on the regression line, given an assigned value of X.

---

[1] The population model describing the specific value of the dependent variable Y_i, corresponding to the ith observation of the independent variable X, where e_i is the (normally distributed) random error, would be $Y_i = \alpha + \beta X_i + \varepsilon_i$.

STEPS IN REGRESSION We can summarize the steps in determining and using the linear regression of Y on X as follows:

1. Collect observations (data) of the independent and dependent variables.
2. Verify the linear nature of the relationship (e.g., via a scatter diagram)
3. Compute the parameters (slope and intercept), and associated error of the regression model.
4. Use the regression model to predict values of the dependent variable.
5. Complete any additional predictions and/or tests as desired.

FINDING THE LINEAR REGRESSION EQUATION

LINE OF BEST FIT Assuming the data have been collected and that a scatter diagram has confirmed the linear nature of the relationship, let us proceed with step 3, the computation of the regression line. In regression, we seek to derive a line through the data points that minimizes the amount of error, or deviation of all points Y from the line, Y_c. Fortunately, we can solve a set of equations that will give us a *line of best fit* through the data. This line (sometimes called the "least squares line of best fit") has three outstanding properties, as illustrated in Figure 11–3:

1. It goes through both $\bar{x}$ and $\bar{y}$.
2. The summation of deviations about the line is zero [i.e., $\Sigma(Y - Y_c) = 0$].
3. The summation of the squared deviations about the line is a minimum [i.e., $\Sigma(Y - Y_c)^2 =$ minimum].

FIGURE 11–3 The least squares line of best fit

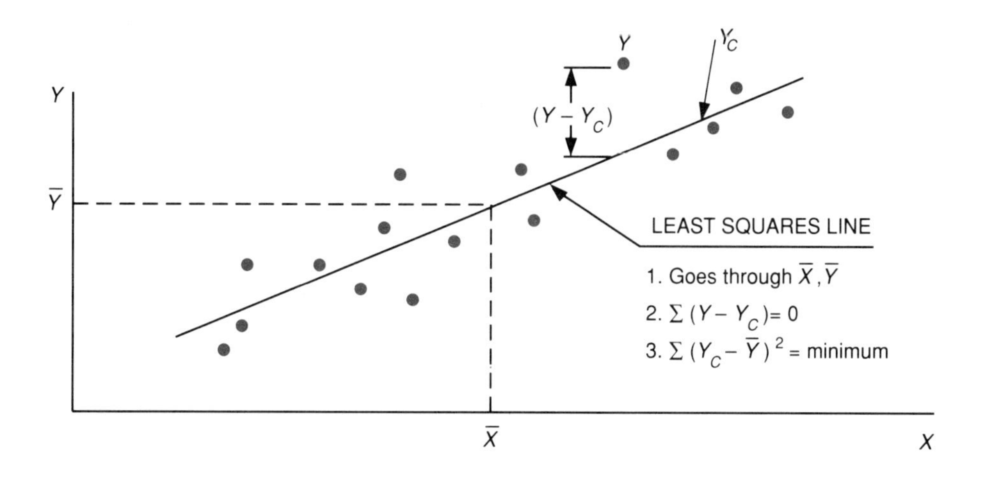

USING THE NORMAL EQUATIONS

The set of equations that are solved to yield the least squares line of best fit are referred to as the *normal equations*.

Normal equations
$$\Sigma Y = na + b\Sigma X$$
$$\Sigma XY = a\Sigma X + b\Sigma X^2 \qquad (11\text{–}2)$$

The name "normal" does not relate to the normal statistical distribution. The normal equations are simply expressions that incorporate X and Y data points that, when solved simultaneously in accordance with the minimizing conditions, will yield values for the constants a and b. (A slightly more complex version is also available for solving three equations for three unknowns.)

Although the normal equations can be solved simultaneously for a and b, the same result can be obtained more directly by rearranging the equations.

slope
$$b = \frac{\Sigma XY - n\bar{x}\bar{y}}{\Sigma X^2 - n\bar{x}^2} \qquad (11\text{–}3)$$

intercept
$$a = \bar{y} - b\bar{x} \qquad (11\text{–}4)$$

Example 11–1 Spaceage Materials Co. produces a new composite material (CM7) that is especially suitable for space applications. Tensile strength is a critical variable, and strength measurements are important, but strength tests destroy the material. However, the product line manager feels that the strength of CM7 material may be predicted from surface hardness measurements. He has collected some data and would like to derive a regression model to explore the relationship between hardness and strength. Test data for ten samples of the CM7 material are as shown below. Determine whether a linear relationship exists and, if so, use the data to develop the regression equation.

TABLE 11–1 Strength and surface hardness scores for CM7 material

SAMPLE NUMBER	1	2	3	4	5	6	7	8	9	10
Hardness *X*	7	9	5	8	6	9	7	4	8	7
Strength *Y*	10	12	6	9	8	11	10	5	10	9

Solution

First plot the observations of hardness and strength on a scatter diagram to verify that the relationship is linear. See Figure 11–4.

FIGURE 11–4 Scatter diagram of hardness (*X*), and strength (*Y*)

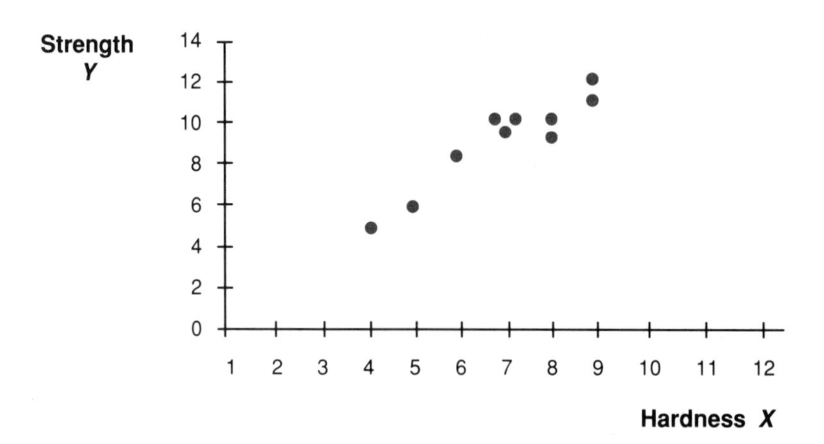

The ten hardness and strength measurements are ordered (X, Y) pairs and are plotted at the intersection of the X and Y scale values. Note that the relationship is apparently linear, and it is positive; strength tends to

increase as hardness increases. However, it is not a perfect relationship because all the points do not fall on a single diagonal straight line.

Next, we must compute the slope and intercept constants of the regression line. For hand calculations, it is helpful to arrange the data into a format that will facilitate calculation of the ΣXY and ΣX^2 value. (Computer programs take care of this automatically.) Table 11–2 shows the values we will need.

TABLE 11–2 Strength and hardness score calculations

SAMPLE NUMBER	Hardness X	Strength Y	X^2	XY	Y^2
1	7	10	49	70	100
2	9	12	81	108	144
3	5	6	25	30	36
4	8	9	64	72	81
5	6	8	36	48	64
6	9	11	81	99	121
7	7	10	49	70	100
8	4	5	16	20	25
9	8	10	64	80	100
10	7	9	49	63	81
Totals	$\Sigma X = 70$	$\Sigma Y = 90$	$\Sigma X^2 = 514$	$\Sigma XY = 660$	$\Sigma Y^2 = 852$

From the ΣX and ΣY values, we can find the sample means for the X and Y variables as:

$$\bar{x} = \frac{\Sigma X}{n} = \frac{70}{10} = 7 \qquad \bar{y} = \frac{\Sigma Y}{n} = \frac{90}{10} = 9$$

Using equations 11–3 and 11–4, we can then compute the slope and intercept constants for our regression line:

$$\text{slope} \qquad b = \frac{\Sigma XY - n\bar{x}\bar{y}}{\Sigma X^2 - n\bar{x}^2} = \frac{(660) - 10(7)(9)}{(514) - 10(7)^2} = \frac{30}{24} = 1.25$$

$$\text{intercept} \qquad a = \bar{y} - b\bar{x} = 9 - (1.25)(7) = .25$$

The regression equation is thus: $Y_c = .25 + 1.25X$.

The regression equation describes a line of best fit through the data points as shown in Figure 11–5. The intercept is at .25 units on the Y (strength) axis. And we estimate that the average strength increases 1.25 units for each unit increase in hardness.

FIGURE 11–5 Regression line for hardness and strength data

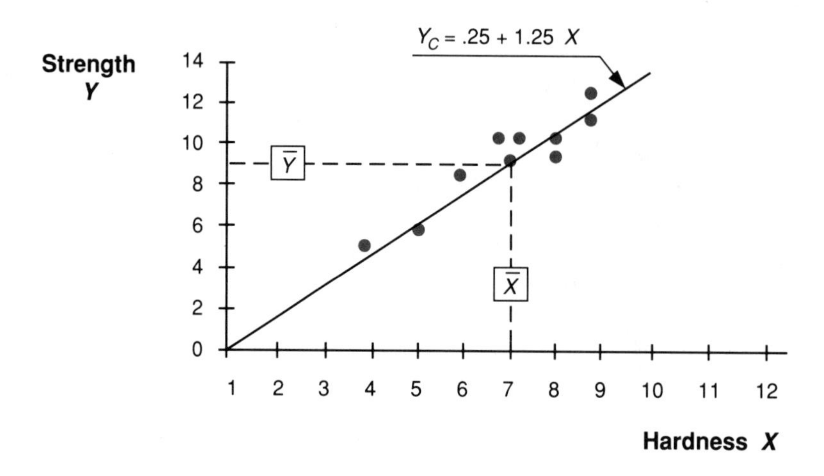

EQUATIONS FOR *b* AND *a* There are several different forms of the equations we used for computing the slope (*b*) and intercept (*a*). We choose not to offer them all here because one method is sufficient for understanding the theory of regression. Moreover, any significant problems will most likely be done on a computer anyway. However, you should be aware that other references may show different equations for these constants.

VALIDITY OF THE RELATIONSHIP Although we have quantified the relationship between the two variables, we should also be aware of two limitations at this point. *First*, regression *does not establish a logical cause and effect* relationship. In other words, hardness is not necessarily the cause of tensile strength—indeed it might even lead to brittleness and fracture. *Second*, we have not yet validated the relationship. That is, the relationship might have come about by chance. In order to assess the validity, we must calculate the error associated with the points about the regression line, which we will do momentarily. But first let us complete the mechanics of using the regression line to make a simple point estimate.

MAKING A POINT ESTIMATE

Once the regression equation is derived, it can be used for estimating the value of the Y_c for a given value of X. If a point estimate is desired, one can simply substitute a value for X (the independent variable) into the equation, and solve it for Y_c (the dependent variable). Any estimates should, however, be *limited to the approximate range within which the data lie*, for the same relationship may not hold outside of this range.

Example 11–2 Using the regression equation $Y_c = .25 + 1.25X$, make a point estimate of the strength of the CM7 material when the hardness measures: (a) 10 and (b) 5. (c) Suppose the hardness measurement were 20. Estimate the strength.

Solution

By substituting the value of X into the regression equation we have:

(a) $Y_c = .25 + 1.25X = .25 + 1.25(10) = 12.75$ strength units

(b) $Y_c = .25 + 1.25X = .25 + 1.25(5) = 6.50$ strength units

(c) The X value of 20 lies too far outside the range to make a valid estimate.

These point estimates do, of course, suffer from the same weaknesses as the univariate data point estimates we worked with in earlier chapters. They are most likely incorrect, and we have no measure of certainty attached to them. To enhance the value of our predictions, therefore, we must obtain a measure of the error of the estimate. In addition, we should be aware of the assumptions that underlie our estimates. We take up these considerations in the next section.

EXERCISES I

1. How does regression differ from correlation?
2. Identify the independent variable in the following:
   (a) Highway funds are allocated to counties on the basis of population.
   (b) Purchase orders are counted to estimate storage space needs.
   (c) A survey of buyer intentions suggests strong Christmas sales of toys.
3. Distinguish between the following terms:
   (a) univariate and bivariate data
   (b) simple and multiple regression
   (c) linear and nonlinear regression
4. What is the purpose of a scatter diagram?
5. How does the general form of the regression model differ from the form we use for simple linear regression? Why is there a difference?
6. Name three properties of the line of best fit?

7. Graph the following data on X, Y coordinates to determine whether they have a linear relationship, a nonlinear relationship, or no apparent relationship.

(a)

X	Y
2	2
6	4
5	2
4	1
5	3
1	4
3	1
1	3
6	3

(b)

X	Y
5	4
3	1
1	3
2	5
4	4
6	2
3	3
2	2
5	3

(c)

X	Y
4	4
1	2
5	4
2	2
6	5
1	1
3	3
5	5
3	2

8. The following data were collected to measure the success of a training program for word processing employees.
   (a) Plot the data on X, Y coordinates.
   (b) Do they appear to have a linear relationship?
   (c) Is the slope positive or negative?

Hrs of training	**X**	6	10	12	2	4	10	3	9	2	8	5	7
No. errors/job	**Y**	14	6	6	16	20	4	18	10	18	6	10	10

9. A retail sales manager would like to determine whether the number of coupons in their newspaper advertisements can be used to help predict sales of small appliances in the following week. Data from past weeks are:

	No. coupons X	Sales ($000) Y	X^2	XY
	4	5	16	20
	8	7	64	56
	10	7	100	70
	3	4	9	12
	2	5	4	10
	6	6	36	36
Totals	$\Sigma X = 33$	$\Sigma Y = 34$	$\Sigma X^2 = 229$	$\Sigma XY = 204$

   (a) Graph the data to confirm that a linear relationship exists.
   (b) Compute the least squares line of best fit.

(c) Make a point estimate of sales when seven coupons are included in the advertisement in the previous week.

10. Compute the line of best fit through the data of problem 8.
11. For the following data, compute the least squares line of best fit: $n = 10$, $\Sigma X = 20$, $\Sigma Y = 62$, $\Sigma X^2 = 54$, $\Sigma Y^2 = 398$, $\Sigma XY = 135$
12. Suppose a line of best fit relating expenditures of selected international companies on research and development (R & D) to their export sales (both in $ millions) is $Y_c = -.60 + 30.40\ X$. Estimate the export sales for companies spending (a) $1.5 million on R & D, (b) $80,000 on R & D.
13. Use the following data to develop a point estimate of the number of material handling (MH) robots in manufacturing companies with (a) 15,000 employees and (b) 50,000 employees:

No. of employees (000)	30	10	20	50	10	60	40	10	50
No. of MH robots	70	30	110	200	0	180	150	50	150

USING THE REGRESSION MODEL FOR INFERENCE

Having established the general procedure for obtaining a regression equation, which is descriptive in nature, let us extend our inquiry to learn how regression is used for making statistical inferences. We will begin by recognizing the assumptions underlying regression and calculating the error associated with a regression.

ASSUMPTIONS UNDERLYING REGRESSION

LEAST SQUARES IS DETERMINISTIC The least squares procedure we used to find the line of best fit is simply a deterministic mathematical procedure. There is no uncertainty associated with it, and it will work for any two variables—they need not have an independent/dependent relationship.

REGRESSION IS PROBABILISTIC Regression uses the least squares procedure, but *it goes beyond* this deterministic relationship. This is because the regression model assumes that the dependent (Y) variable is distributed according to a normal probability distribution. That is, for every given value of X, there are many possible values of Y. These individual Y values, taken together, form a series of probability distributions around the regression line. Figure 11–6 illustrates three such distributions, which are referred to as *conditional probability distributions of Y given X*, or ($Y.X$). As can be seen, the regression line goes through the means (each $\mu_{y \cdot x}$) of this set of distributions.

FIGURE 11–6 Conditional distributions of Y given X

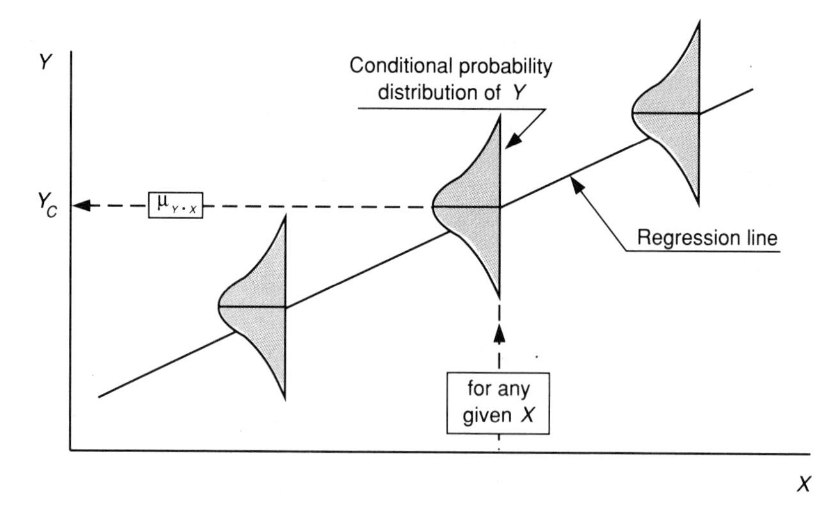

ASSUMPTIONS In summary, three assumptions of regression modeling are:

1. The data are normally distributed (in the Y direction) about the regression line.
2. The means of all of the distributions, $\mu_{y \cdot x}$, lie on the line.
3. The variances of all of the distributions are equal.

The normality assumption is not severely restrictive because variables frequently tend to be distributed about their mean in a normal manner.

BASIS OF INFERENCE It is this assumption about the probabilistic nature of the dependent variable that allows us to make inferential statements when working with regression models. Insofar as we assume that the Y values form a normal distribution about the regression line, we can use the table of areas under the normal curve to help establish prediction intervals within which we can predict Y values to lie (once an X value is given or assumed). This, of course, requires that we know the standard deviation of the conditional probability distribution—which is the topic we take up next.

THE STANDARD DEVIATION OF REGRESSION

In *univariate* analysis, we measured the error, or dispersion of points from their mean by the standard deviation, s. And we computed s by summing the squared deviations of the values from their mean $\Sigma(X - \bar{x})^2$, dividing by the degrees of freedom $(n - 1)$, and taking the square root of this.

With *bivariate* data, the standard deviation is also a useful measure of dispersion. However, the *mean* is the regression line, designated by Y_c, and the deviations are the sum of the squared differences between the points and the regression line $\Sigma(Y - Y_c)^2$. Also, we divide by the number of pairs of data minus two ($n - 2$), because we lose two degrees of freedom in estimating the two constants of the regression line (a and b). The resultant *definitional form* of the equation for the standard deviation of regression, s_e, is:

$$s_e = \sqrt{\frac{\Sigma(Y - Y_c)^2}{n - 2}} \tag{11–5}$$

Note that s_e is a measure of distances in the Y direction, and not perpendicular to the regression line. The standard deviation of regression is also sometimes designated $s_{y \cdot x}$, or $s_{y|x}$, and is commonly referred to as the *standard error of estimate.* This name is unfortunate because s_e is really a standard deviation of points around the regression line—not a standard error in the sense that we have been using the term. Nevertheless, standard error of estimate is a commonly used term, so we should be alert to it.

Example 11–3 Compute the standard deviation of regression for the data in Example 11–2.

Solution

The X and Y data points from Example 11–2 are repeated in the table below, along with the computation of each Y_c value, the deviations, the square of the deviations, and the sum of the squared deviations, SSD. These deviations represent the *unexplained,* or *random, variation* of Y values about the regression line; hence the designation of error, or s_e.

TABLE 11–3 Computation of *SSD* of points about the regression line

SAMPLE NUMBER	Hardness X	Strength Y	Point Estimate $Y_c = .25 + 1.25X$	Deviation $(Y - Y_c)$	Squared Deviation $(Y - Y_c)^2$
1	7	10	9.00	1.00	1.0000
2	9	12	11.50	.50	.2500
3	5	6	6.50	−.50	.2500
4	8	9	10.25	−1.25	1.5625
5	6	8	7.75	.25	.0625
6	9	11	11.50	−.50	.2500
7	7	10	9.00	1.00	1.0000
8	4	5	5.25	−.25	.0625
9	8	10	10.25	−.25	.0625
10	7	9	9.00	.00	.0000
Totals	70 (ΣX)	90 (ΣY)		0.00	4.5000 $\Sigma(Y - Y_c)^2$

$$s_e = \sqrt{\frac{\Sigma(Y - Y_c)^2}{n - 2}} = \sqrt{\frac{4.5000}{10 - 2}} = .75$$

SHORTCUT EQUATION FOR s_e As is evident from the above, the calculation of s_e from equation 11–5 can be tedious. Each Y_c value must first be calculated, then the deviations computed and squared, and so forth. Fortunately, a shortcut equation that utilizes the a and b constants of the regression equation has been derived. Equation 11–6 is mathematically equivalent to the "definitional" equation above, but much easier to compute.

$$s_e = \sqrt{\frac{\Sigma Y^2 - a\Sigma Y - b\Sigma XY}{n - 2}} \tag{11–6}$$

Example 11–4 Use equation 11–6 to compute the standard deviation of regression for the strength and hardness data of Example 11–1. [Note: The regression equation developed there was $Y_c = .25 + 1.25X$, and from Table 11–2 we have $\Sigma Y = 90$, $\Sigma Y^2 = 852$, and $\Sigma XY = 660$.]

Solution

$$s_e = \sqrt{\frac{\Sigma Y^2 - a\Sigma Y - b\Sigma XY}{n - 2}} = \sqrt{\frac{852 - (.25)(90) - (1.25)(660)}{10 - 2}} = .75$$

Once we have computed the standard deviation of regression, s_e, it can be used to help test the validity of the regression equation—which we will take up momentarily. It can also be used for making inferential statements in a manner similar to what we have done before when using other normal distributions. For example, as illustrated in Figure 11–7, if the points about the regression line are normally distributed, we can expect 68.2% of the observations to lie within $\pm$ 1 s_e, 95.5% of them to lie within $\pm$ 2 s_e, and 99.7% of them to lie within $\pm$ 3 s_e. Note, however, that we are referring to the location of *individual Y values*, not to the mean of a sample of Y values. For this reason, we do not refer to these estimates associated with regression as "confidence intervals" (of means). Instead, we refer to them as *prediction intervals* (of individual values).

FIGURE 11–7 Normal distribution around the regression line

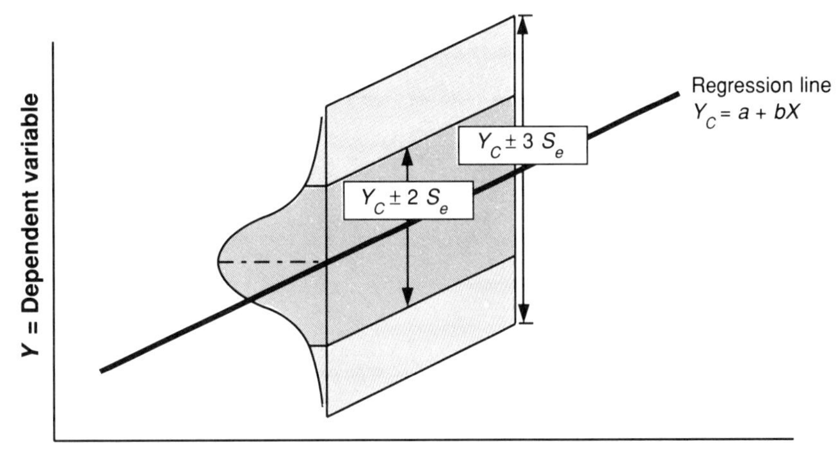

APPROXIMATE PREDICTION INTERVAL FOR INDIVIDUAL VALUES

LARGE SAMPLES We have already seen that we can make a point estimate of the mean of the regression line by substituting any desired value of X into the regression equation $Y_c = a + bX$. If the sample size is large, the Y values are normally distributed about the regression line, and we can establish approximate prediction limits for individual values of Y, which we can designate as Y_{PL}:

Prediction Limits
(large samples): $$Y_{PL} = Y_c \pm Zs_e \quad (11\text{–}7)$$

SMALL SAMPLES Analysts do not agree on what constitutes a "large" sample for making predictions using regression; some stay with the $n \geq 30$ limit we have used before, and others suggest 50 or even 100. But certainly if the sample size is less than 30, one would want to use the t-distribution for establishing the approximate prediction limits:

Prediction Limits
(small samples): $$Y_{PL} = Y_c \pm ts_e \quad (11\text{–}8)$$

As before, the t-distribution is theoretically correct whenever σ is unknown, and the t-value must reflect the loss of two degrees of freedom $(n - 2)$.

Example 11–5 Use the t-distribution to establish approximate 95% prediction limits using the data from Example 11–1. [Recall that the 10 data points resulted in a regression line of $Y_c = .25 \pm 1.25X$, and s_e was later computed to be .75.] Make an interval prediction of strength, Y, when the surface hardness measurement is $X = 8.5$.

Solution

$$Y_{PL} = Y_c \pm ts_e \quad \text{where:} \quad Y_c = .25 + 1.25\,(8.5) = 10.88$$
$$t = t_{\frac{\alpha}{2} n-2} = t_{\frac{.05}{2},\ 10-2} = 2.306$$

$$LPL_Y = 10.88 - 2.306\,(.75) = 9.15$$

$$UPL_Y = 10.88 + 2.306\,(.75) = 12.61$$

Thus we would predict that the strength of a piece of CM7 material that tested 8.5 on the hardness scale would be within the range of 9.15 to 12.61 units of strength. If the actual data conform to the underlying assumptions, this should be a reasonably good interval estimate for an individual item.

MORE ACCURATE PREDICTION INTERVALS FOR INDIVIDUAL VALUES

As you may have noticed, the discussion about prediction intervals has, to this point, been qualified by referring to the intervals as "approximate." This was with good reason. The true (population) regression line is $\mu_{y \cdot x} = \alpha + \beta X$. But we are estimating the true values of the Y-intercept (α) and the slope (β) with the point estimates a and b. We know point estimates are usually wrong, and we have no way of knowing whether our estimates agree with the true values. So we should allow for this uncertainty in our prediction interval—especially if our sample size is small.

There are two major sources of error—the possibility of an intercept error and the possibility of a slope error. Each regression line from a sample passes through its own $\bar{x}$ and $\bar{y}$ values. But these $\bar{x}$ and $\bar{y}$ values are not necessarily the true means. And the values of a and b obtained from different samples will vary from one sample to the next. The estimated Y_c value is least dependent on the true values of α and β, where the X value is equal to the mean of the X variable. As we move away from this mean, we are less and less certain of the Y_c value because the error due to estimating b is magnified. Estimates based on small samples are particularly vulnerable to this error.

ERROR ADJUSTMENT Fortunately, statisticians have derived a correction factor to adjust the standard error of estimate to the additional intercept and slope uncertainties. The corrected standard error of estimate for individual predictions is termed s_{IND}:

$$s_{IND} = s_e \sqrt{1 + \frac{1}{n} + \frac{(X - \bar{x})^2}{\Sigma(X - \bar{x})^2}} \qquad (11\text{–}9)$$

The (corrected) form of s_{IND} includes an allowance for the type of model, the intercept error, and the slope error. The third factor makes allowance for the slope uncertainty. The $(X - \bar{x})^2$ in the numerator squares the distance that a value, X, is from the sample mean, $\bar{x}$. The $\Sigma(X - \bar{x})^2$ in the denominator is the total variation in X. (This is also the denominator of the equation for the slope of the regression line, except that in Equation 11–3 it is expressed in the more convenient form of $\Sigma X^2 - n\bar{x}^2$.) The effect of this is to give the prediction limits a flared effect, as illustrated in Figure 11–8.

FIGURE 11–8 Prediction limits for 95% interval

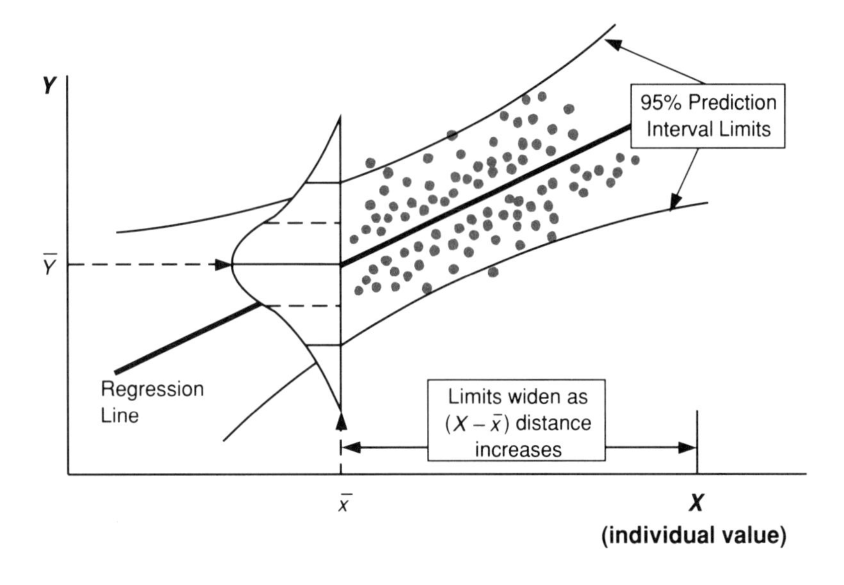

Example 11–6 Repeat the calculation of the 95% prediction limits requested in the previous example, except use the adjusted form of the standard error of estimate from Equation 11–9. [Note: In Example 11–5, we used $s_e = .75$, and computed a prediction interval for strength based upon a hardness measurement of $X = 8.5$. We found that the $Y_c = 10.88$, and used $t_{\frac{.05}{2},\ 10-2} = 2.306$. Also, for this example, $\bar{x} = 7$, and $\Sigma(X - \bar{x})^2 = \Sigma X^2 - n\bar{x}^2 = 24$.]

Solution

Insofar as we will need to use s_{IND}, let us calculate that first. Equation 11–9 is set up in such a way that one can obtain s_{IND} by multiplying the (uncorrected) standard deviation of regression by the adjustment factor.

$$s_{IND} = s_e \sqrt{1 + \frac{1}{n} + \frac{(X - \bar{x})^2}{\Sigma(X - \bar{x})^2}} = (.75) \sqrt{1 + \frac{1}{10} + \frac{(8.5 - 7.0)^2}{24}}$$

$$(.75) \sqrt{1 + .1000 + .0938} = (.75)(1.0926) = .82$$

The corrected prediction limits are thus:

$$Y_{PL} = Y_c \pm ts_{IND} \quad \text{where:} \quad Y_c = 10.88$$

$$t_{\frac{.05}{2},\, 10-2} = 2.306$$

$$LPL_Y = 10.88 - 2.306(.82) = 8.99 \text{ (Uncorrected limit was 9.15)}$$

$$UPL_Y = 10.88 + 2.306(.82) = 12.77 \text{ (Uncorrected limit was 12.61)}$$

Our corrected prediction of strength of an individual piece of CM7 material is thus within the range of 8.99 to 12.77 units. Note that these limits are not substantially different from the earlier limits, even though the sample size is quite small.

CONFIDENCE INTERVAL FOR MEAN VALUES

Thus far we have assumed that the regression model would be used to predict an *individual value* of the dependent variable. Some statistical references also provide an equation that can be used to predict the mean or average of the dependent variable from samples of any given size, say, of $n = 25$. The point estimate Y_c is, of course, the same for both. But the distribution of sample means will tend to be much closer to the regression line than that of individual values.

Estimation of the mean Y value for any given value of X is analogous to the confidence interval estimation procedures we used in earlier chapters. The approximate confidence limits for large samples are $Y_{CL} = Y_c \pm Z \dfrac{s_e}{\sqrt{n}}.$

The corrected interval corresponding to Equation 11–9, except for *means*, is:

Confidence interval (for means)

$$Y_{CL} = Y_c \pm ts_e \sqrt{\frac{1}{n} + \frac{(X - x)^2}{\Sigma(X - \bar{x})^2}} \qquad \text{(11–10)}$$

Inasmuch as the confidence limits for mean values seem to have less application in business decision making than the prediction limits for individual values, we will not carry forward with an example of this. However, the calculation is straightforward. And the interpretation is that one can be confident (at the specified level) that the interval includes the true mean value of Y, when the independent variable represents a mean of values rather than an individual value.

INFERENCE ABOUT THE SLOPE OF THE REGRESSION LINE

The two parameters (and sources of error) of the regression line are the Y-intercept and the slope. Are the intercept and slope values of the sample regression line we computed statistically significant, or might they have come about by chance? Tests are available to answer these questions. The test of the slope of the regression line is more common (and useful) than that of the intercept, so we shall limit our inquiry to the slope coefficient.

TESTING THE SLOPE The question of most interest is usually whether the regression line has any significant slope *at all*. If the X and Y variables are not related (as in Figure 11–2 d), then the slope of the true (population) line through them should be zero, and X would be of no help in predicting Y. Is the slope of the sample line simply due to the chance that certain points—along a line—happened to be included in the sample? Or is it due to a valid relationship between the X and Y variables?

We can test this in a forceful way by hypothesizing that the slope of the population line is zero, H_0: $\beta = 0$. Then if the hypothesis is rejected, there is strong evidence to suggest the slope is significantly different from zero, so the alternative hypothesis is H_1: $\beta \neq 0$. We could also set up the null to test whether the value of β is equal to some specific value other than zero if we have some prior value to use.

The test of a regression slope (or regression coefficient as it is sometimes called) follows the same general format that we have used for other tests. We specify the level of significance and the reject limit from the appropriate distribution—which is frequently the t-distribution. Then the test statistic is computed by dividing the difference between the sample value (b) and hypothesized value (β) by the estimated standard error of the slope, s_b:

$$\text{Test statistic for slope:} \qquad t = \frac{b - \beta}{s_b} \tag{11–11}$$

Here, s_b is the estimated standard deviation of the sampling distribution of the possible slope (b) values, or estimated standard error of the slope:

$$s_b = \frac{s_e}{\sqrt{\Sigma(X - \overline{x})^2}} \tag{11–12}$$

Example 11–7 Test whether the slope of the regression equation developed in Example 11–1 is significantly different from zero. Use the $\alpha = .05$ level. [Note: From previous examples we have $Y_c = .25 + 1.25X$, $s_e = .75$, and $\Sigma(X - \bar{x})^2 = 24$.]

Solution

1. H_0: $\beta = 0$ (There is no linear relationship between X and Y in the population.)

   H_1: $\beta \neq 0$ (The population variables X and Y are linearly related.)
2. $\alpha = .05$
3. Use t-distribution. Reject if $|t_c| > t_{\frac{.05}{2},\,10-2} = 2.306$
4. $t = \dfrac{b - \beta}{s_b}$ where: $s_b = \dfrac{s_e}{\sqrt{\Sigma(X - \bar{x})^2}} = \dfrac{.75}{\sqrt{24}} = .153$

   $= \dfrac{1.25 - 0}{.153} = 8.17$
5. *Conclusion*: Reject H_0: $\beta = 0$ because $8.17 > 2.306$. There does appear to be a positive linear relationship between hardness and strength; on average the harder materials are stronger over the range of values in these data.

The calculated value of 8.17 actually exceeds the t-value required to reject H_0 at the .01 level of 3.355, so the slope is highly significant and quite unlikely to have come about by chance.

CONFIDENCE INTERVAL ESTIMATE OF THE SLOPE Once the hypothesis of zero slope is rejected, one might well ask, "If it isn't zero, what is it?" We do have a point estimate of the true slope ($b = 1.25$), but we may wish to have an interval estimate with a specified level of confidence. The computation of confidence limits for the true slope is similar in form to other confidence intervals we have worked with:

$$CL \text{ for } \beta = b \pm t_{\frac{\alpha}{2},n-2}\, s_b \qquad (11\text{–}13)$$

Example 11–8 Provide a 95% confidence interval estimate for the true slope of the regression line developed in Example 11–1. [Note: From previous examples, $Y_c = .25 + 1.25X$ and $s_b = .153$.]

Solution

$$CL = b \pm t_{\frac{\alpha}{2},n-2}\, s_b \quad \text{where: } b = 1.25, \quad t_{\frac{.05}{2},\,10-2} = 2.306$$

$$= 1.25 \pm 2.306\,(.153)$$

$$= .90 \text{ to } 1.60$$

ANALYSIS The interval is wide because the small sample has yielded a b value that is not very precise. With a larger sample size, s_b would be smaller because the numerator used to calculate it (s_e) would be smaller, and the denominator, which is the total variation of the X variable, $\Sigma(X - \bar{x})^2$, would be larger. Both changes would act to decrease the interval width.

EXERCISES II

14. Someone has defined a regression line as simply a mathematical "fit" of the best line through a series of points, and someone else has defined it as a line that describes the relationship between any given value of X and the mean, $\mu_{y.x}$, of the corresponding probability distribution of Y. Which is correct? Explain.
15. When computing s_e using the long, or *definitional*, form:
    (a) along which axis are the deviations measured?
    (b) how does one find the values for $\bar{x}$ and $\bar{y}$?
    (c) why is the $\Sigma(Y - Y_c)$ always equal to zero?
    (d) why is $(n - 2)$ used in the denominator rather than $(n - 1)$?
16. In regression we compute *prediction intervals* for Y values, whereas in the estimation materials studied earlier, we were concerned with an X variable, and developed *confidence intervals*. How do prediction intervals for Y values in regression differ from the confidence intervals we worked with earlier?
17. How does the *standard error of estimate* differ from (a) the standard errors of one-sample data, and (b) the standard deviations of one-sample data we have worked with previously?
18. Compute s_e for the following:
    (a) a sample of 40 observations where $\Sigma(Y - Y_c)^2 = 26.80$
    (b) a sample of 102 data points where $\Sigma(Y - Y_c)^2 = 288.00$
19. Compute s_e for a sample of 10 observations where: $\Sigma X = 400$, $\Sigma Y = 660$, $\Sigma X^2 = 17{,}600$, $\Sigma XY = 27{,}286$, and $\Sigma Y^2 = 44{,}264$. The regression line for these data is $Y_c = 43.85 + .55X$.
20. For the regression equation $Y_c = -2.75 + 1.70X$, compute the standard error of estimate. Assume $n = 4$ and $\Sigma Y = 23$, $\Sigma Y^2 = 165$, and $\Sigma XY = 132$.
21. For the following data, compute the standard deviation of regression: $n = 10$, $\Sigma X = 20$, $\Sigma Y = 62$, $\Sigma X^2 = 54$, $\Sigma Y^2 = 398$, $\Sigma XY = 13$, $b = .79$.

22. An economist has developed a political index (X) which he hopes to use to predict incremental price index changes (Y). For the four pairs of observations below, (a) compute the regression equation, (b) compute the predicted Y_c values for each X value, and (c) use the "definitional" equation (11–5) to compute the standard deviation of regression, s_e.

Political index X	3	7	4	6
Price index change Y	2	10	5	6

23. Use the t-distribution to establish *approximate* 99% prediction limits for the inventory level of soft drinks at a store in Dallas. The point estimate from a regression equation using 14 stores is $Y_c = 240$ cases and $s_e = 4$ cases.
24. Data were collected from 20 firms to relate their earnings per share (Y in \$) to expenditures on new equipment (X in \$ millions). If the equation was $Y_c = 4.2 + 1.9X$ with a standard error of estimate of .6, make an *approximate* 95% prediction interval for (a) a firm that spends \$5 million on new equipment, and (b) one that spends nothing on new equipment.
25. The marketing manager of an industrial products company has collected the following data relating the number of sales visits to a customer (X) to the annual purchase orders, or bookings (in \$000) from that customer (Y).

X = No. Visits	5	7	3	1	7	8	4	2	5	6	3	5
Y = Orders (\$000)	5.40	6.70	3.70	1.30	6.60	8.70	4.50	2.20	5.90	6.10	2.00	3.70
Predicted value Y_c	5.06	7.04	3.09	1.11	7.04	8.02	4.08	2.10	5.06	6.05	3.09	5.06
Error ($Y - Y_c$)	.34	−.34	.61	.19	−.44	.68	.42	.10	.84	.05	−1.09	−1.36

Also from the data $\Sigma Y^2 = 323.48$, $\Sigma XY = 315.10$, $\Sigma(Y - Y_c)^2 = 5.2776$, and the regression equation is $Y_c = .125 + .987X$. For this data compute the standard deviation of regression using (a) the definitional form (equation 11–5) and (b) the short form (equation 11–6). Then (c) make an approximate 95% prediction interval estimate of the value of orders booked when eight visits are made to a customer (equation 11–8), and (d) make a more accurate prediction interval estimate (using equation 11–9). Finally, (e) comment on any differences in your answers in (a) and (b), and in (c) and (d).

26. An investor is seeking to determine whether it is possible to predict the number of new stocks that go up in price on the basis of the number of new stock offerings handled by that broker. Her research revealed the following:

X Number of Stock Offerings	Y No. that increased in price
8	6
4	2
10	9
3	3
14	11
6	5
10	7

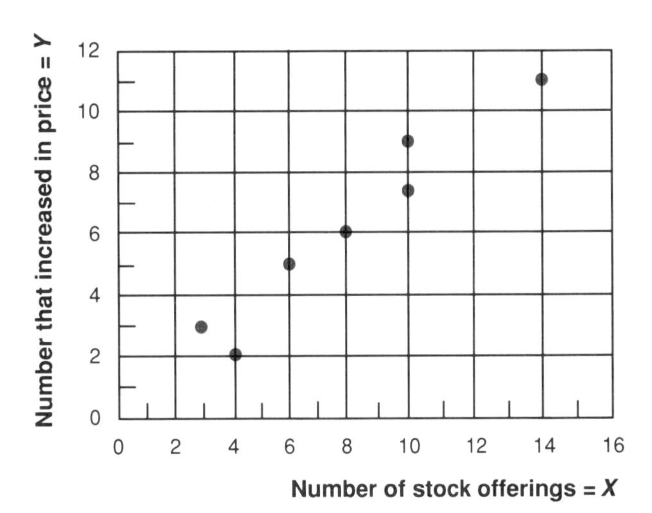

(a) Use simple linear regression to develop the equation that best fits the data. (b) Compute s_e. (c) Develop a point estimate of the number of stocks that will increase in price for a broker that handles 12 stock offerings. (d) Make an approximate 98% individual prediction interval estimate for the number of stocks that increase in price when the firm handles 12 stock offerings.

27. Is population a good predictor of the number of large-scale nuclear power plants in operation? The data below show the estimated population of seven countries (in the year 2000), along with the number of large-scale nuclear power plants either in operation, under construction, or being planned. For these data
    (a) estimate the regression equation
    (b) compute the standard error of estimate
    (c) make (an accurate) 95% prediction interval estimate of the number of nuclear reactors in a country of 150 million people in the year 2000.
    (d) [*optional*] Make a 95% confidence interval estimate of the *mean* number of reactors in countries of 150 million people in the year 2000.

COUNTRY	POPULATION ESTIMATE (Year 2000 in millions)	NUMBER NUCLEAR REACTORS
Canada	32	22
France	56	67
Germany	59	32
Japan	129	46
USA	264	129
UK	55	43
USSR	310	110

[Source: Japan Institute for Social and Economic Affairs]

28. A regression analysis of five observations relating food production to fertilizer usage resulted in an equation with a slope of 1.8 and a standard error of estimate of .73. Also, $\Sigma(X - \bar{x})^2 = 10$.
    (a) Test whether the slope is significantly different from zero.
    (b) Construct a 90% confidence interval for the true slope.
29. (a) Use the data from Problem 22 to test whether the slope of the regression line is significantly different from zero. Let $\alpha = .05$.
    (b) If the slope is significant, provide an accurate 95% confidence interval estimate.
30. A regression study of 12 employees relating hours of quality control training (X) to defect-free production (Y) resulted in the equation $Y_c = 276.2 + 33.4X$, with $s_e = 69.3$, and $\Sigma(X - \bar{x})^2 = 70.0$.
    (a) Calculate the t-value used to test the slope to verify that it is significant at the .05 level.
    (b) What is the 95% confidence interval for the true slope?

SIMPLE LINEAR CORRELATION

HOW REGRESSION AND CORRELATION DIFFER The regression procedures we have worked with to this point have been useful for describing the *nature of the relationship* between the two variables. With regression, we quantified that relationship in terms of the intercept and slope of the linear equation relating the independent (X) and dependent (Y) variables. Then, for any selected value of the X variable, we found that there was a conditional probability distribution of Y values. And observed Y values were assumed to be randomly selected members of those conditional distributions.

Correlation is concerned with the *degree*, or *closeness, of the relationship* between the variables. It makes no difference which variable is labeled X or Y because there is no independent-dependent relationship. Both variables enjoy equal status. We are not directly estimating one variable from another—instead we seek a mathematical measure of the closeness, or strength, of the relationship. Once a close relationship is identified, however, it may in turn be useful for forecasting one of the variables.

The mathematical relationship between the variables is referred to as the *coefficient of correlation*. It is a modified, or quasi measurement of the variation explained by the X and Y relationship compared to the total variation in the data. For the two-variable model discussed in this chapter, we assume the X and Y values form joint probability distributions that are normal in both the X and Y directions, as illustrated in Figure 11–9.[2]

[2] Our discussion here follows the Pearson method. The nonparametric methods chapter describes a method of correlating ranked data (via Spearman's rank correlation method), and techniques are also available for correlating nominal data (via the contingency coefficient).

FIGURE 11–9 Distinction between regression and correlation

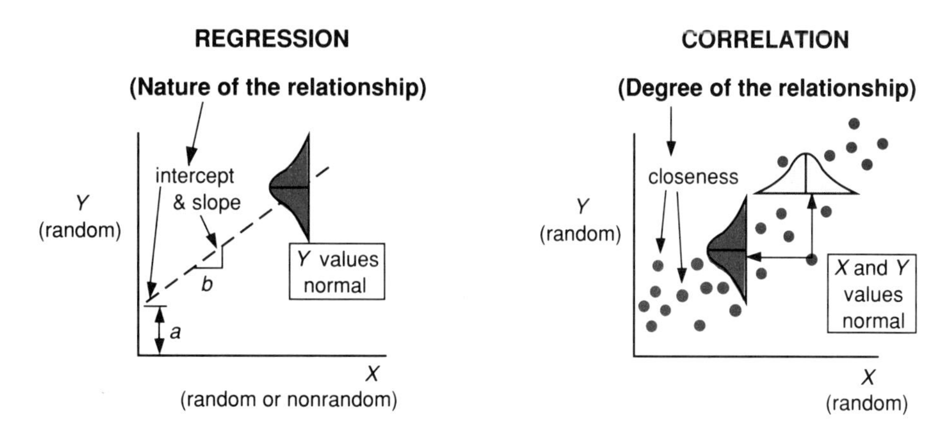

Our initial inquiry will be limited to *simple linear correlation*. That is, we will consider only two variables, X and Y, and we shall assume the underlying relationship is linear. Moreover, we shall explain the meaning of correlation in terms of the deviation of the Y variable, recognizing that we might just as well have used the X variable. First we look to the intuitive meaning and then to the theory and calculation of the coefficient.

CORRELATION COEFFICIENT

We begin with a statement about the correlation coefficient.

> The **simple linear correlation coefficient, *r*,** is a relative number between -1 and $+1$ that is a measure of how closely two variables conform to a linear relationship.

Figure 11–10 utilizes some scatter diagrams to offer an intuitive meaning of the term correlation. Note that the correlation coefficient tells us

1. whether the relationship (line) is positive (+) or negative (−)
2. how (relatively) close the data points are to a line through the data.

When r is positive, the Y values increase as X increases, and when negative, they decrease as X increases. The numerical value of r is a summary expression of the location of the observations relative to the two distributions. It can go as high as $+1$ or as low as -1 if the points all fall on the conditional means of their respective distributions. These extremes would signify perfect relationships where all the points on the XY plane would lie on the line connecting the means of the joint distributions. On the other hand, a correlation coefficient of zero would signify that there is no relationship between the two variables. Nevertheless, as we shall see next, there is a related coefficient that has more intuitive meaning than the coefficient of correlation. It is called the *coefficient of determination*.

FIGURE 11–10 Characteristics of the correlation coefficient

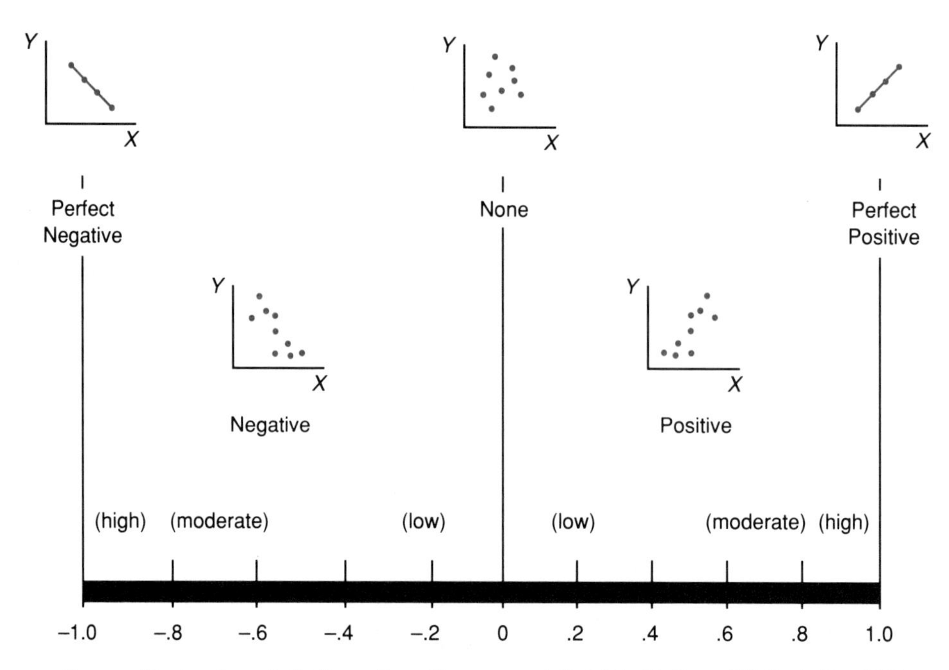

Source: Monks, Joseph G. *Operations Management*, 3rd Ed., McGraw-Hill Book Co., 1987, p. 287.

COEFFICIENT OF DETERMINATION

EXPLAINED AND UNEXPLAINED VARIATION As an aid to understanding the theory underlying correlation, let us assume we have a scatter diagram with a positively sloped regression line, Y_c. Figure 11–11 uses two of many possible points Y to illustrate the variation.

FIGURE 11–11 Explained and unexplained variation

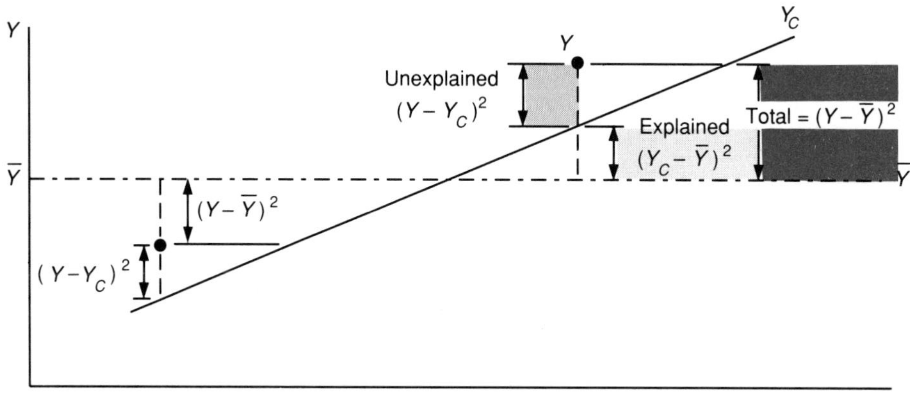

The *total variation* of points in the Y direction is the sum of squared deviations of points about their mean, $\bar{y}$. Part of this total variation can be *explained* by relating the Y values to X, and part of it is simply random, or *unexplained*, variation. In symbolic form we could say:

Total Variation in Y	=	**Explained Variation**	+	**Unexplained Variation**
[*SSD* of all points from their mean $\bar{y}$]		[*SSD* of the regression line value, Y_c, from $\bar{y}$]		[*SSD* of Y values from the regression line value, Y_c]
$\Sigma(Y - \bar{y})^2$	=	$\Sigma(Y_c - \bar{y})^2$	+	$\Sigma(Y - Y_c)^2$

COEFFICIENT OF DETERMINATION The $\Sigma(Y - \bar{y})^2$ is total variation (and would be the variance of Y if we divided by *df*). But the Y values are linked with the values of X. The reason the Y values are not clustered about the mean of y is explained by (or associated with) the values of X. For this particular relationship, high values of X tend to be associated with high values of Y. One might even expect to have all Y values on the regression line if X were a perfect predictor of Y. However, it is not, so the actual points deviate from the line by the distance $(Y - Y_c)$. Those are unexplained deviations.

The **coefficient of determination, r^2**, is a number between zero and one that expresses the explained variation as a percent of the total variation.

$$r^2 = \frac{\text{explained variation}}{\text{total variation}} = 1 - \frac{\text{unexplained variation}}{\text{total variation}}$$

In equation form we can express the coefficient of determination as:

$$r^2 = \frac{\Sigma(Y_c - \bar{y})^2}{\Sigma(Y - \bar{y})^2} = 1 - \frac{\Sigma(Y - Y_c)^2}{\Sigma(Y - \bar{y})^2} \qquad (11\text{–}14)$$

Example 11–9 A city homebuilders association has collected the data shown below in an attempt to relate the number of housing starts per month to the prevailing home loan interest rate at local banks. A graph of the data appears linear—much like the one on the opening page of the chapter.

Interest Rate (X)	No. Housing Starts (Y)	X^2	XY	Y^2	Pt. Est. Y_c	Error $(Y - Y_c)$	$(\text{Error})^2$ $(Y - Y_c)^2$
18	4	324	72	16	3.87	.13	.0169
12	10	144	120	100	11.25	1.25	1.5625
8	20	64	160	400	16.17	3.83	14.6689
7	18	49	126	324	17.40	.60	.3600
14	12	196	168	144	8.79	3.21	10.3041
9	15	81	135	225	14.94	.06	.0036
17	6	289	102	36	5.10	.90	.8100
20	2	400	40	4	1.41	.59	.3481
16	4	256	64	16	6.33	2.33	5.4289
10	8	100	80	64	13.71	5.71	32.6041
$\Sigma X = 131$	$\Sigma Y = 99$	$\Sigma X^2 = 1903$	$\Sigma XY = 1067$	$\Sigma Y^2 = 1329$			$\Sigma = 66.1071$

The data above yield the equation $Y_c = 26.014 - 1.23X$ with $s_e = 2.87$, and the total variation in Y is $\Sigma(Y - \bar{y})^2 = 348.90$. Use the data to compute the coefficient of determination and interpret its meaning.

Solution

Inasmuch as the unexplained variation $\Sigma(Y - Y_c)^2$ is given, we can use it:

$$r^2 = 1 - \frac{\Sigma(Y - Y_c)^2}{\Sigma(Y - \bar{y})^2} = 1 - \frac{66.1071}{348.90} = 1 - .1895 = .8105$$

Thus the sample values have a coefficient of determination of .81. From this we can estimate that approximately 81% of the variation in housing starts is associated with (or explained by) the level of interest rates on home loans.

EQUATION FOR THE COEFFICIENT OF CORRELATION The coefficient of correlation, r, is the square root of the coefficient of determination. In the definitional form of Equation 11–14, we must assign a + or − to the r value, depending upon whether the relationship is positive or negative.

$$r = \pm\sqrt{\frac{\Sigma(Y_c - \bar{y})^2}{\Sigma(Y - \bar{y})^2}} \qquad (11\text{–}15)$$

Example 11–10 Compute and interpret the correlation coefficient for the interest rate and housing start data of the previous example.

Solution

$$r = \sqrt{r^2} = \sqrt{.8105} = -.9003$$

The best interpretation is probably to use the r^2 value and say that an estimated 81% of the variation in housing starts is associated with the level of interest rates. Note that we have assigned a negative value to r because the slope of the regression line is negative.

INTERPRETATION OF r The coefficient of correlation is more widely known and used than the coefficient of determination. However, r^2 is easier to understand and constitutes a more precise interpretation because it shows the proportion of the variation in Y that can be explained by relating Y to X. For example, suppose the r^2 value relating company sales of fast foods to outdoor temperature was .49. This would suggest that just under 50% of sales are associated with the weather. The correlation would be the square root of this, or $r = +.70$—but the percent of sales explained by temperature is still .49. One must be careful. The r measure (alone) can tend to overstate the degree of relationship between the two variables in the minds of many users.

OTHER EQUATIONS FOR r An alternate, and more common, procedure for calculating r is to use the same data table that is used for the regression equation. This form is typically easier to use and determines the sign of r directly. Note the symmetry in the X and Y terms of equation 11–16, confirming the equality of the two variables.

$$r = \frac{n\Sigma XY - \Sigma X \Sigma Y}{\sqrt{[n\Sigma X^2 - (\Sigma X)^2]\,[n\Sigma Y^2 - (\Sigma Y)^2]}} \qquad (11\text{–}16)$$

The correlation coefficient is a widely used measure and several other equations for it are also available. One that utilizes values that might be obtained from other calculations (of regression) is:

(using the a and b from regression)

$$r = \pm\sqrt{\frac{a\Sigma Y + b\Sigma XY - n\bar{y}^2}{\Sigma Y^2 - n\bar{y}^2}} \qquad (11\text{–}17)$$

INFERENCE ABOUT THE CORRELATION COEFFICIENT

Statistical procedures are available to test the hypothesis that the value of the (true) population correlation coefficient is equal to zero, and to set up a confidence interval for the true coefficient of correlation.

HYPOTHESIS TEST OF ρ As with other parameters, we designate the population correlation coefficient with a Greek letter, in this case, ρ (rho). The test to determine whether it is reasonable to believe that $\rho = 0$ follows the usual procedure. The null hypothesis is H_0: $\rho = 0$, and the alternate is H_1: $\rho \neq 0$. After setting the level of significance, we can designate the test statistic as the t-distribution. The calculated value is assumed to come from a bivariate normal population, and is computed as:

$$t_c = \frac{r - \rho_H}{\sqrt{(1 - r^2)/(n - 2)}} \qquad (11\text{–}18)$$

Example 11–12 Test for the significance of the correlation coefficient of Example 11–11 where we calculated an r of $-.90$ from a sample of $n = 10$ observations. Use a significance level of 2%.

Solution

1. H_0: $\rho = 0$ (There is no correlation between the two variables.)
   H_1: $\rho \neq 0$ (There is a significant correlation between the two variables.)
2. $\alpha = .02$
3. Use t, reject if $|t_c| > t_{\frac{\alpha}{2},\, n-2df} = t_{\frac{.02}{2},\, 8} = 2.896$
4. $t_c = \dfrac{r - \rho_H}{\sqrt{[1 - r^2]/(n - 2)}} = \dfrac{-.90 - 0}{\sqrt{[1 - (.90)^2]/(10 - 2)}} = 5.84$
5. *Conclusion*: Reject hypothesis of no correlation between the populations. Because of the large value of t, we may be practically certain that the true coefficient of correlation is *not* zero.

Other methods of testing the significance of r are also available, including some charting techniques that can be found in statistical references. The method illustrated above is limited to tests that $\rho = 0$, but can accommodate any reasonable sample size.

CONFIDENCE INTERVAL ESTIMATE OF ρ When the hypothesis that $\rho = 0$ is rejected, one may wish to set up a confidence interval estimate for the true coefficient of correlation. Unfortunately, r is not normally distributed around the true value ρ. So it is necessary to transform the r values into values which are approximately normally distributed, find the interval from these transformed values, and then convert these figures back to correlation coefficients. Another method of establishing the confidence intervals is to use confidence band charts prepared especially for this purpose. Using the charts, one can enter an axis marked sample r, proceed up to the curves for the appropriate sample size, and then read off the confidence interval limits on the other axis. Because of the relatively minor importance of these procedures, however, they are not included here.

CORRELATION LIMITATIONS

This completes our inquiry into simple linear regression and correlation, except for some notes of caution. Once we have a high correlation coefficient, we may be tempted to think of the two variables in a cause-and-effect relationship. For example, we might want to jump to the conclusion that low interest rates *cause* housing starts. However, low interest rates might really chase funds out of the domestic market into better foreign investments. Maybe the government has a policy of channeling federal funds into housing during recessionary periods—periods when interest rates just happen to be low.

Whatever the case, we should note emphatically that *correlation, per se, implies nothing about cause and effect*. It is simply a statistical procedure that seeks to document a relationship; it doesn't prove one variable is the cause of the action of another. Cause-and-effect conclusions require carefully controlled conditions and/or judgmental inputs that must be made by the manager or decision maker. Correlation might help by supplying some information, but the conclusions from correlation must be limited to statements about the degree of association between the variables.

Sometimes correlations are *spurious*. That is, there is no apparent relationship, but the r value shows a high correlation anyway. For example, you may run across stories about the high correlation between the success of a particular baseball team and the level of the stock market, or the salary of ministers and the consumption of liquor. Very often, the variables we are interested in are associated with other variables that move in the same direction. So it is important to try to *work with variables that have a commonality* and not be misled by the influence of overlapping variables.

Finally, we should always *limit our conclusions* to the range of the variables we are working with. We cannot extrapolate to other times or locations if our sample data are not representative of those populations.

EXERCISES III

31. Compare simple linear correlation with regression in terms of the
    (a) "status" of the variables
    (b) nature or degree of the relationship
    (c) type(s) of statistical distributions assumed
32. What two items of information are conveyed by the correlation coefficient?
33. In considering the coefficient of correlation and the coefficient of determination, (a) how do they differ, and (b) which is more intuitively understandable? Why?
34. Solve for the following:
    (a) Given the coefficient of determination = .80, find r.
    (b) Given $r^2 = .90$, find the coefficient of correlation.
    (c) Given $\Sigma(Y_c - \bar{y})^2 = 340$ and $\Sigma(Y - \bar{y})^2 = 400$, find r^2.
    (d) Given $\Sigma(Y_c - \bar{y})^2 = 86$, and $\Sigma(Y - Y_c)^2 = 14$, find r.
35. A study of the relationship between employment levels ($Y \times 10^5$) and freightcar shipments (X) in a certain community yielded the following data.

    $\bar{x} = 1.4$, $\bar{y} = 30$, $\Sigma(Y_c - \bar{y})^2 = 1{,}000$, $\Sigma(Y - \bar{y})^2 = 1300$, $\Sigma(Y - Y_c)^2 = 300$

    (a) Compute r^2, and (b) explain its meaning in terms of this problem.
36. For the data of the previous problem, (a) compute the coefficient of correlation, and (b) comment on the causal relationship between the two variables.
37. Use equation 11–16 to compute the correlation coefficient for Example 11–9.
38. Use equation 11–17 to compute the correlation coefficient for Example 11–9.
39. Computer programs frequently give correlation as well as regression information for the same problem.
    (a) Use the data from Table 11–2, along with Equation 11–16, to compute the correlation coefficient for the CM7 material hardness (X) and strength (Y) data of Example 11–1.
    (b) For the same data, test the H_0: $\rho = 0$ against H_1: $r \neq 0$ at the .01 level of significance.
40. A random sample of 15 observations revealed a correlation .48. Test whether the correlation is significantly different from zero at the .05 level.

SUMMARY

Simple linear regression and correlation are statistical techniques we use for analyzing *bivariate* (two-variable) data. The first step is to make a scatter diagram of the observed data points to confirm that the relationship is indeed linear. For regression, the X variable is independent and the Y variable—the one being predicted—is termed dependent. Then the normal equations can be solved to yield the slope (b) and intercept (a) of the line of best fit through the sample points. The sample regression equation, $Y_c = a + bX$ goes through $\overline{x},\overline{y}$ and minimizes the variation (or SSD) of the observed points about the regression line.

For every X value, the population regression model $[Y_c = f(X,\epsilon)]$ assumes there is a normal distribution of Y values whose mean lies on the regression line. A regression line thus describes the relationship between any given value of X and the mean of the corresponding conditional probability distribution of Y. We use this mean value of the regression line Y_c as a point estimate of Y for any given value of X. Any deviation of actual Y values from the regression line value Y_c is referred to as error, ϵ. The most common measure of this error is the standard deviation of regression, which is called the standard error of estimate, s_e.

Given the assumption that Y values are normally distributed about the regression line, and the standard deviation of this distribution, s_e, we can establish prediction limits (for individual Y values) and confidence limits (for mean values of Y). The expression $Y_{PL} = Y_c \pm Zs_e$ is only approximately correct for predicting Y from large samples, however. If small samples are used, the t-distribution is appropriate. And for more accurate predictions, the formula should be modified to take account of the slope and intercept error resulting from the use of sample data—even though the correction is often quite small. The effect of this is to widen the prediction limits more as the X values deviate farther from the mean, $\overline{x}$, that the regression line passes through.

The simple linear correlation coefficient, r, is a number between -1 and $+1$ that tells *how closely* two variables conform to a linear relationship. A more understandable measure is the coefficient of determination, r^2. It expresses the explained variation as a percent of the total variation. However, the correlation coefficient is more widely used, and several equations are available to calculate it. In addition, a relatively simple t-test is available to test if the population correlation coefficient $\rho = 0$.

Finally, we are cautioned that even high correlation does not prove a cause-and-effect relationship exists. That is a matter for more controlled study and managerial judgment.

CASE: KATHY DUBOIS, LEGISLATIVE ASSISTANT[3]

The job of legislative assistant to a dynamic senator like Paul Hawkins was anything but dull. Senator Hawkins had said the task of his staff was to "keep him credible." That was not an easy task, especially when one was such a vocal critic of White House policies as Senator Hawkins. But Kathy Dubois was a talented young Ivy League graduate who could accept the challenge. Moreover, each new task enhanced her confidence.

Now the White House was proposing a tax cut that Senator Hawkins intuitively felt would send the U.S. economy into a tailspin. Kathy got the assignment of "analyzing it." She knew that meant examining every facet of the Administration's proposal, for if she overlooked something important, her boss could be mighty embarrassed.

Senator Hawkins objected strongly to a tax cut, and he was looking for evidence to suggest that decreasing taxes would drop the industrial, or nonfarm, productivity in the country. The data he gave to Kathy included some federal tax rate information along with some historical productivity change figures. He hoped it could be presented in a way that would clearly show that lower taxes were associated with slow growth and that higher taxes were needed to provide funds to improve the economy. Hopefully, that would be enough ammunition to sink the White House proposal.

The data Senator Hawkins handed to Kathy had two of the sets of numbers circled. As he gave it to her he said with a commanding grin, "Kathy, let's see if we can use this to send a *loud and clear signal* to the White House."

Tax Rate	Productivity Index	Tax Rate (cont.)	Productivity Index (cont.)	Tax Rate (cont.)	Productivity Index (cont.)
26	.4	22	3.8	28	1.2
23	3.1	30	−2.2	24	2.2
19	3.1	20	3.8	24	3.0
24	2.0	26	2.2	33	−1.0
26	−.3	28	2.0	31	2.0
31	.2	34	−1.4	25	1.5
*32	3.0	23	2.6	22	1.9
22	3.2	23	2.2	28	.9
30	.6	*25	−3.0	26	3.1
28	.7	22	4.0	23	3.5

The next day Kathy Dubois had some bad news for Senator Hawkins.

KATHY DUBOIS: I'm afraid you'll be shooting yourself in the foot if you use these data, Senator.

[3] Suggested by (but modified from) an editorial article in *The Wall Street Journal*, 8/5/85.

SENATOR HAWKINS: You've got to be kidding, Kathy. Didn't you see those numbers I circled? The 32% tax rate and 3.0 productivity index, coupled with the 25% tax rate and −3.0 index! They tell me that high taxes generate higher productivity. The government pumps that tax money right back into the economy, you know—instead of letting it slip out of the country.

KATHY DUBOIS: Well it may appear that way on the surface, Senator. But the rest of the data doesn't bear that out. As a matter of fact, it says the opposite. Moreover, the 25% and −3.0 index point shouldn't even be in there. It was really a "quirk" measurement that reflected the temporary effects of an oil embargo on our economy. Statisticians would classify that as an "outlier," and probably rule it out.

SENATOR HAWKINS: Kathy, you've really made my day! You know I was getting a speech all ready to go on this, don't you?

KATHY DUBOIS: I'm sorry, sir, but that's the way it turned out. Here's a little more detailed analysis of the data.

Case Questions

(a) In general terms, what type of analysis would Kathy most likely have done with the senator's data? Explain how the analysis you mention was expected to support the senator's position.

(b) For a regression analysis, what would be the independent and dependent variables?

(c) What type of graphic aid would most likely have been Kathy's first step in the analysis? Would it (alone) have given Kathy a cause for questioning the outcome the senator hoped to see?

(d) Using all the data, what is the resulting regression line?

(e) Is the slope of this regression line significant? Why might the slope play such a key role in the senator's position?

(f) Using all the data, what percent of the variation in the productivity index is associated with the tax rate?

(g) Is there a significant correlation between the two variables? How do you think the correlation would be affected by discarding the questionable point at $X = 25$, $Y = -3$?

(h) Was Kathy Dubois correct in concluding the data do not support the senator's position?

QUESTIONS AND PROBLEMS

41. Compare and contrast the following terms:
   (a) dependent variable and independent variable
   (b) positive relationship and negative relationship
   (c) correlation and causation
   (d) the standard deviations: s_x, s_y, and s_e
42. Comment on the following:
   (a) Why should a scatter diagram be prepared as the first step in regression analysis?
   (b) Is it possible to mistakenly fit a straight line to data which are curvilinearly related?
   (c) What is the danger in making estimates of Y from X values that are outside the domain of the observed X values?
43. A firm markets an impulse item (M) that is sold near the checkstands of supermarkets. Prepare a scatter diagram to study the relationship between sales taxes collected (in a specified time span) and sales of product M in ten counties selected at random from those served by the company. (Data have been simplified.)

County	Sales Taxes Collected = X	Units of M Sold = Y
A	$ 16	40
B	24	50
C	32	68
D	15	36
E	20	45
F	12	27
G	18	42
H	14	36
I	10	29
J	29	67
Totals	$190	440

44. In problem 43, sales taxes collected was considered the independent variable and was plotted on the x-axis.
   (a) How can an analyst decide which variable should be considered the dependent variable?
   (b) What is the effect on the regression equation if the X and Y variables are reversed and the dependent variable is labeled X?

(c) What is the effect on the correlation coefficient if the wrong variable is labeled Y?

45. Use the data from problem 43 to prepare a table from which the values needed for the regression and correlation equations can be obtained.
    (a) What is the total variation in the dependent variable?
    (b) What is the total variation in the independent variable?
    (c) Compute the slope of the regression line.
    (d) Compute the average value of Y, where $X = 0$.
    (e) Compute the standard deviation around the mean of Y.
    (f) Compute the standard deviation around the regression line.
    (g) Which of the values computed in parts (a) through (f) is called the standard error of estimate?
46. Use the data obtained in problem 45 to compute the following:
    (a) the predicted (point estimate) of Y_c when X is equal to 20
    (b) the 95% prediction interval for the number of units sold in one individual county if the sales taxes collected there are $20
47. Using the data from problem 45, test the hypothesis that the sample described in problem 43 was drawn from a population with a true beta value of 2.000. Use the .01 level, a two-tailed test, and remember that the number of degrees of freedom is equal to $n - 2$.
48. Equation 11–5 for the standard error of estimate is:

$$s_e = \sqrt{\frac{\Sigma(Y - Y_c)^2}{n - 2}}$$

    (a) Use this equation to compute s_e for problem 43.
    (b) Why is this equation seldom used in actual practice?
    (c) In equation 11–5, the $\Sigma(Y - Y_c)^2$ is the unexplained vertical variation from the regression line. Compare the $\Sigma(Y - Y_c)^2$ portion of that equation with the $\Sigma Y^2 - a\Sigma Y - b\Sigma XY$ expression in Equation 11–6. What can you conclude about the nature of the $\Sigma Y^2 - a\Sigma Y - b\Sigma XY$ term?
49. Prepare a diagram to illustrate the slope and location of a regression line equation in which
    (a) both a and b are positive
    (b) a is positive and b is negative
    (c) a is negative and b is positive
    (d) both a and b are negative
50. Draw a chart on which the total deviation in Y, the explained deviation, and the unexplained deviation are illustrated. Explain the meaning of the coefficient of determination in terms of your chart.

51. Compute the following measures of the strength of the relationship between X and Y using the data from problem 43:
    (a) the coefficient of determination
    (b) the coefficient of correlation
52. A highway construction firm has done a study of the relationship between federal gasoline taxes collected and their own sales. The following results were obtained:

    $n = 10 \qquad \Sigma(X - \bar{x})^2 = 300 \qquad s_e = .756 \qquad b = .80 \qquad a = 4.6$

    (a) Establish a 95% confidence interval estimate for the true value of β.
    (b) Explain the meaning of your interval computed in (a).
53. Test scores from a sales aptitude test (X) and actual first-year sales figures (Y) of 100 sales representatives were studied, and the resultant correlation coefficient was .40. Let $\alpha = .05$.
    (a) Test the hypothesis H_0: $\rho = 0$ against the alternative H_1: $\rho \neq 0$
    (b) State your conclusion and explain what it means.
54. Use the electrical and electronic firms in the Corporate Statistical Database (Appendix A).
    (a) Let the current year earnings per share be the independent variable, X, and develop a regression equation to predict the annual dividend, Y.
    (b) Make a point estimate of the dividend for an individual firm that has $5.00 earnings per share.
    (c) Make either: (1) a 95% prediction interval estimate *for an individual firm* that has $5.00 earnings per share, or (2) a 95% confidence interval estimate for *the mean value of* Y (dividends) for firms that have earnings of $5.00 per share.
55. Use the electrical and electronic firms in the Corporate Statistical Database (Appendix A).
    (a) Let X = the current year earnings per share and Y = the previous year earnings per share. Determine the correlation between the two.
    (b) Do you feel the correlation is significant? Explain.
56. Use the electrical and electronic firms in the Corporate Statistical Database (Appendix A). Is there a significant correlation between the annual salary of employees and the state's per capita income?

57. Use all the firms in the Corporate Statistical Database (Appendix A), and let X = number of employees (or number in hundreds) and Y = annual sales (\$m).
    (a) Determine the regression equation relating number of employees to annual sales. Include a scatter diagram if your computer program offers it.
    (b) Using the equation, predict the annual sales for a firm that has 15,000 employees.
    (c) What is the value of the standard error of estimate?
    (d) Is the slope of the equation significant?
    (e) To what extent are the total annual sales explained by the number of employees?

CHAPTER

12 Multiple Regression and Correlation

INTRODUCTION: MULTIPLE REGRESSION AND CORRELATION

FINDING THE MULTIPLE LINEAR REGRESSION EQUATION

USING THE MULTIPLE REGRESSION MODEL FOR INFERENCE
- The Standard Error of Estimate
- Prediction Intervals for Values of Y
- Evaluating the Overall Regression Model
- Results from the Computer
- Testing and Estimation of the Regression Coefficients

THE COEFFICIENT OF MULTIPLE DETERMINATION
- Calculating R^2
- Correcting for Small Sample Size
- Multicollinearity in Multiple Regression Data

COMPUTER APPROACHES TO MULTIVARIATE ANALYSIS
- Micro and Mainframe Programs
- Stepwise Multiple Linear Regression

SUMMARY

CASE: FORECASTING AT SUN BELT EQUIPMENT COMPANY

QUESTIONS AND PROBLEMS

Chapter 12 Objectives

1. Explain the terms multiple regression and multiple correlation.
2. Given some sets of data, be able to find and use the multiple regression equation.
3. Be able to find and interpret the coefficient of multiple determination.
4. Describe the role of computers in multivariate analysis.

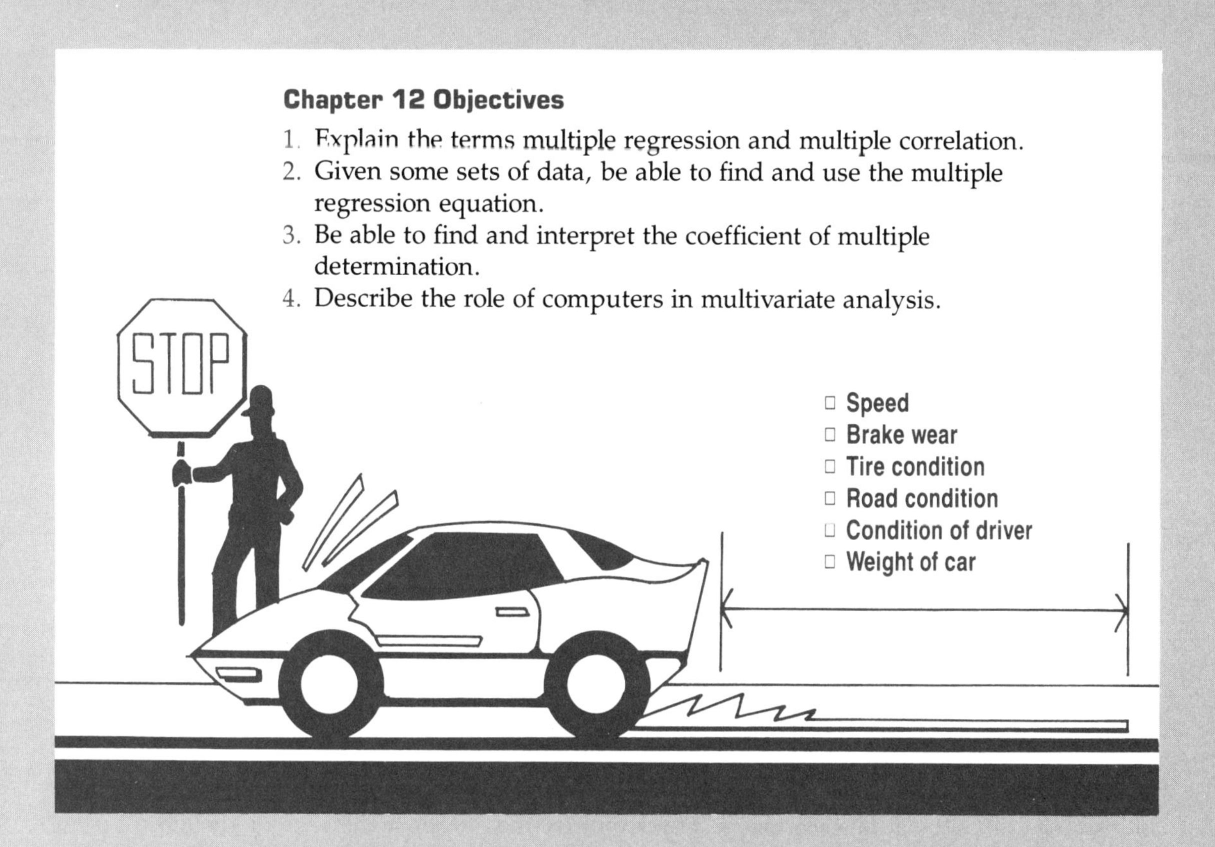

Suppose you worked as an analyst for a state highway department. If you really wanted an accurate regression model of how long it took cars to stop on a freeway, data from one variable (such as speed) probably wouldn't be sufficient. If a car's brakes or tires were badly worn, the car would no doubt take longer to stop. Or if the highway were wet, or the driver had been drinking, the stopping distance would probably be longer.

A regression model that uses more than one independent variable to predict the value of a dependent variable is referred to as a multiple regression *model. Just as we can express the nature of a multivariate relationship by means of a multiple regression model, we can also express the closeness of the relationship among several variables in terms of their* multiple correlation.

INTRODUCTION: MULTIPLE REGRESSION AND CORRELATION

OUR APPROACH The purpose of this chapter is to gain an appreciation of multiple regression and correlation. Inasmuch as this type of analysis can be complex, and hand calculations very tedious, our approach will be to explain the underlying theory so that you understand the reasons for, and logic of, the procedures. Beyond that, we shall rely heavily upon a computer for multivariate analysis. An abundant number of excellent software packages are available. Some of the more widely used mainframe packages are SPSS, BMDP, MINITAB, and SAS. And there are numerous packages for microcomputers—many of which are quite "user friendly."

HYPERPLANES Multiple regression is simply an extension of the two variable procedures we discussed in the previous chapter. With simple regression, we could depict the relationship on a two-dimensional (X, Y) *plane*. But when several independent variables are involved at the same time, the relationships cannot be so easily visualized, for they extend into third and higher level dimensions that mathematicians refer to as *hyperplanes*. Fortunately, the mechanics of finding the necessary constants and coefficients are well documented and quite suitable for computers—which never seem to tire of doing such repetitive calculations.

MULTIPLE INDEPENDENT VARIABLES Figure 12–1 illustrates the type of data we might find useful for the stopping distance prediction problem posed in the opening figure of this chapter. In Figure 12–1, we're taking the liberty of looking at the individual XY relationships of each of several dependent variables—assuming that other effects could be disregarded, or in effect held constant. That may not seem realistic, but it is essentially what we did when working with the simple regression models of the previous chapter. We used one dependent variable only, and lumped any other effects into what we conveniently called "unexplained error." With multiple regression, we use more independent variables in hopes of reducing that unexplained error and developing a better predictive model.

Numerous possible relationships exist, some of which are positive (such as a, b, c, and f in the Figure 12–1) and others negative (d and e). And some are linear (a, b, and e), whereas others may be nonlinear (c, d, and f). We seek to identify those relationships that are statistically significant and incorporate them into our model.

THE MULTIPLE LINEAR REGRESSION MODEL For purposes of illustration, we shall begin by limiting our scope to two or three independent variables (Xs), that each share a linear relationship with the dependent Y variable. The entire relationship of the variables need not be linear, however. If a variable is linear or approximately linear over the range of interest, that variable may be useful in the model.

FIGURE 12–1 Some dependent/independent variable relationships

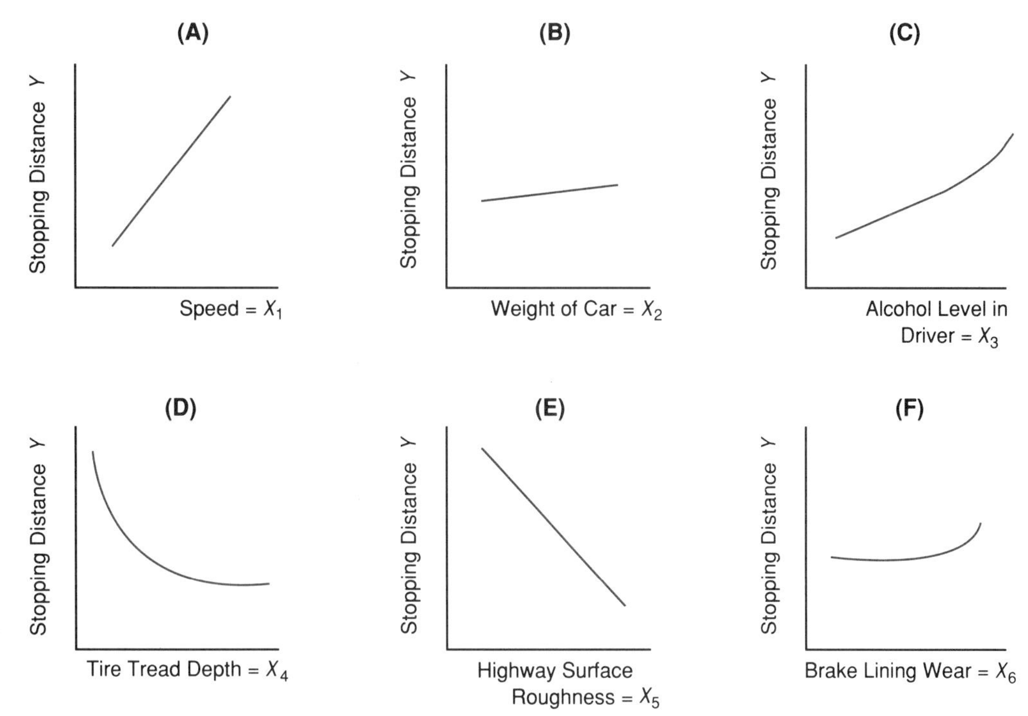

The generalized (theoretical) form of a multiple regression model is $Y = f(X_1, X_2, \ldots, X_n, \varepsilon)$, where $X_1, X_2, \ldots, X_n$ represent the n independent variables and ε again represents the unaccounted-for error. In the stopping distance model, ε might include a driver's age or a response time coefficient that had not otherwise been explicitly factored into the model.

Using the above model as a guide, we can formulate an expression for the multiple linear regression of a population in a manner similar to that used for simple linear regression. For this we let α represent a constant (comparable to the intercept of a simple regression model), and β the influence of each independent variable (ie., the slope effect). The mean of the Y values given each of several X values, is then $\mu_{y.x_1, x_2, \ldots, x_n}$. Condensing this to $\mu_{y..}$ the population, or true multiple linear regression, model for describing the mean value of Y may be expressed as:

$$\mu_{y..} = \alpha + \beta_1 X_1 + \beta_2 X_2 + \ldots + \beta_n X_n$$

Individual values of Y will, of course, differ from $\mu_{y..}$ by a unique error amount, ε_i. However, insofar as we will again be working with sample data and using residual or standard measures of error, our estimating model will take the more operational form of:

$$Y_c = a + b_1 X_1 + b_2 X_2 + \ldots + b_n X_n \qquad (12\text{–}1)$$

MEANING OF THE *a*'s AND *b*'s It is important to have a clear understanding of the meaning of the *a* and *b* coefficients at this point. They signify values derived from sample data. The *a* value (or intercept coefficient) is a constant that represents the expected value of Y when the X values are all equal to zero. It thus constitutes a basis, or starting point, for the model. The *b* values are sample estimators of the true (population) regression coefficients, the βs. Note that each *b* value carries a subscript to indicate the independent variable to which it applies. Thus, b_1 is the estimator of β_1, and b_2 is the estimator of β_2. We refer to b_1 and b_2 as the individual sample regression coefficients, or simply the *regression coefficients*.[1]

The regression coefficients (*b* values) are typically of more interest than the constant (*a* value). They describe the amount of change in the expected value of Y for each unit change in the X variable to which they are attached. That is, each regression coefficient describes the change that might be expected in Y if all the other X variables were held constant.

Let us illustrate the above. Suppose an expression for stopping distance Y, based upon speed X_1, weight of car X_2, and alcohol level X_3, was:

$$\underset{\text{(ft)}}{Y} = 0 + \underset{\text{(mph)}}{10X_1} + \underset{\text{(lbs)}}{.05X_2} + \underset{\text{(units)}}{2.5X_3}$$

The b_1 estimate of 10 would suggest that an increase of one unit of X_1 could be expected to increase Y by 10 units. In other words, going one more mile per hour faster (say from 55 to 56 mph) could be expected to increase the stopping distance by 10 feet. This effect assumes that the weight of the car (X_2) and alcohol level of the driver (X_3) are held constant or removed from consideration. The other regression coefficients have a similar interpretation. Thus, if the car weight is increased by 100 lbs, the stopping distance might be expected to increase by $(100)(.05) = 5$ feet, assuming speed and alcohol levels are unchanged.

Although there is no theoretical limit to the number (n) of independent variables that can be included in a model, practical considerations (and computer capacity) will automatically limit this. In addition, many computer routines are designed to stop admitting variables that contribute little or nothing to the strength of the model (as measured by the proportion of variance they explain).

Once the multiple regression equation is formulated, the dependent variable, Y, can be estimated from the intercept (the constant *a*) and the series of regression coefficients (*b* values) in a manner similar to what we used with simple regression. But before proceeding with an example, let us identify some addtional counterparts in multiple regression models.

OTHER MEASURES We shall also find it useful to develop a standard error of estimate for multiple regression, comparable to the s_e of simple regression.

[1] Some computer programs designate the intercept constant (*a*) as b_0. Others sometimes designate the dependent variable (Y) as X_1, and then the independent variables become X_2, X_3, and so on. Be sure to carefully check the symbols in whatever program you use.

It is a combined measure of the dispersion of actual values from the multiple regression hyperplane. And, on occasion, we may need to compute a measure comparable to the simple coefficient of determination r^2. With multivariate data, the coefficient of multiple determination is designated with a capital letter, R^2, and subscripts. We'll return to these other measures after first seeing how the multiple regression equation is developed.

FINDING THE MULTIPLE LINEAR REGRESSION EQUATION

MODEL ASSUMPTIONS Multiple regression models are based upon similar assumptions to those we used for simple regression models, except more variables come into play. We begin with linear relationships between the dependent variable and each independent variable. Then, instead of developing a line of best fit through the data, our multiple regression equation represents a three-dimensional plane, or n-dimensional "hyperplane of best fit." We assume the errors about the hyperplane are random and normally distributed, with an expected value of zero. Moreover, they have a constant variance for all combined values of the independent variables.

USING THE NORMAL EQUATIONS The solution procedure is to solve a set of equations for the required intercept (a) and slope (b) constants. Inasmuch as we must solve for one intercept plus all the slope coefficients, there must be one more equation than the number of independent variables. For example, the basic normal equations for a model with two independent variables are:

$$\Sigma Y = na \quad + b_1\Sigma X_1 \quad + b_2\Sigma X_2$$

$$\Sigma X_1 Y = a\Sigma X_1 + b_1\Sigma X_1^2 \quad + b_2\Sigma X_1 X_2$$

$$\Sigma X_2 Y = a\Sigma X_2 + b_1\Sigma X_1 X_2 + b_2\Sigma X_2^2 \qquad (12\text{–}2)$$

With only three equations and three unknowns, one might be tempted to do a simultaneous solution by hand if the problem were not too large.[2] However, computer routines are readily available and would most likely be used. Either way, the solution of these equations would yield values for a, b_1, and b_2, which would enable us to write a regression equation of the form: $Y_c = a + b_1X_1 + b_2X_2$. This model minimizes the sum of squared deviations of all actual points about the plane defined by the regression equation.

The equations for models with three or more independent variables are more complex but follow the same general format. Let us illustrate the procedures involved by using a simplified two-independent-variable example, after which we will review the typical output from a computer solution to the same problem.

[2] Either the normal equations or matrix algebra methods can be used. Hand calculations should be carried to about 6 digits to minimize cumulative rounding error—even though they may be reported to 3 or 4 places.

Example 12–1 Automart Products Company supplies car care items such as antifreeze, stereos, and seat covers to retailers in ten sales districts. In an effort to predict their sales better, a market analyst has collected the following data on Automart sales and two associated variables: (1) spendable income and (2) automobile registrations in the market area. (A third independent variable, federal gasoline taxes collected, was also being considered but is not included here.) All data are scaled to simplified units of measurement.

TABLE 12–1 Sales, spendable income, and automobile registrations

District	Sales of Automart Y	Spendable Income X_1	Automobile Registrations X_2
A	15	9	11
B	16	13	8
C	17	14	9
D	18	15	9
E	20	16	12
F	20	16	13
G	22	17	14
H	24	19	15
I	23	20	13
J	25	21	16
Totals	200	160	120

(a) Confirm that the relationships between sales and spendable income (Y,X_1), and between sales and automobile registrations (Y,X_2) are linear, and

(b) Derive the multiple regression equation.

Solution

(a) A graph of each variable against the dependent variable shows that both relationships are indeed linear. Many computer packages have routines to readily display these graphs, such as those at the right.

Other computer programs simply use the specified variables and rely upon sorting routines to discard variables that do not "fit" well, as measured by the error associated with that variable. In addition, many programs have routines that can transform nonlinear data into a relatively good linear fit.

(b) The values needed for solving the three equations for the three unknowns are obtained in the same manner as before. Table 12–2 is for illustrative purposes only. Even with only 10 sets of observations, the number of calculations suggests the advantage of using a computer.

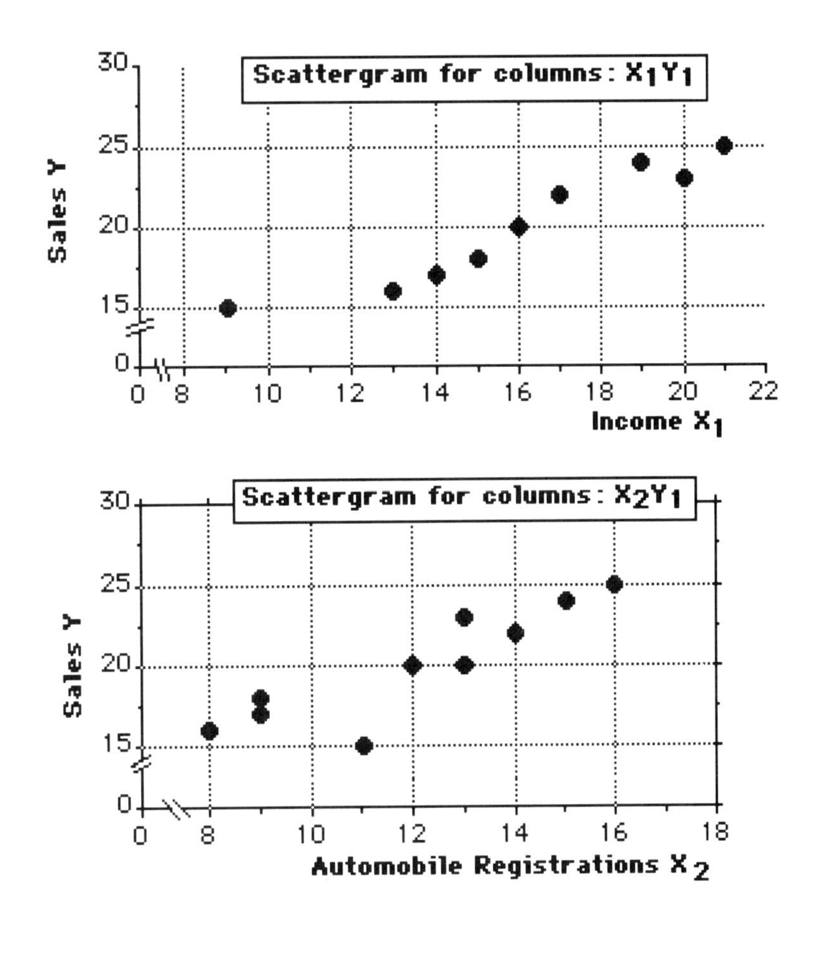

TABLE 12–2 Sales, income, and registration calculations

Dist	Y Sales	X_1 Income	X_2 Regstr.	Y^2	X_1^2	X_2^2	X_1Y	X_2Y	X_1X_2
A	15	9	11	225	81	121	135	165	99
B	16	13	8	256	169	64	208	128	104
C	17	14	9	289	196	81	238	153	126
D	18	15	9	324	225	81	270	162	135
E	20	16	12	400	256	144	320	240	192
F	20	16	13	400	256	169	320	260	208
G	22	17	14	484	289	196	374	308	238
H	24	19	15	576	361	225	456	360	285
I	23	20	13	529	400	169	460	299	260
J	25	21	16	625	441	256	525	400	336
Totals	200	160	120	4,108	2,674	1,506	3,306	2,475	1,983

Using these values in the normal equations, we have the three equations:

$$\begin{aligned} \text{(I)} \quad & 200 = 10a + 160b_1 + 120b_2 \\ \text{(II)} \quad & 3{,}306 = 160a + 2{,}674b_1 + 1{,}983b_2 \\ \text{(III)} \quad & 2{,}475 = 120a + 1{,}983b_1 + 1{,}506b_2 \end{aligned}$$

Solving these three equations simultaneously,[3] we (or the computer) obtain:

$$a = 3.460 \qquad b_1 = .6388 \qquad b_2 = .5266$$

Note that both regression coefficients happen to be positive, which is not always the case. The multiple regression equation derived from this sample data can then be expressed by substituting these values into the general linear model form of Equation 12–1:

$$Y_c = 3.460 + .639X_1 + .527X_2$$

ANALYSIS This equation tells us that a point estimate of Automart sales is 3.460 units, plus .639 times the spendable income factor X_1, plus .527 times the automobile registration factor X_2. Where spendable income and automobile registrations are both equal to zero, the expected sales figure is clearly outside the experience (and data range) of the company. None of the spendable income or automobile registration data points go below a value of 8. Nevertheless, the intercept value (3.460) does provide a starting point for locating the plane of the relationship. Then, for each unit increase in the spendable income factor X_1 *within the range of the data points*, the company can expect sales to increase by .639 sales units. Similarly, each unit increase in automobile registrations should be accompanied by an increase of .527 sales units.

[3] Multiplying equation (I) by (−12) and adding it to equation (III) drops out the a term. Then multiplying (I) by (−16) and adding it to equation (II) yields another equation without a. Solving these two equations with two unknowns then yields $b_2 = .5266$, and substituting this back into one of the two new equations gives us $b_1 = .6388$. The value for a can then be found by substitution or by the expression:

$$a = \overline{Y} - b_1\overline{X}_1 - b_2\overline{X}_2$$

(I) × (−12):	−2400=	− 120 a −	$1920b_1$ −	$1440b_2$	(IV)
(III) + (IV):	75=		$63b_1$ +	$62b_2$	(V)
(I) ×(−16):	−3200=	−160 a −	2560 b_1 −	1920 b_2	(VI)
(II) + (VI) :	106=		$114b_1$ +	$63b_2$	(VII) Then, 114 ÷ 63 = 1.8095, and
(1.8095) × (V):	135.714=		$114b_1$ +	119.429 b_2	(VIII)
(VIII) − (VII):	29.714=			$56.429b_2$	(IX) So b_2 = .5266

Substituting b_2 into (VIII) we have:

106=		$114b_1$ +	63(.5266)	(X) So b_1 = .6388

Then $a = \overline{Y} - b_1\overline{X}_1 - b_2\overline{X}_2 = 20 - .6388(16) - .5266(12) = 3.460$ (12–3)

EXERCISES I

1. How does multiple regression differ from simple regression?
2. Multiple regression is "more complicated" than simple regression.
   (a) Why might a company prefer to use it?
   (b) What is accomplished by using several independent variables instead of one?
3. (a) How many regression coefficients are included in multiple regression models? Explain. (b) How many normal equations would be needed to develop the equation for a multiple linear regression model with six independent variables?
4. A multiple regression model relating net earned income of insurance companies (Y in \$m) to the company's assets (X_1 in \$ billions), and its world-wide employment (X_2 in thousands) is
   $$Y_c = 460 - 8X_1 + 20X_2.$$
   (a) Explain the meaning of each term of the model.
   (b) Why might the intercept term, by itself, be unrealistic?
   (c) Make a point estimate of net income for an insurance firm with \$32 billion in assets and 20 thousand employees.
5. Assume the following data have been summarized from a multiple regression analysis:
   $$b_1 = -2.70 \qquad b_2 = .85 \qquad b_3 = 120.70 \qquad a = 35.48$$
   (a) Use the values to formulate and state the multiple regression model.
   (b) Make a point estimate of Y_c when $X_1 = 9$, $X_2 = 20$, and $X_3 = 2$.
6. Given the following data set:

Y	X_1	X_2
7	4	2
4	3	10
6	4	10
8	5	1
5	4	9

   (a) Arrange the data in a table that will facilitate calculation of the multiple regression coefficients.
   (b) Using the values developed in (a), state the normal equations that would be solved to find the multiple regression coefficients.
   (c) Compute the multiple regression equation.
   (d) Use your equation to predict a value for Y when $X_1 = 5$ and $X_2 = 8$.

7. (Note: This can also be done by hand.) An appliance sales manager, who has been advertising with coupons in a local newspaper, feels he can improve his sales forecast if he uses both X_1 = number of coupons placed in the newspaper and X_2 = newspaper circulation in thousands of copies. Data with sales (in 000) are:

	Y Sales	X_1 Coupons	X_2 Copies	Y^2	X_1^2	X_2^2	X_1Y	X_2Y	X_1X_2
	5	4	7	25	16	49	20	35	28
	7	8	11	49	64	121	56	77	88
	7	10	10	49	100	100	70	70	100
	4	3	8	16	9	64	12	32	24
	5	2	7	25	4	49	10	35	14
	6	6	10	36	36	100	36	60	60
Totals	34	33	53	200	229	483	204	309	314

(a) Determine the multiple regression equation that best fits this data.

(b) Predict units sold when 7 coupons are included in the advertisement in the previous week, and the newspaper circulation is 9,000 copies.

8. Data were collected to help predict the success of a commercial training program for word processors. Two independent variables were measured: X_1 = hrs of training given the employee, and X_2 = age of employee.

Hrs of training X_1	6	10	12	2	4	10	3	9	2	8	5	7
Age of employee X_2	38	25	31	51	49	18	42	33	58	22	37	40
No. errors per job Y	14	6	6	16	20	4	18	10	18	6	10	10

(a) Use a computer program to find the multiple regression equation.

(b) Explain the specific meaning of the regression coefficients in the context of your model.

(c) Make a point estimate of the number of errors per job for a 32-year-old employee with 6 hrs of training.

USING THE MULTIPLE REGRESSION MODEL FOR INFERENCE

In the previous section, we used sample data to find the multiple regression equation relating sales (Y) to spendable income (X_1) and automobile registrations (X_2). In this section we learn how to measure the error associated with a multiple regression equation. Then, knowing the characteristics of the error, we can use it in appropriate tests to make inferences about the relationships that are presumed to exist within the population being studied.

THE STANDARD ERROR OF ESTIMATE

As is the case with simple regression, the most common measure of dispersion in multiple regression is the standard deviation of actual points about the line defined by the regression equation. And this is still referred to as the standard error of estimate, s_e. However, the degrees of freedom are now the number of data points in the sample, n, minus the number of independent variables, k, minus the number of dependent variables (1):

$$s_e = \sqrt{\frac{\Sigma(Y - Y_c)^2}{n - k - 1}} \qquad (12\text{–}4)$$

The standard error of estimate for a specific problem is also sometimes written with subscripts designating the variables involved. For example, given one dependent variable, Y, and two independent variables, X_1 and X_2, the standard error of estimate might be designated $s_{y.12}$. A more intuitive expression for $s_{y.12}$ illustrates how the variation is accounted for:

$$s_{y.12} = \sqrt{\frac{\text{Unexplained or residual variation in } Y}{n - k - 1}}$$

$$= \sqrt{\frac{\text{Total variation in } Y - \left[\text{Variation in } Y \text{ explained by relating } Y \text{ to } X_1 + \text{Variation in } Y \text{ explained by relating } Y \text{ to } X_2\right]}{n - k - 1}}$$

This expression makes use of two measures of *covariation*, or variation of one variable in concert with the other. In (more complex) symbols this is:

$$s_{y.12} = \sqrt{\frac{\Sigma(Y - \bar{y})^2 - [b_1\Sigma(X_1 - \bar{x}_1)(Y - \bar{y}) + b_2\Sigma(X_2 - \bar{x}_2)(Y - \bar{y})]}{n - k - 1}} \qquad (12\text{–}5)$$

Example 12–2 From Example 12–1: $Y_c = 3.460 + .6388X_1 + .5266X_2$ and the total variance in Y could be computed as $\Sigma(Y - \bar{y})^2 = 108$. Using Table 12–1, the three measures of covariation can also be found to be:

Covariation of X_1 and Y	Covariation of X_2 and Y	Covariation of X_1 and X_2
$\Sigma(X_1 - \bar{x}_1)(Y - \bar{y}) = 106$	$\Sigma(X_2 - \bar{x}_2)(Y - \bar{y}) = 75$	$\Sigma(X_1 - \bar{x}_1)(X_2 - \bar{x}_2) = 63$

Compute the standard error of estimate s_e.

Solution

$$s_e = \sqrt{\frac{\Sigma(Y - \bar{y})^2 - [b_1\Sigma(X_1 - \bar{x}_1)(Y - \bar{y}) + b_2\Sigma(X_2 - \bar{x}_2)(Y - \bar{y})]}{n - k - 1}}$$

$$= \sqrt{\frac{108 - [(.6388 \times 106) + (.5266 \times 75)]}{10 - 2 - 1}}$$

$$= \sqrt{\frac{108 - 107.208}{7}} = .3364$$

A similar result would be obtained using Equation 12-4. The purpose of this was, however, more to illustrate how the error is accounted for in the calculation; these values are seldom calculated by hand.

PREDICTION INTERVALS FOR VALUES OF *Y*

Once the value of s_e has been computed, it can be used to establish prediction limits for values of Y in the same manner as with simple linear regression. We illustrate with an *approximate prediction interval* for an individual value.

Example 12–3 Using the multiple regression equation derived in Example 12–1, and the standard error of estimate computed in Example 12–2, make a 95% approximate individual prediction interval estimate of the Automart sales in one district (Y), when spendable income $X_1 = 18$ and car registrations $X_2 = 10$.

Solution

From Example 12–1: $Y_c = 3.460 + .6388X_1 + .5266X_2$

From Example 12–2: $s_e = .3364$.

We use the t-distribution for $n = 10$ and $k = 2$ independent variables

$Y_{PL} = Y_c \pm ts_e$ where: $Y_c = 3.460 + .6388(18) + .5266(10) = 27.365$

$$t = t_{\frac{\alpha}{2},n-k-1} = t_{\frac{.05}{2},10-1} = 2.365$$

$Y_{PL} = 27.365 \pm 2.365(.3364) = 27.365 \pm .7956$
$= 26.569$ to 28.161

These are the 95% *approximate individual limits* based on both X_1 and X_2.

EVALUATING THE OVERALL REGRESSION MODEL

Thus far, we have developed the multiple regression model and used it to predict the value of an independent variable. But how do we know it is a valid prediction? Does our model portray a significant relationship between the dependent and independent variables, or might the relationships be due simply to chance? This is one of the first questions likely to arise after developing a multiple regression model.

TESTING THE MODEL We can test the overall significance of the regression model by using the analysis of variance F-test discussed earlier (Chapter 10). In hypothesis form, our test is:

H_0: There *is no relationship* among the Y and X variables (i.e., the regression model does not explain variation in the dependent variable).

H_1: There *is a relationship* among the Y and X variables (i.e, at least one or more of the regression coefficients are $\neq 0$).

Thus we would ordinarily be looking for a large F value that would enable us to reject the H_0, and conclude that our regression model is significant. It should not seem too surprising to use an F-test here. As you may recall from working with ANOVA, the calculated F ratio was the ratio of an explained to a residual deviation. We could also express that ratio as:

$$F = \frac{\text{Between (or explained) mean square deviation}}{\text{Within (or residual) mean square deviation}} = \frac{MSD_{\text{explained}}}{MSD_{\text{residual}}}$$

In ANOVA, if the explained MSD was substantially larger than the residual MSD, we deemed it significant and attributed the larger variance to the difference between the group means represented in the numerator.

F-TEST FOR REGRESSION A similar logic applies to the analysis of variance F-test for regression. Only in this case, the explained deviation is that explained by the regression plane. Up to this point, we have referred to this as the explained sum of squared deviations, $\Sigma(Y_c - \bar{y})^2$. Computer programs frequently refer to this as the sum of squares due to regression, or SSR. The unexplained deviation is the deviation of points from the regression line, $\Sigma(Y - Y_c)^2$, that we have worked with in computing s_e. Computer programs often refer to this as the residual sum of squares or error sum of squares, SSE. Then the F ratio of explained to error variance that we use to test the hypothesis of no significant overall regression is:

$$F = \frac{SSR/k}{SSE/(n - k - 1)} \qquad (12\text{–}6)$$

In this expression, the df for SSR is equal to the number of independent variables, k, and the df for SSE is the number of observations, n, minus the k independent variables, minus the 1 dependent variable. If the calculated F value exceeds the table value for F, we reject the H_0 and conclude the regression is significant at the specified α level.

Example 12–4 Using the data of Examples 12–1 and 12–2, test whether the overall multiple regression model is significant. Use the F-distribution at the .01 level. Note that there are $n = 10$ sets of observations and $k = 2$ independent variables.

Solution

From Example 12–2, we have both the explained variation (107.208) and the total (108.000), so the unexplained is 108.000 − 107.208 = .792.

1. H_0: There *is no relationship* among the Y and X variables
   H_1: There *is a relationship* among the Y and X variables
2. $\alpha = .01$
3. Use F, reject if $F_{calc} > F_{.01,\ k,\ n-k-1} = F_{.01,2,7} = 9.55$
4. $F = \dfrac{SSR/k}{SSE/(n - k - 1)} = \dfrac{107.208/2}{.792 / (10 - 2 - 1)} = \dfrac{53.604}{.113} = 474$ (rounded)
5. *Conclusion*: Reject H_0, and conclude the regression is significant since $474 > 9.55$. This means that the Automart sales are strongly associated with the level of spendable income and automobile registrations. The independent variables account for nearly all the variability in sales. Moreover, this strong a relationship would have come about by chance less than 1% of the time.

RESULTS FROM THE COMPUTER

The major computer programs for multiple regression problems provide the same type of test described above, although the results may be packaged differently from one program to the next. Figure 12–2 illustrates a typical printout from a microcomputer program for multiple regression. [Note: this output is from STATVIEW.]

FIGURE 12–2 Computer printout for multiple regression

Multiple - Y : SALES (Y) Two X variables

DF:	R-squared:	Std. Err.:	Coef. Var.:
9	.993	.336	1.681

Analysis of Variance Table

Source	DF:	Sum Squares:	Mean Square:	F-test:
REGRESSION	2	107.208	53.604	474.037
RESIDUAL	7	.792	.113	$p \leq .0001$
TOTAL	9	108		

Regression Coefficients Table

Parameter:	Value:	Std. Err.:	t-Value:	Partial F:
INTERCEPT	3.46	.549	6.302	
Spendable Income	.639	.046	13.942	194.386
Car Registration	.527	.06	8.745	76.468

Note that the printout provides much of the same analysis that we have developed in the examples thus far:

Example 12–1 (We derived the equation $Y_c = 3.460 + .6388X_1 + .5266X_2$.) In the Regression Coefficients Table of the computer printout are provided the intercept (3.460), the spendable income coefficient (.639), and the car registration coefficient (.527).

Example 12–2 (We computed the standard error of estimate, $s_e = .3364$.) At the top of the computer printout is the *std. err* value of .336.

Example 12–4 (Our test of significance resulted in an F_{calc} value of 474.) The ANOVA table in the computer printout shows the regression and residual sum of squares, the mean square, and the *F*-test result of 474.037. Moreover, it notes that this *F* value would be significant at the *p* level of .0001, so it is very highly significant.

Concluding that the overall regression model is significant means that at least one of the regression coefficients (*b* values) has a significant (nonzero) slope. It helps explain the variability in the dependent variable. Knowing this, it is worthwhile to proceed with further testing and use of the model.

TESTING AND ESTIMATION OF THE REGRESSION COEFFICIENTS

TESTING THE COEFFICIENTS In addition to s_e, the other error measurements of most interest to analysts—and commonly included in computer solutions to multiple regression problems—are the standard errors associated with the regression coefficients, or b values. These standard errors are used in individual tests of the hypotheses that the slope coefficients equal zero. The tests tell whether each independent variable is a statistically significant explanatory variable for Y. If we let i represent the individual number of the regression coefficient, the test is

H_0: $\beta_i = 0$ (The slope is essentially zero; there is no relationship between y and X_i, holding the other variables constant.)

H_1: $\beta_i \neq 0$ (The slope is not zero; there is a relationship between Y and X_i, holding the other variables constant.)

Each such test of the slope of a regression coefficient is *conditional* to the extent that it automatically accepts the presence of the other independent variables in the model. However, the regression coefficient value itself reflects only the *partial* effect of the one variable, assuming the effect of the others is neutralized. The test itself utilizes the t-distribution with $n - k - 1$ degrees of freedom and follows the familiar format where the test statistic is:

$$t_c = \frac{b_i - \beta_i}{s_{b_i}} \tag{12–7}$$

In this expression, b_i is the sample value, β_i is the hypothesized value (usually zero), and s_{b_i} is the standard error of the slope for regression coefficient i. The standard error calculations follow the same general format as with simple regression. If many variables are to be tested, however, a computer is almost a necessity.

Example 12–5 For the data of Table 12–1, where $Y_c = 3.460 + .6388X_1 + .5266X_2$, the standard errors of the slope coefficients are $s_{b_1} = .0458$ and $s_{b_2} = .0602$. Test whether the slopes are significantly different from zero at the .01 level of significance. (Recall that Table 12–1 has $n = 10$ data points.)

Solution

For coefficient of X_1 (.6388)

H_0: $\beta_1 = 0$

H_1: $\beta_1 \neq 0$
$\alpha = .01$

Use t, reject if
$|t_c| > t_{\frac{.01}{2},7} = 3.499$

$$t_c = \frac{b_1 - \beta_1}{s_{b_1}} = \frac{.6388 - 0}{.0458}$$

$t_c = 13.948$

Conclusion: Reject H_0: $\beta_1 = 0$
because $13.948 > 3.499$

For coefficient of X_2 (.5266)

H_0: $\beta_2 = 0$

H_1: $\beta_2 \neq 0$
$\alpha = .01$

Use t, reject if
$|t_c| > t_{\frac{.01}{2},7} = 3.499$

$$t_c = \frac{b_2 - \beta_2}{s_{b_2}} = \frac{.5266 - 0}{.0602}$$

$t_c = 8.748$

Conclusion: Reject H_0: $\beta_2 = 0$
because $8.748 > 3.499$

The conclusions of the regression coefficients tests show that both the spendable income, X_1, and automobile registrations, X_2, are significant predictors of sales, Y.

COMPUTER RESULTS A review of the Regression Coefficients Table in Figure 12–2 shows that these results are consistent with those provided by the computer program—except for rounding error.

STANDARDIZED COEFFICIENTS Because X_1 and X_2 are not in the same units (X_1 is in dollars of spendable income and X_2 is in numbers of automobiles registered), their effect (per unit) on the dependent variable cannot be directly compared. Some computer programs provide *standardized, or beta, coefficients* that facilitate comparison of the regression coefficients by expressing them in standardized units comparable to standard deviations. These coefficients—which differ from the b_i values we have worked with thus far—measure the number of standard deviations that Y_c changes as a result of a change equal to one standard deviation of X_i. In doing this, the beta coefficients, or weights, take better account of both (1) the effects of an individual X_i value on Y_c and (2) the overlapping effects of two (or more) X_is on Y_c.

CONFIDENCE INTERVALS FOR β_i If the slope of a regression coefficient is significantly different from zero, we may wish to make a confidence interval estimate of the true slope. As in simple regression, we can construct confidence interval estimates of any of the regression coefficients by following the same general procedure:

$$CL \text{ for } \beta_i = b_i \pm t_{\frac{\alpha}{2},\, n-k-1}\, s_{b_i} \quad (12\text{–}8)$$

Example 12–6 For the regression equation $Y_c = 3.460 + .6388X_1 + .5266X_2$, both regression coefficients were found to be statistically significant at the .01 level. Develop a 95% confidence interval estimate of the true slope of the X_1 coefficient (where $s_{b_1} = .0458$).

Solution

$$CL \text{ for } \beta_1 = b_1 \pm t_{\frac{\alpha}{2}, n-k-1} s_{b_1} \quad \text{where: } b_1 = .6388$$

$$t_{\frac{\alpha}{2}, n-k-1} = t_{\frac{.05}{2}, 10-3} = 2.365$$

$$= .6388 \pm 2.365(.0458)$$

$$= .5305 \text{ to } .7471$$

We can be 95% confident that the slope of the true (population) regression coefficient β lies within the range of .5305 to .7471. (That is, intervals derived in this way will include the true value 95% of the time.)

THE COEFFICIENT OF MULTIPLE DETERMINATION

The best measure of the strength of the relationship between a variable being predicted, Y, and the several independent (predictor) variables, Xs, is the coefficient of multiple determination, ρ^2 (rho squared). As with most other statistical measures, we estimate the true population value with the sample coefficient of multiple determination, R^2.

The **coefficient of multiple determination,** R^2, shows the proportion of the total variation of the Y variable in the sample that can be accounted for by the independent, or predictor, variables.

CALCULATING R^2

The calculation of R^2 for many variables is best done on the computer. However, we can grasp the theory underlying the equation by reviewing the equation for a two-independent-variable model. It uses some of the same covariation measures as used in calculating s_e, but instead of calculating "unexplained" variation, our interest is now focused on the percentage of "explained" or associated variation. In the $R^2_{y.12}$ expression, $b_1\Sigma(X_1 - \bar{x}_1)(Y - \bar{y})$ is the variation explained by relating Y to X_1, and $b_2\Sigma(X_2 - \bar{x}_2)(Y - \bar{y})$ is the variation explained by relating Y to X_2. As before, *no causal relationship* is implied.

$$R^2_{y.12} = \frac{\overbrace{b_1\Sigma(X_1 - \bar{x}_1)(Y - \bar{y})}^{|\leftarrow \text{explained by } X_1 \rightarrow|} + \overbrace{b_2\Sigma(X_2 - \bar{x}_2)(Y - \bar{y})}^{|\leftarrow \text{explained by } X_2 \rightarrow|}}{\Sigma(Y - \bar{y})^2} \tag{12–9}$$

The denominator shows the total variation in the dependent variable, Y. When we divide the explained variation by the total variation, we obtain the proportion explained, which always has a positive sign.

Example 12–7 From Example 12–1: $Y_c = 3.460 + .6388X_1 + .5266X_2$, and the total variation in Y is $\Sigma(Y - \bar{y})^2 = 108$. In addition, two measures of covariation are:

Covariation of X_1 and Y	Covariation of X_2 and Y
$\Sigma(X_1 - \bar{x}_1)(Y - \bar{y}) = 106$	$\Sigma(X_2 - \bar{x}_2)(Y - \bar{y}) = 75$

Use this data to compute the coefficient of multiple determination.

Solution

$$R^2_{y.12} = \frac{b_1\Sigma(X_1 - \bar{x}_1)(Y - \bar{y}) + b_2\Sigma(X_2 - \bar{x}_2)(Y - \bar{y})}{\Sigma(Y - \bar{y})^2}$$

$$= \frac{.6388(106) + .5266(75)}{108} = .9927$$

ANALYSIS Slightly over 99% of the variation in the sample sales data is attributable to the two independent variables. This is very high (more than might typically be expected), and reflects the contribution of two statistically significant variables. A simple (one independent variable) regression and correlation analysis using each of these independent variables (separately) would have resulted in the following:

Sales Y and Income	$Y_c = 5.123 + .930X$	$r^2 = .913$
Sales Y and auto registrations	$Y_c = 6.364 + 1.136X$	$r^2 = .789$

Both of the simple coefficients of determination (i.e., between Y and X_1 of .913, and between Y and X_2 of .789) are already relatively high. But including both variables in the same (multivariate) model has increased the strength of the relationship by a substantial amount (to a coefficient of multiple determination of $R^2 = .9927$).

COMPUTER OUTPUT From Figure 12–2, you can readily verify that the computer calculated R^2 value is .993. The ease of making different "runs" on the computer also facilitates the type of comparisons discussed above. Inasmuch as many programs refer to the explained variation as the *SSR* and to the total sum of squares as *SST*, the coefficient of multiple determination is also sometimes expressed as:

$$R^2 = \frac{\text{sum of squares regression}}{\text{sum of squares total}} = \frac{SSR}{SST} \tag{12–10}$$

Inasmuch as the total sum of squares (*SST*) equals the explained or regression amount (*SSR*) plus the error amount (*SSE*), identifying the explained portion frequently accounts for a substantial portion of the total variability in Y.

COEFFICIENT OF MULTIPLE CORRELATION The coefficient of multiple correlation is the square root of R^2 (or .9963 for the example above). Unlike the two-variable case, R is always assigned a positive sign, although some b values may be negative and some positive. Hence the sign of R is not indicative of the direction of the relationship. Some computer programs give R^2, whereas others give R values, so you may have to convert R to R^2 in order to interpret the coefficient.

CORRECTING FOR SMALL SAMPLE SIZE

If the sample size is small relative to the number of independent variables, the relationship between the variables as indicated by the coefficient of determination tends to be overstated. For example, if there were only two variables and two observations, both of the points would fall on the regression line regardless of the observation values. Similarly, for the three-variable example, a plane can be made to conform exactly for the first three sales districts regardless of the values of Y, X_1, and X_2. This means that if there were only three observations for each of three variables, we could expect $R^2_{y.12}$ to be 1.00 regardless of the size of the observations. Of course, we would virtually always have more than one observation of each variable, so this is the extreme case—but is illustrative of the problem.

To compensate for this bias, computer programs frequently include a correction for small sample sizes and offer an *adjusted* R^2.

$$\text{Adjusted } R^2 = R^2_A = 1 - (1 - R^2)\left(\frac{n-1}{n-k-1}\right) \tag{12–11}$$

In this expression, n is again the number of sample observations and k the number of independent variables. This correction should be made if the sample size is small, or if there are many independent variables relative to the number of observations. For the data of Example 12–6, the correction results in a minor adjustment from an R^2 of .9927 to an R^2_A of .9906.

MULTICOLLINEARITY IN MULTIPLE REGRESSION DATA

Although an in-depth analysis of multiple regression models is beyond the scope of our text, we should be aware of one source of problems that can distort the results of a multiple regression model.

Multicollinearity is the condition that occurs when some independent variables are highly correlated with each other. It lessens the reliability of the individual regression coefficients.

For example, if two independent variables, say X_3 and X_4, had a correlation (between themselves) of .70 or more, one might be concerned about multicollinearity. When this happens, the influence of one independent variable overlaps that of another, and it is difficult to say what proportion of the effect is explained by any given independent variable.

The effects of multicollinearity may surface in a number of ways. When multicollinearity is present, the overall model may be significant even though some of the predictor variables, by themselves, may not have significant regression coefficients. This is because the Y variable is really correlated with a weighted combination of the independent variables, some of which are accounting for the same variation. It is not clear which X variable is responsible for a given effect in the predicted variable. And some of the individual regression coefficients may be less precise (and have a higher standard error) than the overall model suggests. These coefficients may even have incorrect signs, or change from significant to insignificant predictors when new variables are added to the model.

Dealing with multicollinearity can be a task for the professional. In many cases, the overall regression model can still yield relatively good predictions even though the validity of the individual effects is questionable. However, it can be very difficult to predict the incremental effects of dependent variable changes when multicollinearity is present, so it should be removed if possible. One of the more obvious ways of reducing it is to delete one or more of the intercorrelated variables from the model. This is easily done on many computer routines that allow the user to "try out" different variables in the model. More mathematical approaches to the problem can also be found in advanced level texts.

CORRELATION MATRIX Some computer programs provide a correlation matrix that is useful for identifying which variables are most helpful in explaining the variation in the independent variable. Table 12–3 illustrates the correlation matrix for the Automart Products Company example. It shows the individual correlations between all of the pairs of variables in the model. For example, the correlation between Y and X_1 is .96, and between Y and X_2 is .89. Note that the matrix is symmetrical about the diagonal and that the correlation of each variable with itself (on the diagonal) is always 1.00.

TABLE 12–3 Correlation matrix for Automart Products Company data

Variable	Y	X_1	X_2
Y	1.00	.96	.89
X_1	.96	1.00	.72
X_2	.89	.72	1.00

When first formulated, a multiple regression equation may include more variables than are necessary or useful to model the situation being studied. The correlation matrix can be used to help select (and retain) those independent variables that have the highest correlation with the dependent variable—that is, the correlations closest to 1.00. The "stepwise" regression programs (discussed in the next section) make use of this information. It can also be helpful for identifying significant amounts of overlap or multicollinearity in the data. For example, in Table 12–3 we see that the correlation between X_1 and X_2 of .72 is relatively high, suggesting that perhaps other variables might be sought out to improve or upgrade the model.

EXERCISES II

9. For a multiple regression model:
   (a) Describe, in words, what s_e attempts to measure.
   (b) How are the degrees of freedom for s_e determined?
   (c) How is s_e used in making predictions?
10. (a) If the overall regression model is significant, why is it necessary to test the individual regression coefficients as well?
    (b) What conclusion would you draw if the null hypothesis about an individual coefficient is not rejected?
11. An F ratio was used in ANOVA. What is the logic for using an F ratio to test a multiple regression model?
12. Is the coefficient of multiple determination the sum of the simple coefficients of determination of the variables involved? How do these two coefficients differ?
13. What is the relationship between
    (a) SSR and $[b_1\Sigma(X_1 - \bar{x}_1)(Y - \bar{y}) + b_2\Sigma(X_2 - \bar{x}_2)(Y - \bar{y})]$?
    (b) SST and $\Sigma(Y - \bar{y})^2$?
    (c) $\Sigma(Y - \bar{y})^2$ and $[b_1\Sigma(X_1 - \bar{x}_1)(Y - \bar{y}) + b_2\Sigma(X_2 - \bar{x}_2)(Y - \bar{y})]$?

14. Using Equation 12–9 as a guide, write the equation for the coefficient of multiple correlation for a multiple regression model with three independent variables.
15. Explain multicollinearity and identify one way of reducing it.
16. When is it important to adjust R^2 to R_A^2?
17. Name two uses of a correlation matrix.
18. For $n = 26$ observations of $k = 4$ independent variables, the unexplained variation was found to be $\Sigma(Y - Y_c)^2 = 68.04$. (a) Compute s_e. (b) Make a 95% approximate prediction interval estimate for Y if $Y_c = 33.50$.
19. A multiple regression equation has been derived from 22 observations of 6 independent variables. If the total variation is 342.8 and the explained variation is 286.8, find (a) s_e and (b) F_{calc}. (c) What would be the F_{reject} value for testing the significance of the model at the .05 level?
20. Given the following data for a multiple regression study where $n = 20$ and $k = 5$ independent variables: $SSR = 1452.38$, $SSE = 904.05$. Find (a) s_e, (b) F_{calc}, and (c) F_{reject} for .01 level. (d) Is the model significant at the .01 level?
21. Problem 7 asked to relate X_1 = number of coupons and X_2 = thousands of newspaper copies to sales of appliances, Y. For this problem, assume that $\Sigma(Y - \bar{y})^2 - [b_1\Sigma(X_1 - \bar{x}_1)(Y - \bar{y}) + b_2\Sigma(X_2 - \bar{x}_2)(Y - \bar{y})] = 1.160$. There are n=6 observations and the total sum of squares is 7.333.
    (a) Compute the standard error of estimate.
    (b) What are the SSE and SSR?
    (c) Test whether the regression model is significant at the .05 level.
    (d) Compute the coefficient of multiple determination.
    (e) Compare the value of R^2 with R_A^2 and comment upon the reason for the difference.
22. Given the following data from Problem 20: $n = 20$ and $k = 5$, independent variables: $SSR = 1452.38$ and $SSE = 904.05$, compute (a) the coefficient of multiple determination and (b) the coefficient of multiple correlation.
23. A multiple regression model designed to determine the time required to fill orders at a large building materials warehouse was based upon $n = 20$ observations of the following variables:

    Y = Order fill time (min)
    X_1 = Number of items in order
    X_2 = Distance traveled to reach farthest item (ft)
    X_3 = Size of load (cubic ft)
    X_4 = Similarity Index for items on order (range of 100 down to zero)

Results from the initial phase of the computer run are as shown below.

Multiple - Y : Time min Four X variables

DF:	R-squared:	Std. Err.:	Coef. Var.:
19	.921	1.893	20.572

Analysis of Variance Table

Source	DF:	Sum Squares:	Mean Square:	F-test:
REGRESSION	4	627.471	156.868	43.794
RESIDUAL	15	53.729	3.582	$p \leq .0001$
TOTAL	19	681.2		

Using the values from the table, show the equations and numbers that were used to calculate (a) s_e, (b) R^2, and (c) the F-test value. (d) Show the steps taken in a test of hypothesis to determine whether the data appear to provide a significant regression model (at the .05 level).

24. The following data were provided along with that shown in the previous problem:

Regression Coefficients Table

Parameter:	Value:	Std. Err.:	t-Value:
INTERCEPT	11.51	4.543	2.534
X1 No. Items	-.259	.315	
X2 Distance ft	.004	.002	
X3 Size cu.ft.	.131	.100	1.314
X4 Sim'l Index	-.11	.045	-2.455

(a) Show the equations, numbers, and steps necessary to test whether the regression coefficients associated with the number of items, X_1, and the distance traveled, X_2, are significant at the .10 level.

(b) Which of the coefficients are significant at the .10 level?

25. Using the data in Problem 24, construct a 95% confidence interval estimate of the true slope of the regression coefficient associated with X_4.

26. The Analysis of Variance and Regression Coefficients Tables shown display some of the results of a multiple regression analysis designed to help predict the occupancy percentage, Y, of an airline on the basis of six predictor variables.

Multiple - Y : Y OCCUPANCY % Six X variables

Analysis of Variance Table

Source	DF:	Sum Squares:	Mean Square:	F-test:
REGRESSION	6	7904.884		
RESIDUAL	20	3008.968		
TOTAL	26	10913.852		

Regression Coefficients Table

Parameter:	Value:	Std. Err.:	t-Value:
INTERCEPT	240.8	80.666	2.985
X1 Ticket Cost	-.452	.327	-1.379
X2 Check-in min	-.763	.642	-1.188
X3 Depart Time	-.224	.998	-.225
X4 Food Items	-3.326	1.443	-2.305
X5 Arrival Time	.32	.337	.948
X6 Bag Retriva...	-.841	1.109	-.758

(a) Write the multiple regression equation.
(b) Compute the standard error of estimate.
(c) Is the regression significant at the .05 level?
(d) What percent of the variation in occupancy rate is explained by the six independent variables?
(e) Which of the predictor variables have a significant relationship to the occupancy rate? Use the .05 level.

27. The correlation matrix for a model with 4 predictor variables is as shown.

Variables	Y	X_1	X_2	X_3	X_4
Y	1.00	.48	−.26	.74	.08
X_1	.48	1.00	.32	.55	.85
X_2	−.26	.32	1.00	.16	.37
X_3	.74	.55	.16	1.00	.18
X_4	.08	.85	.37	.18	1.00

(a) Explain the 1.00 values.
(b) Which variables evidence a potential multicollinearity problem?

28. The operations manager of a ski resort in New England has collected the following sample data in an effort to help predict how many customers must be accommodated at their chairlift and restaurant facilities.

Number of Customers for Ski Lift	Miles of Snowcovered Highway	Average Daily Temp °F	Snow Depth at Top (inches)	Ski Lift Price ($/day)
1778	15	30	62	18
1895	18	29	52	15
1510	31	18	45	28
2040	9	32	54	17
912	28	14	40	28
2057	11	28	58	20
240	22	5	37	28
2904	18	33	66	15
1630	25	19	48	25
2890	10	35	64	18
2107	15	44	59	17
808	40	10	46	24
1100	18	15	52	32
385	34	1	42	30
1825	12	25	54	20
1912	14	35	56	20
180	38	0	36	28
1115	33	12	49	27
1918	12	45	45	16
1426	22	24	46	25

Use a computer program to find

(a) the multiple regression equation

(b) the standard error of estimate

(c) the coefficient of multiple determination

(d) Use your equation to make a point estimate of the number of customers on a day when the highway has 25 miles of snow cover, the temperature is 30 °F, the snow depth is 52 inches, and the chairlift price is $26.

29. For the data of the previous problem, and using the .05 level

(a) test whether the overall regression is significant

(b) identify which of the independent variables are significant

(c) Is there any multicollinearity present?

COMPUTER APPROACHES TO MULTIVARIATE ANALYSIS

The computations required to formulate and analyze multiple regression models can become quite complex as the number of variables and/or the number of observations increases. Hence the widespread availability of calculator and computer programs for both micro and mainframe computers.

MICRO AND MAINFRAME PROGRAMS

MICROCOMPUTER PROGRAMS The past few years have witnessed a proliferation of software programs for doing multiple regression (as well as other complex statistical procedures) on microcomputers. They are widely available for the IBM and IBM-compatible machines, as well as for Apple and other computers. In fact, there are too many such programs to attempt to mention them all. The microcomputer output we illustrated earlier in the chapter is from *Statview*, an integrated and easy-to-use package designed for professional statistical and graphic data analysis. A number of other software packages also offer both graphic and tabular analysis.

MAINFRAME PROGRAMS Some of the mainframe software programs mentioned earlier included SPSS-X, BMDP, MINITAB, and SAS. While space does not permit us to illustrate the outputs from these programs in detail, they might be looked upon as different mechanisms for arriving at similar results—as you might use either a Ford, Chevrolet, or Honda to get to your destination. There are substantial differences in the programs, of course, and the format of the computer inputs and outputs varies from program to program. But for the analysis we have followed in this chapter, they provide comparable results. (Some of these "mainframe" programs, such as MINITAB, are also available in versions that can be used on microcomputers.)

USING A DATABASE With today's integrated information systems, firms frequently maintain one common database file that contains substantially all of the historical and operational statistics for their firm. The *mainframe programs* virtually all allow a user to import data directly from a data base or from files in other programs. This eliminates the need to reenter the data for different types of analysis and enables the user to experiment with various combinations of data.

Many of the *microcomputer programs* also permit the user to work from a database. In fact, IBM PCs and Apple Macintosh computers can readily be networked to share the same data files by using distributed network software which costs only a few hundred dollars. So today's organization enjoys considerably more flexibility and opportunity to use statistical data than firms had just a few years ago.

STEPWISE MULTIPLE LINEAR REGRESSION

In addition to simply offering the capability of accommodating large multiple regression problems, computers can enhance the solution by selectively including the "best" predictors in a model. Multiple regression routines often follow a stepwise procedure aimed at finding the most efficient combination of independent variables to predict the dependent variable. This can reduce the data collection and maintenance costs, while still providing an acceptable (or perhaps better) solution to the problem.

Both step-up and step-down procedures are available. In the step-up approach, the computer begins by selecting the one predictor variable that is most highly correlated with the variable being predicted. The residual or error variation is then determined. Then the variable that explains most of this error is added, and the error is recalculated. Other variables are added in this stepwise fashion until the addition of another variable to the model would not reduce the error by any significant amount. This may mean that several potential variables are not included (or needed) in the model.

The step-down procedure works in reverse. The computer program begins with all the predictor variables in the solution and then eliminates them one by one, starting with those that contribute least to the explanation of the variation. This is continued to the point where removing another variable would significantly increase the amount of unexplained variation in the predicted variable. In both the step-up and step-down procedures, variables that have once been admitted (or removed) from the model are not necessarily "frozen." They can be reevaluated with each iteration and remain candidates for addition or deletion in a later step.

SUMMARY

Multiple regression models use more than one independent variable to predict the value of a dependent variable. The general linear model describes a hyperplane of best fit through n sample data points and is of the form:

$$Y_c = a + b_1X_1 + b_2X_2 + \ldots + b_nX_n$$

The intercept, a, and regression coefficients, b, can be determined by solving a set of normal equations (or using linear algebra). Then the model can be used to predict in much the same manner as a simple regression equation.

The *standard error of estimate* for a multiple regression model, s_e, is based upon the deviation of actual points from the hyperplane defined by the equation. The numerator in the expression for s_e describes the total variation in Y, minus that explained by each of the independent variables. And the denominator $(n-k-1)$ reflects the loss of degrees of freedom from the k independent variables and the (1) dependent variable. Prediction intervals are formulated in much the same way as with simple regression.

The overall *significance of the multiple regression model* can be tested by an F-test of the regression mean square deviation, divided by the residual or error mean square deviation. In equation form, F_{calc} is SSR/k divided by $SSE/(n-k-1)$. Large values of F_{calc} cause us to reject the null hypothesis and conclude that the regression is significant. Multiple regression models have more than one predictor variable, and some of the regression coefficients (the b_is) may not be significant even though the model in total is significant. The individual regression coefficients can be tested using a t-test.

The *coefficient of multiple determination*, R^2, shows the proportion of the total variation in the Y variable that can be accounted for by the predictor variables. It is $R^2 = SSR/SST$. If the sample size is small relative to the number of independent variables, an adjusted R_A^2 value should be used.

Multicollinearity is the situation that occurs when two or more of the predictor variables are correlated with each other. A *correlation table* can be helpful in revealing this. When predictor variable correlations become large, the overlapping influence makes it difficult to identify what proportion of the effects are attributable to what specific variables. The most common method of reducing multicollinearity is to delete one or more of the affected variables from the model.

Nearly all multiple regression problems are done on computer. Stepwise programs allow for the introduction (or withdrawal) of variables on the basis of how much they contribute to explaining the relevant variation in the model.

CASE: FORECASTING AT SUN BELT EQUIPMENT COMPANY[4]

James Ryerson had advanced to the position of Manager of Industrial Products rather quickly. Nevertheless, his lack of experience was more than offset by his creativity. A failure here or there didn't discourage him from trying something new. But his latest assignment from company president Smithmore Stevens was a real challenge.

Although Sun Belt Equipment was only 12 years old, they already had a strong foothold in the food processing equipment market in some southern states. Company business had kept pace with the general level of economic activity across the country, and as a result, the firm was undergoing a multimillion dollar expansion in its manufacturing plant. That's where the challenge came in. With the large investment in new manufacturing facilities, Mr. Stevens felt that it was more important than ever to be able to accurately predict the demand for their food processing equipment. Otherwise Sun Belt would have too little (or too much) inventory of steel, gearmotors, and controls on hand. This, in turn, affected their scheduling of manufacturing activities which dictated how well they would be utilizing their highly automated—and expensive—facilities. Unless business forecasts were improved, Mr. Stevens indicated they would be faced with the possibility of a 20% layoff.

Faced with this problem, Jim Ryerson wasn't afraid to seek advice. And he got plenty. Thomas Sylvester, the regional sales manager, volunteered that his field representatives could probably come up with some fairly accurate estimates, and the production manager suggested that Jim plot the demand history and then just keep extending the trend one month at a time. A clerk on the financial analysis staff recommended that Jim try to make use of some economic indicators.

Along with reviewing these and other suggestions, Jim did some reading on his own. He learned that the federal government published indicators of the nation's economic activity on a monthly basis. A dozen of these indexes were so-called "leading indicators" because they tended to precede the economic activity of industry in general; there were also four coincident indicators and six lagging indicators.

With the help of his assistant, Alison Barrett, Jim Ryerson got some of the national and regional economic data together, plus some demand history for the Sun Belt processing equipment. Alison suggested that maybe they could do a regression of the economic indicators against sales, but "lag" the sales by two months. "For example," she explained, "we might use the predictor values for June, but match them with our actual sales demand values for August—which would be our values from two months later."

The matched data derived by Jim and Alison are summarized in Table 12–4.

---

[4] The values of the indicators used in this case are representative of actual values, but they are not necessarily correct representations of actual data in a specific time period. Use of a multiple regression computer package is required for efficient solution of this case.

TABLE 12–4 Sun Belt demand and selected economic indicators

Sales of Sun Belt (2 month lag $m)	83	105	102	107	104	112	118	113
1 Workweek (hrs)	41.0	40.6	40.7	40.7	40.6	40.6	40.6	40.8
2 Unempl Claims (000)	375	384	393	374	378	378	370	379
3 Mfrs New Orders	89.37	87.76	83.20	87.03	84.06	85.65	84.78	85.33
4 Slower Deliveries (%)	46	48	50	50	55	50	54	51
5 Business Formations	118.4	121.2	121.0	123.2	119.7	119.5	121.2	119.3
6 Plant & Equipt Orders	27.67	34.54	31.85	31.08	31.18	32.32	33.53	32.54
7 Building Permits	152.1	143.8	148.0	150.3	142.6	142.9	140.3	133.4
8 Inventory Change ($b)	18.75	22.94	24.00	33.88	5.25	−4.42	−7.64	−10.28
9 Mtl Price Change (%)	−.13	−.20	−.47	−.60	−.26	.23	.46	−.18
10 Stock Prices (500stk)	208.19	219.37	232.33	237.98	238.46	245.30	240.18	245.00
11 Money Supply ($b)	2257.5	2273.1	2294.9	2328.4	2347.4	2357.4	2381.8	2399.2
12 Credit Change (%)	7.9	3.9	4.8	1.9	6.6	3.1	7.2	7.9

Case Questions

[Note: Some instructors may choose to delete the starred (*) items.]

(a) How could you determine whether this national economic data would be of any value for predicting sales of Sun Belt food processing equipment?

*(b) Do you feel it is necessary (or perhaps even worthwhile) to include all twelve variables? Suppose Jim Ryerson wanted to reduce the number of variables in his model. How might he go about identifying which variables to delete?

*(c) Rank the five best predictor variables according to the value of their simple correlation, r, with Sun Belt sales.

(d) Compare the two models below in terms of R^2, s_e, and the overall significance of the model at the .05 level:

Model (1): Y and X_1, X_3, X_9 **Model (2):** Y and X_6, X_7, X_{11}

(e) For the better of the two models in (d):
i) what percent of the variation in Sun Belt demand is explained by the three predictor variables included in this model?
ii) are the regression coefficients all significant at $\alpha = .05$?

*(f) Prepare a correlation table of the Y variable and the three predictor variables for the best model in (d). Is there any evidence of multicollinearity?

(g) If you were Jim Ryerson, would you recommend use of a model developed from the economic data of Table 12–4? Explain.

QUESTIONS AND PROBLEMS

30. Explain what is meant by "stepwise multiple regression."
31. Explain why computers are so essential to the solution of large multiple regression problems.
32. Given the data set shown:

Y	7	3	8	6	6	4
X_1	10	2	9	5	7	3
X_2	2	20	1	12	8	6

   (a) Use a table format to arrange the data in a way that will facilitate calculation of the multiple regression coefficients.
   (b) Using numbers from your problem, state the normal equations that would be solved to find the multiple regression coefficients.
   (c) Find the multiple regression equation that best fits the data.
   (d) Make a point estimate of Y_c when $X_1 = 6$ and $X_2 = 17$.

33. Using the data in the previous problem, find (a) s_e and (b) R^2. (c) Compute R^2_A and explain any difference in R^2 and R^2_A. (d) Is the overall regression significant? (e) Are both individual regression coefficients significant?
34. Use the data given in Table 12–1 for sales, Y, and spendable income, X_1, but use the following hypothetical values of federal gasoline taxes collected as X_2 (instead of the automobile registration data):

District	A	B	C	D	E	F	G	H	I	J
Taxes Collected	8	7	9	10	11	11	12	13	14	15

   (a) Find the value for b_1 and for b_2.
   (b) Determine the value for the Y intercept.
   (c) State the multiple regression equation and use it to predict a value of Y_c when $X_1 = 15$ and $X_2 = 11$.
   (d) Find the proportion of the variation in Y that is explained by relating Y to X_1 and X_2.
   (e) Find the value of s_e.

35. For the data of Problem 8, find
    (a) the standard error of estimate
    (b) the coefficient of multiple determination
    (c) the standard errors of the two regression coefficients
    (d) Are both regression coefficients significant at the .05 level?
    (e) Make an approximate 95% prediction interval estimate of the number of errors per job for a 35-year-old employee with 6 hrs of training.
36. Use the data of Problem 28, but include only the average daily temperature (as X_1) and snow depth (as X_2) as the predictor variables in your model. Find (a) the multiple regression equation, (b) s_e, and (c) R^2. (d) Is the overall regression equation significant at the .05 level? (e) Make a 95% prediction interval estimate of the number of ski lift customers for a 30 °F day when the mountain has 52 inches of snow.
37. A Los Angeles producer of custom fiber optic and microwave equipment has a factory that uses a number of robots and computer-controlled machines. In an effort to better predict the product cost, Y, a manufacturing analyst has collected data on the number of assembly steps X_1, the amount of a certain metal used X_2, and the batch or lot size in which the products are produced X_3. The results are as shown.

Product Cost $/unit	No. Steps in Assembly	Metal Usage (pounds)	Batch Size Produced
Y	X_1	X_2	X_3
8,010	130	419	6
7,830	162	379	9
7,290	149	441	12
6,900	78	334	14
7,800	180	482	6
6,980	109	324	15
6,600	100	301	17
6,800	131	414	26
6,810	88	384	8
7,100	109	404	14
8,120	163	493	7
8,360	159	496	5
6,170	50	258	23
6,380	71	254	19
7,210	108	422	9
7,800	149	478	12
6,980	83	387	7
7,200	150	437	21

Use a computer to analyze the data and answer the following:

(a) What is the multiple regression equation?

(b) Compute the standard error of estimate.

(c) Is the overall regression significant at the .01 level?

(d) Make a point estimate of the per unit cost (in $) for producing a product that takes 110 steps to assemble, uses 375 pounds of metal, and is produced in batches of 20.

(e) What percent of the variation in product cost is explained by the three variables?

(f) Which of the predictor variables are significantly related to the product cost (at the .05 level)?

38. For the data of problem 37:

(a) *Optional—if graphics are available.* Use a computer to develop individual scatterdiagrams of the Y variable with each of the three predictor variables.

(b) Find the simple linear regression equation resulting from each of the simple linear regressions in (a).

(c) Find r^2 for each of the simple linear regressions in (a).

(d) Formulate a correlation matrix for the Y variable and the three predictor variables. Do the variables evidence any multicollinearity?

39. For the data of problem 37:

(a) Using the point estimate and data of part (d), make an approximate 95% prediction interval estimate.

(b) Delete the X_2 variable and find the revised multiple regression equation, s_e, and R^2.

(c) Make a revised approximate 95% prediction interval estimate using the revised equation developed in (b) above.

(d) Comment on the estimates developed in part (a) and (c) of this problem.

40. Use the electrical and electronic firms of the Corporate Statistical Database (Appendix A). Develop a multiple regression model to predict the annual dividend Y_c on the basis of earnings per share 2 years ago (EPS2), earnings per share 1 year ago (EPS1), and earnings per share current year (EPSC). State

(a) the regression equation

(b) the standard error of estimate

(c) the coefficient of multiple determination

(d) whether the overall regression is significant

(e) *Optional*: Comment on the value of the different variables as useful predictors of dividends, Y. [Note: If you use a stepwise regression program for additional information, use an F value of 4 to enter the model.

41. Use all the firms of the Corporate Statistical Database to develop a multiple regression equation to predict the dividends of a stock, Y_c. For predictor variables, use (1) yrs of operation, (2) age of employees, (3) share price, (4) export percentage, and (5) advertising expense. Show the results of your model along with the relevant statistics that are typically supplied by the computer program you are using. Then comment on the usefulness of your model.

42. Are the years of operation of a firm and the number of employees indicative of annual sales of a company? Use the banking and insurance firms of the Corporate Statistical Database to see if a significant regression model can be formulated to predict sales from these two variables. State your conclusions.

CHAPTER 13

Time Series, Index Numbers, and Forecasting

INTRODUCTION: TIME SERIES AND INDEX NUMBERS

CLASSICAL COMPONENTS OF A TIME SERIES

TREND ESTIMATION
- Using the Equation
- Shifting the Time Series
- Nonlinear Trend Equations

SEASONAL, CYCLICAL, AND IRREGULAR PATTERNS
- What is a Seasonal Index?
- Determining a Moving Average
- Computing the Seasonal Index
- Using the Seasonal Index
- Describing the Cyclical and Irregular Components

INDEX NUMBERS: PURPOSE AND TYPES
- Purpose and Limitations
- Types of Indexes

SIMPLE PRICE, QUANTITY, AND VALUE INDEXES

COMPOSITE INDEX NUMBERS
- Unweighted Aggregative Composites
- Weighted Aggregative Composites
- The Laspeyres and Paasche Indexes

SELECTING AND CHANGING THE BASE PERIOD

USING INDEXES IN BUSINESS
- Commonly Used Indexes
- Using Indexes to Deflate a Time Series

FORECASTING METHODS
- Techniques in Use Today
- A Time Series Forecasting Example

SUMMARY

CASE: WINTER COATS INVENTORY AT FASHIONWORLD

QUESTIONS AND PROBLEMS

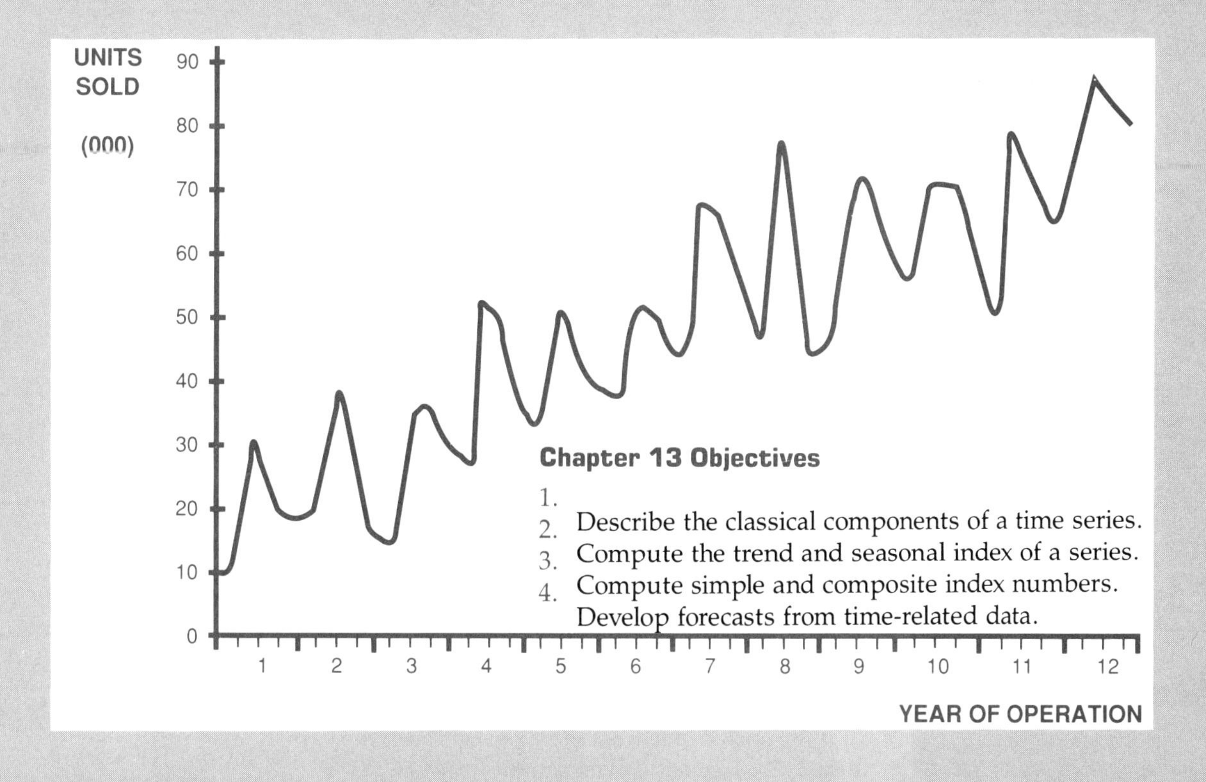

Virtually every organization makes some assessment of their progress over time. Many of these are measurements in the form of tables or charts, like the sales record depicted above. Marketing managers chart their pattern of monthly sales, production supervisors plot their weekly labor costs, and financial managers scrutinize the money markets daily, hoping to anticipate any change in their cost of funds. Time is the one variable that never stands still—always bringing new challenges to managers to use it wisely.

INTRODUCTION: TIME SERIES AND INDEX NUMBERS

A **time series** is a set of successive observations of some variable at regular intervals in time. The purpose of time series analysis is to explain the variation in the data in a manner that may aid in making reasonable projections into the future.

In addition to describing the action of a variable over time, economists and businessmen frequently need to compare the status of a variable (or set of variables) at two different points in time. For example, one may wish to compare the cost of living (for food, shelter, etc.) today with the cost of a similar set of items ten years ago. This can be done quite easily by expressing the two sets of costs as a percentage, or ratio—which we refer to as an index number. Such ratios typically express the price, quantity, or value of the item (or group of items) in a given time period relative to that in a base, or reference, period.

An **index number** is the ratio of a summary measure of data in a given time period to a similar measure of comparable data in a base period.

TIME—THE UMBRELLA VARIABLE Both time series analysis and index number construction use descriptive statistical data from different points in time. In fact, the common denominator of all the topics in the chapter is *time*. Time is a universal measure of change. Our goal in this chapter is to learn how to abstract the trends and pattern of a variable over time, and use that knowledge to better anticipate the future.

In one sense, we will be acting as if time itself were the independent variable, assuming that it incorporates all the explanations of our variable of interest. But instead of using a group of independent explanatory variables (such as $X_1, X_2, X_3, \ldots, X_n$ as we did in a multiple regression), we simply lump them all together under the umbrella of time. Then we measure the dependent variable as it moves through time, and act as though the time period, t, were the single predictor variable.

Why can we do this—what is the justification for using time as such a convenient "catch-all" variable? Because it works! Countless organizations have found that the past is often satisfactory as a predictor variable for the future. This is because the forces affecting business activity (e.g., money supply, weather, unemployment, etc.) exert some pronounced and understandable effects. And it is especially true if we can break the total time series effect down into identifiable elements—elements that seem to have some intuitive justification. We refer to these elements as the classical components of a time series.

CLASSICAL COMPONENTS OF A TIME SERIES

One need only glance at publications such as the *Survey of Current Business*, the *Monthly Labor Review*, *The Wall Street Journal*, brokerage house newsletters, and annual reports of corporations to see the extensive use of time series in business. They describe changes over time in prices, sales, production, inventories, employment, unemployment, hours worked, energy produced, earnings, and so on. Some changes, such as residential construction levels, seem to fluctuate over cycles of two or three years, whereas others, such as heating oil consumption, have definite short-term seasonal patterns.

COMPONENTS One of the most widely accepted methods of analyzing time series data is to break down, or "decompose" the observed time series values into four components that appear to exert effects over different time periods. The time periods vary from long-term growth or decline, through multiyear and one-year periods, to relatively short-term periods. And all time series do not necessarily evidence all these time periods. But the most popular models for analyzing time series data do attempt to explain the variation in terms of four factors that are related to these time intervals.

More succinctly, the classical model of time series analysis assumes that time series data are composed of (1) trend, (2) cyclical, (3) seasonal, and (4) irregular components. These components are defined in Table 13–1.

TABLE 13–1 Classical components of a time series

1. **Trend T** the smooth, long-term increase or decrease in the data due to the consistent growth or decline over time. Trend is often attributed to long-run economic or social factors such as shifts in population, new energy sources, changing modes of transportation.
2. **Cyclical C** the gradual swings or oscillating patterns about the trend line due to changing economic conditions.
3. **Seasonal S** the periodic variations related to seasonal or climatic effects, which are normally yearly occurrences but could be monthly, or even daily.
4. **Irregular I** the remaining short-term or random variations that have no discernible pattern and are not explained by the trend, cyclical, and seasonal components.

Figure 13–1 illustrates, in simplified form, the trend, cyclical, seasonal, and irregular components one might abstract from a time series, such as the one depicted in the opening of this chapter.

FIGURE 13–1 Classical components of a time series

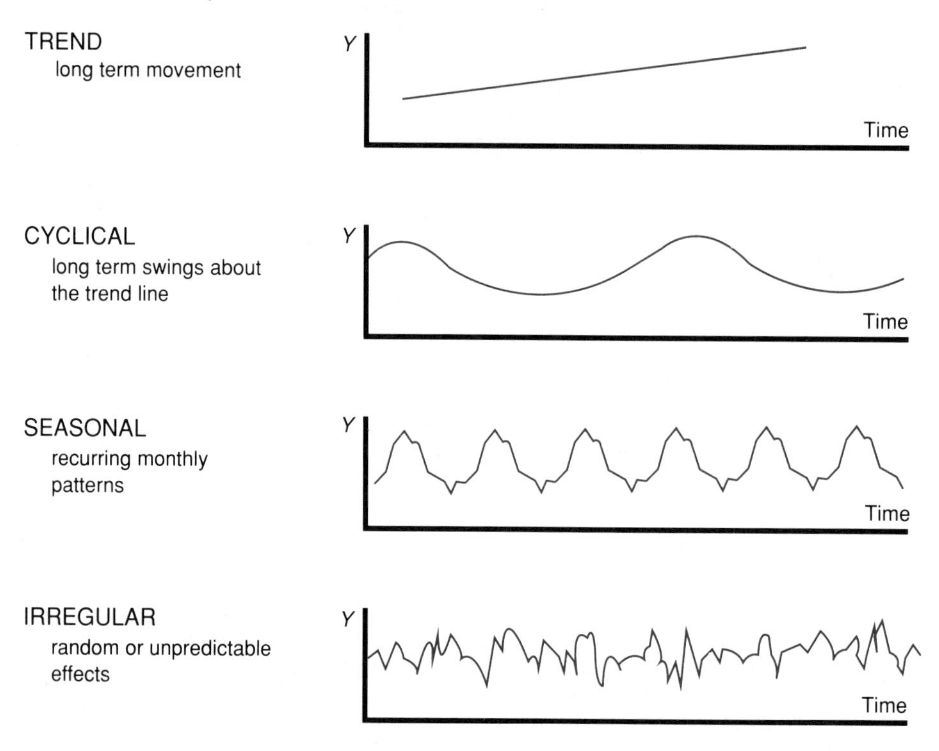

MULTIPLICATIVE MODEL As you might expect, there are different ways of depicting the relationship among the trend, cyclical, seasonal, and irregular components. One model is based upon an additive formulation of these components. Another and more widely used model expresses the time series value, Y, as a product of these four components.

$$Y = T\,C\,S\,I \tag{13–1}$$

We shall concentrate upon this (multiplicative) model in this chapter, recognizing that the procedures used to develop it are descriptive and somewhat subjective in nature.

USES OF THE MODEL Once a valid model has been developed, it can be used as a basis for planning to meet future needs. Thus, we find economic time series models being used to forecast everything from food and heating oil requirements to social needs for transportation, education, water, and electrical energy. Moreover, by recognizing the recurring patterns in a time series, firms are better able to manage the time-related variations they experience in sales, production, labor needs, working capital, and other business-related variables.

With this introduction, we move on now to explore some of the techniques for abstracting the trend equation and a seasonal index from time series data.

TREND ESTIMATION

Trend forecasts are based upon an extrapolation of historical patterns of growth or decline, which may necessitate using 15 or more years of data. If a period of less than 10 years is used, one may be unable to distinguish the trend effects from the long-term cyclical variation. If linear, the resulting trend line value, designated Y_t, would be determined from the standard form:

$$\text{Trend line} \qquad Y_t = a + bX \tag{13–2}$$

If the data exhibit a nonlinear pattern, other equations may be more appropriate, such as exponential or parabolic curves. Procedures for fitting data to these curves can be found in more advanced statistics texts.

FREEHAND Perhaps the easiest way of determining a trend line for much time series data is to simply plot the points and draw the freehand line that best represents the values. In essence, the freehand line becomes a visual "line of best fit" through the data. A minor enhancement on this is to divide the data into two groups, compute their means, plot them, and then draw the trend line as one connecting the two means. Whichever way is used, it is possible to obtain a trend equation by measuring the slope of the line (as the change in Y relative to change in X) and finding the intercept (by identifying where the line crosses the y-axis).

LEAST SQUARES TREND LINE The location of a freehand line is obviously somewhat subjective. A more objective approach is to reduce the line to a mathematical line of best fit upon which everyone can agree. This can be done by using the same normal equations we used for the simple linear regression line. Restating the equations in slightly different form (where they can most readily be solved for the slope, b, and intercept, a), we have:

$$b = \frac{n\Sigma XY - \Sigma X \Sigma Y}{n\Sigma X^2 - (\Sigma X)^2} \tag{13–3}$$

$$a = \frac{\Sigma Y - b\Sigma X}{n} \tag{13–4}$$

In these equations, the time period, t, is designated as the X variable, and the first time period is typically assigned an X value of zero (or sometimes one). The calculations can, of course, be relegated to a computer, using one of the numerous time series or regression programs available. For our first example, however, we shall include the calculations to demonstrate the applicability of the equations.

Example 13–1 Knoxville Toy Co. has produced wooden toys for many years (though only the last eleven years of sales data are shown below to simplify the illustration). (a) Graph the data to confirm a linear fit, and then (b) use equations 13–3 and 13–4 to find the slope and intercept of the time series equation that best expresses the trend of sales over time. [Note: Sales are in $000. Also, the initial year has been coded zero (i.e., 1978 = year 0) to simplify calculations.]

Year *X*	0	1	2	3	4	5	6	7	8	9	10
Sales *Y*	475	535	550	560	555	585	540	525	545	585	650

Solution

(a) Graphing the data is the first, and possibly the most important, step in the solution. The graph below does suggest a linear trend may fit the data, though perhaps some cyclical effect is present. However, more data points would be useful to confirm both of these observations.

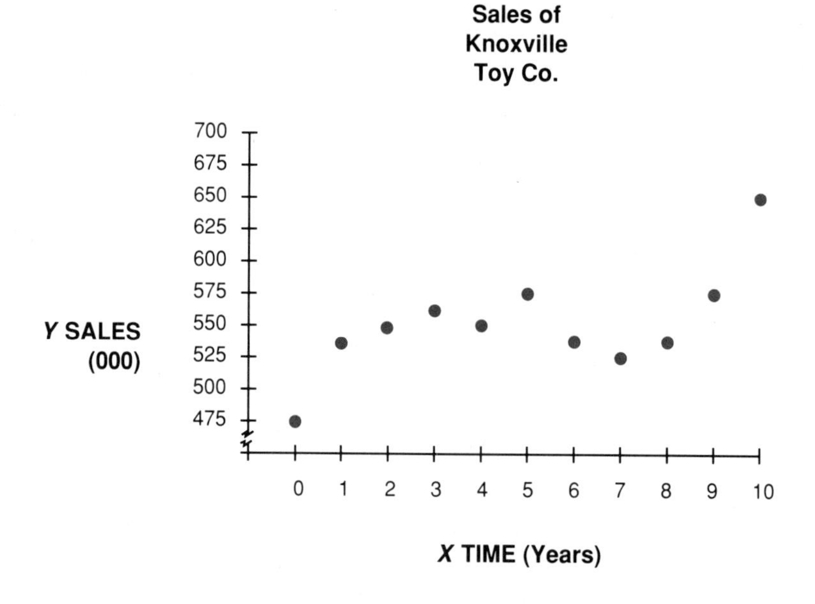

(b)

	Year Code X	Sales Y	XY	X^2
	0	475	0	0
	1	535	535	1
	2	550	1,100	4
	3	560	1,680	9
	4	555	2,220	16
	5	585	2,925	25
	6	540	3,240	36
	7	525	3,675	49
	8	545	4,360	64
	9	585	5,265	81
	10	650	6,500	100
Totals	**55**	**6,105**	**31,500**	**385**

$$b = \frac{n\Sigma XY - \Sigma X \Sigma Y}{n\Sigma X^2 - (\Sigma X)^2} = \frac{(11)(31{,}500) - (55)(6{,}105)}{(11)(385) - (55)^2} = 8.8636$$

Thus the slope coefficient of the time series equation is (8.8636)($1000), or approximately $8,860 per year.

$$a = \frac{\Sigma Y - b\Sigma X}{n} = \frac{6{,}105 - 8.8636(55)}{11} = 510.68$$

The intercept, or value at the origin in year 0, of the time series equation is (510.68)($1,000) = $510,680.

USING THE EQUATION

INFERENTIAL LIMITATIONS Once the slope and intercept coefficients of the time series data have been determined, the trend equation can easily be formulated and used to project the trend into the future. Before doing that, however, we should pause to note that time series projections are not "statistically valid" in the same sense as regression forecasts are valid. That is, we cannot attach a level of confidence to a time series projection in the same way that we can to predictions derived from regression. This is because (1) we have only one Y value for each X, (so we cannot assume that the data points are normally distributed about the trend line) and (2) we are typically projecting beyond the limits of the available data. Thus, the bulk of the estimation and testing statistics we used in regression analysis are (theoretically) not applicable to time series.

On the other hand, time can be treated as an independent variable. Moreover, one can measure the standard error and the explained and unexplained variance associated with a time series. And error measurements, correlations, F ratios, and significance measures are the standard output of computerized regression algorithms that are frequently used for time series analysis. So, as a practical matter, trend forecasts are made, and the accompanying statistical measures are sometimes reported.

Nevertheless, there is little inferential basis for extrapolating linear trends into the future. The economic and market forces that influenced the past may well be different in the future. Trends can and do change over time. That is why one should always bear in mind that time series analysis rests heavily upon the historical past and should be accompanied by sound managerial judgment about how those forces might change in the future.

EQUATION SIGNATURE A second precaution related to time series is that every time series equation should be accompanied by what is sometimes referred to as a *signature*. The *signature* tells what time period has been designated as the origin (or zero period), as well as the units used for X and Y. With this knowledge, the equation can be used to project trend values into the future by substituting any desired values for X.

Example 13–2 (a) Use the coefficients developed in Example 13–1 to express the time series equation—complete with signature. (b) Explain its meaning. (c) Project the toy sales trend value to period 20 (i.e., to the year 1998).

Solution

(a) The equation form is $Y_t = a + bX$, so the time series equation is:

$$Y_t = 510.68 + 8.86X \qquad (1978 = 0,\ X = \text{yrs},\ Y = \$000)$$

(b) This means that the trend value for toy sales was $510,680 in 1978 (when $X = 0$), and it increases by $8,860 for each unit increase in X. That is, sales go up, on average, by $8,860 per year.

(c) $Y_t = 510.68 + 8.86\,(20) = 687.88$. The (point) estimate of the 1998 sales trend is therefore $687,880.

SHIFTING THE TIME SERIES

CENTER OF THE DATA When a time series trend value is stated, an implicit assumption is that the value corresponds with the trend amount in the middle of the time period. Thus, in Example 13–2 the projected trend value of $687,880 would really be centered, or *plotted* in the middle of the year (i.e., on July 1, 1998). Figure 13–2 illustrates.

FIGURE 13–2 Trend values lie in the middle of the period

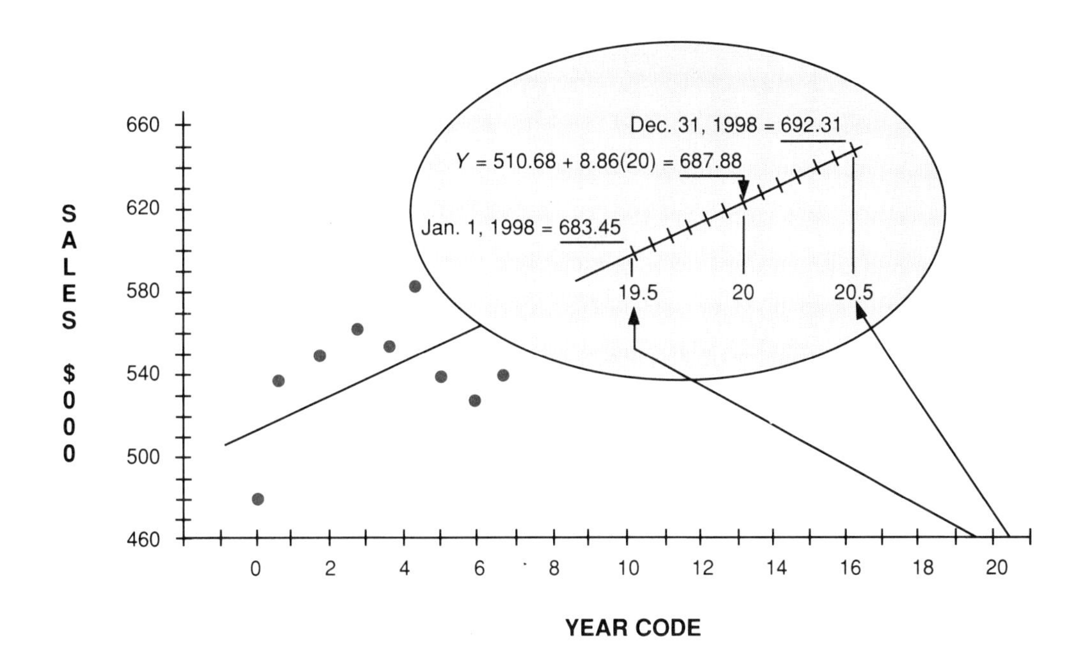

EXPRESSING *X* IN MONTHLY TERMS As is evident from the trend equation $Y_t = 510.68 + 8.86\,X$, toy sales increase by $8,860 per year. This is also apparent from Figure 13–2, where the projected increase over the twentieth year is $692,310 − $683,450 = $8,860. This is equivalent to a monthly increment of $8,860 ÷ 12 or $738.33 per month.

Any time series equation that is expressed in yearly *X* values can be converted to monthly *X* values by dividing the *X* value by 12. When this is done, the signature should be changed to designate the day upon which the origin is centered and to note that *X* is now in months instead of years.

Example 13–3 Given the time series equation:

$Y_t = 510.68 + 8.86X \quad (1978 = 0,\ X = \text{yrs},\ Y = \$000)$

(a) Convert the X values into monthly units, and (b) use your new equation to forecast the trend value for August 1998.

Solution

(a) Dividing the X value by 12, we have the new equation which is centered on July 1, 1978:

$$Y_t = 510.68 + \frac{8.86X}{12}$$

$$Y_t = 510.68 + .7383 \text{ (July 1, 1978} = 0,\ X = \text{months},\ Y = \$000)$$

(b) The trend value for August is located in the center of the month of August, which is 1.5 months away from July 1. Insofar as July 1, 1998 is (20 yrs)(12 mo/yr) = 240 months away from the origin of July 1, 1978, the trend value we seek is 240 + 1.5 = 241.5 months distant from the origin. Thus the trend value for August 1998 is:

$$Y_t = 510.68 + .7383\ (241.5) = 688.99, \text{ or } \$688{,}990.$$

This is consistent with what might be expected from Figure 13–2.

SHIFTING THE ORIGIN Users of time series sometimes wish to update a series to a more recent origin, or base period. Although this does not involve any change in the slope, or b value, of the line, it does necessitate a change of the intercept. However, the shift is easily accomplished by simply adding the designated number of slope increments to the intercept value.

Example 13–4 Given the time series equation:

$$Y_t = 510.68 + 8.86X \quad (1978 = 0,\ X = \text{yrs},\ Y = \$000)$$

(a) Shift the origin to 1988 = 0. (b) Use the new equation to project the toy sales trend value for the year 1998.

Solution

(a) Insofar as the origin is to be shifted 10 years, we add 10 to the X value:

$$Y_t = 510.68 + 8.86(X + 10) = 510.68 + 8.86X + 88.60$$
$$= 599.28 + 8.86X \text{ (1988} = 0,\ X = \text{yrs},\ Y = \$000)$$

(b) $Y_t = 599.28 + 8.86\ (10) = 687.88$, or \$687,880. (Note that this is the same as the trend value computed from the original equation in Example 13–2.)

NONLINEAR TREND EQUATIONS

Of course not all long-term trends are linear. Some time series increase geometrically (at a fairly constant percentage rate), rather than arithmetically (at an absolute rate). Growth in new product sales is frequently exponential. And most products go through a "life cycle" of introduction, rapid growth, leveling off (maturity), and decline. Gompertz growth curves are useful for series that tend to level off and approach a peak after some period of time. Modified exponential curves allow for a diminishing growth rate.

Space does not permit us to explore the many possible trend functions that are used to fit time series data. However, calculator and computer routines are widely available for fitting data to nonlinear functions. And with many of the programs available today, this requires little more than selecting the type of function one wishes to try. For these decisions, we again reemphasize the value of graphing the data early, so as to gain an intuitive understanding of any pattern that may exist in the trend.

EXERCISES I

1. Distinguish between a time series and an index number. In what way are the two related?
2. Which of the classical components of a time series concerns data that
   (a) consist of unexplained, short-term variations
   (b) fluctuate according to weather patterns of a year
   (c) increase or decrease in a constant manner over the long term
3. Suppose you wished to estimate the linear trend line equation for a time series without using the normal equations. (a) What two parameters would you need to estimate, and (b) how might you proceed to do this?
4. The price ($) of an electronics stock on the OTC Exchange over a 13-year period yielded the following values. (The initial year was coded 1976 = 0.)

   $$\Sigma X = 78 \qquad \Sigma X^2 = 650 \qquad \Sigma Y = 178 \qquad \Sigma XY = 1254$$

   (a) Find the slope and intercept of the time series trend line.
   (b) State the time series equation (complete with signature).

5. The trend equation describing the number of jury verdicts of $1 million or more in personal injury cases has been estimated as

   $Y_t = 110 + 60\,X \qquad (1979 = 0,\ X = \text{yrs},\ Y = \text{number of verdicts})$

   (a) Using this equation, what is the trend value for the number of $1 million or more verdicts in 1990?

   (b) Shift the origin to 1990 and state the revised equation.

   (c) Convert the equation from (b) by expressing the number of verdicts in monthly terms. State the equation complete with signature.

6. A personnel manager has developed the following time series equation to describe the trend in average hourly pay of factory workers in his area:

   $$Y_t = 8.85 + .30\,X\ (1983 = 0,\ X = \text{yrs},\ Y = \text{hourly pay in \$})$$

   During a collective bargaining negotiation, the manager finds it necessary to (a) estimate the trend value for 1996. Then, with reference to a specific contract starting date he needs to (b) project the trend value for October 1996. Use the available equation, with whatever shifting or conversion is required, to provide the two projections the manager needs. Show your work.

7. Data on the total number of installed production robots (in hundreds) in an industrialized country is given below (where year 1980 = 0).

Year	**X**	0	1	2	3	4	5	6	7	8
No. Robots	**Y**	2	4	12	20	24	30	35	40	50

   Assuming a linear trend fits the data satisfactorily

   (a) find the linear trend equation.

   (b) explain its meaning.

   (c) use it to project the number of installed robots in 1995.

8. Average gasoline prices (in $ per gallon) in a metropolitan area over a 10-year period were as shown below, where the origin is 1978. Assume a linear trend is to be fit to the data and answer the questions below:

Year X	0	1	2	3	4	5	6	7	8	9
Price Y	.64	.86	1.22	1.33	1.29	1.24	1.20	1.20	.92	.98

   (a) Find the trend equation and state it complete with signature.

   (b) Shift the origin to 1985 and restate the equation.

(c) Use the equation developed in (a) to project the price of gasoline in 1995.

(d) Express the equation developed in (a) in monthly units of X and use it to project the trend value of gasoline prices in March 1998.

9. Overseas expenditures (in $ billion) by U.S. travelers in foreign countries are estimated as follows:

Year	'75	'76	'77	'78	'79	'80	'81	'82	'83	'84	'85	'86	'87
$	6	7	7	8	9	10	11	12	14	16	17	15	17

(a) Does a graph suggest that a linear trend may fit the data?

(b) Find the linear trend equation and state it complete with signature.

(c) Shift the origin to 1990 and restate the equation.

(d) Use the equation developed in (c) to project the overseas expenditures by U.S. travelers in the year 2000.

(e) Explain any reservations you might have about using the projection of expenditures in the year 2000.

10. Per capita expenditures (in $) for beer, liquor, and wine in a region of the U.S. over some recent years are as shown below (1976 = 0). What conclusions can you draw concerning the trend of consumption of these beverages?

Year	X	0	1	2	3	4	5	6	7	8	9	10
Beer	Y_1	39	39	39	40	39	39	38	37	37	37	36
Liquor	Y_2	17	18	19	19	18	18	17	16	16	15	15
Wine	Y_3	9	10	11	11	12	12	12	13	13	13	14

SEASONAL, CYCLICAL, AND IRREGULAR PATTERNS

Nearly every organization is interested in the trend of some aspect of its activity. As we saw above, annual data can be analyzed to reveal the trend. However, sales, production costs, inventory levels, and other measures of business are typically reported in quarterly, monthly, or even daily terms. Managers use this information to make the day-to-day decisions required to plan and control business operations. For example, financial managers must plan for enough working capital to build up inventories for peak demands, and personnel managers must anticipate the variations in employment—especially in construction and other seasonally affected industries.

In this section, we explore some techniques for isolating the cyclical, seasonal, and irregular components of a time series. Our main attention will be directed to procedures for determining a seasonal index, which is probably the most useful of the three. But we shall endeavor to make the analysis more complete by describing one method of identifying and accounting for the cyclical and irregular components of a time series as well. To accomplish these goals, we will use quarterly data that have seasonal and cyclical characteristics. Quarterly data will enable us to illustrate the logic with the least amount of computations. To extend the procedures to work with monthly data, simply change *quarterly* to *monthly* in the following discussion, and use twelve months for the year instead of four quarters.

WHAT IS A SEASONAL INDEX?

We are all familiar with seasonal patterns. Summer brings agricultural, camping, and warm weather activities, whereas winter is associated with school, the Christmas season, and perhaps some snow skiing. While these seasonal changes provide some variety to our lives, they pose some difficult problems for business firms. Managers of both manufacturing and retail firms must contend with the fluctuating seasonal demand for goods and services, while at the same time providing stable employment for their employees.

Firms cope with seasonal demands in a number of ways. Most manufacturers try to balance their product lines with offsetting products (e.g., chain saws and snow blowers). Other techniques call for accumulating inventories during periods of slack demand or contracting extra work to subcontractors during periods of peak demand. Retailers frequently promote special sales or offseason rates.

Underlying all these strategies, however, is the need for good information about the seasonal characteristics of the good or service being produced. This information is commonly summarized in a seasonal index.

A **seasonal index** is a numerical expression of the monthly (quarterly or other) level of a time series variable relative to its longer run average or trend for that same period.

The result of seasonal index computations is an index value for each quarter, or month, of the time series. For example, monthly indexes might be: Jan 85.4%, Feb 97.1%, and Mar 108.6%. This would mean that the actual activity level in January is typically 85.4% of the average or trend value for January. That average value is frequently taken as the *moving average* of data points centered upon the period under consideration. Let us use some quarterly data to illustrate this moving average concept before proceeding with an example of a seasonal index computation.

DETERMINING A MOVING AVERAGE

A moving average (MA) is computed by summing and averaging successive values of a time series. Figure 13–3 illustrates a simple 3-period moving average. We begin by totaling the first 3 values and placing the total opposite the middle value. Then we divide by 3 to obtain the average and show it alongside. Each successive moving average is recalculated by dropping the oldest value in the average and adding the next unused value.

Note that the moving average values are always centered in the middle of the values in the average, and that no moving average values are available for the end points of the series. A graph of the moving average values would obviously be smoother than the actual values. Moreover, this leveling effect increases as the number of periods in the average increases. Thus, a 12-month moving average could be expected to smooth out all of the seasonal effects in a set of monthly data.

FIGURE 13–3 A three-period moving average

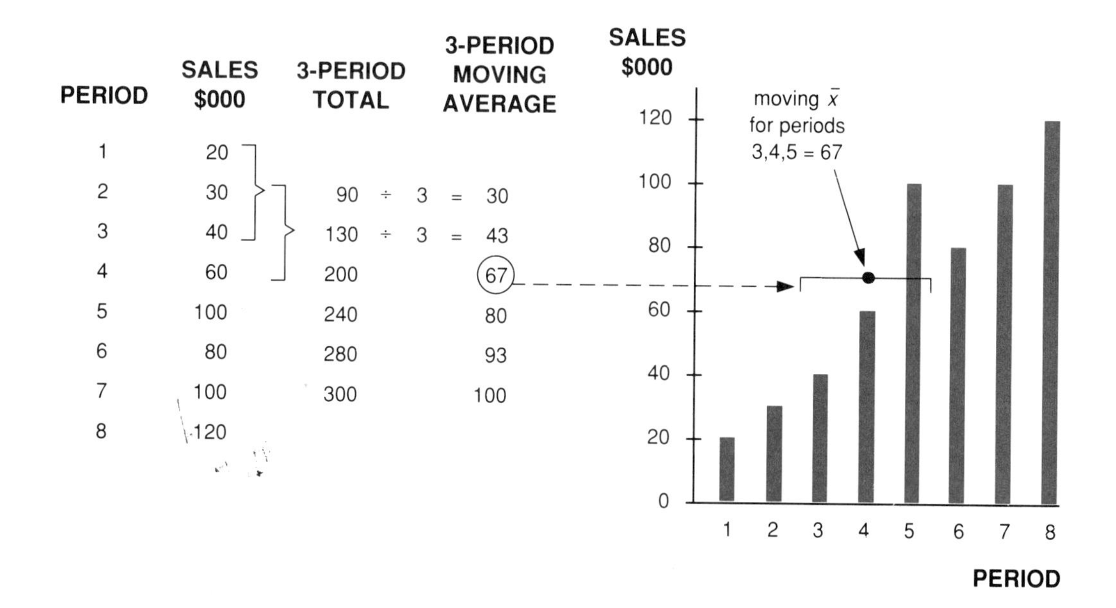

MODIFICATION FOR EVEN NUMBER OF PERIODS Computing the moving average for an even number of periods presents us with a minor problem because the central point for the average lies between the two periods, rather than in the middle of a period. For example, the center of quarters 1, 2, 3, and 4 lies between the second and third quarters (e.g., July 1), and does not correspond with the observed values centered on either quarter. Similarly, the next average of quarters 2, 3, 4, and 5 is centered between quarters three and four (i.e., on October 1).

Figure 13–4 illustrates an easy remedy for this situation. By averaging the two totals (and dividing by 8 instead of 4), the first average is centered opposite the third quarter (i.e., on August 15).

FIGURE 13–4 Centering a moving average

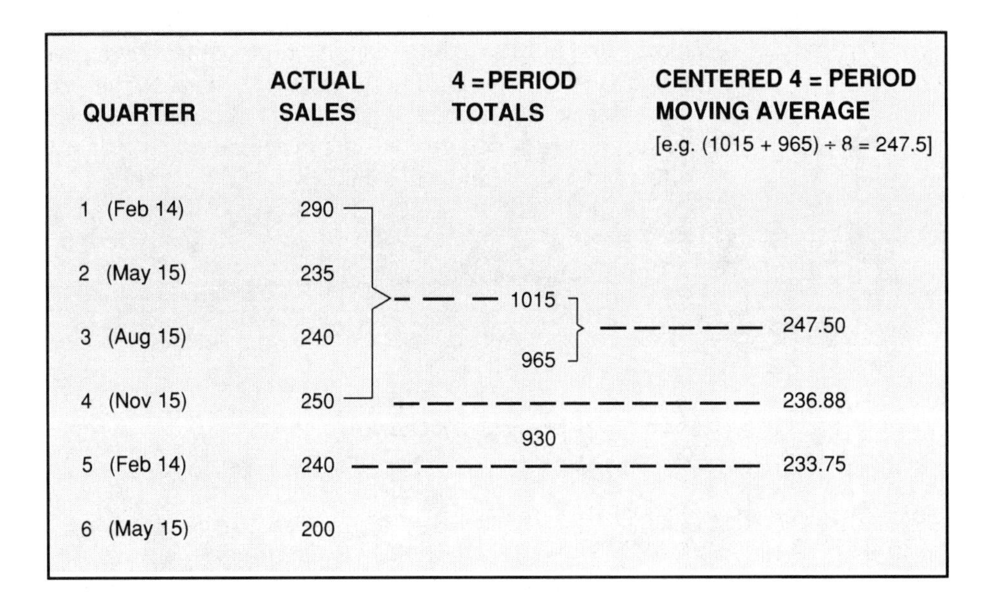

This procedure can be continued until a moving average is centered opposite each quarter, except the two on each end of the series. Each moving average is based on data for five quarters but, as shown, the middle three of each group of five are included twice because they are in both totals. In Figure 13–4, the 247.50 average includes two first quarters (quarters 1 and 5) plus two weights of quarters 2, 3, and 4. Thus, all quarters are included, and each is given equal weight, so the seasonal effect is totally obliterated from each of the moving average values. In addition, the averaging process also cancels out the irregular effects. The assumption is that the seasonal S and irregular I components are effectively removed from the MA values leaving the trend T and cycle C only.

COMPUTING THE SEASONAL INDEX

RATIO-TO-MOVING-AVERAGE METHOD The ratio-to-moving-average is the most widely used seasonal index in business. It expresses each period's activity as a percent of moving average centered on the period. Our illustration of the ratio-to-moving-average index will follow the steps outlined in Table 13–2.

TABLE 13–2 Steps in the ratio-to-moving-average method of computing a seasonal index

(General Model: $Y = T\ C\ S\ I$)

1. Compute a moving average to describe the trend T and cycle C effects. (Use 12 periods for monthly data or 4 periods for quarterly data.)
2. Divide the actual value Y by the moving average centered on that period. (This removes the trend and cyclical effects leaving the seasonal and irregular components. In equation form: $(T\ C\ S\ I) \div (T\ C) = (S\ I)$. These $S\ I$ values are called *specific seasonals*.)
3. Group the specific seasonals for each period in a table and select a representative index for each time period, e.g., a January seasonal. (This may be done by using the median, mean, or some form of modified mean. The purpose here is to help remove irregular effects from the $S\ I$ values leaving only the seasonal S.)
4. Adjust each representative index so that their total is precisely equal to the number of periods ($\times$ 100) in the data. (This is done by multiplying the representative indexes by the ratio of 1200 over the sum of the indexes for monthly data, or 400 over the sum for quarterly data.) The purpose here is to eliminate any remaining irregular effect I, and ensure that the mean for the final seasonal indexes equals 100.0. Each index then reflects what proportion, on average, the actual values are, relative to the trend centered on their respective periods.

Example 13–5 The data shown below represent quarterly sales ($000) of a sportswear line over a 7-year period. Compute a seasonal index for each quarter by using the ratio-to-moving-average method outlined in Table 13–2.

Solution

Computations are shown in Table 13–3. For convenience, column 4 has been included to show the totals of the two 4-period sums, and in column 5 this sum is divided by 8.[1] The specific seasonals are shown in column 6, where the actual Y value from column 2 is divided by the centered moving average from column 5.

TABLE 13–3 Quarterly sales for sport-togs line ($000)

Year	(1) QTR	(2) Actual Sales	(3) 4 Period Total	(4) Two Totals Combined	(5) Centered 4 Period Moving Average	(6) Ratio: Actual to Moving Average
1982	1	290				
	2	235				
			1015			
	3	240		1980	247.50	96.97
			965			
	4	250		1895	236.88	105.54
			930			
1983	1	240		1870	233.75	102.67
			940			
	2	200		1930	241.25	82.90
			990			
	3	250		2080	260.00	96.15
			1090			
	4	300		2280	285.00	105.26
			1190			
1984	1	340		2455	306.88	110.79
			1265			
	2	300		2585	323.13	92.84
			1320			
	3	325		2710	338.75	95.94
			1390			
	4	355		2850	356.25	99.65
			1460			

[1] For monthly data, the first moving total includes one figure for January of the first year, two figures for each month through December, and one figure for January of the following year. This total is divided by 24 to get a moving average which is centered opposite July of the first year.

TABLE 13–3 Continued

Year	(1) QTR	(2) Actual Sales	(3) 4 Period Total	(4) Two Totals Combined	(5) Centered 4 Period Moving Average	(6) Ratio: Actual to Moving Average
1985	1	410		2980	372.50	110.07
			1520			
	2	370		3125	390.63	94.72
			1605			
	3	385		3200	400.00	96.25
			1595			
	4	440		3160	395.00	111.39
			1565			
1986	1	400		3095	386.88	103.39
			1530			
	2	340		3000	375.00	90.67
			1470			
	3	350		2900	362.50	96.55
			1430			
	4	380		2805	350.63	108.38
			1375			
1987	1	360		2700	337.50	106.67
			1325			
	2	285		2610	326.25	87.36
			1285			
	3	300		2580	322.50	93.02
			1295			
	4	340		2645	330.63	102.83
			1350			
1988	1	370		2765	345.63	107.05
			1415			
	2	340		2900	362.50	93.79
			1485			
	3	365				
	4	410				

Next, the specific seasonals are grouped according to the quarter to which they apply, and a representative value is selected. Analysts frequently use a *modified mean* to eliminate any irregular variations. This is accomplished by deleting the highest and lowest (or perhaps the two highest and two lowest) specific seasonals. These values are marked H and L in Table 13–4. Because there were originally six figures in each column and two are deleted to eliminate any atypical seasonal variation, the column totals become the sums of the four "most typical" specific seasonals. The mean for each quarter is then obtained by dividing the column total by the number of specific seasonals added, in this case, four.

TABLE 13–4 Specific seasonals for sport-togs line

Year	First Quarter	Second Quarter	Third Quarter	Fourth Quarter	Adjustment Factor
1982	—	—	H→96.97	105.54	
1983	L→102.67	L→82.90	96.15	105.26	
1984	H→110.79	92.84	95.94	L→ 99.65	
1985	110.07	H→94.72	96.25	H→111.39	Correction $= \dfrac{400}{\Sigma \text{ Indexes}}$
1986	103.39	90.67	96.55	108.38	
1987	106.67	87.36	L→93.02	102.83	
1988	107.05	93.79	—	—	$= \dfrac{400}{399.69} = 1.0008$
Modified Total	427.18	364.66	384.89	422.01	
Modified Mean (M)	106.80	91.17	96.22	105.506	$\Sigma = 399.69$
Final Index (*M*)(1.0008)	106.9	91.2	96.3	105.6	$\Sigma = 400.0$

As a final step, the modified means are multiplied by the adjustment factor, which in this case is greater than one (because the modified mean total is less than 400). The final indexes are shown in the bottom row of Table 13–4.

USING THE SEASONAL INDEX

Seasonal indexes, like those computed in Example 13–5, depict the typical seasonal pattern of the data. Once obtained, they can be used for eliminating the effect of seasonal variation from data and for more accurately predicting future levels of activity. We will be using the seasonal indexes we developed here as we move on to isolate the cyclical and irregular components from the historical data. But let us first note how the seasonal index can be used to analyze current data.

If the major (and only) factors of concern in a data set are trend and seasonal, the actual values should closely approach the trend times the seasonal index S. In other words (and expressing S as a decimal)

$$\text{Actual} = (\text{Trend})\left(\frac{S}{100}\right) \tag{13–5}$$

Economic reports and news broadcasts often refer to data that have been *seasonally adjusted*. For example, *The Wall Street Journal* frequently mentions the seasonally adjusted unemployment or automobile sales. Seasonally adjusted sales are the trend values *after* the seasonal index is taken into account. Rearranging Equation 13–5, we can restate this as:

$$\text{Seasonally Adjusted (Trend) Value} = \frac{\text{Actual Value}}{\text{Seasonal Index}}(100) \tag{13–6}$$

The economic reports are thus projecting what the annual figure of a subject variable, such as unemployment, would be if the seasonally adjusted activity were to continue for the full year. For example, assume auto sales for the first month of a year were one million cars. If the seasonal index for January was 90%, then the seasonally adjusted annual sales would be at a rate of ($1m ÷ .90)(12 months) = 1.3 million cars per year.

Making seasonal adjustments to actual data frequently increases the usefulness of the data for making business decisions. Consider the following.

Example 13–6 The departmental manager of a firm handling a line of fashion sportswear wishes to consider dropping the line just after Christmas, unless sales are on the increase. A review of sales records after Christmas shows a sales increase of from $365,000 in the third quarter to $410,000 in the fourth quarter of the year. If the seasonal indexes for the third and fourth quarters are 96.3 and 105.6, respectively, is there any real (seasonally adjusted) increase in sales—or is the increase due wholly to seasonal factors?

Solution

The seasonally adjusted sales are the trend values obtained after the seasonal index is taken into account.

	Seasonally adjusted sportswear sales 3rd Quarter	 4th Quarter
Seasonally Adjusted Sales $= \frac{\text{Actual}}{S}(100)$	$\frac{\$365{,}000}{96.3}(100) = \$379{,}024$	$\frac{\$410{,}000}{105.6}(100) = \$388{,}258$

There is a seasonally adjusted increase over the one-quarter period of $388,258 − $379,024 = $9,234, so the manager may wish to reconsider dropping the line.

CHANGING SEASONAL PATTERNS As is apparent when one examines the pattern of energy consumption, record sales, and numerous other variables, seasonal patterns do change over time. Thus it behooves the user of economic and business data to do whatever analysis is justified to verify that no significant changes are taking place in the seasonal patterns. One of the best ways to do this is by graphing the actual period values and/or the specific seasonals to see if any changes or discontinuities appear to be present in the data. If so, a single set of index numbers may not be adequate for describing the seasonal pattern.

DESCRIBING THE CYCLICAL AND IRREGULAR COMPONENTS

It is not unusual for a time series analysis to be directed solely toward the trend and seasonal pattern in a set of data; these components have always claimed the important role in time series work. But business activity does advance in cycles, and when cyclical measures can be abstracted from data, they may be useful.

If the economy is in the initial stages of an expansion, or if an analysis shows that recessionary forces will probably exert an effect over the next year or two, managers will want to take that into account when making their decisions about new facilities, marketing campaigns, and other business operations. Similarly, it is advantageous to be able to label the random or irregular variations in a series for what they really are—rather than be misled into believing that they are evidence of some growth trend or cyclical effects. Identifying the irregular component is not always an easy (or fruitful) task, but analytical techniques are frequently useful. The stock market drops in 1929 and 1987 are vivid reminders that businesses must be prepared to deal with more than just trend and seasonal factors.

The following example illustrates one technique for abstracting the cyclical and irregular components of a time series. By using the data of Table 13–3 (and the associated example), we will take advantage of using the seasonal indexes, which have already been calculated.

Example 13–7 The time series diagram below illustrates the trend of the data in Table 13–3. Using that data, isolate the cyclical and irregular components of the time series.

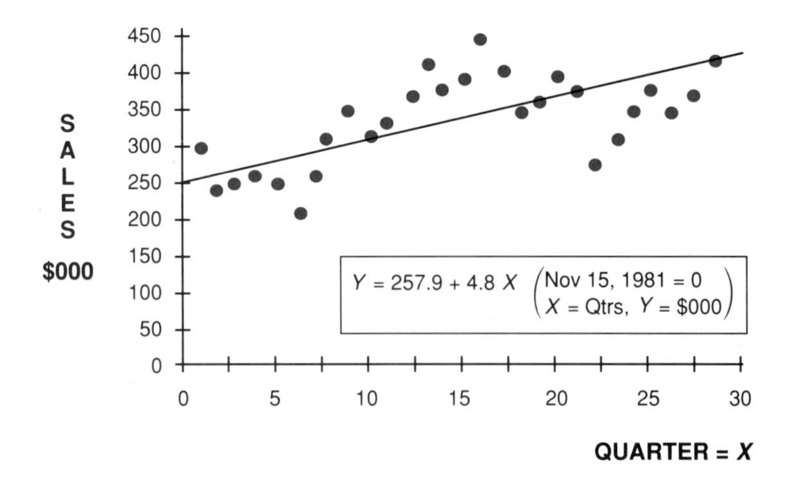

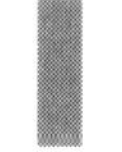

Solution

Table 13–5 shows the calculations required, which are discussed below.

TABLE 13–5 Analysis of cyclical and irregular components

Year and Quarter		(1) Actual Sales Y	(2) Trend Value T	(3) Seasonal Index S	(4) Seasonally Adjusted $\frac{Y}{S}=\frac{TCSI}{S}(100)=TCI$	(5) C & I (4)÷(2) CI	(6) 3-Period MA of (5) C	(7) Irregular (5)÷(6) I
1982	1	290	262.7	106.9	271.3	103.27		
	2	235	267.5	91.2	257.6	96.30	97.03	99.25
	3	240	272.3	96.3	249.2	91.52	91.09	100.47
	4	250	277.1	105.6	236.8	85.46	85.55	99.89
1983	1	240	281.9	106.9	224.6	79.67	80.53	98.93
	2	200	286.7	91.2	219.2	76.46	81.73	93.55
	3	250	291.5	96.3	259.6	89.06	87.13	102.21
	4	300	296.3	105.6	284.1	95.88	96.86	98.98
1984	1	340	301.1	106.9	318.1	105.65	103.02	102.55
	2	300	305.9	91.2	328.9	107.52	107.27	100.23
	3	325	310.7	96.3	337.5	108.63	107.57	100.99
	4	355	315.5	105.6	336.2	106.56	111.65	95.44
1985	1	410	320.3	106.9	383.6	119.76	117.02	102.34
	2	370	325.1	91.2	405.7	124.79	121.91	102.36
	3	385	329.9	96.3	399.8	121.19	123.49	98.14
	4	440	334.7	105.6	416.7	124.50	118.65	104.93
1986	1	400	339.5	106.9	374.2	110.22	114.33	96.40
	2	340	344.3	91.2	372.8	108.28	107.54	100.69
	3	350	349.1	96.3	363.5	104.12	104.70	99.45
	4	380	353.9	105.6	359.9	101.70	99.90	101.80
1987	1	360	358.7	106.9	336.8	93.89	93.85	100.04
	2	285	363.5	91.2	312.5	85.97	88.14	97.54
	3	300	368.3	96.3	311.5	84.58	85.62	98.78
	4	340	373.1	105.6	322.0	86.30	87.50	98.63
1988	1	370	377.9	106.9	346.2	91.61	91.77	99.83
	2	340	382.7	91.2	372.8	97.41	95.61	101.88
	3	365	387.5	96.3	379.0	97.81	98.07	99.73
	4	410	392.3	105.6	388.3	98.98		
1989	1		397.1					
	2		401.9					
	3		406.7					
	4		411.5					

Trend

Column 2 of Table 13–5 shows the trend values depicted by the straight line in the figure. They are determined from the time series trend equation computed as we learned earlier. For these data the equation is:

$$Y = 257.9 + 4.8X \text{ (Nov 15, 1981} = 0,\ X = \text{Qtrs},\ Y = \$000)$$

The intercept value is thus 257.9 (on Nov 15, 1981), and the value for period one of the data (on Feb 15, 1982) is therefore $257.9 + 4.8(1) = 262.7$. Each successive trend value then is increased by the slope of 4.8. Thus, the period-two trend is $262.7 + 4.8 = 267.5$, and the period-three trend is $267.5 + 4.8 = 272.3$.

Seasonal adjustment

Column 3 lists the seasonal indexes developed earlier. In column 4 the seasonally adjusted values are computed by dividing the seasonal index into the actual sales (Y) values. This eliminates seasonal effects so that column 4 reflects only the trend, cyclical, and irregular components of the time series.

Cyclical

In column 5, the trend effects are removed by dividing the TCI values of column 4 by the trend (T) values of column 2. This has the effect of eliminating any of the long-term growth or decline of a series, so that the column 5 values reflect only cyclical and irregular patterns. The irregular effects are then minimized by computing a 3-period moving average, which is shown in column 6. The length of the moving average is a function of the amount of irregular variation in the data and the length of the cycle. (A long moving average runs the risk of canceling out the cyclical effects.) In many cases, a three- or five-period moving average may suffice, especially if the irregular effects are nominal, as is the case in this example.

Irregular

Lastly, column 7 isolates the irregular component. This is achieved by dividing the cyclical and irregular components (the $C\,I$ of column 5) by the moving average (the C of column 6), leaving only the I component.

ANALYSIS The purpose of breaking the T, C, S, and I components out from the actual sales pattern is to (1) understand and explain how the various components contribute to the aggregate Y values, and (2) use that knowledge to better plan for future activities.

We have already dealt extensively with trend and seasonal effects, so let us focus upon the C and I factors here. One of the most effective ways of gaining additional knowledge about the cyclical and irregular factors is to replot the data after the seasonal and trend effects have been removed. Figure 13–5 illustrates two such data plots—(a) is a replot after the seasonal effect has been removed (by using the data from column 4 of Table 13–5), and (b) is a replot after both seasonal and trend are removed (by using column 5 of Table 13 5). Both data plots evidence cyclical effects (and some minor irregular effects), suggesting that further analysis may be useful.

FIGURE 13–5 Replot of sales data with (a) seasonal removed and (b) seasonal and trend removed

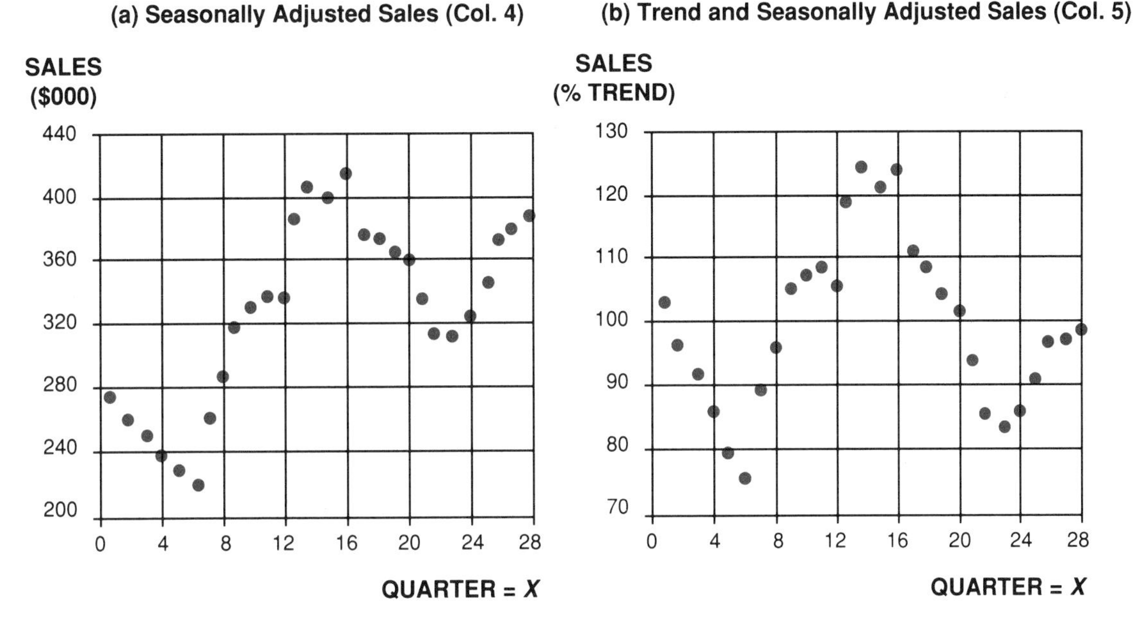

Another, though less vivid, way of identifying cyclical effects is to glance down column 6, which is the "smoothed" cyclical index. Notice that it moves gradually down to a low of 80.53 in the first quarter of 1983, up to a high of 123.49 in the third quarter of 1985, and down again to a low of 85.62 in the third quarter of 1987.

When data such as these confirm the presence of cyclical forces, that knowledge can be used to advantage in the same way seasonal indexes are used. Because the cycle seems to be rising at the end of the series, planners will want to take that into account in planning for inventories, advertising, and other business-related activities. Knowledge of the cyclical pattern for a firm's products also presents the firm with the opportunity of testing for correlations of product sales with other economic indicators, such as building permits or money supply factors.

The irregular indexes in column 7 are all quite close to 100, suggesting that random or irregular forces did not exert a significant effect on the sportswear sales patterns. If those indexes had deviated substantially, they would be signaling management to search for identifiable causes of such irregularities (e.g., strikes) in hopes of better accommodating them in the future.

EXERCISES II

11. Explain what is meant by a *seasonal effect,* and tell how it might be present in statistical data that describe the following:
    (a) monthly sales of toys
    (b) hourly traffic on the New York subway
    (c) monthly workload on employees of public accounting firms
12. Suppose you have 10 years of data and are computing a seasonal index describing lumber production in Oregon. Why would you use a 12-month moving average in the calculation instead of a 3- or 5-month average?
13. What is a *specific seasonal,* and why might the specific seasonals for similar periods differ (e.g., for June 1988, June 1989, and June 1990)?
14. For the economic variable *housing starts,* distinguish between an actual value for March and the seasonally adjusted value in March.
15. Assume you have calculated the trend and seasonal index values for a set of data. Now list the steps you might take to isolate the cyclical and irregular components of the series.
16. Compute (a) a 5-period moving average and (b) a centered 4-period moving average for the data of Figure 13–3.
17. Use the data in Table 13–3, but change the last two years of actual sales to the following:

Quarter	1	2	3	4
Actual sales for 1987	360	210	280	360
Actual sales for 1988	430	230	310	460

    Using the ratio-to-moving-average method, recompute the last eight specific seasonals of the table.
18. Using the data from problem 17 above, assume the specific seasonals for the successive quarters of the recomputed Table 13–3 are

For quarter 1:	102.67	110.79	110.07	103.39	113.83	130.80
For quarter 2:	82.90	92.84	94.72	90.67	68.85	66.67
For quarter 3:	96.97	96.15	95.94	96.25	96.55	89.96
For quarter 4:	105.54	105.26	99.65	111.39	111.36	111.63

    Use the modified-mean method to compute the revised seasonal indexes (using the methodology of Table 13–4).
19. Using the data from problem 17 above, assume the revised time series equation is $Y = 262.70 + 4.15X$ (Nov 15, 1981 = 0, X = Qtrs,

Y = \$000). Use this equation with the new seasonal index values from problem 18 above to determine the revised cyclical and irregular indexes (as originally computed in Table 13–5).

20. Given the following data on number of welding machines sold.

	J	F	M	A	M	J	J	A	S	O	N	D
Yr 1987	4	2	5	7	6	3	1	3	9	6	3	4
Yr 1988	5	4	8	9	9	4	5	7	10	8	6	5
Yr 1989	7	9	12	7	7	6	9	12	11	14	11	8

(a) Graph the data. Does there appear to be any cyclical effect present?

(b) Compute a 3-period moving average and plot it as a solid line.

(c) Compute a centered 12-period moving average and plot it as a dashed line.

(d) Comment on the difference in the 3- and 12-period moving averages.

21. For the data in problem 20, compute the specific seasonals via the ratio-to-moving average method.

22. A time series analysis of sales data from a large retail chain store yielded some specific seasonals from which the following modified means were calculated. Make whatever adjustment is necessary and determine the seasonal indexes for the various months.

Month	J	F	M	A	M	J	J	A	S	O	N	D
Modified Mean	58.2	22.1	37.8	32.2	81.8	113.4	46.7	68.8	119.4	40.1	110.7	422.6

23. Given the following table of specific seasonals. Compute the seasonal indexes for the four quarters of the year. (Use the modified mean approach and discard the highest and lowest values.)

Year	First Quarter	Second Quarter	Third Quarter	Fourth Quarter
1	—	—	72.7	143.6
2	65.6	102.9	88.4	159.8
3	71.2	115.5	74.5	163.2
4	67.7	98.7	72.3	151.8
5	62.1	108.8	84.9	156.1
6	63.9	117.4	81.9	148.3
7	70.8	107.2	—	—

24. A news broadcast reports that wages paid out by employers in a local community were $24.8 million in February, and the February seasonal index for wages is 90.0. Based upon this, at what annual rate are wages flowing into this community?
25. Given that the trend line for data in problem 20 is $Y = 3.1 + .2X$ (Dec 15, 1987 = 0, X = mo, Y = no. machines), and the seasonal indexes are:

J	F	M	A	M	J	J	A	S	O	N	D
89.8	97.3	160.0	124.3	118.5	70.5	47.7	81.4	159.3	112.0	68.8	70.4

Isolate the cyclical and irregular components in the same manner as was done in Table 13–5. Comment on your result.

INDEX NUMBERS: PURPOSE AND TYPES

As noted earlier, index numbers are a convenient mechanism for measuring the change in economic variables, relative to some reference point, over time. Their use is not restricted to economic variables nor to differences in time, however. That is, they could be used to compare social measures such as crime rates or to compare energy production in different countries at the same time. But most applications relate economic comparisons over time—so our focus will be on that. And, in accordance with conventional usage, we will not show the percent sign with the index number, even though the numbers are calculated as percents.

PURPOSE AND LIMITATIONS

Index numbers are used to support managerial decisions by quantitatively summarizing the change of selected variables, such as costs and prices, in one number. The goods or services under study are selected judgmentally and almost never by chance—as in a probability sample. This eliminates the possibility of confidence interval estimates or other inferences based on probability about the results obtained. But it does not invalidate the index as a descriptive statistic to serve the purpose intended.

TYPES OF INDEXES

Numerous schemes exist for classifying index numbers depending upon whether they measure the change in a single item (a *simple* index number) or a group of items (a *composite* index number). Classifications of composite items can be further subdivided into those that consist of unweighted aggregates (e.g., totals of prices of various items) and weighted aggregates (e.g., where prices are weighted by specified quantities).

FIGURE 13–6 One classification for index numbers

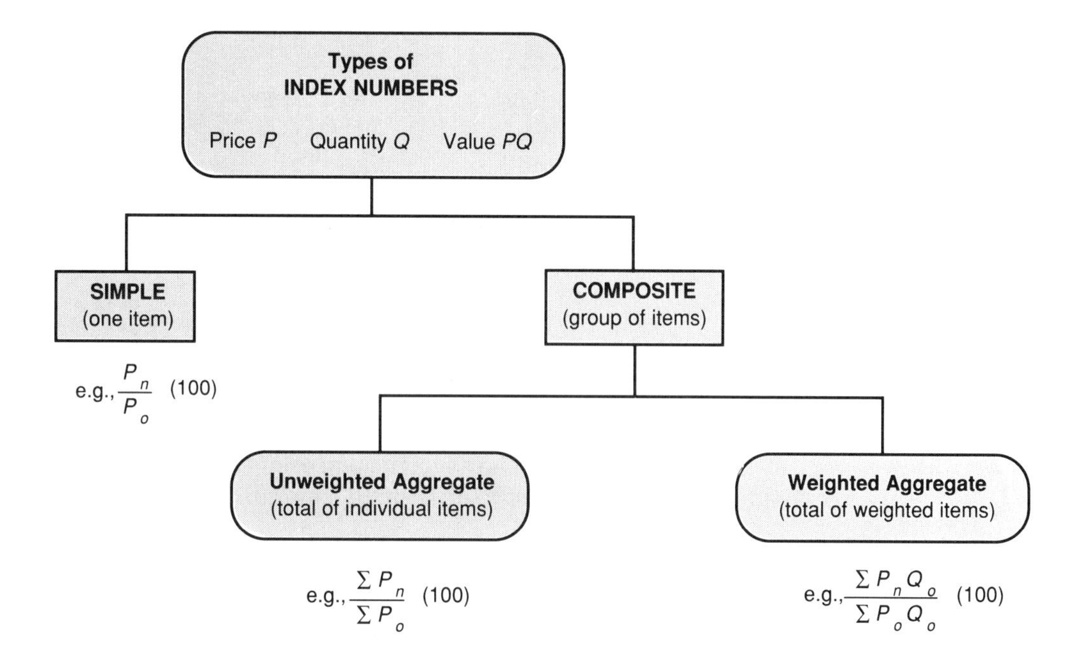

Figure 13–6 illustrates one classification of index numbers based upon the simple versus composite distinction. It also includes some example formulations that we shall develop in the subsequent discussion. For now, however, simply note the groupings of *simple* versus *composite*, and the extension of composite indexes to *unweighted aggregates* and *weighted aggregates*.

A commonly used method of distinguishing index numbers (and one convenient for our purposes) is to identify them in terms of the type of variable they summarize. This leads us to consider briefly three types of indexes: (a) *price indexes*, (b) *quantity indexes*, and (c) *value indexes*. These three types are mentioned in the top box of Figure 13–6. If we let P designate price, and Q designate quantity, then the value of an item is its price times its quantity, or $P \times Q$. Most business index number applications fall into one or more of these three categories. To ensure our understanding of the three types of indexes, we shall first work with an example of a single product, i.e., we will be generating simple index numbers relating to prices, quantities, and values.

SIMPLE PRICE, QUANTITY, AND VALUE INDEXES

A **simple index number** typically describes the relative change in one classification of an item over two time periods. Letting the subscript n refer to the given, or specified, period and the subscript o refer to the base, or reference, period, then the simple relative indexes for price, quantity, and value are:

Simple Price Index: $$SPI = \frac{P_n}{P_o}(100) \quad (13\text{–}7)$$

Simple Quantity Index: $$SQI = \frac{Q_n}{Q_o}(100) \quad (13\text{–}8)$$

Simple Value Index: $$SVI = \frac{P_nQ_n}{P_oQ_o}(100) \quad (13\text{–}9)$$

PRICE INDEXES We begin with a price index computation and use the data in Table 13–6, which reflect the real estate home sales history available from the Board of Realtors in a small community.

TABLE 13–6 Home sales in Lakeview over an 8-year period

(1) Year	(2) Mean Sales Price $ P	(3) Number Homes Sold Q	(2) × (3) Total Value ($m) PQ
1981	50,000	28	1.40
1982	53,100	32	1.70
1983	61,900	21	1.30
1984	58,300	30	1.75
1985	66,700	18	1.20
1986	69,900	37	2.59
1987	74,600	44	3.28
1988	72,000	39	2.81

Example 13–8 The Lakeview Board of Realtors wishes to establish a price index for home sales. (a) Using the data in Table 13–6 with 1981 as the base year, compute the price index. (b) What conclusion can you draw from the index?

Solution

(a) Upon designating 1981 as the base year, the $50,000 price for 1981 is set equal to 100, signifying that the 1981 price is 100%. Then we compute each year's average price as a percentage of the base year price by dividing each subsequent year's price by the base year price, and multiplying by 100 (as in Equation 13-7). Thus for 1982 we have:

$$SPI_{1982} = \frac{P_n}{P_o}(100) = \frac{53{,}100}{50{,}000}(100) = 106.2$$

Index values are frequently reported to one digit beyond the decimal. The price index values for the home sales data with 1981 as the base year are

Year	1981	1982	1983	1984	1985	1986	1987	1988
Price	$50,000	$53,100	$61,900	$58,300	$66,700	$69,900	$74,600	$72,000
Index	100.0	106.2	123.8	116.6	133.4	139.8	149.2	144.0

(b) From the indexes, we can observe that the average price of homes sold in Lakeview in 1988 was 44% higher than the price of those sold in 1981. Note carefully, however, that the comparisons always relate to a base year. Thus we could not say that prices declined by 5.2% from 1987 to 1988, because the percentage decline from 1987 was only ($74,600 − $72,000) ÷ $74,600 = 3.5%.

QUANTITY INDEXES Simple quantity indexes are useful for analyzing the flow of goods and usage patterns of critical raw materials or components. For example, a copper mining firm may use quantity indexes to monitor the amount of ore taken from a mine, or an oil company may use one to help plan for the production rate of petroleum products. Quantity indexes are computed in a manner similar to price indexes.

VALUE INDEXES Value indexes reflect the total change in a variable when both prices and quantities are taken into account. They are used extensively in business—although many people do not recognize them as value indexes. Nevertheless, whenever you multiply a price times a quantity, you get a value.

Example 13–9 Using the data in Table 13–6 , with 1984 as the base year, construct (a) a quantity index for 1988 and (b) a value index for 1988.

Solution

Using 1984 as the base year, both the quantity and value columns (columns 3 and 4 in Table 13–6) would show indexes of 100.0 for 1984. We can calculate the two indexes for 1988 as follows:

$$\text{(a) } SQI_{1988} = \frac{Q_n}{Q_o}(100) = \frac{39}{30}(100) = 130.0$$

$$\text{(b) } SVI_{1988} = \frac{P_nQ_n}{P_oQ_o}(100) = \frac{2.81}{1.75}(100) = 160.6$$

From Example 13–9, one could conclude that the number of homes sold had increased by 30.0% in the last four years and that the total value of the homes sold had increased by just over 60%.

COMPOSITE INDEX NUMBERS

Although simple index numbers can be very useful, you are probably more familiar with composite or aggregative index numbers. Composite indexes are used to describe the relative change in price, quantity, or value of a group of items. The methodology is similar except that several items make up the base-period and given-period numbers. For example, a rental car firm may wish to monitor the prices of repair parts, quantities used, and total parts costs for its fleet of automobiles over a five-year period. Comparative information on the cost indexes associated with different types of automobiles might then influence the firm's selection process when they add new automobiles to their fleet.

UNWEIGHTED AGGREGATIVE COMPOSITES

The unweighted composite index is the most elementary form of an aggregative index. We can illustrate with a composite price index. It is computed by summing the prices of all items in the given year (ΣP_n), dividing by the summation of prices in the base year (ΣP_o), and multiplying by 100.

Unweighted Aggregative Price Index
$$UAPI = \frac{\Sigma P_n}{\Sigma P_o}(100) \tag{13–10}$$

Example 13–10 The cost differences of selected expense account items to make a sales call in a given city were recorded as follows over a five-year interval. Compute the unweighted aggregate price index, using the earlier year as the base year.

	Cost, 1983	Cost, 1988
Airfare	$130	$140
Airport transportation	16	22
Lodging/day	65	120
Meals/day	43	84
Rent-a-car	31	46
	ΣP_o = $285	ΣP_n = $412

Solution

Letting 1983 be the base year and 1988 the given year, we have:

$$UAPI = \frac{\Sigma P_n}{\Sigma P_o}(100) = \frac{\Sigma P_{1988}}{\Sigma P_{1983}}(100) = \frac{412}{285}(100) = 144.5$$

This tells us that the cost of these expense account items has risen by about 45% over the five-year period. We do not know if the cost increase is due to inflationary effects or to a different consumption pattern. The index may not accurately reflect cost changes for a particular situation either, because no adjustment is included to account for the one-time costs (such as airfare) versus the daily costs (of meals and lodging). Insofar as the costs other than airfare have risen about 75% (versus less than 10% for the airfare), the index of 144.5 may be too low if the sales representative stays in the city for more than one day.

WEIGHTED AGGREGATIVE COMPOSITES

As you have no doubt surmised, problems arise when summing the prices of a composite of items. As another example, suppose a maintenance cost index is composed of the prices of a muffler ($40), a headlight ($8), a set of tires ($195), and an engine ($1,600). An unweighted aggregate index may not be very representative—especially if one of each item is included in the index. It will be unduly influenced by price changes in commodities that have the largest price per unit (such as car engines). The prices may not reflect the usage pattern. Moreover, units of measurement are not necessarily the same; some may be in units (a muffler), others in sets (four tires), and others in pounds or gallons (e.g., gasoline). As a result of these measurement problems, composite indexes are almost always weighted.

CHOICE OF WEIGHTS The ideal weights for a composite index would seem to be those that enable the index to represent "typical" quantities consumed or "typical" prices prevailing. In practice, three alternatives usually surface:

1. using base year values (P_o or Q_o)
2. using given year values (P_n or Q_n)
3. using some average or combination of the two.

Numerous weighting schemes exist. The weights used for composite price indexes are frequently quantities from the base year and for quantity indexes are prices from the base year. Value indexes typically reflect the ratio of given year prices and quantities (P_nQ_n) to base year prices and quantities (P_oQ_o). The sales volume of a firm over a several-year period might give rise to a value index, for both prices and quantities change over the years. Table 13–7 summarizes these expressions in equation form.

TABLE 13–7 Equations for weighted composite indexes

TYPE	CHARACTERISTICS	FORMULA
Price	Prices vary (Base year quantities are weights.)	$PI = \frac{\Sigma P_nQ_o}{\Sigma P_oQ_o}(100)$ (13–11)
Quantity	Quantities vary (Base year prices are weights.)	$QI = \frac{\Sigma P_oQ_n}{\Sigma P_oQ_o}(100)$ (13–12)
Value	Prices and quantities vary (Given year values divided by base year values.)	$VI = \frac{\Sigma P_nQ_n}{\Sigma P_oQ_o}(100)$ (13–13)

BASIC DATA To illustrate the computational procedure for one of the composite indexes, we shall use the data presented in Table 13–8. It shows the daily consumption of only four of over one hundred food items monitored periodically by the food service operations manager at a health care facility. The cost data is eventually used for making adjustments in meal charges for residents of the facility. Including all the food commodities would increase the accuracy (and cost) of the computation, but for illustrative purposes, four commodities are sufficient.

TABLE 13–8 Data on food items used at a health care facility

Commodity	Year 1 Quantity/day	Price per Unit ($) Yr 1	Yr 2	Yr 3	Yr 4	Yr 5	Year 5 Quantity/day
A (vegetable)	12 pounds	.30	.33	.36	.36	.39	14 pounds
B (juice)	10 quarts	.25	.24	.30	.32	.30	9 quarts
C (sweet rolls)	20 each	.20	.25	.28	.32	.30	16 each
D (cocoa mix)	1 box	2.00	2.40	2.50	2.50	2.60	1 box
Totals		2.75	3.22	3.44	3.50	3.59	

The table gives the prices of four commodities for five years, and the quantities for years 1 and 5. Notice that the quantities are not in the same units—nor is this necessary. We have deliberately chosen pounds, quarts, each, and boxes to emphasize this. The prices are, however, per unit of quantity expressed. And whereas the units of quantity cannot be added, the prices or values in a given year can be added. However, the price totals do not offer much insight into the meal cost patterns because they simply tell how much it would have cost the facility to purchase one stated unit of each commodity.

Example 13–11 Use the data given in Table 13–8 to compute the composite price index, assuming year 1 is the base year.

Solution

In lieu of any other instructions, we shall assume that the composite price index is to be computed using base year quantities as weights (per Equation 13–11). The price index for year 1 is 100.0, and for year 2 is thus:

$$PI = \frac{\Sigma P_n Q_o}{\Sigma P_o Q_o}(100) = \frac{(.33)(12) + (.24)(10) + (.25)(20) + (2.40)(1)}{(.30)(12) + (.25)(10) + (.20)(20) + (2.00)(1)}(100)$$

$$= \frac{13.76}{12.10}(100) = 113.7$$

Therefore food prices in year 2 were about 13.7 percent higher than they were in the base year (year 1). Table 13–9 shows all the calculation results.

TABLE 13–9 Price index calculation for food items

ITEM	Base Q_o	Price per Unit ($) Yr 1	Yr 2	Yr 3	Yr 4	Yr 5	Price × Base Yr Quantity Yr 1	Yr 2	Yr 3	Yr 4	Yr 5
A	12	.30	.33	.36	.36	.39	3.60	3.96	4.32	4.32	4.68
B	10	.25	.24	.30	.32	.30	2.50	2.40	3.00	3.20	3.00
C	20	.20	.25	.28	.32	.30	4.00	5.00	5.60	6.40	6.00
D	1	2.00	2.40	2.50	2.50	2.60	2.00	2.40	2.50	2.50	2.60
Totals		— Not Required —					12.10	13.76	15.42	16.42	16.28
Price Index							100.0	113.7	127.4	135.7	134.5

The composite price index calculations in Table 13–9 show us that prices for selected quantities of these items have risen by 34.5% over the four-year period. In this calculation, the base year quantities served as a representative weight and as a standard, for they were a constant reference amount over the period.

LASPEYRES AND PAASCHE INDEXES

LASPEYRES INDEX (base year weights) Much of the initial development of weighted indexes is attributed to two individuals named Laspeyres and Paasche—and the equations we use today frequently bear their names.

Etienne Laspeyres introduced the idea of using base period weights in the late 18th century. In recognition of his contribution, equation 13–11 is frequently referred to as the Laspeyres Price Index. Thus the price index we computed in the previous example (i.e., the 113.7) is called a Laspeyres Price Index. Equation (13–12) yields Laspeyres Quantity Index. Note that the price index uses *base year quantities* as weights, and the quantity index uses *base year prices* as weights.

Base year quantities are often a useful historical reference point, and using them as a standard is an effective way of isolating price increases. However, when prices of selected items rise—as often happens in our economy—consumption of those items may decline relative to the less expensive items. The base year quantities of the expensive items do not reflect actual consumption. So, for a given mix of items, the Laspeyres Index may tend to overstate the actual increase in total cost.

PAASCHE INDEX (given year weights) The Paasche Index is also a weighted aggregative composite index. But it differs from the Laspeyres in that it uses given year weights instead of base year weights. Thus the Paasche equation for a price index is:

$$\text{Paasche Price Index} = \frac{\Sigma P_n Q_n}{\Sigma P_o Q_n}(100) \qquad (13\text{–}14)$$

The computational mechanics are similar to those we have already illustrated, except that the weighting scheme differs. Note, however, that the Paasche Index tends to have the opposite effect from the Laspeyres formula. Instead of using base period quantities, the Paasche Index reflects an up-to-date pattern of consumption where the quantities are those currently being purchased. This means that if consumers do indeed reduce consumption of the more expensive items, the Paasche Index may understate the effect of rising prices. Moreover, as usage rates change, new data on quantity values must be collected and entered into the computation. The cost associated with this reassessment and revision is sometimes a deterrent to the use of the Paasche Index, and it is not as commonly used as some other indexes.

OTHER INDEXES Both the Laspeyres and Paasche Indexes have limitations by virtue of using either base year or given year weights. However, other indexes that mitigate some of these limitations are available. For example, Fisher's Ideal Index uses the geometric mean of the Laspeyres and Paasche Price Indexes. This gives it a theoretical advantage over each individual index. It and other indexes are treated in more advanced texts on index numbers.

SELECTING AND CHANGING THE BASE PERIOD

SELECTION As you may have already surmised, selection of the base period can be extremely important. In Example 13–11, the 134.5 index suggests that prices of the food items rose 34.5% over the time period to year 5. If the base period had been year 4 instead of year 1, we would have concluded that prices declined from year 4 to year 5. Such differences can have a significant bearing on business activities. For example, suppose both labor and management were using the Table 13–9 data to support wage negotiations. Each side may seek to reinforce a different bargaining position simply by using a different base period for computing its wage index.

As far as possible, the base period should represent a fairly normal or typical period—although this is not always easy to find. Sometimes an average of three or five periods is used as the base. Whatever the case, periods of excessive expansion or depression should be avoided.

UPDATING THE BASE Organizations usually prefer to keep their index values updated so that the base period is no more than ten to fifteen years distant. As a result, many economic indexes are changed from time to time by moving the base period forward. Once careful consideration is given to the selection of a representative base period, the mechanics of the change are relatively routine. The selected time period (new base) is first assigned an index of 100. Then the old index values are divided by the old index of the newly selected base. The new index values then enjoy the same relative relationship to each other, but expressing them in relation to a current base makes the figures more meaningful.

Example 13–12 Using the data of Table 13–9, (a) change the base to year 3. (b) Then revise the 134.5 price index for year 5 to reflect the new base period.

Solution

From Table 13–9, the old price index for year 3 is 127.4, so the new indexes are:

For year 3: $PI = \frac{127.4}{127.4}(100) = 100.0$ (New Year 3 index is 100.0)

For year 5: $PI = \frac{134.5}{127.4}(100) = 105.6$ (New Year 5 index is 105.6)

USING INDEXES IN BUSINESS

Although individual firms often maintain internal data on their own industry, a number of external data sources are available. The largest source of statistical data in the world is probably the U.S. federal government. Federal and state agencies publish a myriad of indexes that chart the health of the U.S. and world economy—and these are widely available. Publications such as the *Survey of Current Business* and the *Statistical Abstract of the United States* are available in most every library. Other governments also add to the wealth of data available. For example, the Japan Institute for Social and Economic Affairs publishes the very compact but comprehensive *An International Comparison*. It covers a wide range of topics from bank deposits and steel production to religion and life expectancy figures.

COMMONLY USED INDEXES

Three of the most frequently quoted indexes in the U.S. are the (1) consumer price index, (2) producer price index, and (3) industrial production index.

CONSUMER PRICE INDEX The U.S. Department of Labor, Bureau of Labor Statistics, publishes the *consumer price index*, or CPI. This index, commonly referred to as the cost-of-living index, measures the relative change in prices of a standard group of approximately 400 goods and services, ranging from food and housing expense to transportation and health care. Two indexes are published, one specifically weighted for *wage earners and clerical workers* and the second for *all urban households*. Weights are based upon expenditure patterns from sample areas, and the items included are kept current by substitution and linking procedures. Since the mid-1930s, the percentage of costs attributed to food and alcoholic beverages in the all urban households index has been reduced from about 35% to under 20%. Meanwhile, the transportation component of the index has increased its share from about 8% to nearly 18%.

In addition to the national CPI, major cities and regional areas frequently publish consumer price index information geared to their specific locality. This helps local residents appraise their specific living costs even more precisely. These indexes are also used extensively in labor negotiations and wage contracts. In fact, the salaries of approximately 8 million U.S. workers are now tied to some type of consumer price index. It should not be surprising that the CPI and localized versions of it are among the most closely watched governmental statistics by U.S. wage earners.

PRODUCER PRICE INDEX The Bureau of Labor Statistics also publishes the *producer price index*, which covers prices of over 2,000 commodities that American manufacturers use extensively in their production operations. Separate indexes are available for each item, each major group of items, and the entire list. Individual firms sometimes use these listings to construct their own indexes which include items relevant to them at appropriate weights so that they accurately describe price changes in materials for their own firm or industry.

INDUSTRIAL PRODUCTION INDEX The *industrial production index*, published by the Federal Reserve Board, is a barometer of business conditions, for it measures the change in industrial production. This index includes measures of manufacturing, mining, and utility output.

OTHER INDEXES Numerous other indexes cover specific areas, such as employment, department store sales, personal consumption expenditures, freight car loadings, new housing starts, and many more. Among the indexes tracking activity of the stock market are the *Dow-Jones Averages*, the *Standard and Poor's 500*, and the *New York Stock Exchange Composite Index of All Stocks*. Hardly a day goes by that business people are not making reference to some kind of index number.

USING INDEXES TO DEFLATE A TIME SERIES

We saw earlier in the chapter that dividing an actual time series value by the seasonal index essentially removed the seasonal effect from the data and left us with a "trend" value. This process of removing the effect of one variable from another is one of the most common uses of index numbers. The most widespread application of this is undoubtedly in deflation, or removing the effect of price level variation from a time series. (If the price index is less than 100.0, the process of deflation actually results in inflation, but the process is still known as deflation.) Let us illustrate with an example.

Example 13–13 Labor negotiators of a manufacturing firm are seeking a wage hike to compensate for increases in the cost of living. In an attempt to substantiate their case, the following pay and consumer price index information for the local area has been introduced. Use the index to deflate the series and reveal the "real" wages (i.e., after allowing for the cost of living increases).

	1983	1984	1985	1986	1987	1988
Weekly take-home pay	$328.50	336.60	349.20	375.20	406.40	414.30
Local area price index	185.1	193.0	198.8	206.7	219.2	235.4

Solution

Dividing each take-home pay value by the price index, we have the real wages.

	1983	1984	1985	1986	1987	1988
Real (adjusted) weekly pay	$177.47	174.40	175.65	181.52	185.40	176.00

In real (price-adjusted) terms, the weekly pay is fairly close to what it was five years ago. However, there has been a drop in purchasing power in the latest year because of the 7.3% increase in the local consumer price index during the last year of the data.

EXERCISES III

26. Distinguish between a simple index number and a composite (or aggregate) index number.
27. How do the base year weights differ in a composite price index and a composite quantity index?
28. Average gasoline prices (in $ per gallon) in a metropolitan area over a 10-year period were as shown below. Using 1978 as the base year, compute the price index for gasoline. [Note: Problem 8 called for fitting a time series to this same data.]

Year X	1978	1979	1980	1981	1982	1983	1984	1985	1986	1987
Price Y	.64	.86	1.22	1.33	1.29	1.24	1.20	1.20	.92	.98

29. Prices and sales volumes for prerecorded video cassettes in a southern metropolitan area followed the pattern shown above:

	Average Price $/unit	Units Sold (000)
1981	65	.2
1982	61	.5
1983	57	.9
1984	49	2.0
1985	45	4.1
1986	41	7.7
1987	36	13.0

Using 1981 as the base year, compute the simple (a) price index, (b) quantity index, and (c) value index for video cassette sales in this area.

30. Research and development spending in the U.S. (in $ billions) has been estimated as follows

	1965	1975	1985
Basic research	2.6	4.6	13.3
Applied research	4.3	7.9	23.9
Development	13.2	22.7	71.6
Total	20.1	35.2	108.8

Source: National Science Foundation

(a) Using 1965 as the base year, compute the price indexes for total spending on research and development.

(b) Use index numbers to determine which area grew at a faster rate, proportionally (over the 20-year period), basic research or applied research?

31. Using the data of Table 13–8, compute the composite price indexes, assuming year 5 is the base year.

32. Use the quantities given below, along with the price data of Table 13–8, to compute the (composite) value indexes for the food items shown. Let year 1 be the base year.

	Year 1	Quantities/day				
Commodity	**Price**	**Yr 1**	**Yr 2**	**Yr 3**	**Yr 4**	**Yr 5**
A (vegetable)	.30	12	12	13	13	14
B (juice)	.25	10	11	10	9	9
C (sweet rolls)	.20	20	19	18	17	16
D (cocoa mix)	2.00	1	1	1	1	1

33. Computer Centers Southwest has recorded the following business in software sales over the past five years. Prices are averages for the

particular type of software package. Using year 3 as the base year, compute the composite (a) price, (b) quantity, and (c) value indexes of software sales for the firm. (d) Use your data to comment upon any significant change in the sales volume of software products.

ITEM	Price per Unit ($)					Number of Packages Sold				
	Yr 1	Yr 2	Yr 3	Yr 4	Yr 5	Yr 1	Yr 2	Yr 3	Yr 4	Yr 5
A (accounting)	775	650	545	425	390	3	5	18	20	28
B (spreadsheet)	600	395	320	275	295	12	18	30	38	41
C (word processing)	385	240	200	185	175	27	45	143	220	205
D (data base)	—	560	425	310	365	—	4	10	14	35

34. A research foundation has estimated that college students in a selected region of the country have the average spending patterns shown. Compute the unweighted aggregate price index using 1980 as the base year, and discuss the meaning of your result.

Nonschool item	Cost in 1980	Cost in 1989
Clothing	$ 76	$ 193
Snacks & refreshments	130	220
Health & beauty items	38	54
Local transportation	42	67

35. For the data of problem 33, let year 2 be the base year and year 5 the given year. Then compute the (a) Laspeyres Price Index and (b) Paasche Price Index. (c) Are the results comparable? Explain.

36. The materials manager at an electronics plant in Phoenix has been asked whether the "real" dollar value of her finished goods inventory has increased over the past several years, or whether the increase was simply due to inflated prices. The inventory balances and corresponding price indexes for these items are shown below. Compute the "deflated" inventory balances and determine if there has been any real increase in the value of the finished goods inventory.

Year	Inventory Balance ($ 000)	Price Index (Year 1 = 100)
Yr 10	500.0	128.2
Yr 11	520.2	131.4
Yr 12	548.6	136.2
Yr 13	573.0	142.1
Yr 14	589.5	144.5
Yr 15	620.2	147.2

FORECASTING METHODS

Planning for the future is one of the major responsibilities of management. Wise planning means that managers must anticipate the economic and, perhaps, social forces that will be acting upon their organization in the future. These uncertainties may pertain to sales, raw-material prices, labor costs, interest rates, and a host of other economic and technological variables. The more accurately managers can anticipate such effects, the better job they can do directing and controlling their organization.

Forecasts are estimates of uncertain future events or levels of activity. Forecasts typically concern demand, or technological changes, or the social, political, or economic status of the environment.

Forecasts based strictly upon past data assume that what happened in the past will continue on into the future. The key to good forecasting is, of course, to use past data as a guide but to also anticipate changes in the future as much as possible.

TECHNIQUES IN USE TODAY

Although there are numerous forecasting techniques, they might be conveniently grouped into three categories: (1) opinion and judgmental methods, (2) time series methods, and (3) associative methods.

The *opinion and judgmental methods* are based largely upon experience, and include estimates from field salespeople, market surveys, and opinions of executives. Comparisons with similar situations (i.e., historical analogies) are also useful. Although opinion methods can incorporate experience, they are somewhat subjective, and not as statistically rigorous as some other methods. Thus, they are not always amenable to analysis and improvement over time.

Time series methods include the standard trend, cyclical, and seasonal analyses we discussed earlier in this chapter, plus other methods, such as exponential smoothing and the more advanced Box-Jenkins methods. These latter techniques are beyond our scope here, but we will illustrate a basic time series forecast in this section.

Associative methods include the regression and correlation techniques we discussed in earlier chapters, so we will not repeat examples of regression forecasts here. More sophisticated regression approaches, such as econometrics, are also used, however. Econometric forecasts use the simultaneous solution of multiple regression equations that relate to a wide range of economic activity. Essentially all of these types of forecasts require large-scale computer models to manipulate the data.

A TIME SERIES FORECASTING EXAMPLE

Table 13–10 describes the general procedure for making a time series forecast. We can illustrate the procedure using the trend, seasonal, and cyclical values already provided in Table 13–5.

TABLE 13–10 General procedure for time series forecast

1. Collect and plot historical data to confirm the type of relationship.
2. Develop the trend equation, seasonal index, and cyclical index (if all are relevant).
3. Project the trend into the future.
4. Multiply the monthly (or quarterly) trend value by the seasonal index.
5. Modify the projected values for any cyclical or irregular effects.

Example 13–14 Use the data from the sportswear Example 13–5 (as summarized in Table 13–5) to determine what the sales forecast value would have been for the first quarter of 1989. [Note: The trend equation is $Y = 257.9 + 4.8\,X$ (Nov 15, 1981 = 0, X = Qtrs, Y = \$000).]

Solution

Steps 1 and 2 of the time series forecasting procedure have already been completed, and in Table 13–5 the first quarter trend has been projected as:

Trend: $Y = 257.9 + 4.8\,(29) = 397.1$ or \$397,100.

From Table 13–5, the seasonal index for Quarter 1 is 106.9. Multiplying by this index modifies the trend for seasonal effects:

Trend and seasonal: $Y = 397.1\,(1.069) = 424.5$ or \$424,900.

From Column 6 in Table 13–5, we note that the cyclical index has ranged from a low of 80.53 to a high of 123.49. During quarter 4 of 1988, it appears to be on an upswing where the next value would perhaps be in the range of from 99 to 104. Insofar as this involves some judgment about the strength and duration of the business cycle, this is a place to factor in some experienced judgment of managers and business conditions analysts. Assuming that business conditions are strong, and sales prospects look reasonably good, we will apply an index of 103.00 for illustrative purposes.

Trend, seasonal, and cyclical: $Y = 424.5\,(1.030) = 437.2$ or \$437,200.

Finally, if any unusual or irregular factors are anticipated (e.g., weather, labor, or competitive conditions), these would be factored in at this time. In their absence, the first quarter forecast would be \$437,200.

SUMMARY

A *time series* is a set of successive observations of a variable at regular intervals in time. One approach to analyzing a time series is to decompose it into its trend (T), cyclical (C), seasonal (S), and irregular (I) components, using the multiplicative model $Y = T\ C\ S\ I$.

For linear models, the trend can be estimated by the least squares approach. Then the *trend equation* of the form $Y = a + bX$ is stated with a *signature* that specifies the origin, as well as the units of X and Y. If desired, the origin can be shifted by adding (or subtracting) a designated number of slope increments to the intercept value. An annual trend equation can also be expressed in quarterly or monthly units by dividing the slope, or b value, by 4 or 12 respectively.

A *seasonal index* is a numerical expression of the monthly or quarterly level of a time series, relative to its longer run average or trend centered on that same period. The most common way of computing the seasonal index (i.e., the ratio-to-moving-average method) requires that a moving average first be computed—and centered on the period. Dividing actual Y values by the moving average yields specific seasonals, which contain the seasonal and irregular components. The specific seasonals for each time period are then averaged (e.g., by using a modified mean) and adjusted so the total equals the number of periods times 100. When these seasonal indexes are divided into actual values, they result in seasonally adjusted trend values which are widely used as comparative measures of business activity.

Cyclical components are isolated by dividing seasonally adjusted (TCI) values by trend (T) values to get a CI result and taking a moving average of that (to smooth out the irregular component). Gradual swings in the values of this index indicate the presence of cyclical effects. If the resulting cyclical index is divided into the (CI) value, one can isolate the irregular component and assess its magnitude. However, the irregular component is inherently random—with no discernible pattern—so it remains largely unpredictable anyway.

An *index number* is the ratio of a summary measure of data in a given time period to a similar measure of comparable data in a base period. Most interest centers upon (1) price, (2) quantity, and (3) value indexes. *Simple index numbers* describe the relative change in one classification of an item from a base period, o, to a given period, n. The three simple indexes are:

Simple Price Index	*Simple Quantity Index*	*Simple Value Index*
$SPI = \frac{P_n}{P_o}(100)$	$SQI = \frac{Q_n}{Q_o}(100)$	$SVI = \frac{P_nQ_n}{P_oQ_o}(100)$

Composite index numbers describe the relative change in a group of items and can be either unweighted or weighted. Weighted composites are computed by weighting and then summing the items of the group. The Laspeyres Index uses base year weights. The Laspeyres Price Index uses *base year quantities* as weights and the Laspeyres Quantity Index uses *base year prices* as weights. Their equations, along with a composite value index equation, are illustrated below.

Composite Price Index	*Composite Quantity Index*	*Composite Value Index*
$PI = \frac{\Sigma P_n Q_o}{\Sigma P_o Q_o}(100)$	$QI = \frac{\Sigma P_o Q_n}{\Sigma P_o Q_o}(100)$	$VI = \frac{\Sigma P_n Q_n}{\Sigma P_o Q_o}(100)$

The Paasche Index is somewhat similar to the Laspeyres except that it uses given year weights. As a result, the conclusions from use of the two indexes may differ.

If possible, the base period should represent a normal period and be no more than 10 or 15 years distant. Old index values can be updated by dividing the old index values by the old index of the newly selected base. Among the most commonly used index numbers are the consumer price index, the producer price index, the industrial production index, and the stock market indexes.

Forecasts are estimates of uncertain future events or levels of activity. They have been classified into (1) opinion and judgmental methods, (2) time series methods, and (3) associative methods. A time series can be used for forecasting by projecting the (quarterly or monthly) trend into the future, and multiplying the calculated trend value by the seasonal index. The forecast can be further refined by multiplying by a measure of the cyclical index, if the cyclical effect is identifiable. Finally, an adjustment should be made for any irregularities that can be anticipated.

CASE: WINTER COATS INVENTORY AT FASHIONWORLD

Sheryl Miller had hoped to move up the ladder to become a full-time buyer for the Minnesota-based department store, but it was looking more like she had blown her chances for that. Fashionworld Stores Ltd. was a major department store chain with branches in several midwestern cities. Sheryl had started out as a floor clerk three years ago and had recently been given the opportunity to work as an assistant buyer in the Minneapolis headquarters.

Although she lacked formal training in merchandising, she had a good intuitive feel for what was in style, what would sell, and roughly how much. But now the new vice president of sales, Maryanne Jackson, had cast a shadow on her prospects. Mrs. Jackson had sent a short memo to Sheryl's supervisor, questioning Sheryl's order of winter coats for the upcoming third quarter of the year. There was no problem with the style, but Mrs. Jackson felt that "way too many" coats had been ordered for fall. And if they didn't sell, the whole chain would suffer. In her words, "We've only sold over 100,000 coats in one quarter once in our history—and that was five years ago. Margins are already too thin to have to absorb a loss on 20,000 coats!"

Sheryl's supervisor, Bill Vogel, called her into his office to talk about the order.

BILL: I'm with you, Sheryl. I have a feeling that you've hit it right on this order. But who are we to question Maryanne Jackson? She may be new in that job, but she carries a lot of weight—I mean people listen to her.

SHERYL: She might be right, Bill. You know I figured we'd sell about 123,000 coats in the third quarter. I must admit I had some anxiety about sending that order in. But it's too late to rescind it now. Any thoughts on where to go from here?

BILL: I've had my assistant go back over several years' sales records and get some data on how we've done quarter by quarter. You might take a look at it as a way of bolstering our position. By the way, the group down in data processing might be able to help you pull some meaning out of this too. They've been real helpful to me in the past.

SHERYL: Thanks, Bill. I sure hope so. I hope I haven't stuck my neck out too far on this one. I'll be in touch.

Sheryl looked over the data. Then she took it down to data processing and asked them to do "whatever computers do with this kind of stuff." The printout below was in her mail the next morning with a note that the computer had "crashed" before all the data were printed out, but they hoped this would help.

TIME SERIES COMPONENTS FOR ANNUAL COAT SALES

$Y = 14.263 + 1.463\,X$ (Nov 15, Yr 0 = 0, X = Qtr, Y = Coats in 000)

Year &	Quarter	(1) Actual Units Y	(2) Trend Value T	(3) Seasonal Index S	(4) Seasonally Adjusted $\frac{Y}{S} = \frac{TCSI}{S}(100) = TCI$	(5) C & I (4)÷(2) CI	(6) 3-Period MA of (5) C	(7) Irregular (5)÷(6) I
1	1	12.0	15.72	86.04	13.95	88.74		
	2	9.5	17.18	77.40	12.27	71.44	95.67	74.67
	3	29.8	18.64	126.03	23.64	126.84	102.55	123.69
	4	24.3	20.11	110.53	21.99	109.36	107.83	101.41
2	1	16.2	21.58	86.04	18.83	87.30	98.65	88.49
	2	17.7	23.03	77.40	22.87	99.29	103.23	96.19
	3	38.0	24.49	126.03	30.15	123.10	107.24	114.78
	4	28.5	25.96	110.53	25.78	99.34	96.75	102.68
3	1	16.0	27.42	86.04	18.60	67.82	78.98	85.87
	2	15.6	28.88	77.40	20.15	69.78	81.34	85.79
	3	40.7	30.35	126.03	32.29	106.42	99.12	107.36
	4	42.6	31.81	110.53	38.54	121.17	113.59	106.67
4	1	32.4	33.27	86.04	37.66	113.18	113.33	99.87
	2	28.4	34.74	77.40	36.69	105.63	112.39	93.99
	3	54.0	36.20	126.03	42.85	118.37	113.91	103.92
	4	49.0	37.66	110.53	44.33	117.72	113.55	103.67
5	1	35.2	39.12	86.04	40.91	104.57	105.61	99.01
	2	29.7	40.59	77.40	38.37	94.54	95.68	98.80
	3	46.6	42.05	126.03	36.98	87.93	87.37	100.64
	4	38.3	43.51	110.53	34.65	79.64	83.42	95.47
6	1	32.0	44.98	86.04	37.19	82.69	80.54	102.67
	2	28.5	46.44	77.40	36.82	79.29	75.08	105.60
	3	38.2	47.90	126.03	30.31	63.28	75.56	83.74
	4	45.9	49.36	110.53	41.53	84.13	78.63	106.99
7	1	38.7	50.83	86.04	44.98	88.49	92.29	95.88
	2	42.2	52.29	77.40	54.52	104.26	98.69	105.64
	3	70.0	53.75	126.03	55.54	103.33	106.34	97.17
	4	68.0	55.22	110.53	61.52	111.42	116.69	95.48
8	1	66.0	56.68	86.04	76.71	135.33	125.58	107.77
	2	58.5	58.14	77.40	75.58	129.99	138.76	93.68
	3	113.4	59.61	126.03	89.98	150.95	132.81	113.66
	4	79.3	61.07	110.53	71.75	117.49	122.19	96.15
9	1	52.8	62.53	86.04	61.37	98.14	108.15	90.74
	2	53.9	64.00	77.40	69.64	108.81	97.83	111.23
	3	71.4	65.46	126.03	56.65	86.55	98.47	87.89
	4	74.0	66.92	110.53	66.95	100.05	99.02	101.04
10	1	65.0	68.38	86.04	75.55	110.47	101.25	109.11
	2	50.4	69.85	77.40	65.11	93.22	93.97	99.20
	3	70.3	71.31	126.03	55.78	78.22	83.22	94.00
	4	62.9	72.77	110.53	56.91	78.20	84.08	93.01

TIME SERIES COMPONENTS FOR ANNUAL COAT SALES Continued

$Y = 14.263 + 1.463X$ (Nov 15, Yr 0 = 0, X = Qtr, Y = Coats in 000)

Year &	Quarter	(1) Actual Units Y	(2) Trend Value T	(3) Seasonal Index S	(4) Seasonally Adjusted $\frac{Y}{S}=\frac{TCSI}{S}(100)=TCI$	(5) C & I (4)÷(2) CI	(6) 3-Period MA of (5) C	(7) Irregular (5)÷(6) I
11	1	61.2	74.24	86.04	71.13	95.81	85.54	112.01
	2	48.4	75.70	77.40	62.53	82.60	87.03	94.91
	3	80.4	77.16	126.03	63.79	82.67	84.25	98.14
	4	76.0	78.62	110.53	68.76	87.46	89.12	98.13
		- - - - - - - - - - - - - computer	- - - - - failed	- - - - - at	- - - - - this	- - - - - point	- - - - - -	- - - - - -
12	1	67.0						
	2	78.8						
	3	92.4						
	4	93.5						
13	1	96.0						
	2	98.4						

Case Questions

(a) Is there any apparent growth trend pattern in the data? If so, how much?

(b) Does the past history of demand evidence any seasonal effects? If so, what are they?

(c) Comment on the existence of any cyclical effects.

(d) Complete the calculations for the last year and a half of the data.

(e) Assuming the trend and seasonal factors are reliable, what value of a cyclical index does Sheryl's forecast of 123,000 coats implicitly assume? In view of the data, does this seem reasonable?

(f) Do you feel Sheryl's projection of 123,000 coats for the third quarter was on target? What response would you recommend Sheryl and Bill Vogel offer to Maryanne Jackson?

QUESTIONS AND PROBLEMS

37. Distinguish between cyclical and seasonal components of a time series. Which can be more accurately predicted and why?

38. In economic data, what are the principal causes of (a) long-term trends, (b) regularly recurring patterns, (c) cyclical variation?

39. Suppose you have some actual demand values that you want to analyze via classical time series analysis.

    (a) In what order (sequence) are the causes (or sources) of change in the demand values removed from the actual values?

    (b) How are the trend and seasonal components eliminated?

    (c) Which factor is eliminated by smoothing?

40. Compare the freehand trend line with that obtained using the least squares procedure from the standpoint of (a) effort required, (b) accuracy, (c) credibility, (d) skill required, and (e) objectivity.

41. Maintenance expenses ($000) for a Salt Lake City trucking company over a 15-year period were as shown.

Year	Expense	Year	Expense	Year	Expense
1	21	6	19	11	21
2	34	7	26	12	49
3	11	8	30	13	27
4	18	9	44	14	39
5	31	10	35	15	47

    (a) Graph the data to judge (visually) whether a linear trend is present.

    (b) Using your graph, estimate maintenance expenses in year 20.

42. Using the data from the problem above, (a) compute the least squares trend line for the maintenance expense data. (b) Plot your trend line on a graph of the data and use it to project expenses in year 20. (c) Use your time series equation to estimate expenses in year 20. Does it agree with your estimate from (b)?

43. Assume the data shown represent the number of new residential housing starts in a small community over a seven-year period.

Year	Year Code X	Housing Starts Y	XY	X^2
19X0	0	25	0	0
19X1	1	28	28	1
19X2	2	32	64	4
19X3	3	36	108	9
19X4	4	34	136	16
19X5	5	43	215	25
19X6	6	47	282	36
Totals	**21**	**245**	**833**	**91**

(a) Compute the slope and intercept of the least squares trend line and state the trend equation.

(b) Project the trend value for the number of housing starts in 19X9.

44. Assume the data shown below represent the number of new businesses seeking a license over a five-year period in a city in Missouri.

(a) Develop a least squares trend line and state it complete with signature. (b) Use your equation to project the number of license requests in the year 1997. (c) Shift the origin of the equation to 1990 and restate the equation.

Year	1985	1986	1987	1988	1989
No. businesses	10	20	15	30	25

45. The population of the United States has grown as shown in the following table (where population values are in millions):

Year	Population
1890	63
1900	76
1920	106
1940	132
1960	179
1980	227

The U.S. Bureau of Census forecast for the year 2000 is 267 million. How does their forecast compare with what would be derived from a linear trend?

46. The manager of Desertpalace Motor Inn in Dallas has developed a time series model to predict room occupancy. The trend equation is:

$$Y = 30{,}600 + 2{,}400\,X \;(1986 = 0,\ X = \text{yrs},\ Y = \text{occupants/yr})$$

(a) Project the occupancy rate for 1996.

(b) Shift the origin to 1990 and state the equation complete with signature.

(c) Restate the original equation, except with X = months.

(d) Use your equation for (c) to forecast the annual occupancy rate as of mid-September 1996.

47. Jersey Chemical Products Co. produces a plastic powder used in toys and housewares, and they "book" orders over a year in advance of production. Their quarterly sales of one product ($000) are estimated as follows:

1983		1984		1985		1986		1987		1988		1989	
Qtr	**Sales**	**Qtr**	**Sales**	**Qtr**	**Sales**	**Qtr**	**Sales**	**Qtr**	**Sales**	**Qtr**	**Sales**	**Qtr**	**Sales**
1	290	1	320	1	340	1	370	1	370	1	355	1	360
2	280	2	305	2	321	2	360	2	350	2	336	2	341
3	285	3	310	3	320	3	362	3	355	3	335	3	342
4	310	4	330	4	340	4	380	4	370	4	350	4	359

(a) Compute the specific seasonals.

(b) Compute the seasonal index using the modified mean by eliminating the highest and lowest specific seasonals.

(c) Using the trend equation $Y_t = 303.33 + 2.346\,X$ (Nov 15, 1982 = 0, X = quarters, Y = sales in $000), compute the trend and the cyclical and irregular indexes. Use the same format as Table 13–5.

(d) Do the data evidence a strong cyclical component?

48. (a) Contrast price index numbers and quantity index numbers.

(b) How does a value index utilize price and quantity numbers?

49. Explain the difference between the selection of the items to be included in a price index and a random sample intended as a basis for inference about some characteristic of a production process.

50. (a) Why do firms update the base period of their indexes on a regular basis? (b) What kind of period should be used for the base?

51. Given are the following data on personal computer prices and sales during the mid-1980s. Using year 1982 as the base year, compute the simple (a) price, (b) quantity, and (c) value indexes.

	1982	1983	1984	1985
Average price ($)	4,200	4,060	3,690	3,910
Quantity (millions)	1.0	1.6	2.6	3.4

Source: The Wall Street Journal, 9/16/85, p.7c)

52. Records of transactions of a merchant in Chicago reveal the following:

Commodity and Unit	Quantities Sold				Prices Charged			
	1986	1987	1988	1989	1986	1987	1988	1989
A pound	10	12	13	15	2.00	2.40	2.50	2.60
B gallon	40	44	48	42	3.00	3.30	3.42	3.54
C box	20	24	22	26	6.00	5.40	6.12	6.24
D dozen	100	108	116	120	.50	.52	.58	.62

(a) If the quantities are to be used as weights in a composite price index, must they all be in the same units?

(b) Compute the composite price indexes for each year using 1986 as the base year and 1986 quantities as weights.

53. Given the transactions in the problem above and using 1986 as the base year, compute the composite value index for each year.

54. Assume the figures below represent the net sales of an office products company for the past five years. Also given is a price index appropriate for the types of products they sell.

Year	Sales	Price Index
19X1	$3,264,545	125.0
19X2	3,395,600	132.1
19X3	4,256,800	136.7
19X4	5,310,600	138.4
19X5	5,623,755	143.5

(a) Convert the price index to a 19X1 base.

(b) Use your new price index to tell what each year's sales would have been if there had been no change in price level.

(c) Was there a real increase in sales (i.e., after price level changes are taken into account)?

55. Computer records at the Vineyard Mountain Pump Co. show that the inventory of parts on hand over four years was as shown. The table shows the average inventory value and the index of prices paid for the parts. These data are used to evaluate the company's inventory control procedures.

Year	Average Inventory	Index of prices paid
1985	$386,000	123.8
1986	402,000	132.0
1987	416,000	134.5
1988	440,000	138.0
1989	462,000	143.6

(a) Prepare an index of deflated inventory balances to help in analyzing changes in the average (dollar) size of the inventory.

(b) Shift the index to 1987 and show the revised indexes.

56. The employees of Specialty Steel Company of Ohio have presented the following data in support of their contention that they are entitled to a wage adjustment. Dollar amounts shown represent the average weekly take-home pay of the group, and the index is a locally constructed price index.

Year:	19X6	19X7	19X8	19X9
Take-home Pay:	$960.00	$969.96	$1,008.28	$1,041.77
Area Price Index:	126.8	129.5	136.2	141.2

(a) Compute the real wages based on the take-home pay and the price indexes given.

(b) Compute the amount of pay needed in 19X9 to provide the amount of buying power equal to that enjoyed in 19X6.

(c) What percent increase in pay over the 19X9 level (if any) is required to provide buying power equal to that of 19X6?

57. A time series analysis has revealed the seasonal indexes and trend equation as shown:

$$Q_1 = 84.5 \quad Q_2 = 96.0 \quad Q_3 = 114.4 \quad Q_4 = 105.1$$

$$Y_t = 280.0 + 6.2X \text{ (Nov 15, 1985} = 0,\ X = \text{Qtrs},\ Y = \$000)$$

If the analysis suggests a cyclical index of 118 for the fourth quarter of 1992, use it and the above data to develop a forecast for that quarter.

58. The only time series data in Appendix A are the three years of EPS values. In order to work with a longer data set, for this problem, assume that the sales of the 30 electronic firms represent 30 years of sales data for one firm. Use these data to develop the trend equation for a time series with origin in year zero (corresponding to firm number 71).
    (a) Use a time series (or regression) program to develop the trend equation and state your equation.
    (b) What is the trend value for year zero?
    (c) Use the trend equation to develop a point estimate for when $X = 35$.

CHAPTER

Nonparametric Statistical Methods

INTRODUCTION

RUNS TEST FOR RANDOMNESS

SIGN TESTS FOR CENTRAL TENDENCY AND DIFFERENCES

One-Sample Tests

Paired-Samples Tests

RANK SUM TEST OF MEANS

RANK CORRELATION

Computing the Rank Correlation Coefficient

Testing the Correlation Coefficient

SUMMARY

CASE: GLOBE PERSONNEL SERVICES—WE'LL FIND YOU A JOB!

QUESTIONS AND PROBLEMS

Throughout the book, we have accepted some assumptions that enabled us to apply selected statistical techniques to business decision problems. For example, we have often assumed that we had numerical measurements from populations that were normally distributed. Unfortunately, many decision situations do not meet conditions assumed earlier in the text. Perhaps the data are only name (nominal) or rank (ordinal) ordered. Or maybe the samples are small, and we cannot assume the data are normally distributed. Then many of the inference procedures discussed earlier do not apply.

The figure above illustrates a business decision situation that is difficult to quantify. Labor and management are bargaining over a new contract and several issues are on the table. Do the two parties essentially agree on the relative priorities of the issues? If management proposes a new contract, will they have correctly addressed the issues that must be resolved? If not, a critical opportunity for settlement of a labor dispute may have been lost.

Unfortunately the issues here concern categories, such as work rules, that are not so easily quantified. So the managers may not be able to base their decisions on traditional parameters like means and standard deviations. Nevertheless, decision making does not come to a screeching halt when these less tangible decision problems arise. Decisions must still be made—often quickly. In many cases, nonparametric methods can help. They are less efficient techniques (and typically less powerful in a statistical sense) than the parametric methods we have studied earlier in the text. But they have a broad range of applicability because there are so few restrictions upon their use.

INTRODUCTION

Thus far, we have concerned ourselves with making inferences about population parameters. The tests we used, which assumed the existence of a specific (often normal) distribution, relied upon the use of parameters (such as μ and σ), and are referred to as parametric tests. Nonparametric methods do not make as much use of these parameters and are less restrictive.

> **Nonparametric methods** are statistical procedures that do not assume any underlying distribution parameters (i.e., are "distribution free"), and generally require only minimal assumptions about the population under study.

Research into nonparametric methods has yielded such an extensive array of statistical tests, that whole books (and courses) are available on nonparametric methods alone.[1] Most of the techniques are used to test hypotheses about some characteristic of the population—as opposed to estimating a population parameter.

We shall make no attempt to survey the multitude of nonparametric techniques in this chapter. Our inquiry will be limited to four of the more widely used methods—methods which happen to use counts, signs, and ranks.

1. *Runs test* This test uses *counts* of similar outcomes to determine whether a sample has been randomly selected from a population.
2. *Sign test* This test uses positive and negative *signs* to test the central tendency of a data set or the significance of the difference in two data sets.
3. *Rank sum test* This test uses overall *ranks* of observations to test whether there is a significant difference between two sets of data.
4. *Rank correlation* This is a method of measuring the correlation in a set of data by using only the *ranks* of the observations.

ADVANTAGES AND DISADVANTAGES As suggested above, nonparametric methods are widely applicable because the assumptions underlying them are less constraining than parametric methods. Moreover, the computations are typically easier to perform. They often use signs and

[1] For example, see W. J. Conover, *Practical Nonparametric Statistics*, 2nd ed., (New York: John Wiley & Sons, Inc., 1980).

ranking information in contrast to specific numerical values of the observations. By doing this, however, they may ignore or discard some potentially useful information—which explains why they are less precise (and less powerful) than standard parametric tests.

RUNS TEST FOR RANDOMNESS

In many of the problems we have illustrated in this text, we have assumed that samples were randomly selected from a population. The one-sample runs test is a method of testing the validity of this assumption. It is based upon a comparison between the actual sequence in which two symbols occur and what would occur under conditions of complete randomness.

A **run** is a succession of identical occurrences, preceded or followed by a different sequence of identical occurrences.

For example, suppose a safety study required that cars be randomly selected without regard to whether they were of American (A) or foreign (F) design. If the actual sequence recorded were either one of the following, you might well question whether the selection was really random:

SEQUENCE I A A A A A A A A A A F F F F F F F F F F

SEQUENCE II A F A F A F A F A F A F A F A F A F A F

Both sequences have a total of $n = 20$ observations. But sequence I has only one run of As and one run of Fs for two runs total. On the other hand, sequence II has twenty runs. Either way, the clustering or perfect alternation is not what would be expected from a random process. The runs test will reveal if too few or too many runs exist.

Suppose we let n_1 equal the number of occurrences of one type (A) and n_2 the number of the other (F). Then the total number of observations is the sum of n_1 and n_2.

$$\text{For Runs Test:} \quad n = n_1 + n_2 \qquad (14\text{–}1)$$

TEST HYPOTHESIS Like other hypotheses, the null hypothesis for a runs test is a claim about a population. It is typically a statement that the number of runs is consistent with what observations would be, if taken randomly from a process. Alternatively, the observations are not randomly mixed. Insofar as either too few or too many runs could cause rejection of the hypothesis, a two-tailed test is common. (However, lower- and upper-tailed tests can be used if the null hypothesis is phrased in such a way as to conclude that there are too few or too many runs to be considered random.)

TEST STATISTIC If the n observations are selected by a random process, the mean of the sampling distribution of the number of runs (or expected number of runs), μ_r, is:

$$\mu_r = \frac{n + 2n_1n_2}{n} \qquad (14\text{–}2)$$

Also, the standard error of the runs statistic is

$$\sigma_r = \sqrt{\frac{2n_1n_2(2n_1n_2 - n)}{n^2(n - 1)}} \qquad (14\text{–}3)$$

The test statistic for a runs test takes a form similar to those we have used before, except the symbols are different. Letting r = the number of runs, we have:

$$Z_{calc} = \frac{r - \mu_r}{\sigma_r} \qquad (14\text{–}4)$$

Tables of the sampling distribution of r for small samples are available in reference texts on nonparametric statistics. However, if either n_1 or n_2 is 20 or more (and n is 30 or more), the test statistic is approximately normally distributed with a mean of zero and standard error of 1. In fact, if n_1 and n_2 are roughly equal, the approximation is reasonably good whenever n_1 and n_2 are both more than 10.

Example 14–1 Forty-five cars were used in a test conducted to measure the stopping distance for cars going 60 mph. The sequence in which the American (A) and foreign (F) design cars were selected is given below. Test whether the cars were randomly selected at the .20 level of significance.

A A F F F F F A A F F A A A F A A F F F F

A A A F F F A A A F A A A F F F F F A A A F F

Solution

There are $n_1 = 21$ American cars, and $n_2 = 24$ foreign cars for $n = 45$ observations total. And, as marked above by the groupings, there are 16 runs in the data. We can use equation (14-4) to compute our test statistic.

1. H_0: the cars are randomly occurring
   H_1: the cars are not randomly occurring
2. $\alpha = .20$
3. Use runs test. Reject if $Z_{calc} < -1.28$ or > 1.28

4. Calculation:

$$Z_{calc} = \frac{r - \mu_r}{\sigma_r} \quad \text{where: } r = \text{number of runs} = 16$$

$$\mu_r = \frac{n + 2n_1n_2}{n} = \frac{45 + 2(21)(24)}{45} = 23.40$$

$$\sigma_r = \sqrt{\frac{2n_1n_2(2n_1n_2 - n)}{n^2(n - 1)}} = \sqrt{\frac{2(21)(24)[2(21)(24) - 45]}{(45)^2(45 - 1)}} = 3.30$$

$$Z_{calc} = \frac{r - \mu_r}{\sigma_r} = \frac{16 - 23.40}{3.30} = -2.24$$

5. *Conclusion*: Reject hypothesis that observations are random because $-2.24 < -1.28$. The limited number of runs (16) suggests that the sequence in which the cars were selected was not random. This small number of runs would have occurred by chance less than 20 % of the time.

WHY THE TEST IS NONPARAMETRIC The "nonparametric" designation of this and other tests does *not* mean that they avoid the use of statistical distributions (such as the normal) to arrive at their conclusions. As is evident, the runs test actually takes advantage of knowledge that for samples of 30 or more, the test statistic approaches a normal distribution.

On the other hand, we *made no prior assumptions about the data*. The runs test, like other nonparametric tests, does avoid, negate, or lessen the need to make certain qualifying assumptions about the nature of the data being tested. Moreover, we *used the similarity of names* (American and foreign) as the only basis of classification—the data in this particular example were not even characterized by a numerical measure.

As might be expected, the runs could be identified by any symbols, including numbers if we choose to use them. The requirement is simply that there be *one sample* that can be *portioned into two distinguishable categories*, so that the number of runs and number of elements in each run can be counted. We needed 20 or more observations of one type to ensure normality (although tables are available for smaller sample sizes). The sign test, which we take up next, can be used for analysis of both one-sample and two-sample situations.

SIGN TESTS FOR CENTRAL TENDENCY AND DIFFERENCES

The sign test is a test concerning the central tendency of a data set that uses a measure of direction of the data points (that is, a plus or minus sign) as the basis for inference. Although the sign test is an alternative to the one-sample (and paired-sample) t-tests of means, *it does not require that the population be normally distributed.* The one-sample sign test often uses the median, or some assumed (symmetrical) division of the population as the basis for comparison. Two-sample sign tests do not presuppose any symmetry in the data. They simply use differences in direction for their test.

ONE-SAMPLE TESTS

One-sample sign tests are quite easy to conduct once a null hypothesis is established. The hypothesis proposes some central value for the population. Then the sample values are observed and assigned a + or a − sign depending upon whether they are larger or smaller than the hypothesized value. If the hypothesized value is correct, we would expect that roughly half of the sample values would be above it and half below it, as illustrated in Figure 14–1.

FIGURE 14–1 Logic of a one-sample sign test

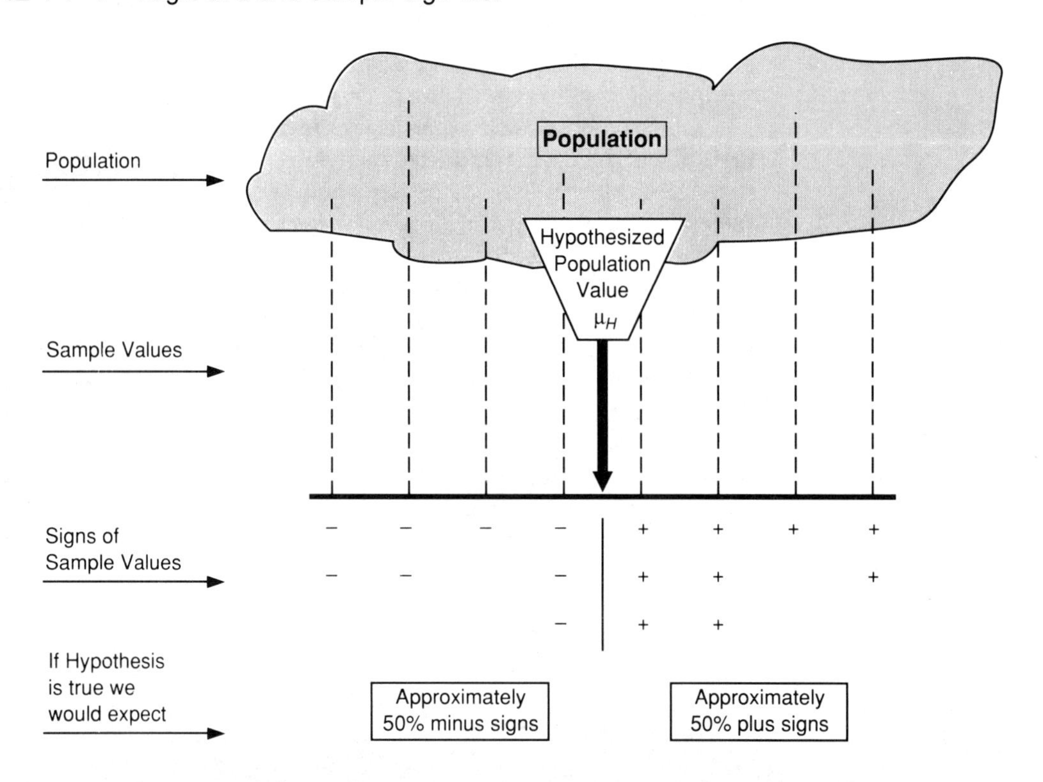

The one-sample sign test consists of comparing the number of + or − signs to what would randomly occur by chance if the proportions should be equal at $\pi = .5$. We base the chance occurrence upon the binomial distribution for small samples (of $n < 20$) and upon the normal approximation to the binomial for larger samples of up to 30 or so. (Beyond 30, the standard parametric test of means can be used because the central limit theorem provides us with some assurance that the distribution is normal. However, the sign test may still be used if desired.)

We shall limit our first example to a small sample case where we can use the binomial distribution to establish the probability that the actual number of plus (or minus) signs would have occurred by chance. If the probability we calculate is less than our specified alpha level, the chance of getting that proportion of plus (or minus) signs is less than the alpha level—i.e., if the hypothesis is true. If the observations are not fairly equally apportioned around the hypothesized value, we reject the hypothesis that our chosen value represents the central tendency of the population.

Example 14–2

An insurance firm is considering relocating their headquarters to a new community. According to a Chamber of Commerce publication, secretaries in the proposed area make approximately $23,000 per year. A sample of 15 secretaries revealed the wages shown below. Use a sign test to determine whether the salaries of the population of secretaries in the community are really $23,000—or are they < $23,000? Use $\alpha = .05$.

$18,000	$24,400	$21,600	$28,000	$19,750
26,350	22,200	20,250	18,100	31,450
23,000	27,700	19,650	21,400	21,370

Solution

The sample size is relatively small ($n = 15$), and population distributions of income are likely to be skewed, so a parametric test may be unsatisfactory. Therefore we will use the sign test as an alternative to the t-test of means. For the sign test, we assign − (and +) signs for values smaller (or larger) than the $23,000. Disregarding the value equal to $23,000, we obtain 9 minus signs (below $23,000), and 5 plus signs (above $23,000), so now our $n = 5 + 9 = 14$.

Value	sign	Value	sign	Value	sign	Value	sign	Value	sign
$18,000	−	$24,400	+	$21,600	−	$28,000	+	$19,750	−
26,350	+	22,200	−	20,250	−	18,100	−	31,450	+
23,000	(delete)	27,700	−	19,650	−	26,400	+	21,370	−

1. H_0: $\mu = \$23{,}000$
   H_1: $\mu < \$23{,}000$
2. $\alpha = .05$
3. Test criterion: Use the sign test with the binomial distribution for $n = 14$ and $\pi = .50$. Because the alternative hypothesis has a $<$ sign, we will count the number of negative signs and reject H_0 if the probability of getting that many (9 or more) is $< \alpha = .05$. That is, we want to reject the H_0 if the chance of getting 9 or more negative signs is so small that we could not reasonably expect that many from a population that is equally divided around $23,000.
4. Calculation:

$$P\,(X \geq 9 \mid n = 14,\ \pi = .50) = P(X = 9 \text{ through } X = 14)$$
$$= .1222 + .0611 + .0222 + .0056 + .0009 + .0001 = .2121$$

5. *Conclusion*: Do not reject the H_0: $\mu = \$23{,}000$ because the binomial probability of getting five salaries below $23,000 is .2121. This is more than the .05 probability (α limit) that serves as our test criterion, so the result is not sufficiently unlikely as to cause us to question the $23,000 average.

ANALYSIS Note that the hypothesis might just as well have been that the median was equal to $23,000, for no use was made of the individual salary values other than their direction (+ or −) in relation to the hypothesized value. Thus, the *test is not powerful enough to reject* the null hypothesis that the salaries average $23,000—even though 9 of the 14 salaries were below $23,000. [And with good reason! It is interesting to note that because of two or three high salaries, the mean salary is actually more than $23,000 ($\bar{x} = \$23{,}215$ and $s = \$4{,}019$). Thus a t test would not have rejected the hypothesis either—partially because the data is positively skewed as a result of the few high values.]

PAIRED-SAMPLES TESTS

The paired-samples or paired difference sign test is similar to the one-sample test except it is used to determine whether a significant difference exists between two sets of paired data. It does not require that the data values be divided around some central value (as the one-sample test assumes), or even that the data be numerical, for directional information will suffice. However, the variable being tested should be continuous.

We typically begin the paired-samples sign test with the claim of no difference, so the null hypothesis is stated in a way that claims the proportion of pluses and minuses is equal to .50. The differences are measured, and inference is again based upon the direction of the difference in scores (i.e., + or −, excluding ties). Then, for small samples, the binomial probability of getting the actual result is calculated and compared with the specified α level, and a conclusion is drawn. For larger samples, the normal approximation may be used, as illustrated next.

Example 14–3 A market research firm has randomly asked 24 shoppers in a supermarket to test two new soft drinks being imported from foreign countries and score them on a 1 (poor) to 5 (excellent) scale. The scores are shown below:

SCORES ASSIGNED TO TWO SOFT DRINKS BY SUPERMARKET SHOPPERS

Shopper I.D. Ltr	A	B	C	D	E	F	G	H	I	J	K	L	M	N	O	P	Q	R	S	T	U	V	W	X
Asparagus Juice	2	3	3	3	4	2	5	4	3	3	2	4	1	4	2	4	2	3	5	5	3	2	4	2
Garden Delite	3	4	3	2	5	1	4	3	4	2	2	1	2	2	1	3	2	2	4	3	4	4	2	1

Use the paired difference sign test to determine whether there is a real difference in customer preference between the two types of soft drinks. Use the 5% level of significance.

Solution

Designating the change in preference from asparagus juice to Garden Delite as + (for a higher score) or − (for a lower score), we have:

Shopper No.	A	B	C	D	E	F	G	H	I	J	K	L	M	N	O	P	Q	R	S	T	U	V	W	X
Sign	+	+	0	−	+	−	−	−	+	−	0	−	+	−	−	−	0	−	−	−	+	+	−	−

In summary, we have 7 (+) signs and 14 (−) signs = 21 total, plus 3 zeros (which are discarded). We are looking for a difference in any direction so this will be a two-tailed test, and we can work with either the proportion of + or of − signs. (We'll let p = proportion of + signs.)

1. H_0: $\pi = .50$ (i.e., proportions equal; no difference in preference)
   H_1: $\pi \neq .50$
2. $\alpha = .05$
3. Use sign test. Reject if $Z_{calc} < -1.96$ or $Z_{calc} > 1.96$
4. *Calculation*:

$$Z_{calc} = \frac{p - \pi}{\sigma_p} \quad \text{where: } p = \text{observed proportion of plus signs} = 7/21 = .3333$$

$$\sigma_p = \sqrt{\frac{\pi(1 - \pi)}{n}} = \sqrt{\frac{(.50)(.50)}{21}} = .1091$$

$$= \frac{.33 - .50}{.11} = -1.52$$

5. *Conclusion*: Do not reject H_0. Using the sign test, the data are not sufficient to conclude that customers in general preferred one type of soft drink over the other (even though twice as many customers preferred the asparagus juice).

APPLICATION The sign test is easy to use and has a wide variety of applications. This is primarily because it can be used for ranked scores and one need only designate the sign of the difference in a matched pair. Moreover, it can also be used for pairs of quantitative measurements, if one wishes to do so. However, as suggested earlier, it ignores the size of the differences, so it is not as powerful as a parametric test. That is, it will not be as likely to cause us to reject a false hypothesis as would a parametric test. Thus, the probability of committing a type II error in a given situation (accepting the null hypothesis when false), would likely be higher when using this nonparametric test.

EXERCISES I

1. Distinguish between parametric and nonparametric methods.
2. Inasmuch as nonparametric methods are often easier to use than parametric methods, why are they not used more extensively than parametric methods?
3. Which nonparametric test would be used to (a) test whether there is a significant difference between two sets of paired data, and (b) determine whether some sample data has been randomly selected from a population?
4. What is (a) a run and (b) the null hypothesis in a runs test?
5. A sportswear manufacturing plant has two large sewing rooms (A and B) where cloth goods are sewed together. The output is marked (A or B) and randomly placed onto a common conveyor belt that transports it into room C. A production control analyst in room C monitors the output on the conveyor belt to ensure it is coming steadily from both rooms. If the output is not randomly mixed, it signals him that a problem exists with the sewing machines in one of the rooms.

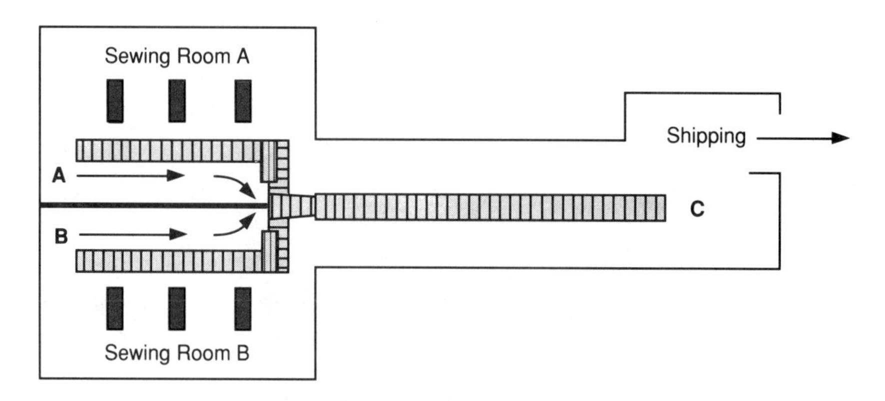

A recent check of two dozen items on the conveyor in room C yielded the following sequence.

B	B	B	A	A	A	B	A	A	A	B	B	B	B	A	B	B	A	A	A	B	A	A	A

Test whether the cloth goods are being randomly placed on the conveyor. Use the .25 level of significance and assume the normal approximation is adequate.

6. A receptionist at the Midvale city council office has recorded the following sequence of phone calls by residents expressing their favor (F) or opposition (O) to a real estate tax designed to support a new sports center.

F	F	O	O	O	O	O	F	F	F	F	O	O	O	O	F	F	O	O	F	F	F	F	F	O	O	O	O	O	O

The city manager wishes to learn if the calls are coming in randomly or if perhaps some special interest groups are sponsoring the calls. Use the runs test at the $\alpha = .20$ level to determine whether the preference of the callers is being expressed on a random basis or not.

7. A pharmaceutical company has developed a diet they claim will cause an overweight person to lose at least 10 pounds in one month. A sample of 14 subscribers to the plan reported the following weight losses:

12	8	14	11	9	4	18	22	15	13	0	12	19	11

Use the sign test at the .15 level to determine whether clients really lose an average of 10 pounds (or more), against the alternative that they lose less than 10 pounds

8. The median score of applicants on a qualifying exam for positions with an electronics company has been 82. However, the personnel director has recently revised a few of the questions on the exam and results from the latest group of applicants are:

Number scoring less than 82	5
Number scoring 82	4
Number scoring greater than 82	13

Use the sign test at the $\alpha = .02$ level to determine whether the median is still 82.

9. A business periodical surveyed twenty-five stock brokers to assess their feelings about the direction the interest rates would take depending upon whether a Republican or Democrat was appointed as chairman of the Federal Reserve Board. Results were as shown below where U represents interest rates will go up, D means they will go down.

DIRECTION PROJECTED FOR INTEREST RATES BY STOCK BROKERS

BROKER NO.	1	2	3	4	5	6	7	8	9	10	11	12	13	14	15	16	17	18	19	20	21	22	23	24	25
Republican	U	U	D	U	D	D	U	U	D	U	U	U	U	D	U	U	U	D	D	U	D	U	U	D	D
Democrat	D	D	D	D	U	D	D	U	U	D	D	D	D	U	D	D	D	U	U	D	D	D	D	U	U

Use the paired difference sign test to determine whether there is a real difference in broker expectations, depending upon whether a Republican or Democrat is appointed. Use the 8% level of significance.

10. Ten employees of a telemarketing concern were given a sales ability test before and after a two-day training program. Results were as shown below:

Employee	Score Before	Score After
1	74	76
2	62	77
3	88	85
4	76	92
5	59	64
6	73	73
7	90	94
8	83	82
9	84	95
10	78	86

Use the paired difference sign test to determine whether there was a real improvement in performance, at the .10 level of significance. (*Hint*: Use the binomial distribution.)

RANK SUM TEST OF MEANS

Thus far we have explored a nonparametric test for randomness (runs test), one for central tendency (one-sample sign test), and a test for differences in paired data (two-sample sign test). We now turn to a useful nonparametric method of testing whether two independent samples have been drawn from populations having the same mean.

The *rank sum test* is the nonparametric equivalent to the small sample t-test, and the corresponding large sample normal distribution test of the difference in means. As you may recall, the (parametric) t-tests assumed two normal populations with equal variances. The rank sum test makes no assumptions about the underlying distribution, but it does assume the variables are continuous and that the samples are independent.

A **rank sum test** is a test of the difference between the means (or medians) of two populations and uses rankings of the data as a basis for calculating the test statistic.

LOGIC OF TEST As with the parametric tests, we typically hypothesize that the two samples were drawn from the same population. Thus our hypothesis is that the means (or medians) are equal. For example:

Null hypothesis H_0: $\mu_1 = \mu_2$ (populations have same mean)

Alt. hypothesis H_1: $\mu_1 \neq \mu_2$ (populations have different means)

As with other tests, an α level is specified, and random samples are taken. To carry on from here, we use a form of the *Wilcoxon rank sum test*, which is one of several rank sum tests. For this test, the samples need not be the same size. But the data (i.e., the values obtained) from both samples must be ranked as if they were in a single group. All values are ranked from lowest to highest without regard to which group they are from—although their group identity is preserved. If the null hypothesis is true, the locations (means) of the two distributions are equal and they have the same shape (e.g., same skewness). We would expect to see both samples relatively equally represented in the high and low rankings. Figure 14–2 illustrates this situation.

FIGURE 14–2 Logic for a rank sum test of means

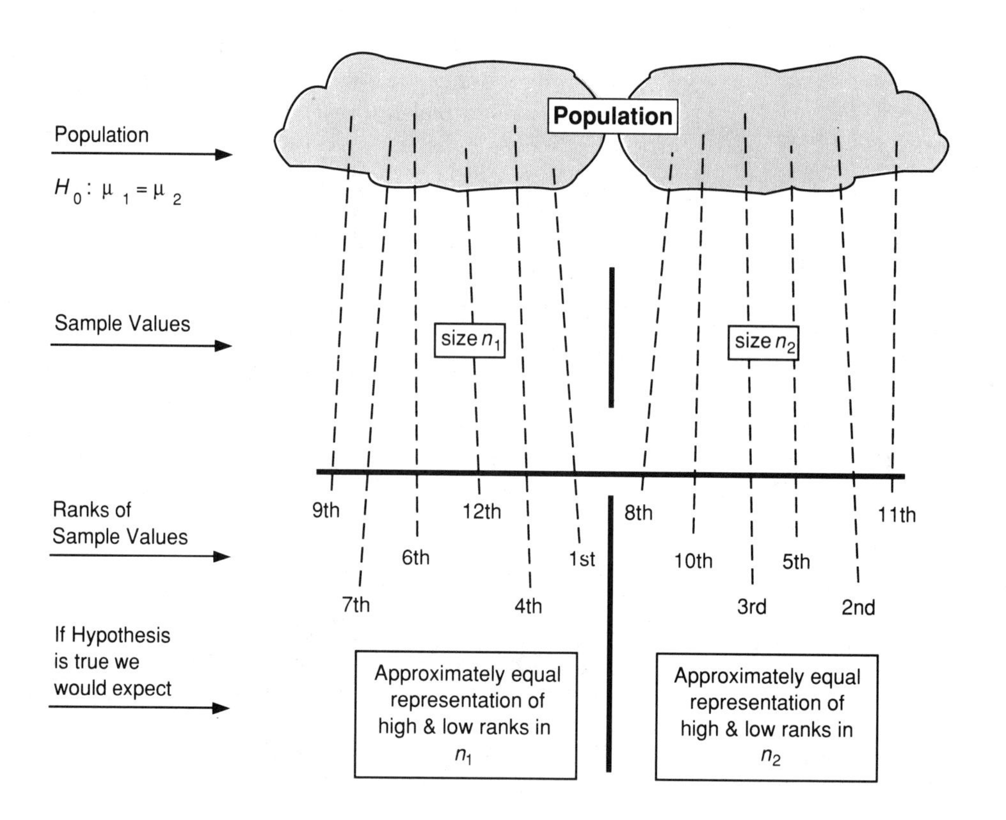

In Figure 14–2, both samples have some high and low rankings, which seems to support the null hypothesis that they are from the same population. If one of the samples has significantly more high (or low) rankings than the other, the rank sum test procedure will cause a rejection of the null hypothesis. The specific test procedure differs slightly for small versus large samples.

SMALL SAMPLES Suppose we let n_1 represent the number in the smaller sample, and n_2 the number in the larger one. The ranks assigned to elements from sample 1 are summed to yield ΣR_1, and the ranks of the data in sample 2 yield $\Sigma R_{\_}$. These ranked sums can then be used to calculate a statistic that can be compared with exact sampling distribution values that have previously been determined by accounting for all the equally likely combinations. For small samples, these predetermined values are available in reference texts on nonparametric statistics. The hypothesis is then rejected (or not) on the basis of this comparison.

LARGER SAMPLES If both n_1 and n_2 are larger than 10, the rank sum test statistic is approximately normally distributed with a mean of zero and standard deviation of one. The test statistic, Z_{calc}, is:

$$\text{Test Statistic:} \quad Z_{calc} = \frac{\Sigma R_1 - \mu_R}{\sigma_R} \tag{14–5}$$

In this expression, ΣR_1 is the sum of the ranks from sample 1, and μ_R and σ_R are the mean and standard deviation, respectively, of the sampling distribution of the rank sum ΣR_1. They are computed as follows:

$$\text{Mean of statistic:} \quad \mu_R = \frac{n_1(n_1 + n_2 + 1)}{2} \tag{14–6}$$

$$\text{Standard deviation:} \quad \sigma_R = \sqrt{\frac{n_1\, n_2(n_1 + n_2 + 1)}{12}} \tag{14–7}$$

The test is completed by calculating the test statistic (Z_{calc}) and determining whether it lies outside the normal distribution limits as specified by the level of significance, α.

Note that we have chosen to use ΣR_1 values for our equations as a matter of convention. The difference between ΣR_1 and the expected mean (μ_R) is the same as that between ΣR_2 and μ_R, because by whatever amount ΣR_1 is higher than the mean, then ΣR_2 is that much lower. For example, if the ΣR_1 is 35 units above μ_R, then the ΣR_2 must be 35 units below μ_R.

STRENGTH OF THE TEST Insofar as we are using information about the relative position of observations in a sample, the rank sum test is more powerful than a nonranking test such as the sign test, which takes only the direction (+ or −) into account. However, the rank sum test still carries a higher β risk than the parametric *t*-test, because the *t*-test makes use of additional information by taking specific values into account. We illustrate with an example from the personnel/production area.

Example 14–4 A North Carolina furniture factory has 30 workers in a workcenter, producing hand-finished maple chairs. To evaluate two proposed incentive pay plans, 15 workers have been randomly selected to participate under plan A, and the other 15 under plan B. The weekly production of each group (measured to the nearest tenth of a chair) was as shown below. Is there a difference in average production under the two programs? Use $\alpha = .01$

Production totals of Group A

70.4	81.5	69.0	55.8
44.5	78.0	56.9	62.3
52.9	66.0	70.5	48.6
62.0	45.4	50.0	

Production totals of Group B

61.0	72.5	70.7	64.8
68.5	76.4	69.0	57.0
57.0	73.5	77.0	69.0
76.0	48.5	59.4	

Solution

Before setting out the steps of the test, it is useful to arrange each group into an array (say from lowest to highest), so that ranks can be more easily assigned. Then assign ranks using data from both groups. If any ties occur, assign each a score equal to the mean of the ranks that would otherwise be used for the affected scores.

Group A		Group B	
Production	Rank	Production	Rank
44.5	1	48.5	3
45.4	2	57.0	9.5
48.6	4	57.0	9.5
50.0	5	59.4	11
52.9	6	61.0	12
55.8	7	64.8	15
56.9	8	68.5	17
62.0	13	69.0	19
62.3	14	69.0	19
66.0	16	70.7	23
69.0	19	72.5	24
70.4	21	73.5	25
70.5	22	76.0	26
78.2	29	76.4	27
81.5	30	77.0	28
Σ Ranks	197		268

1. H_0: $\mu_A = \mu_B$
   H_1: $\mu_A \neq \mu_B$
2. $\alpha = .01$
3. Use rank sum test. Reject if $Z_{calc} < -2.58$ or > 2.58

4. $Z_{calc} = \dfrac{\Sigma R_1 - \mu_R}{\sigma_R}$

where ΣR_1 will be the sum of ranks in group A = 197

$$\mu_R = \frac{n_1(n_1 + n_2 + 1)}{2} = \frac{15(15 + 15 + 1)}{2} = 232.5$$

$$\sigma_R = \sqrt{\frac{n_1\, n_2(n_1 + n_2 + 1)}{12}} = \sqrt{\frac{15(15)(15 + 15 + 1)}{12}} = 24.11$$

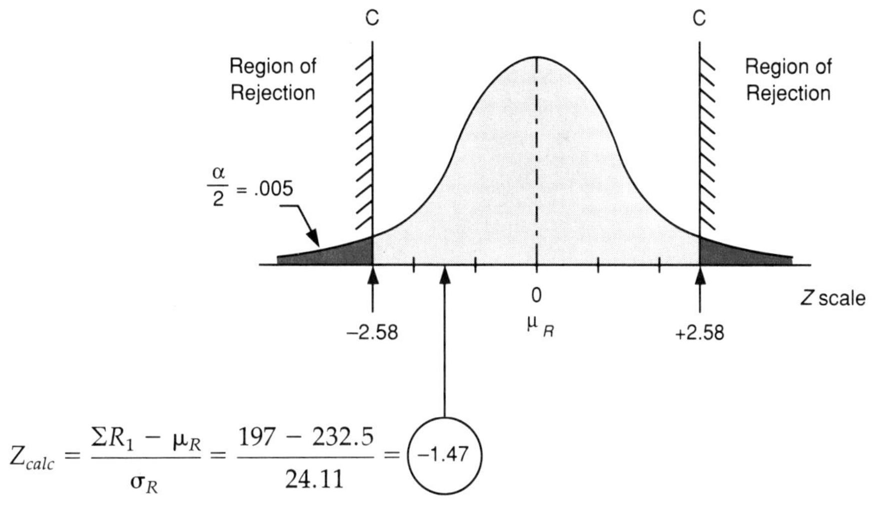

$$Z_{calc} = \frac{\Sigma R_1 - \mu_R}{\sigma_R} = \frac{197 - 232.5}{24.11} = -1.47$$

5. *Conclusion*: Do not reject H_0 of equal means. Using the rank sum test, the sums of the ranks are not sufficiently different to reject the hypothesis at the .01 level. [Note, however, that there is some evidence that pay plan B may be more motivating than plan A. That is, we could reject the hypothesis of equal means at the .10 level, where the critical Z value is -1.28.]

EXERCISES II

11. Concerning the rank sum test, (a) to what type of parametric test is it most similar—in terms of function, and (b) what assumptions are made about the underlying distribution?
12. Explain why the rank sum test is not as powerful a test as the two-sample *t*-test.
13. The financial manager of a chemical company must select a broker to manage the company's retirement funds. Two firms under consideration have agreed to give the manager some profitability data from a random sample of a dozen of their clients' accounts. Shown below are the profits gained over the past year per $1,000 invested for the 12 accounts from each firm. Use the rank sum test at the .05 level to determine whether there is a significant difference in the profitability obtained by the two firms.

Adkins-Stevens Brokers Inc.

Customer No.	1	2	3	4	5	6	7	8	9	10	11	12
Profit	80	206	120	45	308	−170	120	27	833	215	184	−88

Pfafner, Jeffrey, and Jones Investment Company

Customer No.	1	2	3	4	5	6	7	8	9	10	11	12
Profit	110	−30	220	78	155	389	120	327	295	67	156	145

14. Preliminary calculations for a rank sum test comparing the mean operating hours before failure of two competitive products revealed the following:

    $n_1 = 13 \qquad n_2 = 16 \qquad \Sigma R_1 = 153.0 \qquad \mu_R = 201.5 \qquad \sigma_R = 23.9$

    Complete the test at the .10 level and state whether the hypothesis of equal means should be rejected or not.
15. Each year, the American Cancer Society reports on cancer death rates in several countries as compiled in the World Health Statistics Annual a few years earlier. In *Cancer Statistics, 1985*, the estimated lung cancer death rates per 100,000 population in selected countries was estimated as shown below. Rank the data and use the rank sum test to test whether males have a significantly higher lung cancer death rate than females at the .01 level. (Hint: Use a one-tailed test.)

LUNG CANCER DEATH RATES IN SELECTED COUNTRIES

Country	Male	Female
Australia	61.6	10.3
Austria	67.6	8.7
Belgium	91.5	7.2
Canada	62.7	12.5
Denmark	62.0	15.2
England & Wales	96.5	19.4
Germany D.R.	67.0	6.0
Hong Kong	65.6	30.4
Hungary	65.0	11.2
Ireland	61.8	16.9
Netherlands	97.7	6.2
New Zealand	65.6	13.5
Philippines	8.7	3.6
Romania	37.7	7.2
Scotland	108.5	23.1
Spain	18.7	4.8
Thailand	4.9	1.9
United States	68.1	17.2

RANK CORRELATION

In Chapter 11 on regression and correlation, we saw that correlation was a technique for describing the strength of a relationship between two variables. In that chapter, we assumed a (bivariate) normal distribution of the X and Y variables, and used numerical values measured on an interval or ratio scale. Using equation (11-16), we were able to summarize the association between the variables in terms of a coefficient referred to as the Pearson coefficient of correlation, or more simply, r.

Unfortunately, as we have already noted, business situations do not always closely approximate the parametric assumptions such as those mentioned above. Observations frequently consist of judgmental, attitudinal, or behavioral measures that yield only a measure of preference (or rank), as opposed to a numerical score. For example, a purchasing agent may be unable to assign numerical values to each of her suppliers, although she may be able to rank them by order of preference (1, 2, 3, etc.). Or a financial manager seeking to invest some long-term funds may at best be able to compare his priority listing of stocks or bonds with the rankings recommended by his banker. These situations do, however, lend themselves to the use of nonparametric methods. More specifically, the rank correlation coefficient is a widely used measure of the degree of association between two sets of rankings.

Spearman's rank correlation coefficient is a measure of the closeness of the relationship between the ranked values (as opposed to the numerical values) of two variables.

Named after the statistician who developed it in the early 1900s, the rank correlation coefficient, r_r, relies only upon the squared difference in ranks, d, of each of the pairs of n observations. So the computation of r_r is really quite straightforward.

$$r_r = 1 - \frac{6\Sigma d^2}{n(n^2 - 1)} \tag{14–8}$$

As with the standard (Pearson) r, the rank correlation coefficient has the same range (−1.00 to +1.00) and meaning. We shall illustrate its computation by using the same example as used in Chapter 11, so that we can compare the resultant values. The difference here will be that instead of using the original X and Y values for our computation, we will use the ranks of the X and Y values. This means that the rank correlation coefficient will *reflect only the relative position of each observation*, and not the individual observations.

COMPUTING THE RANK CORRELATION COEFFICIENT

For purposes of illustration, we repeat the data from Example 11–9 below.

Example 14–5 A city homebuilders association has collected the data shown below in an attempt to relate the number of housing starts/month to the prevailing home loan interest rate at local banks. Use the data to compute the rank correlation coefficient.

Month	A	B	C	D	E	F	G	H	I	J
Interest Rate % (X)	18	12	8	7	14	9	17	20	16	10
No. Housing Starts (Y)	4	10	20	18	12	15	6	2	4	8

Solution

For r_r, we need ranked data only, so we shall assign ranks from 1 (lowest) to 10 (highest) for each of the two variables. Thus the rankings for month D will be first with respect to interest rates and ninth with respect to housing starts. The tie of four housing starts in months A and J will be handled by the same averaging procedure used in the rank sum tests. (A special procedure is available—in nonparametric texts—for situations with a large percentage of ties.) The computations are shown in Table 14–1.

TABLE 14–1 Rank correlation computation

Month	(1) Interest rate rank (X)	(2) Housing starts rank (Y)	(3) Difference (1) − (2)	(4) (Difference)² [(1) − (2)]²
A	9	2.5	6.5	42.25
B	5	6	−1	1
C	2	10	−8	64
D	1	9	−8	64
E	6	7	−1	1
F	3	8	−5	25
G	8	4	4	16
H	10	1	9	81
I	7	2.5	4.5	20.25
J	4	5	−1	1
			$\Sigma d^2 =$	315.50

$$r_r = 1 - \frac{6\Sigma d^2}{n(n^2 - 1)} = 1 - \frac{6(315.50)}{10(100 - 1)} = -.9121$$

The negative value suggests that high interest rates are associated with a low number of housing starts, and vice versa.

COMPARISON OF r AND r_r Note that the calculated r_r of $-.9121$ compares favorably with the $-.9003$ value of r, calculated from the actual values (and reported in the Chapter 11 example). And r_r is much simpler to calculate.

One reason the results are so close is because there were no extreme values, or "outliers" in the original data of Example 14–5. Unusually large or small values (say, two housing start values of 50 or 60) tend to have a distorting effect on the standard (Pearson) correlation coefficient. On the other hand, because the rank correlation coefficient uses relative position only, it is not affected by such extremes. Thus r_r can provide a close estimate of r in many situations, and is a preferable choice for situations with some outliers. When applied to ranked data only, however, the r and r_r equations result in the same value of the correlation coefficient. So if only ranked data are available, computing r_r would be preferred because of its ease of calculation.

TESTING THE CORRELATION COEFFICIENT

The r_r calculated from sample data can be used to test a hypothesis of no correlation between the two populations in much the same manner as was done with the sample correlation coefficient, r. Special tables are available (in reference texts) for small numbers of paired observations. However, if ten or more pairs of observations are available, a statistic for testing the H_0: $\rho_r = 0$ can be computed from the equation:

Test statistic for r_r

$$t_c = \frac{r_r - 0}{\sqrt{(1 - r_r^2) / (n - 2)}} \qquad (14\text{–}9)$$

This statistic has a t-distribution with $n - 2$ degrees of freedom. You may recognize that this is the same calculation as used in equation (11-18) for testing H_0: $\rho = 0$, except that in equation (14–9), r_r is substituted for r.

Example 14–6 Test the hypothesis of no correlation in the ranked data of Example (14–5) at the .02 level. [Note: For samples of $n = 10$, r_r was found to be $-.91$]

Solution

1. H_0: $\rho_r = 0$ (There is no correlation in the ranked data of the population.)
   H_1: $\rho_r \neq 0$ (There is significant correlation in the ranked data of the population.)
2. $\alpha = .02$
3. Use t, reject if $|t_c| > t_{\frac{.02}{2},8} = 2.896$
4. $t_c = \dfrac{r_r - 0}{\sqrt{(1 - r_r^2) / (n - 2)}} = \dfrac{-.91 - 0}{\sqrt{(1 - .91^2) / (10 - 2)}} = -6.21$
5. *Conclusion*: Reject hypothesis of no correlation in the ranked data. The ranked data strongly evidence a correlation.

ANALYSIS Rejecting the hypothesis of no correlation leads us to conclude that there is a close relationship between the number of housing starts per month and the prevailing home loan interest rate at local banks. Moreover, insofar as the sample correlation coefficient is negative, the two variables move in opposite directions. That is, low interest rates are associated with higher levels of housing starts and high interest rates with lower levels of housing starts.

EXERCISES III

16. How does the rank correlation coefficient r_r differ from the standard (Pearson) coefficent of correlation, r?
17. The following data show the rank of 15 job candidates on two employment test criteria. Compute the rank correlation coefficient, r_r.

Candidate no.	1	2	3	4	5	6	7	8	9	10	11	12	13	14	15
Dexterity	12	5	15	2	4	9	13	10	3	11	7	14	1	6	8
Math ability	2	8	7	13	10	4	1	6	15	9	5	3	12	14	11

18. The purchasing agent of a firm producing electronic components for a space shuttle for NASA would like to reduce the number of suppliers from a dozen down to half that many. She has asked one of her buyers to rate the 12 suppliers on the basis of quality of their components, and another to rate them on the basis of promptness in meeting delivery requirements. Using the results shown below, compute the rank correlation coefficient and test it at the .01 level of significance.

Supplier no.	**1**	**2**	**3**	**4**	**5**	**6**	**7**	**8**	**9**	**10**	**11**	**12**
Rank on Quality	4	2	7	9	10	3	8	12	11	6	1	5
Rank on Delivery	2	5	9	8	9	4	10	9	12	11	3	6

19. The product development department of a dairy products company has engaged two panels to help select which of ten new frozen dessert products to market. One panel consists of company personnel and the other of typical consumers. Each panel is asked to evaluate the market potential of the product on a score of 1 to 100. Results are as shown.

	Score (1 – 100) of Dessert from Panel of	
New Product	**Company Personnel**	**Typical Consumers**
A	68	74
B	92	75
C	77	58
D	45	52
E	75	48
F	83	63
G	90	68
H	86	68
I	72	30
J	95	82

(a) Calculate the rank correlation coefficient and comment on its meaning.

(b) Conduct a two-tailed test of the hypothesis that there is no correlation in the ranks assigned by the two panels at the .10 level of significance.

SUMMARY

Nonparametric methods are statistical procedures for testing and decision making that do not assume any underlying distributions and are generally much less restrictive than parametric methods. They typically use signs and ranking information only, which makes them easy to use and widely applicable. On the other hand, they are less powerful than parametric techniques because they do not make use of the interval scale measurements that lend more precision to the data.

Runs tests use counts of similar outcomes to test hypotheses that the sequence in which the data appear is truly random. The data must first be apportioned into groupings of two distinguishable categories so that the number of runs (of identical occurrences) can be determined. Then the actual test is usually based upon comparison to a normal distribution, because if n_1 and n_2 are both more than 10, the sampling distribution of the number of runs is approximately normal.

The *sign test* uses + and − signs to test the central tendency of a data set or the significance of the difference between two data sets. *One-sample tests* hypothesize some central value for the population, and assign a + or − sign to the sample values depending upon whether they are larger or smaller than the hypothesized value. The test consists of comparing the number of + or − signs to what would randomly occur, if the proportions should be equal at $\pi = .5$. For samples less than 20, the binomial distribution serves as the criteria, whereas for larger samples, the normal distribution is used. *Paired-samples tests* are similar except the + or − sign is assigned on the basis of whether one value of the pair is greater (+) or less (−) than the other.

The *rank sum test* is the nonparametric equivalent to the test of the difference between two means. However, it uses rankings of the data as a basis for calculating the test statistic instead of interval scale measurements. The data in both groups are ranked as if they were in a single group. Then the test statistic is computed by subtracting the mean of the rank sums from the rank sum of one of the groups, and dividing by the standard deviation of the sampling distribution. If this value lies outside the specified normal distribution limits, the hypothesis of equal means is rejected.

The *rank correlation test* is a method of measuring the correlation in a set of data by using only the ranks of the observations. It is especially useful for judgmental and behavioral measures. The test statistic computation is quite straightforward, with a result that is comparable to what would be obtained from the Pearson coefficient of correlation if its use were restricted to ranked data.

CASE: GLOBE PERSONNEL SERVICES—WE'LL FIND YOU A JOB!

It was only a part-time job at a local personnel services firm to help pay her college tuition. So Belinda Gregory didn't expect to contribute (or learn) a whole lot. The first day seemed to confirm that. She spent it stuffing envelopes and typing up some multiple choice questions for her supervisor, Dan Cruthers—the purpose of which she couldn't possibly imagine. For example, the first two questions were:

1. Managers get most of their exercise by:
(a) running the show
(b) jogging their memory
(c) jumping to conclusions
(d) skipping meetings
(e) none of the above

2. If you want a managerial position, you should first:
(a) change your age
(b) change your appearance
(c) become more assertive
(d) mind your own business
(e) more than one of the above

But Belinda knew better than to question Mr. Cruthers, especially on her first day. On day two he came to her with a photocopy in his hand.

DAN CRUTHERS: Those questions you did yesterday are part of an aptitude test I'm working on. I didn't mean to scare you off on your first day.

BELINDA: Oh, you didn't. But I must admit I was a little curious about them. Do you have something for me to do today?

DAN CRUTHERS: As a matter of fact I do. I got ahold of this survey published some time back in one of the journals we take here at the office.[2] It reports on 257 respondents in 55 firms ranging from manufacturing to public service. The employers were all asked to rank the characteristics they considered important in securing employment. And included with that is a ranking of the same characteristics from 250 students who were close to graduation—one year or less away.

BELINDA: Unfortunately I'm not in that category!

DAN CRUTHERS: I know—but I wonder if you might look it over to see if you can come up with some measure of how strongly both of these parties agree. If there is some validity to this, I'd like to build some of those terms into my aptitude quiz. This could help us get our clients into the right slots if there's anything to it. Does the word "nonparametric" mean anything to you?

BELINDA: Sure, I'll get right on it.

[2] Source: *Personnel Administrator*, March 1983.

The material Dan gave Belinda was as shown below.

Employer Rank	Characteristic	Student Rank
22	Age	19
10	Appearance	4
4	Assertiveness	9
21	Community Involvement	21
14	Disposition	13
8	Enthusiasm	3
15	Extroversion	18
25	Fraternal Organization	25
13	Grades	16
24	Hobbies	23
3	Initiative	9
20	Knowledge of Company	17
6	Leadership	12
5	Loyalty	9
16	Mannerisms	15
26	Marital Status	24
7	Maturity	7
2	Motivation	2
1	Oral Communication	1
9	Punctuality	11
18	School Reputation	20
19	Social Activities	22
23	Sports Activities	26
17	Willingness to Relocate	14
12	Work Experience	5
11	Written Communication	6

Case Questions

(a) Why would some form of nonparametric analysis be most appropriate for analyzing this data?

(b) What kind of nonparametric analysis would you recommend?

(c) Compute a measure of the closeness of the association between the employer and student rankings.

(d) Are the ranks assigned by the employers and by the students closer than what might be expected to occur by chance? Determine whether there is a significant association and state your results.

(e) Of the 26 characteristics mentioned, which two appear to be most important? How important to securing employment are fraternal memberships, hobbies, and marital status?

(f) For which characteristic are the employers and students in most disagreement?

(g) Suppose this same data were analyzed by its counterpart parametric technique. Would the result have been the same?

QUESTIONS AND PROBLEMS

20. Why are nonparametric tests referred to as "less powerful" than parametric tests?
21. Applicants for a summer travel scholarship to India are supposed to be evaluated and selected in order "without regard to race, sex, or religious preference." The thirty winners are announced in the following order where C = caucasian, N = noncaucasian, M = male, F = female. Test the race sequence for randomness at the .05 level.

Race	N	C	C	N	C	C	C	N	N	N	N	C	C	N	N	N	N	C	C	N	N	N	C	C	C	C	C	N	N	N
Sex	M	F	F	M	F	M	M	M	F	M	F	F	F	M	M	F	F	M	F	M	M	F	F	F	M	M	F	M	M	F

22. Using the data from the previous problem, test for whether the applicants were selected without regard to whether they were male or female. Use the .01 level of significance.
23. Negotiators at a collective bargaining session are attempting to agree on a cost of living allowance to be incorporated into the next wage agreement. Company officials claim that an historical average of a 3.1% increase would adequately represent the several items the firm uses to chart living expenses. However, union representatives say that their recent sample data suggests the increase should be more than 3.1%. Using a sign test at the .05 level, test whether the data below are sufficient to refute the claim that the average is 3.1%, against the alternative that it should be higher.

PERCENT INCREASE IN SAMPLE OF ITEMS

3.8	4.1	3.1	3.9	2.0	3.2	3.4	1.9	3.3	3.6	3.4	3.1	2.2	3.5	3.3

24. A computer chip manufacturing process has one machine that has been producing an average of 35 defective chips per day. The most recent sampling of defectives from this line is as shown:

45	33	17	53	28	61	35	48	37	21	44	37	31	49
42	22	47	19	61	50	38	44	52	20	36	45	35	52

Test whether the average is still 35 or whether it has changed. Let $\alpha = .20$.

25. The Operations Department of a major airline company has developed a 3-minute film about airline safety that it proposes to show all passengers on overseas flights. The film is designed to convince passengers that the airline is extremely safety conscious. To test the effectiveness of the film, 25 passengers were randomly selected to answer a questionnaire before seeing the film, watch the film, and then answer similar questions after seeing the film. The questionnaire results, where 0 = low and 10 = high concern for safety, are as shown below.

Passenger No	**1**	**2**	**3**	**4**	**5**	**6**	**7**	**8**	**9**	**10**	**11**	**12**
Score before film	6	7	9	5	3	8	8	5	6	7	2	8
Score after film	8	7	10	9	7	5	9	5	4	9	3	9

Passenger No	**13**	**14**	**15**	**16**	**17**	**18**	**19**	**20**	**21**	**22**	**23**	**24**	**25**
Score before film	7	6	8	4	3	7	6	5	7	9	6	8	4
Score after film	9	9	10	8	9	5	9	6	8	9	8	9	5

Is there a significant change in the passengers' opinions about the airline's safety practices as a result of watching the film? Use the $\alpha = .01$ level.

26. The marketing research department of an advertising agency is preparing an advertisement for a breakfast cereal. As part of their research, they have randomly selected several boxes of raisin bran cereal from two competitive products and measured the ounces of raisins per box with results as shown.

OUNCES OF RAISINS PER SERVING

Brand A	2.9	4.3	3.3	3.6	2.9	4.4	3.7	4.2	3.5	3.0	4.3	4.8		
Brand B	3.2	2.5	4.0	3.3	2.8	3.5	3.8	2.4	2.6	3.1	2.7	3.6	2.7	3.9

Conduct a two-tailed rank sum test of the hypothesis that the mean weights of raisins are equal in the two brands at the .10 level.

27. Economists in the international division of a major bank have devised an index to measure the level of productivity of firms in different locations but in comparable industries. For one segment of the paper manufacturing industry, a random sample revealed the following indexes for firms in the U.S. and a foreign country.

PRODUCTIVITY INDEX FOR INDIVIDUAL FIRMS

U.S. Locations	86.4	75.4	94.7	88.9	80.7	65.1	82.1	86.0	66.8	73.5	78.6
Foreign Locations	84.3	72.7	71.5	92.7	94.5	88.9	89.2	92.5	85.3	73.0	87.7

Use the Wilcoxon rank sum method to test whether the means are equal at the .025 level.

28. Use the rank sum method to test the equality of the mean scores assigned by company personnel and by consumers in problem 19. Let $\alpha = .05$.
29. Distinguish between the rank sum test and the test of the rank correlation coefficient.
30. A staff analyst working at the headquarters of a food processing company is evaluating the company's maintenance activities. She has collected the following data concerning the number of equipment breakdowns, and the preventive maintenance costs at several of the company's plants. Breakdowns are per month and maintenance costs are in $000 per month.

Plant	Maintenance Cost ($000)	Number of Breakdowns
Farmdale	2.8	8
Centerville	6.3	11
Battleground	5.0	6
Lester	2.7	13
Altona	12.4	4
Red Bluff	9.0	6
Opportunity	11.5	5
Mead Center	3.8	10
Indian Hills	7.8	6
Gold Creek	2.0	9
Sun River	12.0	2

(a) Calculate the rank correlation coefficient for these observations.

(b) Comment on the sign of the correlation coefficient.

(c) Is the correlation significant at the .05 level?

31. [Note: Requires computer program for rank sum test or can be done manually. Helps to have a sorting routine that can be applied to the database to isolate two groups of firms and rank the variables.] Use the data on number of employees in the retail and chemical industry firms as provided in the Corporate Statistical Database (Appendix A) for this problem. Assuming the data represent a random sample of firms in these industries, use the Wilcoxon rank sum test to test whether there is a difference in the average age of employees for firms in the retail versus the chemical and drug industries. Let $\alpha = .05$.
32. [Note: Requires computer program for rank correlation.] Using the data from the Corporate Statistical Database, determine whether there is a significant correlation between the average annual salary of employees (Slry), and the state's per capita income (ST $). Use all 100 firms of the database and a .05 level of significance.

Appendixes

A. CORPORATE STATISTICAL DATABASE
B. AREAS UNDER THE NORMAL CURVE
C. THE BINOMIAL DISTRIBUTION
D. THE POISSON DISTRIBUTION
E. STUDENT'S *t*-DISTRIBUTION
F. THE CHI-SQUARE DISTRIBUTION
G. THE *F*-DISTRIBUTION
H. ANSWERS TO ODD-NUMBERED PROBLEMS

APPENDIX A: Corporate Statistical Database

This database provides information on 100 business organizations as follows:

U = 20 utilities
T = 5 transportation firms
R = 15 retail firms
C = 20 chemical, drug firms
B = 10 banking, insurance firms
E = 30 electronics firms

The data have been disguised and/or modified and are not intended to reflect actual corporate statistics, although they are representative of a cross section of firms. The content of each column of the database is as listed below:

Col. No.	Designation	Content
1	No	Number assigned to firm
2	Corp. Name	Corporate Name
3	I	Industry Code (as designated above)
4	OP	Years of Operation
5	Sales	Annual Sales ($m)
6	Exp	Exports as % of Sales
7	Sh	Price per Share ($)
8	EPS2	Earnings per Share 2 years ago
9	EPS1	Earnings per Share 1 year ago
10	EPSC	Earnings per Share current year
11	PE	Price/Earnings Ratio = Sh ÷ EPSC
12	Div$	Annual Dividend ($)
13	P	Profit Sharing? (yes or no)
14	Ad Ex	Advertising and Public Relations Expense ($m)
15	Slry	Average Annual Salary of Employees ($000)
16	AG	Average Employee Age
17	No Emp	Number of Employees
18	HQ	Corporate State Headquarters
19	ST $	State's per Capita Income ($ annual)

No	Corp Name	I	OP	Sales	Exp	Sh	EPS2	EPS1	EPSC	PE	Div$	P	Ad Ex	Slry	AG	No Emp	HQ	ST $
1	Mountain Gas	U	22	399	0	29	4.07	3.90	4.04	7	2.34	N	2.79	20	51	2577	ND	6400
2	Mohawk Power	U	44	489	0	27	3.97	4.22	3.13	9	2.60	N	4.89	22	38	2575	MT	6600
3	Mountain P & L	U	60	567	0	24	2.78	3.04	2.98	8	1.50	N	4.82	20	45	3040	UT	6300
4	Portland Electric	U	54	585	0	15	2.54	2.31	2.21	7	1.80	N	2.93	22	41	3240	OR	7600
5	Puget Electric	U	24	519	0	15	2.29	1.69	1.92	8	1.50	N	3.12	25	49	2380	WA	8100
6	Utah Gas & Elec	U	56	855	0	23	2.38	2.46	2.26	10	0.90	N	9.49	20	45	4575	UT	6300
7	Centre Electric	U	61	715	0	17	1.81	2.33	1.95	9	1.61	N	5.72	25	42	2900	IL	8100
8	Cleveland Gas	U	47	1450	0	15	2.75	2.94	2.35	6	2.16	N	13.05	22	47	4800	OH	7300
9	Common Electric	U	71	4900	0	27	3.75	4.39	4.30	6	3.00	Y	54.88	25	41	17757	IL	8100
10	Day Power & Lt	U	73	1040	0	16	2.65	2.80	2.20	7	2.00	N	6.80	23	39	3000	OH	7300
11	Denton Electric	U	83	2530	0	15	1.75	2.21	2.15	7	1.68	N	50.60	22	36	11152	MI	7700
12	Great States Util	U	59	1560	0	13	1.95	2.13	2.30	6	1.60	N	12.48	31	38	4958	TX	7200
13	Hector Industries	U	8	4200	0	21	3.77	3.54	3.42	6	2.25	N	46.20	26	29	10700	TX	7200
14	Interlake Power	U	61	1290	0	22	3.04	3.80	4.25	5	2.50	N	14.95	23	34	3969	IL	8100
15	Isoquant Power	U	52	460	0	31	3.42	4.17	4.60	7	3.00	N	4.58	24	46	2300	IN	7100
16	Kanton City Gas	U	75	430	0	27	3.00	3.08	3.20	8	2.27	N	5.16	22	31	2015	KS	7400
17	Lakeville Gas	U	71	700	0	35	2.21	2.87	2.95	12	2.32	N	7.20	21	35	3499	KY	6000
18	Northeast Power	U	62	1800	0	42	4.79	5.60	5.85	7	4.00	N	25.20	23	47	6450	MN	7500
19	Oklahoma P & L	U	82	1040	0	28	2.57	2.62	2.55	11	1.85	N	10.38	20	34	3850	OK	6900
20	Texas Group	U	19	3900	0	26	3.85	3.90	3.21	8	2.85	N	43.30	21	39	16240	TX	7200
21	Alkan Airlines	T	50	4110	0	28	–0.99	4.79	4.00	7	0.00	Y	120.87	21	41	35500	TX	7200
22	Overland Airways	T	41	480	0	9	0.77	0.17	1.45	6	0.00	Y	9.12	23	29	3980	MO	6900
23	Northern Air Lines	T	4	2050	0	38	0.23	–2.00	5.00	8	0.90	N	34.76	21	37	11225	MN	7500
24	Continental Truck	T	55	1200	0	23	2.04	2.43	2.70	9	0.88	N	24.13	22	35	24400	CA	8300
25	Burgandy Rail	T	23	4508	0	21	2.28	2.08	3.70	6	1.10	Y	85.65	23	47	35721	WA	8100
26	Norsteads	R	45	788	0	35	1.50	1.51	2.15	16	0.50	Y	32.31	17	36	8400	WA	8100
27	Pacific Stores	R	53	833	0	21	1.54	1.07	0.81	26	0.10	N	30.80	17	35	5900	WA	8100
28	Pay Little Stores	R	79	1202	0	22	1.52	2.02	1.25	18	0.55	N	57.69	16	37	11543	WA	8100
29	Albi Dept Stores	R	55	4040	0	48	4.41	6.15	6.90	7	1.80	Y	161.60	17	38	62540	NY	7500
30	Angus Stores	R	22	800	0	26	1.00	1.54	2.00	13	0.14	Y	29.20	18	40	9200	CT	8500
31	The Trend Setter	R	12	525	0	20	2.33	2.52	2.30	9	0.40	N	26.25	17	31	8700	CA	8300
32	Hadrians	R	25	480	0	10	0.60	1.03	0.95	11	0.37	N	30.65	17	33	6800	WV	6100
33	Jasper Fashions	R	18	475	22	42	1.20	1.77	2.00	21	0.00	Y	19.47	18	35	5300	NJ	8100
34	Magic-City Inc	R	74	4700	0	40	3.25	4.32	4.95	8	1.30	Y	202.15	19	35	58000	MO	6900
35	Merchants Leigh	R	65	1765	0	53	4.74	5.65	6.20	9	1.00	N	72.36	16	39	21100	NY	7500
36	Lucy May Stores	R	65	4065	0	49	2.74	3.72	4.10	12	1.00	N	154.00	17	43	49760	NY	7500
37	Pepperbox Stores	R	52	920	0	35	2.10	2.29	2.65	13	1.10	N	3.55	17	31	9010	NJ	8100
38	Pier Five Imports	R	6	255	0	9	0.67	2.82	2.25	4	0.00	Y	10.46	16	33	2530	TX	7200

No	Corp Name	I	OP	Sales	Exp	Sh	EPS2	EPS1	EPSC	PE	Div$	P	Ad Ex	Slry	AG	No Emp	HQ	ST $
39	Stop & Go Shops	R	24	3200	0	50	3.29	4.10	5.00	10	1.30	N	121.60	17	28	35470	MA	7500
40	Xron Stores	R	25	2985	0	47	2.05	3.19	4.05	12	2.00	N	120.70	16	31	38580	MA	7500
41	Allied Products	C	58	4750	30	51	3.59	4.00	4.30	12	2.64	N	140.92	16	46	54680	NY	7500
42	Boundary Mining	C	84	4200	30	50	2.59	3.00	3.45	14	1.50	Y	130.20	22	37	14560	NY	7500
43	Eastern Chem	C	83	3100	33	63	5.42	6.13	9.65	7	3.00	N	96.10	24	39	29200	IN	7100
44	Millwood Ltd	C	23	3535	43	89	5.61	6.10	6.75	13	3.00	Y	98.98	23	40	32620	NJ	8100
45	Periodic Chem Co	C	42	3900	52	40	2.13	2.73	3.10	13	1.35	Y	128.76	21	37	40750	NY	7500
46	Rainland Mines	C	4	1282	30	29	2.74	2.10	2.69	11	1.50	Y	39.90	25	40	11123	CT	8500
47	Rome & Richards	C	16	530	28	29	1.82	2.02	2.05	14	1.00	Y	20.14	21	42	4580	PA	7100
48	Sand Petroleum	C	76	1235	0	59	2.76	2.39	3.25	18	0.50	Y	46.93	23	45	10200	IL	8100
49	Shoshone Falls Co	C	17	1885	38	52	3.01	3.31	3.70	14	2.40	Y	60.32	21	37	24100	NY	7500
50	Union City Co	C	82	2170	15	66	4.18	5.28	5.40	12	4.00	Y	71.61	20	36	21425	NJ	8100
51	Warner-Massey	C	64	3180	33	35	2.05	2.51	2.85	12	1.50	Y	98.58	22	39	42380	NJ	8100
52	Hammer Powder	C	72	2600	35	34	1.97	2.76	4.90	7	1.46	N	8.32	21	43	24200	DE	7500
53	Mocha Chemical	C	51	6700	27	45	4.24	4.72	5.75	8	2.18	Y	194.31	20	44	42330	MO	6900
54	Ortega Products	C	95	2060	40	34	2.66	3.01	3.75	9	1.40	N	63.58	24	47	18705	CT	8500
55	Stone Mtn Chem	C	14	1506	12	17	3.06	-0.03	1.12	15	0.00	Y	28.62	28	42	9752	CT	8500
56	Regency Group	C	7	2035	35	61	2.92	5.33	6.86	9	5.00	N	50.87	23	43	11450	PA	7100
57	Bauer Chemical	C	27	310	15	31	1.93	2.08	2.35	13	1.10	N	8.68	21	44	3890	PA	7100
58	Chem Corp Am	C	14	350	13	29	1.79	2.15	2.45	12	1.48	Y	10.01	20	36	4252	OH	7300
59	Crater Lake Chem	C	36	245	19	20	1.05	1.81	2.40	8	1.00	Y	0.65	28	38	2400	NY	7500
60	Dewey Labs	C	60	704	26	25	2.44	2.40	2.69	9	1.25	N	18.58	21	39	6713	MN	7500
61	Chem Trust Co	B	17	1350	0	49	4.02	4.56	3.70	13	2.14	N	14.85	19	29	7600	NJ	8100
62	Cork Trust Co	B	16	2600	0	32	2.04	-0.07	1.45	22	2.60	Y	31.20	18	31	19200	NY	7500
63	Safeguard Insur	B	55	153	0	34	2.86	3.56	2.86	12	1.10	N	3.21	19	44	10305	WA	8100
64	Radium Bank	B	57	47	0	31	4.12	4.90	5.75	5	2.20	N	0.42	17	34	5422	WA	8100
65	Penny Bank	B	95	6	0	20	3.33	1.56	1.50	13	1.00	N	0.07	18	35	2209	WA	8100
66	Fire Insur Am	B	50	873	0	63	3.67	4.48	6.05	10	0.75	N	5.98	20	45	4743	DC	9000
67	Ohio Insurance	B	15	842	0	42	4.91	4.85	2.50	17	2.50	N	7.57	18	41	5300	OH	7300
68	St. Louis Bank	B	95	1744	0	48	9.23	6.03	2.80	17	2.00	N	6.97	19	43	9870	MN	7500
69	Central Bank Corp	B	65	41	0	23	3.04	2.67	3.00	8	1.80	N	0.49	17	32	3518	MO	6900
70	American Trust	B	12	120	0	51	6.05	5.49	7.85	6	2.88	N	1.22	16	39	3906	OH	7500
71	Adv Computer	E	12	235	16	19	0.43	0.98	1.60	12	0.32	Y	2.35	23	29	4300	NY	7500
72	Wesland Labs	E	18	360	42	13	-0.45	1.05	1.50	9	0.00	N	4.32	25	31	1965	CA	8300
73	Advtech	E	15	1025	0	32	0.39	1.23	2.85	11	0.00	N	9.15	25	33	10680	CA	8300
74	Antigua Lazer	E	19	313	40	12	0.41	0.73	1.35	9	0.00	N	2.96	23	37	1050	MA	7500
75	Arrow Edwards	E	38	765	0	15	-1.20	0.85	2.10	7	0.20	N	9.18	30	40	1985	CT	8500

No	Corp Name	I	OP	Sales	Exp	Sh	EPS2	EPS1	EPSC	PE	Div$	P	Ad Ex	Slry	AG	No Emp	HQ	ST $
76	Ashton Software	E	19	155	11	22	0.60	0.63	0.90	24	0.00	Y	1.24	28	31	2230	CA	8300
77	Battle Creek Inc	E	32	312	0	24	1.70	1.50	2.21	11	0.35	N	1.25	23	42	4005	CA	8300
78	Fort Wayne Instr	E	61	1060	16	22	3.34	1.16	1.40	16	1.00	Y	4.51	21	31	26700	NY	7500
79	Genesee Ind	E	69	260	24	16	0.63	1.25	1.15	14	0.40	Y	0.08	21	41	3059	MA	7500
80	Geologic Systems	E	36	1550	0	23	2.10	1.75	1.90	12	0.60	Y	10.82	25	43	20550	IL	8100
81	International Chip	E	14	1650	28	58	-0.33	1.02	2.15	27	0.00	Y	14.85	22	41	21500	CA	8300
82	National Chip Co	E	7	1655	30	12	-0.20	-0.16	0.66	18	0.00	N	24.82	25	33	32700	CA	8300
83	Techdyne	E	24	385	35	26	0.24	1.01	1.75	15	0.00	Y	2.86	24	43	3900	MA	7500
84	Thomas Electric	E	60	330	0	36	1.52	1.74	2.50	14	1.00	N	39.60	22	39	3420	NJ	8100
85	Vinyard Mtn Ltd	E	3	929	20	39	1.48	2.01	2.72	14	0.00	Y	8.73	21	29	13670	CA	8300
86	Zag Electric	E	61	1685	0	24	-1.10	2.11	2.45	10	1.00	N	18.54	23	40	30242	IL	8100
87	Tarawa Electron	E	7	1331	55	65	4.25	2.57	4.44	15	3.50	N	0.88	22	38	20693	OR	7600
88	Apolo Electronics	E	17	598	10	17	2.23	0.72	0.91	19	0.27	N	6.15	21	35	2700	IL	8100
89	Badger Electric	E	64	185	5	19	1.33	1.20	1.55	12	0.30	N	2.04	19	36	3985	AR	5600
90	Cosmos Systems	E	48	1750	60	62	1.92	3.50	4.95	13	2.00	Y	24.50	18	36	27600	NY	7500
91	Designtronics	E	39	625	0	10	1.37	1.00	0.95	11	0.00	N	0.85	21	31	10560	VA	7500
92	Ellington	E	94	4200	24	70	4.37	4.42	6.10	11	2.50	Y	54.60	21	34	48760	MO	6900
93	Erromatics	E	15	260	0	14	1.42	0.78	1.20	12	0.80	N	2.86	23	39	3118	IL	8100
94	Federal Signal Co	E	31	210	23	26	1.85	3.45	5.34	5	0.00	N	0.95	31	36	1850	WA	8100
95	General Service	E	80	1810	10	47	3.85	3.16	3.85	12	2.10	N	15.80	25	31	23530	CT	8500
96	Grandview Ltd	E	56	1065	25	57	3.50	3.56	4.80	12	1.25	N	12.78	24	29	5970	IL	8100
97	Guild Industries	E	16	150	15	15	-1.20	1.33	1.85	8	0.60	Y	0.62	23	33	2669	NJ	8100
98	Jordan Controls	E	32	1425	30	42	3.83	4.17	4.71	9	1.80	Y	13.68	23	43	20700	WI	7200
99	Mayfield Electric	E	58	2315	0	35	3.09	1.92	3.75	9	2.00	N	19.68	24	39	28000	IL	8100
100	Scotch Electric	E	81	1395	9	38	2.61	2.12	3.80	10	1.00	Y	12.55	24	41	21320	IL	8100

APPENDIX B: Areas Under The Normal Curve

Values in the table represent the proportion of area under the normal curve between the mean ($\mu = 0$) and a positive value of z.

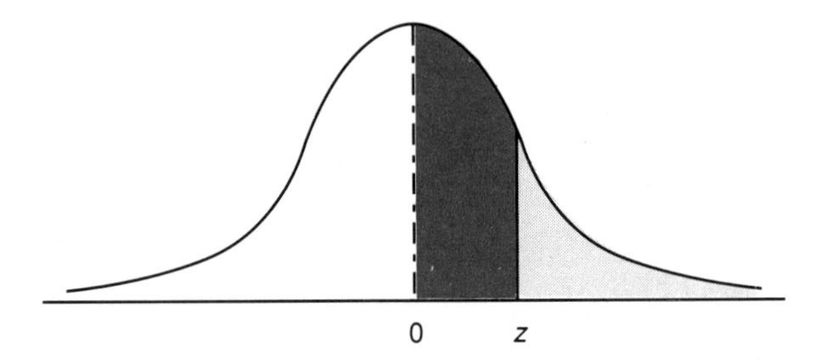

z	.00	.01	.02	.03	.04	.05	.06	.07	.08	.09
.0	.0000	.0040	.0080	.0120	.0160	.0199	.0239	.0279	.0319	.0359
.1	.0398	.0438	.0478	.0517	.0557	.0596	.0636	.0675	.0714	.0753
.2	.0793	.0832	.0871	.0910	.0948	.0987	.1026	.1064	.1103	.1141
.3	.1179	.1217	.1255	.1293	.1331	.1368	.1406	.1443	.1480	.1517
.4	.1554	.1591	.1628	.1664	.1700	.1736	.1772	.1808	.1844	.1879
.5	.1915	.1950	.1985	.2019	.2054	.2088	.2123	.2157	.2190	.2224
.6	.2257	.2291	.2324	.2357	.2389	.2422	.2454	.2486	.2517	.2549
.7	.2580	.2611	.2642	.2673	.2703	.2734	.2764	.2794	.2823	.2852
.8	.2881	.2910	.2939	.2967	.2995	.3023	.3051	.3078	.3106	.3133
.9	.3159	.3186	.3212	.3238	.3264	.3289	.3315	.3340	.3365	.3389
1.0	.3413	.3438	.3461	.3485	.3508	.3531	.3554	.3577	.3599	.3621
1.1	.3643	.3665	.3686	.3708	.3729	.3749	.3770	.3790	.3810	.3830
1.2	.3849	.3869	.3888	.3907	.3925	.3944	.3962	.3980	.3997	.4015
1.3	.4032	.4049	.4066	.4082	.4099	.4115	.4131	.4147	.4162	.4177
1.4	.4192	.4207	.4222	.4236	.4251	.4265	.4279	.4292	.4306	.4319
1.5	.4332	.4345	.4357	.4370	.4382	.4394	.4406	.4418	.4429	.4441
1.6	.4452	.4463	.4474	.4484	.4495	.4505	.4515	.4525	.4535	.4545
1.7	.4554	.4564	.4573	.4582	.4591	.4599	.4608	.4616	.4625	.4633
1.8	.4641	.4649	.4656	.4664	.4671	.4678	.4686	.4693	.4699	.4706
1.9	.4713	.4719	.4726	.4732	.4738	.4744	.4750	.4756	.4761	.4767

z	.00	.01	.02	.03	.04	.05	.06	.07	.08	.09
2.0	.4772	.4778	.4783	.4788	.4793	.4798	.4803	.4808	.4812	.4817
2.1	.4821	.4826	.4830	.4834	.4838	.4842	.4846	.4850	.4854	.4857
2.2	.4861	.4864	.4868	.4871	.4875	.4878	.4881	.4884	.4887	.4890
2.3	.4893	.4896	.4898	.4901	.4904	.4906	.4909	.4911	.4913	.4916
2.4	.4918	.4920	.4922	.4925	.4927	.4929	.4931	.4932	.4934	.4936
2.5	.4938	.4940	.4941	.4943	.4945	.4946	.4948	.4949	.4951	.4952
2.6	.4953	.4955	.4956	.4957	.4959	.4960	.4961	.4962	.4963	.4964
2.7	.4965	.4966	.4967	.4968	.4969	.4970	.4971	.4972	.4973	.4974
2.8	.4974	.4975	.4976	.4977	.4977	.4978	.4979	.4979	.4980	.4981
2.9	.4981	.4982	.4982	.4983	.4984	.4984	.4985	.4985	.4986	.4986
3.0	.4987	.4987	.4987	.4988	.4988	.4989	.4989	.4989	.4990	.4990

Source: Table available from National Bureau of Standards, *Tables of Normal Probability Functions*, Applied Mathematics Series 23, U.S. Department of Commerce, 1953.

APPENDIX C: The Binomial Distribution

$$P(X \mid n, \pi) = \frac{n!}{X!\,(n-X)!}\pi^{x}(1-\pi)^{n-x}$$

For sample data replace π with p. For values of $\pi > .5$, find $P(X)$ by substituting $(1 - \pi)$ for π. Then look up $P(n - X)$. For example:

$$P(X = 2 \mid n = 8, \pi = .6) = P[(n - X) \mid n, (1 - \pi) = P(6|8, .4) = .0413$$

						π					
n	**x**	**.05**	**.10**	**.15**	**.20**	**.25**	**.30**	**.35**	**.40**	**.45**	**.50**
1	0	.9500	.9000	.8500	.8000	.7500	.7000	.6500	.6000	.5500	.5000
	1	.0500	.1000	.1500	.2000	.2500	.3000	.3500	.4000	.4500	.5000
2	0	.9025	.8100	.7225	.6400	.5625	.4900	.4225	.3600	.3025	.2500
	1	.0950	.1800	.2550	.3200	.3750	.4200	.4550	.4800	.4950	.5000
	2	.0025	.0100	.0225	.0400	.0625	.0900	.1225	.1600	.2025	.2500
3	0	.8574	.7290	.6141	.5120	.4219	.3430	.2746	.2160	.1664	.1250
	1	.1354	.2430	.3251	.3840	.4219	.4410	.4436	.4320	.4084	.3750
	2	.0071	.0270	.0574	.0960	.1406	.1890	.2389	.2880	.3341	.3750
	3	.0001	.0010	.0034	.0080	.0156	.0270	.0429	.0640	.0911	.1250
4	0	.8145	.6561	.5220	.4096	.3164	.2401	.1785	.1296	.0915	.0625
	1	.1715	.2916	.3685	.4096	.4219	.4116	.3845	.3456	.2995	.2500
	2	.0135	.0486	.0975	.1536	.2109	.2646	.3105	.3456	.3675	.3750
	3	.0005	.0036	.0115	.0256	.0469	.0756	.1115	.1536	.2005	.2500
	4	.0000	.0001	.0005	.0016	.0039	.0081	.0150	.0256	.0410	.0625
5	0	.7738	.5905	.4437	.3277	.2373	.1681	.1160	.0778	.0503	.0312
	1	.2036	.3280	.3915	.4096	.3955	.3602	.3124	.2592	.2059	.1562
	2	.0214	.0729	.1382	.2048	.2637	.3087	.3364	.3456	.3369	.3125
	3	.0011	.0081	.0244	.0512	.0879	.1323	.1811	.2304	.2757	.3125
	4	.0000	.0004	.0022	.0064	.0146	.0284	.0488	.0768	.1128	.1562
	5	.0000	.0000	.0001	.0003	.0010	.0024	.0053	.0102	.0185	.0312
6	0	.7351	.5314	.3771	.2621	.1780	.1176	.0754	.0467	.0277	.0156
	1	.2321	.3543	.3993	.3932	.3560	.3025	.2437	.1866	.1359	.0938
	2	.0305	.0984	.1762	.2458	.2966	.3241	.3280	.3110	.2780	.2344
	3	.0021	.0146	.0415	.0819	.1318	.1852	.2355	.2765	.3032	.3125
	4	.0001	.0012	.0055	.0154	.0330	.0595	.0951	.1382	.1861	.2344
	5	.0000	.0001	.0004	.0015	.0044	.0102	.0205	.0369	.0609	.0938
	6	.0000	.0000	.0000	.0001	.0002	.0007	.0018	.0041	.0083	.0516

The Binomial Distribution (cont.)

		π									
n	*x*	.05	.10	.15	.20	.25	.30	.35	.40	.45	.50
7	0	.6983	.4783	.3206	.2097	.1335	.0824	.0490	.0280	.0152	.0078
	1	.2573	.3720	.3960	.3670	.3115	.2471	.1848	.1306	.0872	.0547
	2	.0406	.1240	.2097	.2753	.3115	.3177	.2985	.2613	.2140	.1641
	3	.0036	.0230	.0617	.1147	.1730	.2269	.2679	.2903	.2918	.2734
	4	.0002	.0026	.0109	.0287	.0577	.0972	.1442	.1935	.2388	.2734
	5	.0009	.0002	.0012	.0043	.0115	.0250	.0466	.0774	.1172	.1641
	6	.0000	.0000	.0001	.0004	.0013	.0036	.0084	.0172	.0320	.0547
	7	.0000	.0000	.0000	.0000	.0001	.0002	.0006	.0016	.0037	.0078
8	0	.6634	.4305	.2725	.1678	.1001	.0576	.0319	.0168	.0084	.0039
	1	.2793	.3826	.3847	.3355	.2670	.1977	.1373	.0896	.0548	.0312
	2	.0515	.1488	.2376	.2936	.3115	.2965	.2587	.2090	.1569	.1094
	3	.0054	.0331	.0839	.1468	.2076	.2541	.2786	.2787	.2568	.2188
	4	.0004	.0046	.0815	.0459	.0865	.1361	.1875	.2322	.2627	.2734
	5	.0000	.0004	.0026	.0092	.0231	.0467	.0808	.1239	.1719	.2188
	6	.0000	.0000	.0002	.0011	.0038	.0100	.0217	.0413	.0703	.1094
	7	.0000	.0000	.0000	.0001	.0004	.0012	.0033	.0079	.0164	.0312
	8	.0000	.0000	.0000	.0000	.0000	.0001	.0002	.0007	.0017	.0039
9	0	.6302	.3874	.2316	.1342	.0751	.0404	.0207	.0101	.0046	.0020
	1	.2985	.3874	.3679	.3020	.2253	.1556	.1004	.0605	.0339	.0176
	2	.0629	.1722	.2597	.3020	.3003	.2668	.2162	.1612	.1110	.0703
	3	.0077	.0446	.1069	.1762	.2336	.2668	.2716	.2508	.2119	.1641
	4	.0006	.0074	.0283	.0661	.1168	.1715	.2194	.2508	.2600	.2461
	5	.0000	.0008	.0050	.0165	.0389	.0735	.1181	.1672	.2128	.2461
	6	.0000	.0001	.0006	.0028	.0087	.0210	.0424	.0743	.1160	.1641
	7	.0000	.0000	.0000	.0003	.0012	.0039	.0098	.0212	.0407	.0703
	8	.0000	.0000	.0000	.0000	.0001	.0004	.0013	.0035	.0083	.0716
	9	.0000	.0000	.0000	.0000	.0000	.0000	.0001	.0003	.0008	.0020
10	0	.5987	.3487	.1969	.1074	.0563	.0282	.0135	.0060	.0025	.0010
	1	.3151	.3874	.3474	.2684	.1877	.1211	.0725	.0403	.0207	.0098
	2	.0746	.1937	.2759	.3020	.2816	.2335	.1757	.1209	.0763	.0439
	3	.0105	.0574	.1298	.2013	.2503	.2668	.2522	.2150	.1665	.1172
	4	.0010	.0112	.0401	.0881	.1460	.2001	.2377	.2508	.2384	.2051
	5	.0001	.0015	.0085	.0264	.0584	.1029	.1536	.2007	.2340	.2461
	6	.0000	.0001	.0012	.0055	.0162	.0368	.0689	.1115	.1596	.2051
	7	.0000	.0000	.0001	.0008	.0031	.0090	.0212	.0425	.0746	.1172
	8	.0000	.0000	.0000	.0001	.0004	.0014	.0043	.0106	.0229	.0439
	9	.0000	.0000	.0000	.0000	.0000	.0001	.0005	.0016	.0042	.0098
	10	.0000	.0000	.0000	.0000	.0000	.0000	.0000	.0001	.0003	.0010

The Binomial Distribution (cont.)

		π									
n	*x*	.05	.10	.15	.20	.25	.30	.35	.40	.45	.50
11	0	.5688	.3138	.1673	.0859	.0422	.0198	.0088	.0036	.0014	.0005
	1	.3293	.3835	.3248	.2362	.1549	.0932	.0518	.0266	.0125	.0054
	2	.0867	.2131	.2866	.2953	.2581	.1998	.1395	.0887	.0513	.0269
	3	.0137	.0710	.1517	.2215	.2581	.2568	.2254	.1774	.1259	.0806
	4	.0014	.0158	.0536	.1107	.1721	.2201	.2428	.2365	.2060	.1611
	5	.0001	.0025	.0132	.0388	.0803	.1321	.1830	.2207	.2360	.2256
	6	.0000	.0003	.0023	.0097	.0268	.0566	.0985	.1471	.1931	.2256
	7	.0000	.0000	.0003	.0017	.0064	.0173	.0379	.0701	.1128	.1611
	8	.0000	.0000	.0000	.0002	.0011	.0037	.0102	.0234	.0462	.0806
	9	.0000	.0000	.0000	.0000	.0001	.0005	.0018	.0052	.0126	.0269
	10	.0000	.0000	.0000	.0000	.0000	.0000	.0002	.0007	.0021	.0054
	11	.0000	.0000	.0000	.0000	.0000	.0000	.0000	.0000	.0002	.0005
12	0	.5404	.2824	.1422	.0687	.0317	.0138	.0057	.0022	.0008	.0002
	1	.3413	.3766	.3012	.2062	.1267	.0712	.0368	.0174	.0075	.0029
	2	.0988	.2301	.2924	.2835	.2323	.1678	.1088	.0639	.0339	.0161
	3	.0173	.0852	.1720	.2362	.2581	.2397	.1954	.1419	.0923	.0537
	4	.0021	.0213	.0683	.1329	.1936	.2311	.2367	.2128	.1700	.1208
	5	.0002	.0038	.0193	.0532	.1032	.1585	.2039	.2270	.2225	.1934
	6	.0000	.0005	.0040	.0155	.0401	.0792	.1281	.1766	.2124	.2256
	7	.0000	.0000	.0006	.0033	.0115	.0291	.0591	.1009	.1489	.1934
	8	.0000	.0000	.0001	.0005	.0024	.0078	.0199	.0420	.0762	.1208
	9	.0000	.0000	.0000	.0001	.0004	.0015	.0048	.0125	.0277	.0537
	10	.0000	.0000	.0000	.0000	.0000	.0002	.0008	.0025	.0068	.0161
	11	.0000	.0000	.0000	.0000	.0000	.0000	.0001	.0003	.0010	.0029
	12	.0000	.0000	.0000	.0000	.0000	.0000	.0000	.0000	.0001	.0002
13	0	.5133	.2542	.1209	.0550	.0238	.0097	.0037	.0013	.0004	.0001
	1	.3512	.3672	.2774	.1787	.1029	.0540	.0259	.0113	.0045	.0016
	2	.1109	.2448	.2937	.2680	.2059	.1388	.0836	.0453	.0220	.0095
	3	.0214	.0997	.1900	.2457	.2517	.2181	.1651	.1107	.0660	.0349
	4	.0028	.0277	.0838	.1535	.2097	.2337	.2222	.1845	.1350	.0873
	5	.0003	.0055	.0266	.0691	.1258	.1803	.2154	.2214	.1989	.1571
	6	.0000	.0008	.0063	.0230	.0559	.1030	.1546	.1968	.2169	.2095
	7	.0000	.0001	.0011	.0058	.0186	.0442	.0833	.1312	.1775	.2095
	8	.0000	.0000	.0001	.0011	.0047	.0142	.0336	.0656	.1089	.1571
	9	.0000	.0000	.0000	.0001	.0009	.0034	.0101	.0243	.0495	.0873
	10	.0000	.0000	.0000	.0000	.0001	.0006	.0022	.0065	.0162	.0349
	11	.0000	.0000	.0000	.0000	.0000	.0001	.0003	.0012	.0036	.0095
	12	.0000	.0000	.0000	.0000	.0000	.0000	.0000	.0001	.0005	.0016
	13	.0000	.0000	.0000	.0000	.0000	.0000	.0000	.0000	.0000	.0001

The Binomial Distribution (cont.)

		π									
n	x	.05	.10	.15	.20	.25	.30	.35	.40	.45	.50
14	0	.4077	.2288	.1028	.0440	.0178	.0068	.0024	.0008	.0002	.0001
	1	.3593	.3559	.2539	.1539	.0832	.0407	.0181	.0073	.0027	.0009
	2	.1229	.2570	.2912	.2501	.1802	.1134	.0634	.0317	.0141	.0056
	3	.0259	.1142	.2056	.2501	.2402	.1943	.1366	.0845	.0462	.0222
	4	.0037	.0348	.0998	.1720	.2202	.2290	.2022	.1549	.1040	.0611
	5	.0004	.0078	.0352	.0860	.1468	.1963	.2178	.2066	.1701	.1222
	6	.0000	.0013	.0093	.0322	.0734	.1262	.1759	.2066	.2088	.1833
	7	.0000	.0002	.0019	.0092	.0280	.0618	.1082	.1574	.1952	.2095
	8	.0000	.0000	.0003	.0020	.0082	.0232	.0510	.0918	.1398	.1833
	9	.0000	.0000	.0000	.0003	.0018	.0066	.0183	.0408	.0762	.1222
	10	.0000	.0000	.0000	.0000	.0003	.0014	.0049	.0136	.0312	.0611
	11	.0000	.0000	.0000	.0000	.0000	.0002	.0010	.0033	.0093	.0222
	12	.0000	.0000	.0000	.0000	.0000	.0000	.0001	.0005	.0019	.0056
	13	.0000	.0000	.0000	.0000	.0000	.0000	.0000	.0001	.0002	.0009
	14	.0000	.0000	.0000	.0000	.0000	.0000	.0000	.0000	.0000	.0001
15	0	.4633	.2059	.0874	.0352	.0134	.0047	.0016	.0005	.0001	.0000
	1	.3658	.3432	.2312	.1319	.0668	.0305	.0126	.0047	.0016	.0005
	2	.1348	.2669	.2856	.2309	.1559	.0916	.0476	.0219	.0090	.0032
	3	.0307	.1285	.2184	.2501	.2252	.1700	.1110	.0634	.0318	.0139
	4	.0049	.0428	.1156	.1876	.2252	.2186	.1792	.1268	.0780	.0417
	5	.0006	.0105	.0449	.1032	.1651	.2061	.2123	.1859	.1404	.0916
	6	.0000	.0019	.0132	.0430	.0917	.1472	.1906	.2066	.1914	.1527
	7	.0000	.0003	.0030	.0138	.0393	.0811	.1319	.1771	.2013	.1964
	8	.0000	.0000	.0005	.0035	.0131	.0348	.0710	.1181	.1647	.1964
	9	.0000	.0000	.0001	.0007	.0034	.0116	.0298	.0612	.1048	.1527
	10	.0000	.0000	.0000	.0001	.0007	.0030	.0096	.0245	.0515	.0916
	11	.0000	.0000	.0000	.0000	.0001	.0006	.0024	.0074	.0191	.0417
	12	.0000	.0000	.0000	.0000	.0000	.0001	.0004	.0016	.0052	.0139
	13	.0000	.0000	.0000	.0000	.0000	.0000	.0001	.0003	.0010	.0032
	14	.0000	.0000	.0000	.0000	.0000	.0000	.0000	.0000	.0001	.0005
	15	.0000	.0000	.0000	.0000	.0000	.0000	.0000	.0000	.0000	.0000

The Binomial Distribution (cont.)

		π									
n	x	.05	.10	.15	.20	.25	.30	.35	.40	.45	.50
16	0	.4401	.1853	.0743	.0281	.0100	.0033	.0010	.0003	.0001	.0000
	1	.3706	.3294	.2097	.1126	.0535	.0228	.0087	.0030	.0009	.0002
	2	.1463	.2745	.2775	.2111	.1336	.0732	.0353	.0150	.0056	.0018
	3	.0359	.1423	.2285	.2463	.2079	.1465	.0888	.0468	.0215	.0085
	4	.0061	.0514	.1311	.2001	.2252	.2040	.1553	.1014	.0572	.0278
	5	.0008	.0137	.0555	.1201	.1802	.2099	.2008	.1623	.1123	.0667
	6	.0001	.0028	.0180	.0550	.1101	.1649	.1982	.1983	.1684	.1222
	7	.0000	.0004	.0045	.0197	.0524	.1010	.1524	.1889	.1969	.1746
	8	.0000	.0001	.0009	.0055	.0197	.0487	.0923	.1417	.1812	.1964
	9	.0000	.0000	.0001	.0012	.0058	.0185	.0442	.0840	.1318	.1746
	10	.0000	.0000	.0000	.0002	.0014	.0056	.0167	.0392	.0755	.1222
	11	.0000	.0000	.0000	.0000	.0002	.0013	.0049	.0142	.0337	.0667
	12	.0000	.0000	.0000	.0000	.0000	.0002	.0011	.0040	.0115	.0278
	13	.0000	.0000	.0000	.0000	.0000	.0000	.0002	.0008	.0029	.0085
	14	.0000	.0000	.0000	.0000	.0000	.0000	.0000	.0001	.0005	.0018
	15	.0000	.0000	.0000	.0000	.0000	.0000	.0000	.0000	.0001	.0002
	16	.0000	.0000	.0000	.0000	.0000	.0000	.0000	.0000	.0000	.0000
17	0	.4181	.1668	.0631	.0225	.0075	.0023	.0007	.0002	.0000	.0000
	1	.3741	.3150	.1893	.0957	.0426	.0169	.0060	.0019	.0005	.0001
	2	.1575	.2800	.2673	.1914	.1136	.0581	.0260	.0102	.0035	.0010
	3	.0415	.1556	.2359	.2393	.1893	.1245	.0701	.0341	.0144	.0052
	4	.0076	.0605	.1457	.2093	.2209	.1868	.1320	.0796	.0411	.0182
	5	.0010	.0175	.0668	.1361	.1914	.2081	.1849	.1379	.0875	.0472
	6	.0001	.0039	.0236	.0680	.1276	.1784	.1991	.1839	.1432	.0944
	7	.0000	.0007	.0065	.0267	.0668	.1201	.1685	.1927	.1841	.1484
	8	.0000	.0001	.0014	.0084	.0279	.0644	.1134	.1606	.1883	.1855
	9	.0000	.0000	.0003	.0021	.0093	.0276	.0611	.1070	.1540	.1855
	10	.0000	.0000	.0000	.0004	.0025	.0095	.0263	.0571	.1008	.1484
	11	.0000	.0000	.0000	.0001	.0005	.0026	.0090	.0242	.0525	.0944
	12	.0000	.0000	.0000	.0000	.0001	.0006	.0024	.0021	.0215	.0472
	13	.0000	.0000	.0000	.0000	.0000	.0001	.0005	.0021	.0068	.0182
	14	.0000	.0000	.0000	.0000	.0000	.0000	.0001	.0004	.0016	.0052
	15	.0000	.0000	.0000	.0000	.0000	.0000	.0000	.0001	.0003	.0010
	16	.0000	.0000	.0000	.0000	.0000	.0000	.0000	.0000	.0000	.0001
	17	.0000	.0000	.0000	.0000	.0000	.0000	.0000	.0000	.0000	.0000

The Binomial Distribution (cont.)

		π									
n	x	.05	.10	.15	.20	.25	.30	.35	.40	.45	.50
18	0	.3972	.1501	.0536	.0180	.0056	.0016	.0004	.0001	.0000	.0000
	1	.3763	.3002	.1704	.0811	.0338	.0126	.0042	.0012	.0003	.0001
	2	.1683	.2835	.2556	.1723	.0958	.0458	.0190	.0069	.0022	.0006
	3	.0473	.1680	.2406	.2297	.1704	.1046	.0547	.0246	.0095	.0031
	4	.0093	.0700	.1592	.2153	.2130	.1681	.1104	.0614	.0291	.0117
	5	.0014	.0218	.0787	.1507	.1988	.2017	.1664	.1146	.0666	.0327
	6	.0002	.0052	.0301	.0816	.1436	.1873	.1941	.1655	.1181	.0708
	7	.0000	.0010	.0091	.0350	.0820	.1376	.1792	.1892	.1657	.1214
	8	.0000	.0002	.0022	.0120	.0376	.0811	.1327	.1734	.1864	.1669
	9	.0000	.0000	.0004	.0033	.0139	.0386	.0794	.1284	.1694	.1855
	10	.0000	.0000	.0001	.0008	.0042	.0149	.0385	.0771	.1248	.1669
	11	.0000	.0000	.0000	.0001	.0010	.0046	.0151	.0374	.0742	.1214
	12	.0000	.0000	.0000	.0000	.0002	.0012	.0047	.0145	.0354	.0708
	13	.0000	.0000	.0000	.0000	.0000	.0002	.0012	.0044	.0134	.0327
	14	.0000	.0000	.0000	.0000	.0000	.0000	.0002	.0011	.0039	.0117
	15	.0000	.0000	.0000	.0000	.0000	.0000	.0000	.0002	.0009	.0031
	16	.0000	.0000	.0000	.0000	.0000	.0000	.0000	.0000	.0001	.0006
	17	.0000	.0000	.0000	.0000	.0000	.0000	.0000	.0000	.0000	.0001
	18	.0000	.0000	.0000	.0000	.0000	.0000	.0000	.0000	.0000	.0000
19	0	.3774	.1351	.0456	.0144	.0042	.0011	.0003	.0001	.0000	.0000
	1	.3774	.2852	.1529	.0685	.0268	.0093	.0029	.0008	.0002	.0000
	2	.1787	.2852	.2428	.1540	.0803	.0358	.0138	.0046	.0013	.0003
	3	.0533	.1796	.2428	.2182	.1517	.0869	.0422	.0175	.0062	.0018
	4	.0112	.0798	.1714	.2182	.2023	.1491	.0909	.0467	.0203	.0074
	5	.0018	.0266	.0907	.1636	.2023	.1916	.1468	.0933	.0497	.0222
	6	.0002	.0069	.0374	.0955	.1574	.1916	.1844	.1451	.0949	.0518
	7	.0000	.0014	.0122	.0443	.0974	.1525	.1844	.1797	.1443	.0961
	8	.0000	.0002	.0032	.0166	.0487	.0981	.1489	.1797	.1771	.1442
	9	.0000	.0000	.0007	.0051	.0198	.0514	.0980	.1464	.1771	.1762
	10	.0000	.0000	.0001	.0013	.0066	.0220	.0528	.0976	.1449	.1762
	11	.0000	.0000	.0000	.0003	.0018	.0077	.0233	.0532	.0970	.1442
	12	.0000	.0000	.0000	.0000	.0004	.0022	.0083	.0237	.0529	.0961
	13	.0000	.0000	.0000	.0000	.0001	.0005	.0024	.0085	.0233	.0518
	14	.0000	.0000	.0000	.0000	.0000	.0001	.0006	.0024	.0082	.0222
	15	.0000	.0000	.0000	.0000	.0000	.0000	.0001	.0005	.0022	.0074
	16	.0000	.0000	.0000	.0000	.0000	.0000	.0000	.0001	.0005	.0018
	17	.0000	.0000	.0000	.0000	.0000	.0000	.0000	.0000	.0001	.0003
	18	.0000	.0000	.0000	.0000	.0000	.0000	.0000	.0000	.0000	.0000
	19	.0000	.0000	.0000	.0000	.0000	.0000	.0000	.0000	.0000	.0000

The Binomial Distribution (cont.)

		π									
n	**x**	**.05**	**.10**	**.15**	**.20**	**.25**	**.30**	**.35**	**.40**	**.45**	**.50**
20	0	.3585	.1216	.0388	.0115	.0032	.0008	.0002	.0000	.0000	.0000
	1	.3774	.2702	.1368	.0576	.0211	.0068	.0020	.0005	.0001	.0000
	2	.1887	.2852	.2293	.1369	.0669	.0278	.0100	.0031	.0008	.0002
	3	.0596	.1901	.2428	.2054	.1339	.0716	.0323	.0123	.0040	.0011
	4	.0133	.0898	.1821	.2182	.1897	.1304	.0738	.0350	.0139	.0046
	5	.0022	.0319	.1028	.1746	.2023	.1789	.1272	.0746	.0365	.0148
	6	.0003	.0089	.0454	.1091	.1686	.1916	.1712	.1244	.0746	.0370
	7	.0000	.0020	.0160	.0545	.1124	.1643	.1844	.1659	.1221	.0739
	8	.0000	.0004	.0046	.0222	.0609	.1144	.1614	.1797	.1623	.1201
	9	.0000	.0001	.0011	.0074	.0271	.0654	.1158	.1597	.1771	.1602
	10	.0000	.0000	.0002	.0020	.0099	.0308	.0686	.1171	.1593	.1762
	11	.0000	.0000	.0000	.0005	.0030	.0120	.0336	.0710	.1185	.1602
	12	.0000	.0000	.0000	.0001	.0008	.0039	.0136	.0355	.0727	.1201
	13	.0000	.0000	.0000	.0000	.0002	.0010	.0045	.0146	.0366	.0739
	14	.0000	.0000	.0000	.0000	.0000	.0002	.0012	.0049	.0150	.0370
	15	.0000	.0000	.0000	.0000	.0000	.0000	.0003	.0013	.0049	.0148
	16	.0000	.0000	.0000	.0000	.0000	.0000	.0000	.0003	.0013	.0046
	17	.0000	.0000	.0000	.0000	.0000	.0000	.0000	.0000	.0002	.0011
	18	.0000	.0000	.0000	.0000	.0000	.0000	.0000	.0000	.0000	.0002
	19	.0000	.0000	.0000	.0000	.0000	.0000	.0000	.0000	.0000	.0000
	20	.0000	.0000	.0000	.0000	.0000	.0000	.0000	.0000	.0000	.0000

Source: National Bureau of Standards, *Tables of the Binomial Probability Distribution,* Applied Mathematics Series, U.S. Department of Commerce, 1950.

APPENDIX D: The Poisson Distribution

$$P(x \mid \lambda) = \frac{\lambda^x e^{-\lambda}}{x!}$$

					λ						
x	**0.1**	**0.2**	**0.3**	**0.4**	**0.5**	**0.6**	**0.7**	**0.8**	**0.9**	**1.0**	**x**
0	.9048	.8187	.7408	.6703	.6065	.5488	.4966	.4493	.4066	.3679	0
1	.0905	.1637	.2222	.2681	.3033	.3293	.3476	.3595	.3659	.3679	1
2	.0045	.0164	.0333	.0536	.0758	.0988	.1217	.1438	.1647	.1839	2
3	.0002	.0011	.0033	.0072	.0126	.0198	.0284	.0383	.0494	.0613	3
4	.0000	.0001	.0002	.0007	.0016	.0030	.0050	.0077	.0111	.0153	4
5	.0000	.0000	.0000	.0001	.0002	.0004	.0007	.0012	.0020	.0031	5
6	.0000	.0000	.0000	.0000	.0000	.0000	.0001	.0002	.0003	.0005	6
7	.0000	.0000	.0000	.0000	.0000	.0000	.0000	.0000	.0000	.0001	7
x	**1.1**	**1.2**	**1.3**	**1.4**	**1.5**	**1.6**	**1.7**	**1.8**	**1.9**	**2.0**	**x**
0	.3329	.3012	.2725	.2466	.2231	.2019	.1827	.1653	.1496	.1353	0
1	.3662	.3614	.3543	.3452	.3347	.3230	.3106	.2975	.2842	.2707	1
2	.2014	.2169	.2303	.2417	.2510	.2584	.2640	.2678	.2700	.2707	2
3	.0738	.0867	.0998	.1128	.1255	.1378	.1496	.1607	.1710	.1804	3
4	.0203	.0260	.0324	.0395	.0471	.0551	.0636	.0723	.0812	.0902	4
5	.0045	.0062	.0084	.0111	.0141	.0176	.0216	.0260	.0309	.0361	5
6	.0008	.0012	.0018	.0026	.0035	.0047	.0061	.0078	.0098	.0120	6
7	.0001	.0002	.0003	.0005	.0008	.0011	.0015	.0020	.0027	.0034	7
8	.0000	.0000	.0001	.0001	.0001	.0002	.0003	.0005	.0006	.0009	8
9	.0000	.0000	.0000	.0000	.0000	.0000	.0001	.0001	.0001	.0002	9
x	**2.1**	**2.2**	**2.3**	**2.4**	**2.5**	**2.6**	**2.7**	**2.8**	**2.9**	**3.0**	**x**
0	.1225	.1108	.1003	.0907	.0821	.0743	.0672	.0608	.0550	.0498	0
1	.2472	.2438	.2306	.2177	.2052	.1931	.1815	.1703	.1596	.1494	1
2	.2700	.2681	.2652	.2613	.2565	.2510	.2450	.2384	.2314	.2240	2
3	.1890	.1966	.2033	.2090	.2138	.2176	.2205	.2225	.2237	.2240	3
4	.0992	.1082	.1169	.1254	.1336	.1414	.1488	.1557	.1622	.1680	4
5	.0417	.0476	.0538	.0602	.0668	.0735	.0804	.0872	.0940	.1008	5
6	.0146	.0174	.0206	.0241	.0278	.0319	.0362	.0407	.0455	.0504	6
7	.0044	.0055	.0068	.0083	.0099	.0118	.0139	.0163	.0188	.0216	7
8	.0011	.0015	.0019	.0025	.0031	.0038	.0047	.0057	.0068	.0081	8
9	.0003	.0004	.0005	.0007	.0009	.0011	.0014	.0018	.0022	.0027	9
10	.0001	.0001	.0001	.0002	.0002	.0003	.0004	.0005	.0006	.0008	10
11	.0000	.0000	.0000	.0000	.0000	.0001	.0001	.0001	.0002	.0002	11
12	.0000	.0000	.0000	.0000	.0000	.0000	.0000	.0000	.0000	.0001	12

λ

x	3.1	3.2	3.3	3.4	3.5	3.6	3.7	3.8	3.9	4.0	x
0	.0450	.0408	.0369	.0334	.0302	.0273	.0247	.0224	.0202	.0183	0
1	.1397	.1304	.1217	.1135	.1057	.0984	.0915	.0850	.0789	.0733	1
2	.2165	.2087	.2008	.1929	.1850	.1771	.1692	.1615	.1539	.1465	2
3	.2237	.2226	.2209	.2186	.2158	.2125	.2087	.2046	.2001	.1954	3
4	.1734	.1781	.1823	.1858	.1888	.1912	.1931	.1944	.1951	.1954	4
5	.1075	.1140	.1203	.1264	.1322	.1377	.1429	.1477	.1522	.1563	5
6	.0555	.0608	.0662	.0716	.0771	.0826	.0881	.0936	.0989	.1042	6
7	.0246	.0278	.0312	.0348	.0385	.0425	.0466	.0508	.0551	.0595	7
8	.0095	.0111	.0129	.0148	.0169	.0191	.0215	.0241	.0269	.0298	8
9	.0033	.0040	.0047	.0056	.0066	.0076	.0089	.0102	.0116	.0132	9
10	.0010	.0013	.0016	.0019	.0023	.0028	.0033	.0039	.0045	.0053	10
11	.0003	.0004	.0005	.0006	.0007	.0009	.0011	.0013	.0016	.0019	11
12	.0001	.0001	.0001	.0002	.0002	.0003	.0003	.0004	.0005	.0006	12
13	.0000	.0000	.0000	.0000	.0001	.0001	.0001	.0001	.0002	.0002	13
14	.0000	.0000	.0000	.0000	.0000	.0000	.0000	.0000	.0000	.0001	14
x	**4.1**	**4.2**	**4.3**	**4.4**	**4.5**	**4.6**	**4.7**	**4.8**	**4.9**	**5.0**	**x**
0	.0166	.0150	.0136	.0123	.0111	.0101	.0091	.0082	.0074	.0067	0
1	.0679	.0630	.0583	.0540	.0500	.0462	.0427	.0395	.0365	.0337	1
2	.1393	.1323	.1254	.1188	.1125	.1063	.1005	.0948	.0894	.0842	2
3	.1904	.1852	.1798	.1743	.1687	.1631	.1574	.1517	.1460	.1404	3
4	.1951	.1944	.1933	.1917	.1898	.1875	.1849	.1820	.1789	.1755	4
5	.1600	.1633	.1662	.1687	.1708	.1725	.1738	.1747	.1753	.1755	5
6	.1093	.1143	.1191	.1237	.1281	.1323	.1362	.1398	.1432	.1462	6
7	.0640	.0686	.0732	.0778	.0824	.0869	.0914	.0959	.1002	.1044	7
8	.0328	.0360	.0393	.0428	.0463	.0500	.0537	.0575	.0614	.0653	8
9	.0150	.0168	.0188	.0209	.0232	.0255	.0280	.0307	.0334	.0363	9
10	.0061	.0071	.0081	.0092	.0104	.0118	.0132	.0147	.0164	.0181	10
11	.0023	.0027	.0032	.0037	.0043	.0049	.0056	.0064	.0073	.0082	11
12	.0008	.0009	.0011	.0014	.0016	.0019	.0022	.0026	.0030	.0034	12
13	.0002	.0003	.0004	.0005	.0006	.0007	.0008	.0009	.0011	.0013	13
14	.0001	.0001	.0001	.0001	.0002	.0002	.0003	.0003	.0004	.0005	14
15	.0000	.0000	.0000	.0000	.0001	.0001	.0001	.0001	.0001	.0002	15

λ

x	5.1	5.2	5.3	5.4	5.5	5.6	5.7	5.8	5.9	6.0	x
0	.0061	.0055	.0050	.0045	.0041	.0037	.0033	.0030	.0027	.0025	0
1	.0311	.0287	.0265	.0244	.0225	.0207	.0191	.0176	.0162	.0149	1
2	.0793	.0746	.0701	.0659	.0618	.0580	.0544	.0509	.0477	.0446	2
3	.1348	.1293	.1239	.1185	.1133	.1082	.1033	.0985	.0938	.0892	3
4	.1719	.1681	.1641	.1600	.1558	.1515	.1472	.1428	.1383	.1339	4
5	.1753	.1748	.1740	.1728	.1714	.1697	.1678	.1656	.1632	.1606	5
6	.1490	.1515	.1537	.1555	.1571	.1584	.1594	.1601	.1605	.1606	6
7	.1086	.1125	.1163	.1200	.1234	.1267	.1298	.1326	.1353	.1377	7
8	.0692	.0731	.0771	.0810	.0849	.0887	.0925	.0962	.0998	.1033	8
9	.0392	.0423	.0454	.0486	.0519	.0552	.0586	.0620	.0654	.0688	9
10	.0200	.0220	.0241	.0262	.0285	.0309	.0334	.0359	.0386	.0413	10
11	.0093	.0104	.0116	.0129	.0143	.0157	.0173	.0190	.0207	.0225	11
12	.0039	.0045	.0051	.0058	.0065	.0073	.0082	.0092	.0102	.0113	12
13	.0015	.0018	.0021	.0024	.0028	.0032	.0036	.0041	.0046	.0052	13
14	.0006	.0007	.0008	.0009	.0011	.0013	.0015	.0017	.0019	.0022	14
15	.0002	.0002	.0003	.0003	.0004	.0005	.0006	.0007	.0008	.0009	15
16	.0001	.0001	.0001	.0001	.0001	.0002	.0002	.0002	.0003	.0003	16
17	.0000	.0000	.0000	.0000	.0000	.0001	.0001	.0001	.0001	.0001	17

x	6.1	6.2	6.3	6.4	6.5	6.6	6.7	6.8	6.9	7.0	x
0	.0022	.0020	.0018	.0017	.0015	.0014	.0012	.0011	.0010	.0009	0
1	.0137	.0126	.0116	.0106	.0098	.0090	.0082	.0076	.0070	.0064	1
2	.0417	.0390	.0364	.0340	.0318	.0296	.0276	.0258	.0240	.0223	2
3	.0848	.0806	.0765	.0726	.0688	.0652	.0617	.0584	.0552	.0521	3
4	.1294	.1249	.1205	.1162	.1118	.1076	.1034	.0992	.0952	.0912	4
5	.1579	.1549	.1519	.1487	.1454	.1420	.1385	.1349	.1314	.1277	5
6	.1605	.1601	.1595	.1586	.1575	.1562	.1546	.1529	.1511	.1490	6
7	.1399	.1418	.1435	.1450	.1462	.1472	.1480	.1486	.1489	.1490	7
8	.1066	.1099	.1130	.1160	.1188	.1215	.1240	.1263	.1284	.1304	8
9	.0723	.0757	.0791	.0825	.0858	.0891	.0923	.0954	.0985	.1014	9
10	.0441	.0469	.0498	.0528	.0558	.0588	.0618	.0649	.0679	.0710	10
11	.0245	.0265	.0285	.0307	.0330	.0353	.0377	.0401	.0426	.0452	11
12	.0124	.0137	.0150	.0164	.0179	.0194	.0210	.0227	.0245	.0264	12
13	.0058	.0065	.0073	.0081	.0089	.0098	.0108	.0119	.0130	.0142	13
14	.0025	.0029	.0033	.0037	.0041	.0046	.0052	.0058	.0064	.0071	14
15	.0010	.0012	.0014	.0016	.0018	.0020	.0023	.0026	.0029	.0033	15
16	.0004	.0005	.0005	.0006	.0007	.0008	.0010	.0011	.0013	.0014	16
17	.0001	.0002	.0002	.0002	.0003	.0003	.0004	.0004	.0005	.0006	17
18	.0000	.0001	.0001	.0001	.0001	.0001	.0001	.0002	.0002	.0002	18
19	.0000	.0000	.0000	.0000	.0000	.0000	.0000	.0001	.0001	.0001	19

	λ										
x	7.1	7.2	7.3	7.4	7.5	7.6	7.7	7.8	7.9	8.0	x
0	.0008	.0007	.0007	.0006	.0006	.0005	.0005	.0004	.0004	.0003	0
1	.0059	.0054	.0049	.0045	.0041	.0038	.0035	.0032	.0029	.0027	1
2	.0208	.0194	.0180	.0167	.0156	.0145	.0134	.0125	.0116	.0107	2
3	.0492	.0464	.0438	.0413	.0389	.0366	.0345	.0324	.0305	.0286	3
4	.0874	.0836	.0799	.0764	.0729	.0696	.0663	.0632	.0602	.0573	4
5	.1241	.1204	.1167	.1130	.1094	.1057	.1021	.0986	.0951	.0916	5
6	.1468	.1445	.1420	.1394	.1367	.1339	.1311	.1282	.1252	.1221	6
7	.1489	.1486	.1481	.1474	.1465	.1454	.1442	.1428	.1413	.1396	7
8	.1321	.1337	.1351	.1363	.1373	.1382	.1388	.1392	.1395	.1396	8
9	.1042	.1070	.1096	.1121	.1144	.1167	.1187	.1207	.1224	.1241	9
10	.0740	.0770	.0800	.0829	.0858	.0887	.0914	.0941	.0967	.0993	10
11	.0478	.0504	.0531	.0558	.0585	.0613	.0640	.0667	.0695	.0722	11
12	.0283	.0303	.0323	.0344	.0366	.0388	.0411	.0434	.0457	.0481	12
13	.0154	.0168	.0181	.0196	.0211	.0227	.0243	.0260	.0278	.0296	13
14	.0078	.0086	.0095	.0104	.0113	.0123	.0134	.0145	.0157	.0169	14
15	.0037	.0041	.0046	.0051	.0057	.0062	.0069	.0075	.0083	.0090	15
16	.0016	.0019	.0021	.0024	.0026	.0030	.0033	.0037	.0041	.0045	16
17	.0007	.0008	.0009	.0010	.0012	.0013	.0015	.0017	.0019	.0021	17
18	.0003	.0003	.0004	.0004	.0005	.0006	.0006	.0007	.0008	.0009	18
19	.0001	.0001	.0001	.0002	.0002	.0002	.0003	.0003	.0003	.0004	19
20	.0000	.0000	.0001	.0001	.0001	.0001	.0001	.0001	.0001	.0002	20
21	.0000	.0000	.0000	.0000	.0000	.0000	.0000	.0000	.0001	.0001	21
x	8.1	8.2	8.3	8.4	8.5	8.6	8.7	8.8	8.9	9.0	x
0	.0003	.0003	.0002	.0002	.0002	.0002	.0002	.0002	.0001	.0001	0
1	.0025	.0023	.0021	.0019	.0017	.0016	.0014	.0013	.0012	.0011	1
2	.0100	.0092	.0086	.0079	.0074	.0068	.0063	.0058	.0054	.0050	2
3	.0269	.0252	.0237	.0222	.0208	.0195	.0183	.0171	.0160	.0150	3
4	.0544	.0517	.0491	.0466	.0443	.0420	.0398	.0377	.0357	.0337	4
5	.0882	.0849	.0816	.0784	.0752	.0722	.0692	.0663	.0635	.0607	5
6	.1191	.1160	.1128	.1097	.1066	.1034	.1003	.0972	.0941	.0911	6
7	.1378	.1358	.1338	.1317	.1294	.1271	.1247	.1222	.1197	.1171	7
8	.1395	.1392	.1388	.1382	.1375	.1366	.1356	.1344	.1332	.1318	8
9	.1256	.1269	.1280	.1290	.1299	.1306	.1311	.1315	.1317	.1318	9
10	.1017	.1040	.1063	.1084	.1104	.1123	.1140	.1157	.1172	.1186	10
11	.0749	.0776	.0802	.0828	.0853	.0878	.0902	.0925	.0948	.0970	11
12	.0505	.0530	.0555	.0579	.0604	.0629	.0654	.0679	.0703	.0728	12
13	.0315	.0334	.0354	.0374	.0395	.0416	.0438	.0459	.0481	.0504	13
14	.0182	.0196	.0210	.0225	.0240	.0256	.0272	.0289	.0306	.0324	14
15	.0098	.0107	.0116	.0126	.0136	.0147	.0158	.0169	.0182	.0194	15
16	.0050	.0055	.0060	.0066	.0072	.0079	.0086	.0093	.0101	.0109	16
17	.0024	.0026	.0029	.0033	.0036	.0040	.0044	.0048	.0053	.0058	17
18	.0011	.0012	.0014	.0015	.0017	.0019	.0021	.0024	.0026	.0029	18
19	.0005	.0005	.0006	.0007	.0008	.0009	.0010	.0011	.0012	.0014	19
20	.0002	.0002	.0002	.0003	.0003	.0004	.0004	.0005	.0005	.0006	20
21	.0001	.0001	.0001	.0001	.0001	.0002	.0002	.0002	.0002	.0003	21
22	.0000	.0000	.0000	.0000	.0001	.0001	.0001	.0001	.0001	.0001	22

	λ										
x	**9.1**	**9.2**	**9.3**	**9.4**	**9.5**	**9.6**	**9.7**	**9.8**	**9.9**	**10**	**x**
0	.0001	.0001	.0001	.0001	.0001	.0001	.0001	.0001	.0001	.0000	0
1	.0010	.0009	.0009	.0008	.0007	.0007	.0006	.0005	.0005	.0005	1
2	.0046	.0043	.0040	.0037	.0034	.0031	.0029	.0027	.0025	.0023	2
3	.0140	.0131	.0123	.0115	.0107	.0100	.0093	.0087	.0081	.0076	3
4	.0319	.0302	.0285	.0269	.0254	.0240	.0226	.0213	.0201	.0189	4
5	.0581	.0555	.0530	.0506	.0483	.0460	.0439	.0418	.0398	.0378	5
6	.0881	.0851	.0822	.0793	.0764	.0736	.0709	.0682	.0656	.0631	6
7	.1145	.1118	.1091	.1064	.1037	.1010	.0982	.0955	.0928	.0901	7
8	.1302	.1286	.1269	.1251	.1232	.1212	.1191	.1170	.1148	.1126	8
9	.1317	.1315	.1311	.1306	.1300	.1293	.1284	.1274	.1263	.1251	9
10	.1198	.1210	.1219	.1228	.1235	.1241	.1245	.1249	.1250	.1251	10
11	.0991	.1012	.1031	.1049	.1067	.1083	.1098	.1112	.1125	.1137	11
12	.0752	.0776	.0799	.0822	.0844	.0866	.0888	.0908	.0928	.0948	12
13	.0526	.0549	.0572	.0594	.0617	.0640	.0662	.0685	.0707	.0729	13
14	.0342	.0361	.0380	.0399	.0419	.0439	.0459	.0479	.0500	.0521	14
15	.0208	.0221	.0235	.0250	.0265	.0281	.0297	.0313	.0330	.0347	15
16	.0118	.0127	.0137	.0147	.0157	.0168	.0180	.0192	.0204	.0217	16
17	.0063	.0069	.0075	.0081	.0088	.0095	.0103	.0111	.0119	.0128	17
18	.0032	.0035	.0039	.0042	.0046	.0051	.0055	.0060	.0065	.0071	18
19	.0015	.0017	.0019	.0021	.0023	.0026	.0028	.0031	.0034	.0037	19
20	.0007	.0008	.0009	.0010	.0011	.0012	.0014	.0015	.0017	.0019	20
21	.0003	.0003	.0004	.0004	.0005	.0006	.0006	.0007	.0008	.0009	21
22	.0001	.0001	.0002	.0002	.0002	.0002	.0003	.0003	.0004	.0004	22
23	.0000	.0001	.0001	.0001	.0001	.0001	.0001	.0001	.0002	.0002	23
24	.0000	.0000	.0000	.0000	.0000	.0000	.0000	.0001	.0001	.0001	24

	λ										
x	11	12	13	14	15	16	17	18	19	20	x
0	.0000	.0000	.0000	.0000	.0000	.0000	.0000	.0000	.0000	.0000	0
1	.0002	.0001	.0000	.0000	.0000	.0000	.0000	.0000	.0000	.0000	1
2	.0010	.0004	.0002	.0001	.0000	.0000	.0000	.0000	.0000	.0000	2
3	.0037	.0018	.0008	.0004	.0002	.0001	.0000	.0000	.0000	.0000	3
4	.0102	.0053	.0027	.0013	.0006	.0003	.0001	.0001	.0000	.0000	4
5	.0224	.0127	.0070	.0037	.0019	.0010	.0005	.0002	.0001	.0001	5
6	.0411	.0255	.0152	.0087	.0048	.0026	.0014	.0007	.0004	.0002	6
7	.0646	.0437	.0281	.0174	.0104	.0060	.0034	.0018	.0010	.0005	7
8	.0888	.0655	.0457	.0304	.0194	.0120	.0072	.0042	.0024	.0013	8
9	.1085	.0874	.0661	.0473	.0324	.0213	.0135	.0083	.0050	.0029	9
10	.1194	.1048	.0859	.0663	.0486	.0341	.0230	.0150	.0095	.0058	10
11	.1194	.1144	.1015	.0844	.0663	.0496	.0355	.0245	.0164	.0106	11
12	.1094	.1144	.1099	.0984	.0829	.0661	.0504	.0368	.0259	.0176	12
13	.0926	.1056	.1099	.1060	.0956	.0814	.0658	.0509	.0378	.0271	13
14	.0728	.0905	.1021	.1060	.1024	.0930	.0800	.0655	.0514	.0387	14
15	.0534	.0724	.0885	.0989	.1024	.0992	.0906	.0786	.0650	.0516	15
16	.0367	.0543	.0719	.0866	.0960	.0992	.0963	.0884	.0772	.0646	16
17	.0237	.0383	.0550	.0713	.0847	.0934	.0963	.0936	.0863	.0760	17
18	.0145	.0256	.0397	.0554	.0706	.0830	.0909	.0936	.0911	.0844	18
19	.0084	.0161	.0272	.0409	.0557	.0699	.0814	.0887	.0911	.0888	19
20	.0046	.0097	.0177	.0286	.0418	.0559	.0692	.0798	.0866	.0888	20
21	.0024	.0055	.0109	.0191	.0299	.0426	.0560	.0684	.0783	.0846	21
22	.0012	.0030	.0065	.0121	.0204	.0310	.0433	.0560	.0676	.0769	22
23	.0006	.0016	.0037	.0074	.0133	.0216	.0320	.0438	.0559	.0669	23
24	.0003	.0008	.0020	.0043	.0083	.0144	.0226	.0328	.0442	.0557	24
25	.0001	.0004	.0010	.0024	.0050	.0092	.0154	.0237	.0336	.0446	25
26	.0000	.0002	.0005	.0013	.0029	.0057	.0101	.0164	.0246	.0343	26
27	.0000	.0001	.0002	.0007	.0016	.0034	.0063	.0109	.0173	.0254	27
28	.0000	.0000	.0001	.0003	.0009	.0019	.0038	.0070	.0117	.0181	28
29	.0000	.0000	.0001	.0002	.0004	.0011	.0023	.0044	.0077	.0125	29
30	.0000	.0000	.0000	.0001	.0002	.0006	.0013	.0026	.0049	.0083	30
31	.0000	.0000	.0000	.0000	.0001	.0003	.0007	.0015	.0030	.0054	31
32	.0000	.0000	.0000	.0000	.0001	.0001	.0004	.0009	.0018	.0034	32
33	.0000	.0000	.0000	.0000	.0000	.0001	.0002	.0005	.0010	.0020	33
34	.0000	.0000	.0000	.0000	.0000	.0000	.0001	.0002	.0006	.0012	34
35	.0000	.0000	.0000	.0000	.0000	.0000	.0000	.0001	.0003	.0007	35
36	.0000	.0000	.0000	.0000	.0000	.0000	.0000	.0001	.0002	.0004	36
37	.0000	.0000	.0000	.0000	.0000	.0000	.0000	.0000	.0001	.0002	37
38	.0000	.0000	.0000	.0000	.0000	.0000	.0000	.0000	.0000	.0001	38
39	.0000	.0000	.0000	.0000	.0000	.0000	.0000	.0000	.0000	.0001	39

Source: This table is reprinted from Table VII of Burington and May, *Handbook of Probability and Statistics with Tables,* second edition, 1970, by permission of the author's trustee.

APPENDIX E: Student's *t*-Distribution

One-Tail Probability	.10	.05	.025	.01	.005
Two-Tail Probability	.20	.10	.05	.02	.01
DEGREES OF FREEDOM 1	3.078	6.314	12.706	31.821	63.657
2	1.886	2.920	4.303	6.965	9.925
3	1.638	2.353	3.182	4.541	5.841
4	1.533	2.132	2.776	3.747	4.604
5	1.476	2.015	2.571	3.365	4.032
6	1.440	1.943	2.447	3.143	3.707
7	1.415	1.895	2.365	2.998	3.499
8	1.397	1.860	2.306	2.896	3.355
9	1.383	1.833	2.262	2.821	3.250
10	1.371	1.812	2.228	2.764	3.169
11	1.363	1.796	2.201	2.718	3.106
12	1.356	1.782	2.179	2.681	3.055
13	1.350	1.771	2.160	2.650	3.012
14	1.345	1.761	2.145	2.624	2.977
15	1.341	1.753	2.131	2.602	2.947
16	1.337	1.746	2.120	2.583	2.921
17	1.333	1.740	2.110	2.567	2.898
18	1.330	1.734	2.101	2.552	2.878
19	1.328	1.729	2.093	2.539	2.861
20	1.325	1.725	2.086	2.528	2.845
21	1.323	1.721	2.080	2.518	2.831
22	1.321	1.717	2.074	2.508	2.819
23	1.319	1.714	2.069	2.500	2.807
24	1.318	1.711	2.064	2.492	2.797
25	1.316	1.708	2.060	2.485	2.787
26	1.315	1.706	2.056	2.479	2.779
27	1.314	1.703	2.052	2.473	2.771
28	1.313	1.701	2.048	2.467	2.763
29	1.311	1.699	2.045	2.462	2.756
30	1.310	1.697	2.042	2.457	2.750
40	1.303	1.684	2.021	2.423	2.704
60	1.296	1.671	2.000	2.390	2.660
120	1.289	1.658	1.980	2.358	2.617
∞	1.282	1.645	1.960	2.326	2.576

Source: From Table III of Fisher & Yates': *Statistical Tables for Biological, Agricultural and Medical Research* published by Longman Group UK Ltd. London (previously published by Oliver and Boyd Ltd, Edingburgh) and by permission of the authors and publishers.

APPENDIX F: The Chi-Square Distribution

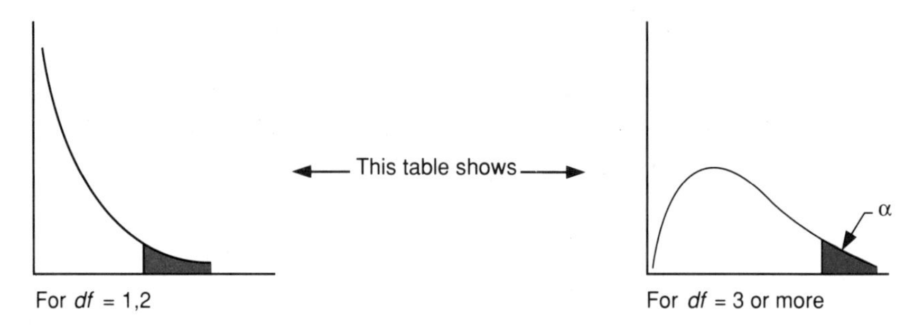

Degrees of Freedom *DF*	Level of Significance						(right tail area)		
	.99	.975	.95	.50	.10	.05	.025	.02	.01
1	.0001	.0009	.0039	.455	2.706	3.841	5.02	5.412	6.635
2	.0201	.0506	.103	1.386	4.605	5.991	7.38	7.824	9.210
3	.115	.216	.352	2.366	6.251	7.815	9.35	9.837	11.345
4	.297	.484	.711	3.357	7.779	9.488	11.14	11.668	13.277
5	.554	.831	1.145	4.351	9.236	11.000	12.83	13.388	15.086
6	.872	1.237	1.635	5.348	10.645	12.592	14.45	15.033	16.812
7	1.239	1.690	2.167	6.346	12.017	14.067	16.01	16.622	18.475
8	1.646	2.18	2.733	7.344	13.362	15.507	17.53	18.168	20.090
9	2.088	2.70	3.325	8.343	14.684	16.919	19.02	19.679	21.666
10	2.558	3.25	3.940	9.342	15.987	18.307	20.5	21.161	23.209
11	3.053	3.82	4.575	10.341	17.275	19.675	21.9	22.618	24.725
12	3.571	4.40	5.226	11.340	18.549	21.026	23.3	24.054	26.217
13	4.107	5.01	5.892	12.340	19.812	22.362	24.7	25.472	27.688
14	4.660	5.63	6.571	13.339	21.064	23.685	26.1	26.873	29.141
15	5.229	6.26	7.261	14.339	22.307	24.996	27.5	28.259	30.578
16	5.812	6.91	7.962	15.338	23.542	26.296	28.8	29.633	32.000
17	6.408	7.56	8.672	16.338	24.769	27.587	30.2	30.995	33.409
18	7.015	8.23	9.390	17.338	25.989	28.869	31.5	32.346	34.805
19	7.633	8.91	10.117	18.338	27.204	30.144	32.9	33.687	36.191
20	8.260	9.59	10.851	19.337	28.412	31.410	34.2	35.020	37.566
21	8.897	10.28	11.591	20.337	29.615	32.671	35.5	36.343	38.932
22	9.542	10.98	12.338	21.337	30.813	33.924	36.8	37.659	40.289
23	10.196	11.69	13.091	22.337	32.007	35.172	38.1	38.968	41.638
24	10.856	12.40	13.848	23.337	33.196	36.415	39.4	40.270	42.980
25	11.524	13.12	14.611	24.337	34.382	37.652	40.6	41.566	44.314
26	12.198	13.84	15.379	25.336	35.563	38.885	41.9	42.856	45.642
27	12.879	14.57	16.161	26.336	36.741	40.113	43.2	44.140	46.963
28	13.565	15.31	16.928	27.336	37.916	41.337	44.5	45.419	48.278
29	14.256	16.05	17.708	28.336	39.087	42.557	45.7	46.693	49.588
30	14.953	16.79	18.493	29.336	40.256	43.773	47.0	47.962	50.892

Source: From Table IV of Fisher & Yates': *Statistical Tables for Biological, Agricultural and Medical Research* published by Longman Group UK Ltd. London (previously published by Oliver and Boyd Ltd, Edingburgh) and by permission of the authors and publishers.

APPENDIX G: The *F*-Distribution

Values of *F*

Right tail of the distribution for $P = .05$ (light-face type), .01 (bold-face type)

df_1 = Degrees of Freedom for Numerator

df_2—Degrees of Freedom for Denominator

df_2	1	2	3	4	5	6	7	8	9	10	11	12
1	161	200	216	225	230	234	237	239	241	242	243	244
	4,052	**4,999**	**5,403**	**5,625**	**5,764**	**5,859**	**5,928**	**5,981**	**6,022**	**6,056**	**6,082**	**6,106**
2	18.51	19.00	19.16	19.25	19.30	19.33	19.36	19.37	19.38	19.39	19.40	19.41
	98.49	**99.01**	**99.17**	**99.25**	**99.30**	**99.33**	**99.34**	**99.36**	**99.38**	**99.40**	**99.41**	**99.42**
3	10.13	9.55	9.28	9.12	9.01	8.94	8.88	8.84	8.81	8.78	8.76	8.74
	34.12	**30.81**	**29.46**	**28.71**	**28.24**	**27.91**	**27.67**	**27.49**	**27.34**	**27.23**	**27.13**	**27.05**
4	7.71	6.94	6.59	6.39	6.26	6.16	6.09	6.04	6.00	5.96	5.93	5.91
	21.20	**18.00**	**16.69**	**15.98**	**15.52**	**15.21**	**14.98**	**14.80**	**14.66**	**14.54**	**14.45**	**14.37**
5	6.61	5.79	5.41	5.19	5.05	4.95	4.88	4.82	4.78	4.74	4.70	4.68
	16.26	**13.27**	**12.06**	**11.39**	**10.97**	**10.67**	**10.45**	**10.27**	**10.15**	**10.05**	**9.96**	**9.89**
6	5.99	5.14	4.76	4.53	4.39	4.28	4.21	4.15	4.10	4.06	4.03	4.00
	13.74	**10.92**	**9.78**	**9.15**	**8.75**	**8.47**	**8.26**	**8.10**	**7.98**	**7.87**	**7.79**	**7.72**
7	5.59	4.74	4.35	4.12	3.97	3.87	3.79	3.73	3.68	3.63	3.60	3.57
	12.25	**9.55**	**8.45**	**7.85**	**7.46**	**7.19**	**7.00**	**6.84**	**6.71**	**6.62**	**6.54**	**6.47**
8	5.32	4.46	4.07	3.84	3.69	3.58	3.50	3.44	3.39	3.34	3.31	3.28
	11.26	**8.65**	**7.59**	**7.01**	**6.63**	**6.37**	**6.19**	**6.03**	**5.91**	**5.82**	**5.74**	**5.67**
9	5.12	4.26	3.86	3.63	3.48	3.37	3.29	3.23	3.18	3.13	3.10	3.07
	10.56	**8.02**	**6.99**	**6.42**	**6.06**	**5.80**	**5.62**	**5.47**	**5.35**	**5.26**	**5.18**	**5.11**
10	4.96	4.10	3.71	3.48	3.33	3.22	3.14	3.07	3.02	2.97	2.94	2.91
	10.04	**7.56**	**6.55**	**5.99**	**5.64**	**5.39**	**5.21**	**5.06**	**4.95**	**4.85**	**4.78**	**4.71**
11	4.84	3.98	3.59	3.36	3.20	3.09	3.01	2.95	2.90	2.86	2.82	2.79
	9.65	**7.20**	**6.22**	**5.67**	**5.32**	**5.07**	**4.88**	**4.74**	**4.63**	**4.54**	**4.46**	**4.40**
12	4.75	3.88	3.49	3.26	3.11	3.00	2.92	2.85	2.80	2.76	2.72	2.69
	9.33	**6.93**	**5.95**	**5.41**	**5.06**	**4.82**	**4.65**	**4.50**	**4.39**	**4.30**	**4.22**	**4.16**
13	4.67	3.80	3.41	3.18	3.02	2.92	2.84	2.77	2.72	2.67	2.63	2.60
	9.07	**6.70**	**5.74**	**5.20**	**4.86**	**4.62**	**4.44**	**4.30**	**4.19**	**4.10**	**4.02**	**3.96**
14	4.60	3.74	3.34	3.11	2.96	2.85	2.77	2.70	2.65	2.60	2.56	2.53
	8.86	**6.51**	**5.56**	**5.03**	**4.69**	**4.46**	**4.28**	**4.14**	**4.03**	**3.94**	**3.86**	**3.80**
15	4.54	3.68	3.29	3.06	2.90	2.79	2.70	2.64	2.59	2.55	2.51	2.48
	8.68	**6.36**	**5.42**	**4.89**	**4.56**	**4.32**	**4.14**	**4.00**	**3.89**	**3.80**	**3.73**	**3.67**
16	4.49	3.63	3.24	3.01	2.85	2.74	2.66	2.59	2.54	2.49	2.45	2.42
	8.53	**6.23**	**5.29**	**4.77**	**4.44**	**4.20**	**4.03**	**3.89**	**3.78**	**3.69**	**3.61**	**3.55**
17	4.45	3.59	3.20	2.96	2.81	2.70	2.62	2.55	2.50	2.45	2.41	2.38
	8.40	**6.11**	**5.18**	**4.67**	**4.34**	**4.10**	**3.93**	**3.79**	**3.68**	**3.59**	**3.52**	**3.45**

df_1 = Degrees of Freedom for Numerator												
14	16	20	24	30	40	50	75	100	200	500	∞	df_2
245	246	248	249	250	251	252	253	253	254	254	254	1
6,142	**6,169**	**6,208**	**6,234**	**6,258**	**6,286**	**6,302**	**6,323**	**6,334**	**6,352**	**6,361**	**6,366**	
19.42	19.43	19.44	19.45	19.46	19.47	19.47	19.48	19.49	19.49	19.50	19.50	2
99.43	**99.44**	**99.45**	**99.46**	**99.47**	**99.48**	**99.48**	**99.49**	**99.49**	**99.49**	**99.50**	**99.50**	
8.71	8.69	8.66	8.64	8.62	8.60	8.58	8.57	8.56	8.54	8.54	8.53	3
26.92	**26.83**	**26.69**	**26.60**	**26.50**	**28.41**	**26.35**	**26.27**	**26.23**	**26.18**	**26.14**	**26.12**	
5.87	5.84	5.80	5.77	5.74	5.71	5.70	5.68	5.66	5.65	5.64	5.63	4
14.24	**14.15**	**14.02**	**13.93**	**13.83**	**13.74**	**13.69**	**13.61**	**13.57**	**13.52**	**13.48**	**13.46**	
4.64	4.60	4.56	4.53	4.50	4.46	4.44	4.42	4.40	4.38	4.37	4.36	5
9.77	**9.68**	**9.55**	**9.47**	**9.38**	**9.29**	**9.24**	**9.17**	**9.13**	**9.07**	**9.04**	**9.02**	
3.96	3.92	3.87	3.84	3.81	3.77	3.75	3.72	3.71	3.69	3.68	3.67	6
7.60	**7.52**	**7.39**	**7.31**	**7.23**	**7.14**	**7.09**	**7.02**	**6.99**	**6.94**	**6.90**	**6.88**	
3.52	3.49	3.44	3.41	3.38	3.34	3.32	3.29	3.28	3.25	3.24	3.23	7
6.35	**6.27**	**6.15**	**6.07**	**5.98**	**5.90**	**5.85**	**5.78**	**5.75**	**5.70**	**5.67**	**5.65**	
3.23	3.20	3.15	3.12	3.08	3.05	3.03	3.00	2.98	2.96	2.94	2.93	8
5.56	**5.48**	**5.36**	**5.28**	**5.20**	**5.11**	**5.06**	**5.00**	**4.96**	**4.91**	**4.88**	**4.86**	
3.02	2.98	2.93	2.90	2.86	2.82	2.80	2.77	2.76	2.73	2.72	2.71	9
5.00	**4.92**	**4.80**	**4.73**	**4.64**	**4.56**	**4.51**	**4.45**	**4.41**	**4.36**	**4.33**	**4.31**	
2.86	2.82	2.77	2.74	2.70	2.67	2.64	2.61	2.59	2.56	2.55	2.54	10
4.60	**4.52**	**4.41**	**4.33**	**4.25**	**4.17**	**4.12**	**4.05**	**4.01**	**3.96**	**3.93**	**3.91**	
2.74	2.70	2.65	2.61	2.57	2.53	2.50	2.47	2.45	2.42	2.41	2.40	11
4.29	**4.21**	**4.10**	**4.02**	**3.94**	**3.86**	**3.80**	**3.74**	**3.70**	**3.66**	**3.62**	**3.60**	
2.64	2.60	2.54	2.50	2.46	2.42	2.40	2.36	2.35	2.32	2.31	2.30	12
4.05	**3.98**	**3.86**	**3.78**	**3.70**	**3.61**	**3.56**	**3.49**	**3.46**	**3.41**	**3.38**	**3.36**	
2.55	2.51	2.46	2.42	2.38	2.34	2.32	2.28	2.26	2.24	2.22	2.21	13
3.85	**3.78**	**3.67**	**3.59**	**3.51**	**3.42**	**3.37**	**3.30**	**3.27**	**3.21**	**3.18**	**3.16**	
2.48	2.44	2.39	2.35	2.31	2.27	2.24	2.21	2.19	2.16	2.14	2.13	14
3.70	**3.62**	**3.51**	**3.43**	**3.34**	**3.26**	**3.21**	**3.14**	**3.11**	**3.06**	**3.02**	**3.00**	
2.43	2.39	2.33	2.29	2.25	2.21	2.18	2.15	2.12	2.10	2.08	2.07	15
3.56	**3.48**	**3.36**	**3.29**	**3.20**	**3.12**	**3.07**	**3.00**	**2.97**	**2.92**	**2.89**	**2.87**	
2.37	2.33	2.28	2.24	2.20	2.16	2.13	2.09	2.07	2.04	2.02	2.01	16
3.45	**3.37**	**3.25**	**3.18**	**3.10**	**3.01**	**2.96**	**2.89**	**2.86**	**2.80**	**2.77**	**2.75**	
2.33	2.29	2.23	2.19	2.15	2.11	2.08	2.04	2.02	1.99	1.97	1.96	17
3.35	**3.27**	**3.16**	**3.08**	**3.00**	**2.92**	**2.86**	**2.79**	**2.76**	**2.70**	**2.67**	**2.65**	

df_2—Degrees of Freedom for Denominator

Values of F

Right tail of the distribution for P = .05 (light-face type), .01 (bold-face type)

df_2—Degrees of Freedom for Denominator	df_1 = Degrees of Freedom for Numerator											
df_2	1	2	3	4	5	6	7	8	9	10	11	12
18	4.41	3.55	3.16	2.93	2.77	2.66	2.58	2.51	2.46	2.41	2.37	2.34
	8.28	**6.01**	**5.09**	**4.58**	**4.25**	**4.01**	**3.85**	**3.71**	**3.60**	**3.51**	**3.44**	**3.37**
19	4.38	3.52	3.13	2.90	2.74	2.63	2.55	2.48	2.43	2.38	2.34	2.31
	8.18	**5.93**	**5.01**	**4.50**	**4.17**	**3.94**	**3.77**	**3.63**	**3.52**	**3.43**	**3.36**	**3.30**
20	4.35	3.49	3.10	2.87	2.71	2.60	2.52	2.45	2.40	2.35	2.31	2.28
	8.10	**5.85**	**4.94**	**4.43**	**4.10**	**3.87**	**3.71**	**3.56**	**3.45**	**3.37**	**3.30**	**3.23**
21	4.32	3.47	3.07	2.84	2.68	2.57	2.49	2.42	2.37	2.32	2.28	2.25
	8.02	**5.78**	**4.87**	**4.37**	**4.04**	**3.81**	**3.65**	**3.51**	**3.40**	**3.31**	**3.24**	**3.17**
22	4.30	3.44	3.05	2.82	2.66	2.55	2.47	2.40	2.35	2.30	2.26	2.23
	7.94	**5.72**	**4.82**	**4.31**	**3.99**	**3.76**	**3.59**	**3.45**	**3.35**	**3.26**	**3.18**	**3.12**
23	4.28	3.42	3.03	2.80	2.64	2.53	2.45	2.38	2.32	2.28	2.24	2.20
	7.88	**5.66**	**4.76**	**4.26**	**3.94**	**3.71**	**3.54**	**3.41**	**3.30**	**3.21**	**3.14**	**3.07**
24	4.26	3.40	3.01	2.78	2.62	2.51	2.43	2.36	2.30	2.26	2.22	2.18
	7.82	**5.61**	**4.72**	**4.22**	**3.90**	**3.67**	**3.50**	**3.36**	**3.25**	**3.17**	**3.09**	**3.03**
25	4.24	3.38	2.99	2.76	2.60	2.49	2.41	2.34	2.28	2.24	2.20	2.16
	7.77	**5.57**	**4.68**	**4.18**	**3.86**	**3.63**	**3.46**	**3.32**	**3.21**	**3.13**	**3.05**	**2.99**
26	4.22	3.37	2.98	2.74	2.59	2.47	2.39	2.32	2.27	2.22	2.18	2.15
	7.72	**5.53**	**4.64**	**4.14**	**3.82**	**3.59**	**3.42**	**3.29**	**3.17**	**3.09**	**3.02**	**2.96**
27	4.21	3.35	2.96	2.73	2.57	2.46	2.37	2.30	2.25	2.20	2.16	2.13
	7.68	**5.49**	**4.60**	**4.11**	**3.79**	**3.56**	**3.39**	**3.26**	**3.14**	**3.06**	**2.98**	**2.93**
28	4.20	3.34	2.95	2.71	2.56	2.44	2.36	2.29	2.24	2.19	2.15	2.12
	7.64	**5.45**	**4.57**	**4.07**	**3.76**	**3.53**	**3.36**	**3.23**	**3.11**	**3.03**	**2.95**	**2.90**
29	4.18	3.33	2.93	2.70	2.54	2.43	2.35	2.28	2.22	2.18	2.14	2.10
	7.60	**5.42**	**4.54**	**4.04**	**3.73**	**3.50**	**3.33**	**3.20**	**3.08**	**3.00**	**2.92**	**2.87**
30	4.17	3.32	2.92	2.69	2.53	2.42	2.34	2.27	2.21	2.16	2.12	2.09
	7.56	**5.39**	**4.51**	**4.02**	**3.70**	**3.47**	**3.30**	**3.17**	**3.06**	**2.98**	**2.90**	**2.84**
32	4.15	3.30	2.90	2.67	2.51	2.40	2.32	2.25	2.19	2.14	2.10	2.07
	7.50	**5.34**	**4.46**	**3.97**	**3.66**	**3.42**	**3.25**	**3.12**	**3.01**	**2.94**	**2.86**	**2.80**
34	4.13	3.28	2.88	2.65	2.49	2.38	2.30	2.23	2.17	2.12	2.08	2.05
	7.44	**5.29**	**4.42**	**3.93**	**3.61**	**3.38**	**3.21**	**3.08**	**2.97**	**2.89**	**2.82**	**2.76**
36	4.11	3.26	2.86	2.63	2.48	2.36	2.28	2.21	2.15	2.10	2.06	2.03
	7.39	**5.25**	**4.38**	**3.89**	**3.58**	**3.35**	**3.18**	**3.04**	**2.94**	**2.86**	**2.78**	**2.72**
38	4.10	3.25	2.85	2.62	2.46	2.35	2.26	2.19	2.14	2.09	2.05	2.02
	7.35	**5.21**	**4.34**	**3.86**	**3.54**	**3.32**	**3.15**	**3.02**	**2.91**	**2.82**	**2.75**	**2.69**
40	4.08	3.23	2.84	2.61	2.45	2.34	2.25	2.18	2.12	2.07	2.04	2.00
	7.31	**5.18**	**4.31**	**3.83**	**3.51**	**3.29**	**3.12**	**2.99**	**2.88**	**2.80**	**2.73**	**2.66**
42	4.07	3.22	2.83	2.59	2.44	2.32	2.24	2.17	2.11	2.06	2.02	1.99
	7.27	**5.15**	**4.29**	**3.80**	**3.49**	**3.26**	**3.10**	**2.96**	**2.86**	**2.77**	**2.70**	**2.64**

df_1 = Degrees of Freedom for Numerator												
14	**16**	**20**	**24**	**30**	**40**	**50**	**75**	**100**	**200**	**500**	∞	df_2
2.29	2.25	2.19	2.15	2.11	2.07	2.04	2.00	1.98	1.95	1.93	1.92	18
3.27	**3.19**	**3.07**	**3.00**	**2.91**	**2.83**	**2.78**	**2.71**	**2.68**	**2.62**	**2.59**	**2.57**	
2.26	2.21	2.15	2.11	2.07	2.02	2.00	1.96	1.94	1.91	1.90	1.88	19
3.19	**3.12**	**3.00**	**2.92**	**2.84**	**2.76**	**2.70**	**2.63**	**2.60**	**2.54**	**2.51**	**2.49**	
2.23	2.18	2.12	2.08	2.04	1.99	1.96	1.92	1.90	1.87	1.85	1.84	20
3.13	**3.05**	**2.94**	**2.86**	**2.77**	**2.69**	**2.63**	**2.56**	**2.53**	**2.47**	**2.44**	**2.42**	
2.20	2.15	2.09	2.05	2.00	1.96	1.93	1.89	1.87	1.84	1.82	1.81	21
3.07	**2.99**	**2.88**	**2.80**	**2.72**	**2.63**	**2.58**	**2.51**	**2.47**	**2.42**	**2.38**	**2.36**	
2.18	2.13	2.07	2.03	1.98	1.93	1.91	1.87	1.84	1.81	1.80	1.78	22
3.02	**2.94**	**2.83**	**2.75**	**2.67**	**2.58**	**2.53**	**2.46**	**2.42**	**2.37**	**2.33**	**2.31**	
2.14	2.10	2.04	2.00	1.96	1.91	1.88	1.84	1.82	1.79	1.77	1.76	23
2.97	**2.89**	**2.78**	**2.70**	**2.62**	**2.53**	**2.48**	**2.41**	**2.37**	**2.32**	**2.28**	**2.26**	
2.13	2.09	2.02	1.98	1.94	1.89	1.86	1.82	1.80	1.76	1.74	1.73	24
2.93	**2.85**	**2.74**	**2.66**	**2.58**	**2.49**	**2.44**	**2.36**	**2.33**	**2.27**	**2.23**	**2.21**	
2.11	2.06	2.00	1.96	1.92	1.87	1.84	1.80	1.77	1.74	1.72	1.71	25
2.89	**2.81**	**2.70**	**2.62**	**2.54**	**2.45**	**2.40**	**2.32**	**2.29**	**2.23**	**2.19**	**2.17**	
2.10	2.05	1.99	1.95	1.90	1.85	1.82	1.78	1.76	1.72	1.70	1.69	26
2.86	**2.77**	**2.66**	**2.58**	**2.50**	**2.41**	**2.36**	**2.28**	**2.25**	**2.19**	**2.15**	**2.13**	
2.08	2.03	1.97	1.93	1.88	1.84	1.80	1.76	1.74	1.71	1.68	1.67	27
2.83	**2.74**	**2.63**	**2.55**	**2.47**	**2.38**	**2.33**	**2.25**	**2.21**	**2.16**	**2.12**	**2.10**	
2.06	2.02	1.96	1.91	1.87	1.81	1.78	1.75	1.72	1.69	1.67	1.65	28
2.80	**2.71**	**2.60**	**2.52**	**2.44**	**2.35**	**2.30**	**2.22**	**2.18**	**2.13**	**2.09**	**2.06**	
2.05	2.00	1.94	1.90	1.85	1.80	1.77	1.73	1.71	1.68	1.65	1.64	29
2.77	**2.68**	**2.57**	**2.49**	**2.41**	**2.32**	**2.27**	**2.19**	**2.15**	**2.10**	**2.06**	**2.03**	
2.04	1.99	1.93	1.89	1.84	1.79	1.76	1.72	1.69	1.66	1.64	1.62	30
2.74	**2.66**	**2.55**	**2.47**	**2.38**	**2.29**	**2.24**	**2.16**	**2.13**	**2.07**	**2.03**	**2.01**	
2.02	1.97	1.91	1.86	1.82	1.76	1.74	1.69	1.67	1.64	1.61	1.59	32
2.70	**2.62**	**2.51**	**2.42**	**2.34**	**2.25**	**2.20**	**2.12**	**2.08**	**2.02**	**1.98**	**1.96**	
2.00	1.95	1.89	1.84	1.80	1.74	1.71	1.67	1.64	1.61	1.59	1.57	34
2.66	**2.58**	**2.47**	**2.38**	**2.30**	**2.21**	**2.15**	**2.08**	**2.04**	**1.98**	**1.94**	**1.91**	
1.98	1.93	1.87	1.82	1.78	1.72	1.69	1.65	1.62	1.59	1.56	1.55	36
2.62	**2.54**	**2.43**	**2.35**	**2.26**	**2.17**	**2.12**	**2.04**	**2.00**	**1.94**	**1.90**	**1.87**	
1.96	1.92	1.85	1.80	1.76	1.71	1.67	1.63	1.60	1.57	1.54	1.53	38
2.59	**2.51**	**2.40**	**2.32**	**2.22**	**2.14**	**2.08**	**2.00**	**1.97**	**1.90**	**1.86**	**1.84**	
1.95	1.90	1.84	1.79	1.74	1.69	1.66	1.61	1.59	1.55	1.53	1.51	40
2.56	**2.49**	**2.37**	**2.29**	**2.20**	**2.11**	**2.05**	**1.97**	**1.94**	**1.88**	**1.84**	**1.81**	
1.94	1.89	1.82	1.78	1.73	1.68	1.64	1.60	1.57	1.54	1.51	1.49	42
2.54	**2.46**	**2.35**	**2.26**	**2.17**	**2.08**	**2.02**	**1.94**	**1.91**	**1.85**	**1.80**	**1.78**	

df_2—Degrees of Freedom for Denominator

Values of F

Right tail of the distribution for P = .05 (light-face type), .01 (bold-face type)

df_2 — Degrees of Freedom for Denominator	df_1 = Degrees of Freedom for Numerator											
df_2	1	2	3	4	5	6	7	8	9	10	11	12
44	4.06 **7.24**	3.21 **5.12**	2.82 **4.26**	2.58 **3.78**	2.43 **3.46**	2.31 **3.24**	2.23 **3.07**	2.16 **2.94**	2.10 **2.84**	2.05 **2.75**	2.01 **2.68**	1.98 **2.62**
46	4.05 **7.21**	3.20 **5.10**	2.81 **4.24**	2.57 **3.76**	2.42 **3.44**	2.30 **3.22**	2.22 **3.05**	2.14 **2.92**	2.09 **2.82**	2.04 **2.73**	2.00 **2.66**	1.97 **2.60**
48	4.04 **7.19**	3.19 **5.08**	2.80 **4.22**	2.56 **3.74**	2.41 **3.42**	2.30 **3.20**	2.21 **3.04**	2.14 **2.90**	2.08 **2.80**	2.03 **2.71**	1.99 **2.64**	1.96 **2.58**
50	4.03 **7.17**	3.18 **5.06**	2.79 **4.20**	2.56 **3.72**	2.40 **3.41**	2.29 **3.18**	2.20 **3.02**	2.13 **2.88**	2.07 **2.78**	2.02 **2.70**	1.98 **2.62**	1.95 **2.56**
55	4.02 **7.12**	3.17 **5.01**	2.78 **4.16**	2.54 **3.68**	2.38 **3.37**	2.27 **3.15**	2.18 **2.98**	2.11 **2.85**	2.05 **2.75**	2.00 **2.66**	1.97 **2.59**	1.93 **2.53**
60	4.00 **7.08**	3.15 **4.98**	2.76 **4.13**	2.52 **3.65**	2.37 **3.34**	2.25 **3.12**	2.17 **2.95**	2.10 **2.82**	2.04 **2.72**	1.99 **2.63**	1.95 **2.56**	1.92 **2.50**
65	3.99 **7.04**	3.14 **4.95**	2.75 **4.10**	2.51 **3.62**	2.36 **3.31**	2.24 **3.09**	2.15 **2.93**	2.08 **2.79**	2.02 **2.70**	1.98 **2.61**	1.94 **2.54**	1.90 **2.47**
70	3.98 **7.01**	3.13 **4.92**	2.74 **4.08**	2.50 **3.60**	2.35 **3.29**	2.23 **3.07**	2.14 **2.91**	2.07 **2.77**	2.01 **2.67**	1.97 **2.59**	1.93 **2.51**	1.89 **2.45**
80	3.96 **6.96**	3.11 **4.88**	2.72 **4.04**	2.48 **3.56**	2.33 **3.25**	2.21 **3.04**	2.12 **2.87**	2.05 **2.74**	1.99 **2.64**	1.95 **2.55**	1.91 **2.48**	1.88 **2.41**
100	3.94 **6.90**	3.09 **4.82**	2.70 **3.98**	2.46 **3.51**	2.30 **3.20**	2.19 **2.99**	2.10 **2.82**	2.03 **2.69**	1.97 **2.59**	1.92 **2.51**	1.88 **2.43**	1.85 **2.36**
125	3.92 **6.84**	3.07 **4.78**	2.68 **3.94**	2.44 **3.47**	2.29 **3.17**	2.17 **2.95**	2.08 **2.79**	2.01 **2.65**	1.95 **2.56**	1.90 **2.47**	1.86 **2.40**	1.83 **2.33**
150	3.91 **6.81**	3.06 **4.75**	2.67 **3.91**	2.43 **3.44**	2.27 **3.14**	2.16 **2.92**	2.07 **2.76**	2.00 **2.62**	1.94 **2.53**	1.89 **2.44**	1.85 **2.37**	1.82 **2.30**
200	3.89 **6.76**	3.04 **4.71**	2.65 **3.88**	2.41 **3.41**	2.26 **3.11**	2.14 **2.90**	2.05 **2.73**	1.98 **2.60**	1.92 **2.50**	1.87 **2.41**	1.83 **2.34**	1.80 **2.28**
400	3.86 **6.70**	3.02 **4.66**	2.62 **3.83**	2.39 **3.36**	2.23 **3.06**	2.12 **2.85**	2.03 **2.69**	1.96 **2.55**	1.90 **2.46**	1.85 **2.37**	1.81 **2.29**	1.78 **2.23**
1,000	3.85 **6.66**	3.00 **4.62**	2.61 **3.80**	2.38 **3.34**	2.22 **3.04**	2.10 **2.82**	2.02 **2.66**	1.95 **2.53**	1.89 **2.43**	1.84 **2.34**	1.80 **2.26**	1.76 **2.20**
∞	3.84 **6.64**	2.99 **4.60**	2.60 **3.78**	2.37 **3.32**	2.21 **3.02**	2.09 **2.80**	2.01 **2.64**	1.94 **2.51**	1.88 **2.41**	1.83 **2.32**	1.79 **2.24**	1.75 **2.18**

df_1 = Degrees of Freedom for Numerator												df_2—Degrees of Freedom for Denominator
14	**16**	**20**	**24**	**30**	**40**	**50**	**75**	**100**	**200**	**500**	∞	df_2
1.92	1.88	1.81	1.76	1.72	1.66	1.63	1.58	1.56	1.52	1.50	1.48	44
2.52	**2.44**	**2.32**	**2.24**	**2.15**	**2.06**	**2.00**	**1.92**	**1.88**	**1.82**	**1.78**	**1.75**	
1.91	1.87	1.80	1.75	1.71	1.65	1.62	1.57	1.54	1.51	1.48	1.46	46
2.50	**2.42**	**2.30**	**2.22**	**2.13**	**2.04**	**1.98**	**1.90**	**1.86**	**1.80**	**1.76**	**1.72**	
1.90	1.86	1.79	1.74	1.70	1.64	1.61	1.56	1.53	1.50	1.47	1.45	48
2.48	**2.40**	**2.28**	**2.20**	**2.11**	**2.02**	**1.96**	**1.88**	**1.84**	**1.78**	**1.73**	**1.70**	
1.90	1.85	1.78	1.74	1.69	1.63	1.60	1.55	1.52	1.48	1.46	1.44	50
2.46	**2.39**	**2.26**	**2.18**	**2.10**	**2.00**	**1.94**	**1.86**	**1.82**	**1.76**	**1.71**	**1.68**	
1.88	1.83	1.76	1.72	1.67	1.61	1.58	1.52	1.50	1.46	1.43	1.41	55
2.43	**2.35**	**2.23**	**2.15**	**2.06**	**1.96**	**1.90**	**1.82**	**1.78**	**1.71**	**1.66**	**1.64**	
1.86	1.81	1.75	1.70	1.65	1.59	1.56	1.50	1.48	1.44	1.41	1.39	60
2.40	**2.32**	**2.20**	**2.12**	**2.03**	**1.93**	**1.87**	**1.79**	**1.74**	**1.68**	**1.63**	**1.60**	
1.85	1.80	1.73	1.68	1.63	1.57	1.54	1.49	1.46	1.42	1.39	1.37	65
2.37	**2.30**	**2.18**	**2.09**	**2.00**	**1.90**	**1.84**	**1.76**	**1.71**	**1.64**	**1.60**	**1.56**	
1.84	1.79	1.72	1.67	1.62	1.56	1.53	1.47	1.45	1.40	1.37	1.35	70
2.35	**2.28**	**2.15**	**2.07**	**1.98**	**1.88**	**1.82**	**1.74**	**1.69**	**1.62**	**1.56**	**1.53**	
1.82	1.77	1.70	1.65	1.60	1.54	1.51	1.45	1.42	1.38	1.35	1.32	80
2.32	**2.24**	**2.11**	**2.03**	**1.94**	**1.84**	**1.78**	**1.70**	**1.65**	**1.57**	**1.52**	**1.49**	
1.79	1.75	1.68	1.63	1.57	1.51	1.48	1.42	1.39	1.34	1.30	1.28	100
2.26	**2.19**	**2.06**	**1.98**	**1.89**	**1.79**	**1.73**	**1.64**	**1.59**	**1.51**	**1.46**	**1.43**	
1.77	1.72	1.65	1.60	1.55	1.49	1.45	1.39	1.36	1.31	1.27	1.25	125
2.23	**2.15**	**2.03**	**1.94**	**1.85**	**1.75**	**1.68**	**1.59**	**1.54**	**1.46**	**1.40**	**1.37**	
1.76	1.71	1.64	1.59	1.54	1.47	1.44	1.37	1.34	1.29	1.25	1.22	150
2.20	**2.12**	**2.00**	**1.91**	**1.83**	**1.72**	**1.66**	**1.56**	**1.51**	**1.43**	**1.37**	**1.33**	
1.74	1.69	1.62	1.57	1.52	1.45	1.42	1.35	1.32	1.26	1.22	1.19	200
2.17	**2.09**	**1.97**	**1.88**	**1.79**	**1.69**	**1.62**	**1.53**	**1.48**	**1.39**	**1.33**	**1.28**	
1.72	1.67	1.60	1.54	1.49	1.42	1.38	1.32	1.28	1.22	1.16	1.13	400
2.12	**2.04**	**1.92**	**1.84**	**1.74**	**1.64**	**1.57**	**1.47**	**1.42**	**1.32**	**1.24**	**1.19**	
1.70	1.65	1.58	1.53	1.47	1.41	1.36	1.30	1.26	1.19	1.13	1.08	1,000
2.09	**2.01**	**1.89**	**1.81**	**1.71**	**1.61**	**1.54**	**1.44**	**1.38**	**1.28**	**1.19**	**1.11**	
1.69	1.64	1.57	1.52	1.46	1.40	1.35	1.28	1.24	1.17	1.11	1.00	∞
2.07	**1.99**	**1.87**	**1.79**	**1.69**	**1.59**	**1.52**	**1.41**	**1.36**	**1.25**	**1.15**	**1.00**	

Source: Reprinted by permission from *Statistical Methods,* Seventh Edition by George W. Snedecor and William G. Cochran © 1980 by the Iowa State University Press.

APPENDIX H: Answers to Odd-Numbered Problems

CHAPTER 1

1. **(a)** Statistics is the body of methods for the collection, analysis, presentation, and interpretation of quantitative data, and for the use of such data.
   (b) A population is the whole collection of data under study.
   (c) A sample is any part of a population.
3. Management is **(a)** an organized body of knowledge, and **(b)** data is available for analysis, **(c)** mathematical and statistical methods can be used, and **(d)** results are repeatable.
5. **(a)** deductive, **(b)** inductive, **(c)** deductive, **(d)** inductive
7. Sampling error is the difference between the sample statistic and the population parameter. It exists because of the "luck of the draw" of what elements are included in a sample. Bias is a consistent error due to an assignable cause.
9. The demand for bread is uncertain, so the baking company is operating in the realm of inferential statistics. The single previous week may not be a sufficiently large sample to yield a representative average so this enhances the sampling error. In addition, they apply their own intuition, which could be biased. However, they may need intuition and judgment to account for such things as holidays and weather conditions, so this may be the best they can do. A point estimate is almost always going to be incorrect, so it is understandable that the number of loaves is never just right.
11. Several approaches are possible. You might conduct a formal statistical study using the steps outlined in the text. External data may come from trade associations or competitive literature, and internal data from one's own firm. Surveys and field studies would yield customer opinions on the two products, but you want to be sure they are objective and unbiased.
13. **(a)** 20%, **(b)** 37%.

CHAPTER 2

1. **(a)** chronological, **(b)** geographical, **(c)** qualitative, **(d)** quantitative
3. **(a)** continuous, **(b)** 27, 32, 32, 37, 38, 42, 45, 46, 54, 65, **(c)** 42
5. **(a)** Stated class limits are the limits printed on charts or graphs; they do not necessarily encompass every point on a continuum. Real class limits are the points that constitute the upper end of one class and the lower end of the next higher class. They are stated to one more level of accuracy than the data points.
   (b) An absolute frequency distribution is a frequency distribution where the data frequency in each class interval is stated in number of observations, whereas in relative frequency distributions the frequency in each interval is stated as a percentage of the total.
7. **(a)** nearest hours, **(b)** 10 hr (lower) and 14 hr (upper),
   (c) 2nd class = 9.5 and 14.5 hrs, 3rd class = 14.5 and 19.5 hrs, **(d)** 17 hrs.
9. **(a)** Absolute cumulative values of $F >$ LCB should be 100, 86, 62, 30, 11, and 3.
   (b) Relative cumulative values of $\% F >$ LCB should be 100, 86, 62, 30, 11, and 3.
11. Histogram should have frequency on Y-axis and Number of Customers in Store on X-axis with at least 6 classes. Maximum frequency of 27 is in $25 < 35$ class, and minimum of 5 is in $55 < 65$ class.
13. Ogive should have cumulative frequency on y-axis and Battery Lifetime (hr) on x-axis. Cumulative points should be plotted at nine upper class boundaries beginning with 0 at 4.5 hrs and going upward and to the right. For example, 68 observations are < 29.5 hrs and all 100 are < 44.5 hrs.
15. **(a)** Vertical axis of ogive is Y = Percent of Households $\geqslant$ LCB and horizontal axis is X = No. of VCR Tapes Rented. Graph begins at $X = 0$, $Y = 100$ and slopes down and to right where last data point is $X = 55$, $Y = 15$. However, line could be continued down to axis at $X = 65$ if one could assume that no more than 65 tapes were rented to any household.
    (b) median = 22 **(c)** The last class is open ended so we cannot be sure what the maximum is.

17. Pie chart should be round with largest area (47%) for accidents and smallest (8%) for heart disease.
19. Any appropriate type of area, bar, line, or pie chart is satisfactory but should be neatly done.
21. **(a)** Discrete variables are countable, whereas continuous variables are measureable.
(b) Raw data are unarranged observations listed in the order received. Arrays are arranged according to size.
(c) Class limits are the listed intervals in a frequency distribution, whereas class boundaries are the "real" limits that touch each other and are stated to one more level of accuracy.
23. There are several possible solutions to this, but the stated limits of the first class for one acceptable solution are **(a)** 126.0 − 126.9, **(b)** 16 and under 24, **(c)** 220 − 224. Other intervals would have to be added.
25. Histogram should show number of people on Y-axis and age group on X-axis with six classes and largest frequency (328) in the 25–44 age group.
27. Ogive should show Number > LCB on Y-axis and Age Classification on X-axis and should slope down and to the right. Chart begins at $X = 0$, $Y = 1024$ and goes down to $X = 65$ & UP and $Y = 89$.
29. **(a)** 19.5, 19.95, 20.0, 19.5, **(b)** 22, 22.45, 22.5, 24.5, **(c)** 5, 5, 5, 10,
(d) nearest full, nearest tenth, last full, nearest full **(e)** yes, yes, yes **(f)** no.
31. **(a)** Histogram should show Number of Boxes on Y-axis and Weight of Boxes (pounds) on X-axis. It has 7 class intervals with largest frequency (20) in 30–39 class.
(b) Frequency polygon has same name designation on the axis, but X-scale would more likely show class midpoints beginning with zero at 4.5 to highest frequency at 34.5 and down to zero at 84.5. Line connects midpoints of classes.
(c) Ogive has $F <$ UCB on Y-axis and Weight of Boxes (pounds) on X-axis. Values are plotted at upper class boundaries beginning with $Y = 0$ at 9.5 and going to $Y = 80$ at 79.5.
33. **(a)** 10, **(b)** interval $10 < 20$, **(c)** 49,
(d) 5 intervals beginning with $0 < 20$ and ending with $80 < 100$,
(e) The point 50 is within the interval $40 < 60$ so we are uncertain exactly where it is.
(f) There should be enough classes to show general patterns that exist within the data.
35. **(a)** First class is $0 < 10$ with cumulative frequency of 8, and second is $10 < 20$ with 28. Other cumulative values are 38, 45, 51, 67, 81, 89, 96, and 100. **(b)** 28%, **(c)** 89%
37. **(a)** Use Percent Exports on y-axis and Firm Number on x-axis. Then firm #41 is a bar up to 30%, firm #42 is a bar up to 30%, etc.
(b) Pie chart should clearly show different proportions corresponding to industry classifications.

CHAPTER 3

1. **(a)** 5, **(b)** 4, **(c)** 2, **(d)** 2 **3.** **(a)** 16, **(b)** 78, **(c)** 256, **(d)** 42, **(e)** 14
5. **(a)** 5.3, **(b)** 5, **(c)** 5
7. **(a)** 207, **(b)** $Q_1 = 165$, $Q_3 = 396$,
(c) Q_3 is the value in the data set such that 75% of the data points are at or below it.
9. Doris, Jim, and Randy can service 10, 4, and 6 customers/hr, respectively, for 20 total. The mean (harmonic) is $20 \div 3 = 6.67$ customers/hr.
11. **(a)** *MAD* would give the "arithmetic average," but the standard deviation would also yield an average.
(b) Either that the boxes all weigh the same, or the measuring instruments are not sensitive enough to measure any differences.
13. After arranging in an array, the range is 8, and interquartile range is 4.
15. **(a)** $MAD = 2$, **(b)** $s = 2.65$
17. **(a)** 4.60,
(b) 91 and 86. The summation about the true (population) mean is larger and would result in a larger standard deviation if only n were used to compute it. Using $n - 1$ in the sample standard deviation compensates for this downward bias and makes the sample standard deviation larger.

19. **(a)** 15, **(b)** 14.7, **(c)** second class of 10 < 20 beds. **21.** 11 min
23. **(a)** Range is from 7 to 17 minutes, or 10. **(b)** Modal class is the 9 < 11 min class.
(c) Distribution is skewed to the right (positively). **25.** 3.14 min squared
27. **(a)** Number of significant digits cannot exceed the number of significant digits in the number with the fewest significant digits.
(b) Align the decimal points of the numbers and draw a vertical line down after the column where the first digit ceases to be significant. Only those digits to the left of the line are significant.
29. **(a)** mean, median, mode, **(b)** range, *MAD*, standard deviation, variance,
(c) skewness, kurtosis.
31. Positively skewed data have extreme values on the right and taper off to the high end. Negatively skewed data have extremes on the low end.
33. **(a)** 184, **(b)** 18.5, **(c)** 17 and 20 **35.** 1.37
37. **(a)** 5, **(b)** 4.5, **(c)** 2.16, **(d)** 4.67
39. **(a)** 27.67, **(b)** 5.26 **41.** **(a)** 34.2, **(b)** 12.61 **43.** **(a)** $ 1.30, **(b)** $ 1,401.84
45. **(a)** 7.4%, **(b)** Each rate is based upon the preceding year, so the base is not common.
47. **(a)** $11.13, **(b)** $11.00, **(c)** $12.00
49. **(a)** Using the suggested intervals, the frequencies are 22, 34, 31, 8, 2, and 3, respectively.
(b) 11.72, **(c)** 4.63
51. **(a)** $11,000 per year, **(b)** $2,661 per year.

CHAPTER 4

1. Classical probabilities are based upon equally likely outcomes and can be calculated prior to an event. Relative frequency probabilities are based upon historical or empirical evidence. Subjective probabilities are based upon personal assessment or judgment.
3. **(a)** 1/6, **(b)** 26/52, **(c)** .75, **(d)** near zero
5. **(a)** Y = {monday, tuesday, wednesday, thursday, friday},
(b) Z = {c | c is a company car less than 2 years old.}
7. Mutually exclusive events preclude the existence of each other. Collectively exhaustive events are ones that include all possible combinations.
9. **(a)** not mutually exclusive, **(b)** mutually exclusive,
(c) not mutually exclusive because one can talk on a conference call to both at the same time.
11. **(a)** .08, **(b)** .02 **13.** **(a)** .13, **(b)** .024, **(c)** .88 **15.** .90
17.

(a)

	Male	Female	Total
Accounting (A)	120	100	220
Economics (E)	80	200	280
Total	200	300	500

(b)

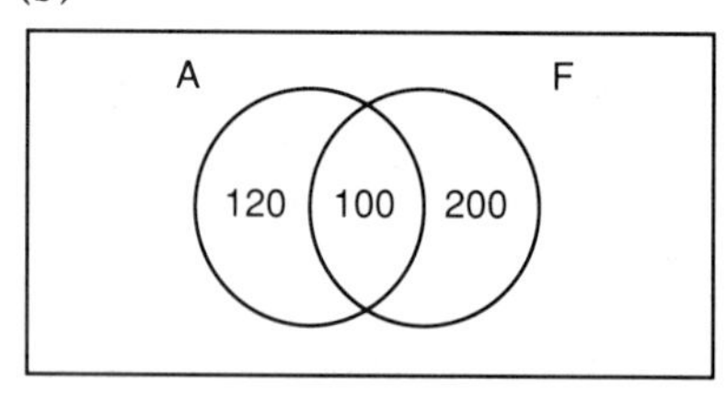

u = 500

(c) .56, **(d)** .33, **(e)** .20, **(f)** .60, **(g)** .40, **(h)** .76,
(i) Male and female are mutually exclusive, as are accounting and economics. **(j)** no
19. **(a)** .07, **(b)** .83
21. **(a)** Yes, because the two are in different locations and industries. **(b)** .48
23. **(a)** .0004, **(b)** .0396, **(c)** .000,008,
(d) Decision is subjective, but the probabilities of failure are 4 in 10,000 flights and airlines do make thousands of flights.

25. Prior probabilities are probability statements made beforehand without the benefit of empirical evidence. Posterior probability statements are made after some data are collected and represent revisions of the prior probabilities. The experimental data is incorporated as a conditional probability, which, when multiplied by the prior probability, becomes a joint probability of the condition and the prior. This ultimately represents a revised, or a "new," prior probability.
27. .96 **29.** .73 **31.** **(a)** 24, **(b)** 1, **(c)** 1, **(d)** 144, **(e)** 625 **33.** 72
35. **(a)** 24, **(b)** 6, **(c)** 90, **(d)** 10, **(e)** 66, **(f)** 1
37. 720 **39.** **(a)** 125, **(b)** 60, **(c)** 10 **41.** 60,480
43. **(a)** subjective, **(b)** classical, **(c)** relative frequency **45.** **(a)** .015, **(b)** .165
47. **(a)** Venn diagram shows A as a circle with designation .65 and remainder is non-A, marked .35. **(b)** .35 **49.** **(a)** .20, **(b)** .15, **(c)** .089, **(d)** .07, **(e)** .46
51. **(a)** 0, 2/29, 8/56,
(b) Industry class and dividend rates are not independent because the conditional probabilities are all different and do not equal the marginal $P(B)$ of 10/100.
53. **(a)** .6, **(b)** .70, **(c)** .08, **(d)** .50
55. **(a)** .08, **(b)** .52,
(c) Yes, because $P(Y) = .40$ and $P(Y \mid X) = .40$, so the probability of Y is not conditioned by whether or not X occurs.
57. .97 **59.** .59
61. **(a)** Tree diagram shows branch for Machine A with $P(A) = .50$ and for Machine B with $P(B) = .50$. Then the A branch divides into conditional $P(Good \mid A) = .90$ and $P(Not\ Good \mid A) = .10$ and the B branch into $P(Good \mid B) = .80$ and $P(Not\ Good \mid B) = .20$. The right hand side values (going down) are .45, .05, .40, and .10. **(b)** .33
63. **(a)** 252, **(b)** 120

CHAPTER 5

1. A random variable assigns values to the events of a sample space, whereas a probability distribution shows the values that the random variable can take on and the probability of taking on each value.
3. **(a)** uniform, **(b)** discrete, **(c)** 1/5 or .20
5.

Relays that work (X)	Probability $P(X)$
W $\overline{W}$ (none)	.06
$\overline{W}$ W (2 and not 1)	.14
W $\overline{W}$ (1 and not 2)	.24
$\overline{W}$ $\overline{W}$ (both)	.56
	1.00

7. **(a)** The discrete uniform distribution is formed by a discrete, or countable, variable, whereas the continuous uniform distribution is formed by a continuous, or measurable, variable.
(b) For the discrete, the mean is $\mu = (n + 1)/2$, where n is the sample size, and for continuous it is $\mu = (a + b)/2$, where a and b are the low and high values. **9.** **(a)** 6.5, **(b)** 11.9
11. **(a)** only two possible outcomes, **(b)** probability constant from trial to trial,
(c) trials independent **13.** **(a)** .8131, **(b)** .0108
15. Probability distribution should show $P(X)$ on Y-axis and X = Number of Successes on X-axis, with largest probability (.356) for $X = 1$, and going down to a probability of .0002 at $X = 6$.
17. **(a)** .5404, **(b)** .3413, **(c)** .1183 **19.** .9445
21. **(a)** There are many possible (discrete) outcomes over a continuous space.
(b) The probabilities of outcomes are low and constant from trial to trial.
(c) Each trial is independent of any previous trial.
23. **(a)** .0954, **(b)** .0244, **(c)** .9600 **25.** **(a)** .0183, **(b)** .2707

27. **(a)** Graph should show $P(X)$ on y-axis and values of X on x-axis, with largest probability (.2240) at $X = 2$ and $X = 3$, going down to a probability of .0001 at $X = 12$.
(b) 3.00 and $\sqrt{3.00}$, **(c)** .4232

29. **(a)** .2169, **(b)** .1205 **31.** Using a Poisson approximation, $P(X \leq 10 \mid \lambda = 6)$ it is .9574.

33. **(a)** zero, **(b)** .6826, **(c)** .9544

35. **(a)** .5000, **(b)** .3413, **(c)** .1587, **(d)** .9974

37. **(a)** .3413, **(b)** .4772, **(c)** .4938, **(d)** .4750

39. **(a)** 210.8 to 289.2, **(b)** 198.5 to 301.5

41. **(a)** when $\pi = .50$,
(b) Binomial applies when the probabilities are constant and independent. It can be easily applied when n is small (< 20) and we are interested in a specific outcome. It can be approximated by the normal if n is larger and $n\pi$ or $n(1 - \pi)$ is > 5, and is better as π approaches .5 and n gets large. If $n\pi$ is < 5 and π is small (i.e., below .05 or .10), then the Poisson can be used.
(c) No, because $n\pi = 20\,(.05) = 1.0$, which is < 5 and too low. **43.** .0078

45. Problem involves first using normal distribution to estimate the proportion (.1056) and then the normal approximation to find the binomial $P(X > 25 \mid n = 50, \pi = .1056)$, which is .9591.

47. **(a)** Probability distribution has uniform $P(X)$ values of .125. Multiplying them by the individual tourist revenue values from 13 through 20 tourists gives $X \cdot P(X)$ values ranging from 8.125 (for 13 tourists) to 12.500 (for 20 tourists). **(b)** \$82.50, **(c)** \$82.50 and \$ 10.00

49. **(a)** .2182, **(b)** Answer is the same as obtained from binomial table.
(c) Proportion is still 20% so the answer is still .2182 **51.** **(a)** .0111, **(b)** .1708, **(c)** .4679

53. **(a)** .0098 (from binomial table), **(b)** .1404 from Poisson table, **(c)** .0918.
(d) This requires continuity correction, for value of .1747.

55. $P(X \geq 45 \mid n = 50, \pi = .96)$ is the same as $P(X \leq 4 \mid n = 50, \pi = .04)$. However, the Poisson approximation may be used because $n\pi = 2$, and using it the answer is .9473.

57. **(a)** .6826, **(b)** .0475, **(c)** 847 hr, **(d)** 880 hr and 1120 hr

59. **(a)** Proportion not paying dividends is 2 in 5 or .40. Therefore $P(X \geq 12 \mid \pi = .40) = .0566$.
(b) .1118, **(c)** Poisson is not a good estimator here because π is too large, and n is only 20.

CHAPTER 6

1. A survey is a field inquiry concerning some condition of interest. Observations are counts or measurements made to estimate specific properties of a population. Experiments are controlled tests that are often used to compare an experimental group with a control group.

3. **(a)** judgment sampling, **(b)** probability sampling, **(c)** judgment sampling

5. The target population is the entire population of the U.S., and a notice in *The Wall Street Journal* is an attempt to obtain respondents who chance to be reading the paper. But only a small proportion read it, and its readers do not represent a cross-section of the population. So any sample may not accurately reflect the population's preference. Moreover, there is bound to be some bias due to self-selection of respondents. In summary, this is a very poor sampling plan.

7. **(a)** 5 digits,
(b) Obtain a sequential listing of all employees numbered from 1 through 21,200. Then beginning at some arbitrary point in a random number table, identify a series of five-digit random numbers. Select 50 numbers in the range of 00,001 to 21,200, and contact the employees corresponding with the numbers selected.

9. 561-80-5114, 576-39-8216, 620-17-8493, 549-78-0414, 537-66-5285, 489-28-7089

11. **(a)** convenience sample, **(b)** purpose sample, **(c)** quota sample

13. Stratified sampling requires that the population be divided into *homogenous* groups, whereas cluster sampling requires that it be divided into *heterogenous* groups. Then random samples are taken of the groups. Thus, the strata (in stratified sampling) should have relatively little variation within the group, whereas the clusters (in cluster sampling) should ideally have the same variation as in the population.
15. Infinite populations contain an infinite number of items, whereas finite populations contain a fixed or limited number of items. In discrete populations, the items are countable, so they would not be infinite. However, if the number is very large, we act as though the population is infinite—even though it (theoretically) is not. **17.** 38,760
19. Every sampling distribution of the means shows (1) all possible values of $\bar{x}$ that can occur from samples of a given size, and (2) the probabilities of each value of $\bar{x}$.
21. **(a)** Frequency distribution should show values of $\bar{x}$ of 2, 3, 4, 5, 6, 8 along with frequencies of 1, 2, 1, 2, 2, and 1. **(b)** 1.76 and 1.76.
23. As n increases, the extreme values in a sample have less representation (weight) in the mean because the extreme values must be combined with more values that are closer to the mean.
25. **(a)** The t-distribution is theoretically correct when samples are drawn from a normally distributed population where σ is unknown.
    (b) As n approaches 30, the sampling distribution of t approaches the normal, so the normal may be used for samples of ≥ 30 even if σ is unknown.
27. **(a)** 150, **(b)** 5, **(c)** .50 **29.** **(a)** \$ 600, **(b)** − 2.063, **(c)** .05
31. **(a)** .017, **(b)** 8.52
33. **(a)** population that should be studied,
    (b) listing of all members of the population,
    (c) displacement error due to a flaw in the design, testing, or recording procedure of a study,
    (d) random error due to the chance that any one sample may not accurately reflect the characteristics of the population.
35. **(a)** judgment, **(b)** probability, **(c)** probability, **(d)** stratified and random sampling.
37. A sampling distribution of the means shows all the sample means that could be obtained from samples of a given size from a population, whereas a sampling distribution of the proportions shows all the sample proportions that could be obtained. Both also show the probability of the statistic.
39. 1,947,792 **41.** 17.31 (Note: Finite population correction factor is required.)
43. The central limit theorem states that the sampling disribution of $\bar{x}$ is approximately normal if the simple random sample size is sufficiently large. It enables statisticians to use the normal distribution for inference about population values regardless of the type of distribution in the population.
45. **(a)** 4.0, **(b)** 2.0, **(c)** the standard deviation and the sample size.
47. **(a)** 2.056, **(b)** − 1.706. **(c)** 10
49. The equation for σ_{np} gives the standard deviation of the sampling distribution of the *number* expected in a sample of size n whereas the equation for σ_p gives the standard deviation of the sampling distribution of the *proportion* expressed as a decimal fraction. Also, $\sigma_p \times n = \sigma_{np}$.

CHAPTER 7

1. Point estimates are rarely correct, and they do not contain any measure of error, telling how close they are likely to be to the population value.
3. **(a)** We estimate the population mean, not the sample mean; the sample is already known.
   (b) The population parameter is not a random variable that "falls" into an interval. Thus we usually refer to the chance of an estimate being correct as confidence, instead of probability.

5. When σ is not known, we must use the standard deviation of the sample instead of the population standard deviation. Then the t is theoretically correct to describe the sampling distribution of sample means. However, if $n \geq 30$, the normal is a satisfactory approximation to the t-distribution.
7. CLs are 4.423 to 4.577 **9.** CLs are 14.04 hr to 15.16 hr **11.** **(a)** 2.33, **(b)** 1.75
13. **(a)** 188, **(b)** 97 **15.** 44
17. Probably not. The standard error of proportion varies with the sample we happen to get, but the standard error of the mean does not differ, so is likely to provide a better estimate.
19. **(a)** 10.6% to 19.4%, **(b)** 11.9% to 18.1% **21.** 21.6% to 34.4% **23.** 1056
25. A point estimate is a single value that is almost certain to be wrong, and it does not include any indication of its precision. An interval estimate is a pair of values within which μ is expected to lie, and it includes a statement of the likelihood or percentage of such estimates that should prove to be correct. We *know* what the sample mean is, but need to *estimate* μ.
27. **(a)** the Z value, width of the confidence interval, an estimate of the standard deviation of the population, and the number of secretaries in the city. **(b)** 384, **(c)** Because n is less than 5% of N, **(d)** 278; The sample is a relevant portion of N, thus reducing $s_{\bar{x}}$.
29. **(a)** .06, **(b)** .0367 to .0833,
    (c) Because the distribution is not symmetrical, and because the standard error of proportion depends on the true value of π, and is not a constant as used in the equation.
    (d) Because sample proportions cannot be estimated from small samples.
31. **(a)** 2400, **(b)** 1536,
    (c) Because the maximum value of pq is .2500, but there is no maximum value for s^2 if s is unknown.
33. CLs = 3.194 ± 1.645 (.345) = \$ 2.62 to \$ 3.76

CHAPTER 8

1. Hypothesis testing is a statistical procedure used to help decide whether a statement or claim about something is true or not true.
3. **(a)** One-sample tests use one sample mean to accept or reject a hypothesis about one population. Two-sample tests use two sample means to test if there is a difference between the two hypothesized population values.
    (b) Tests of means are tests that rely upon a continuous (normal) sampling distrubition of means to test the hypothesized value of a population mean. Tests of proportions also use the normal approximation, but they are designed to test the hypothesized value of a population proportion.
    (c) The null hypothesis is a statement saying that the parameter in question has the same value that it has had in the past—that there is no change. The alternate hypothesis is the position we tend to accept if the null is rejected; it is the "other" position.
    (d) Type I error is rejecting the null hypothesis when it is true. Type II error is accepting the null hypothesis when it is false.
5. Two conclusions are correct: (1) accepting (or not rejecting) a true hypothesis and (2) rejecting a false hypothesis. Two conclusions are erroneous: (3) rejecting a true hypothesis is a type I error and (4) accepting a false hypothesis is a type II error.
7. The level of significance of a test is the risk that one is willing to take of rejecting a hypothesis that is correct. This is also the risk of a type I error, which is designated α.
9. For the test of proportions, assume the proportion is really .20 and a sample of size n is taken.
    (a) Suppose the null hypothesis is H_0: $\pi = .20$. A type I error could arise from obtaining a sample proportion significantly different from .20 (say $p = .12$), causing us to reject the null hypothesis when it really should be accepted.
    (b) Suppose the null hypothesis is H_0: $\pi = .12$. A type II error could arise from obtaining a sample proportion that agrees with the hypothesis (say $p = .12$), causing us to accept the hypothesis when it really should be rejected.

11. **(a)** two: C_1 = the lower limit, and C_2 = the upper limit, **(b)** lower side,
    (c) Limits can be stated in terms of number of standard errors, or units of the variable being tested.
13. **(a)** H_0: μ = \$860, H_1: $\mu >$ \$860, **(b)** H_0: $\mu = .50$, H_1: $\mu \neq 50$
15. The Z_c test statistic may be used on small samples when they are drawn randomly from a normally distributed population. Otherwise the t_c test statistic should be used (assuming the population is normally distributed).
17. Do not reject H_0: $\mu = 1{,}000$ because $-2.05 < -1.67 < +2.05$
19. **(a)** State H_0: $\mu = 14.0$ and H_1: $\mu > 14.0$, α level = .05, reject if $Z_c > 1.645$, $Z_c = 2.0$, and conclusion, which is: Reject H_0: $\mu = 14.0$ min because $2.0 > 1.645$. **(b)** 14.823 min.
21. **(a)** $\pm$ 1.96, reject, **(b)** -1.28, do not reject, **(c)** -2.492, reject, **(d)** + 1.833, do not reject
23. **(a)** No. Do not reject H_0: $\mu = 3{,}000$ because $-1.33 > -2.05$,
    (b) Involves value judgment, but agency should not advertise 3,000 hr life if the bulbs do not last that long.
25. **(a)** H_0: $\pi = .45$, H_1: $\pi \neq .45$,
    (b) Reject if $Z_c < -1.75$ or $> +1.75$, or stated another way, reject if the absolute value $|Z_c| > 1.75$
    (c) Yes, reject the claim because $3.29 > 1.75$.
27. **(a)** 3% level, **(b)** 2.5% level
29. **(a)** Confidence intervals are constructed around sample statistics (means or proportions), whereas rejection limits are established around the hypothesized population parameter.
    (b) We use estimation to arrive at an initial estimate of the population parameter. Tests of hypotheses are used to verify the parameter value or show that it has changed. If a hypothesis is rejected, then confidence interval estimation would be the next step in trying to determine what the value of the parameter might be.
31. The null hypothesis is not true so you cannot commit a type I error.
33. **(a)** t-distribution, **(b)** $-$ 2.064 and + 2.064, **(c)** no,
    (d) -2.40, **(e)** yes, **(f)** from t table = .02.
35. **(a)** Reject H_0: $\mu = 750$ because $-3.00 < -1.96$,
    (b) Do not reject H_0: $\mu = 750$ because $-1.28 > -2.306$,
    (c) The first test has a large enough sample size to permit us to use the normal approximation, whereas the second requires use of the t-distribution. Preference is given to the null. Even though the sample mean values were the same, the sample of $n = 10$ was not sufficient to reject H_0 but the sample of 49 was. Larger samples lend more certainty to the conclusions.
37. **(a)** 1.4, **(b)** Do not reject H_0 at any useful level of significance, **(c)** 5.6,
    (d) Do not reject—there is even less reason to reject it here.
    (e) because σ is not known and n is small. **(f)** the last, or ∞ row.
    (g) the line for $df = 24$, (h) because t approaches the normal distribution as n gets larger and the table covers only a limited number of df values.
39. **(a)** .2266, **(b)** $\approx$ .500, **(c)** .9104, **(d)** no error because the null is not false, **(e)** .6902
41. **(a)** Curve starts on top left at $P(\text{reject } H_0) = 1.0$, drops to type I error of .05 at $\mu_H = 600$, and goes back up to top.
    (b) Curve starts on top left at $P(\text{reject } H_0) = 1.0$, drops to type I error of .05 at $\mu_H = 600$, and continues on down to X-axis.
    (c) Curve starts on X-axis, reaches type I error amount at $\mu_H = 600$, and continues on up to top where $P(\text{reject } H_0) = 1.0$.
    (d) In all cases, the type I error is shown below and the type II error above. However, must compute limits to identify the correct conclusion areas. For (a) limits are 594.12 and 605.88, for (b) the limit is 595.065, and for (c) it is 603.489.

43. **(a)** H_0: $\mu = 20$, H_1: $\mu < 20$, **(b)** 19.30,
(c) no, because Z_c value of -1.092 is not outside the reject value of -1.28,
(d) accepting that the mean is 20.0 when it is not, **(e)** .2912,
(f)

If true μ is	18.0	18.5	19.0	19.3	19.6	20.6
P(accept) is	.0089	.0721	.2912	.5000	.7088	.9911

(g) *OC* curve has possible values of true mean μ on *X*-axis and *P*(Accept) on *Y*-axis. Curve begins at $Y = 0$ when $X = 17.5$, and goes up to .0089 at $X = 18.0$, and .0721 at $X = 18.5$, etc. It is up to $Y = .9911$ at $X = 20.6$. Curve shows probability of accepting hypothesis given various values of μ.

45. Conclusion is do not reject H_0: $\pi = .40$ because $-.6124 > -1.28$

CHAPTER 9

1. The null would state there is no difference: H_0: $\mu_1 = \mu_2$, and the alternative would state there is a difference: H_1: $\mu_1 \neq \mu_2$.
3. To compute the standard error requires knowledge of π_1 and π_2, but if they are both known there is no reason for a test to see if there is a difference in them. Simply observe them to see if they are different.
5. Reject H_0: $\mu_1 = \mu_2$ because $-1.927 < -1.645$. Pay is different.
7. **(a)** Reject H_0: $\mu_1 = \mu_2$ because $4.18 > 2.58$. Methods are different. **(b)** .14 to .39 min/piece.
9. Reject H_0: $\mu_1 = \mu_2$ because $-4.06 < -2.33$. Performance of students is significantly different.
11. Must first compute pooled estimate $s_{pl}^2 = 2.40$. Do not reject H_0: $\mu_1 = \mu_2$ because $-2.205 < |2.228|$. (Result is very close to being significant at 5% level and could be considered significant at the 10% level.)
13. **(a)** H_0: $\mu_1 - \mu_2 = 0$ (which is equivalent to H_0: $\mu_1 = \mu_2$) and H_1: $\mu_1 - \mu_2 \neq 0$.
(b) Include all five steps and end with conclusion: Do not reject because $1.17 < 2.036$.
15. Must first compute $\bar{d} = 4$ and $s_{\bar{d}} = 1.308$. Reject H_0: $\mu_1 - \mu_2 = 0$ because $3.058 > 2.821$. Difference in instructional methods is significant at the 2% level.
17. 1.04 to 6.96 points **19.** Do not reject H_0: $\pi_1 = \pi_2$ because $-1.645 < Z_c$ of $.585 < +1.645$.
21. Reject hypothesis of equal proportions of wealth held by this segment of the population before and after expansion (H_0: $\pi_1 = \pi_2$), because $4.55 > 2.33$.
23. We can be 95% confident that the true mean increase in consumer acceptance is within the interval of 3.69% and 14.31%. **25. (a)** normal, **(b)** *t*-distribution, **(c)** *t*-distribution
27. **(a)** The test for a difference between two independent samples assumes the selection of samples for one group is not influenced by that for the other. With paired samples there is a before/after relationship, or both elements are similar in all other respects except the one treatment under investigation.
(b) For testing hypotheses about the difference in proportions, the standard error of difference is computed from a pooled estimate of the true population proportion. However, for a confidence interval estimate, we use only the sample data p_1 and p_2.
29. The pooled variance is $s_{pl}^2 = 10.00$ and $t_c = 3.76$.
31. The $\bar{d} = 1.2$ and $s_{\bar{d}} = 5.43$. Thus for $n = 15$, $s_{\bar{d}} = 1.40$. Do not reject the hypothesis of no difference in training methods (H_0: $\mu_1 - \mu_2 = 0$) because t_c value of $.857 <$ reject value of $t = 2.624$. The difference in instructional methods is not significant at the 2% level.
33. Using the standard error of difference of 1.1 yr, we calculate $Z_c = 1.82$ which corresponds to an area of .4656 (in one tail). So the two-tailed level of significance is 6.88%.
35. **(a)** H_0: $\pi_1 = \pi_2$, H_1: $\pi_1 < \pi_2$, **(b)** .10, **(c)** .35, **(d)** Yes, because $6.62 > 2.236$. **(e)** no

37. **(a)** Take a sample of 400 and determine the *number* defective. If the number is 62 or more, reject the shipment; otherwise accept it.
    (b) Take a sample of 400 and determine the *percent* defective. If this percent is greater than 15.35, reject the shipment; otherwise accept it.
39. As shown below, the t_c value is .132, which is $<$ the $t_{.10/2,\ 23}$ value of 1.714 , so we cannot reject the H_0: $\mu_{retail} = \mu_{banking}$.

Unpaired t-Test X : Retail P/E Y : Bank-Insur P/E

DF:	X Count:	Y Count:	Mean X:	Mean Y:	Unpaired t Value:
23	15	10	12.6	12.3	.132
					p > .4

41. As shown below, for this paired difference test the H_0: $\mu_{EPS2} = \mu_{EPS1}$ and the t_{calc} value is -1.309. This compares with a table value for $t_{.05/2,\ 19} = 2.093$, so we cannot reject the H_0 at the .05 level.

Paired t-Test X : EPS2 Y : EPS1

DF:	Mean X - Y:	Paired t value:
19	-.292	-1.309
		.1 < p ≤ .375

CHAPTER 10

1. **(a)** 50%, **(b)** 10%, **(c)** 1% 3. **(a)** continuous, **(b)** 23,
   (c) same as might be expected from sampling from a normal (0,1) distribution.
5. Reject H_0: $\pi_1 = \pi_2 = \ldots = \pi_{20}$ at $\alpha = .10$ level because $29.375 > 27.204$
7. Reject H_0: proportions distributed as .90 and .10 because $7.11 > 5.412$
9. Do not reject H_0: $\pi_1 = \pi_2 = \ldots = \pi_{10}$ because $13.40 < 14.684$
11. Do not reject H_0: Data are Poisson distributed, because $7.64 < 11.00$
13. Do not reject H_0: Education level is independent of attitude, because $1.38 < 7.38$
15. Reject H_0: Accident rate is independent of smoking habits, because $18.63 > 9.210$
17. **(a)** 2.42, **(b)** 2.77
19. ANOVA does use means to test for differences in means. But it uses the means to compute squared differences and variances. Because we are comparing two variances, we use the *F* distribution as the standard of comparison.
21. Reject equality of means because $7.76 > F_{.01,\ 5,\ 30}$ of 3.70.
23. ANOVA table should show Between $SSD = 1513.24$ and Within $SSD = 3776.00$. Conclusion is do not reject equality of means because $2.27 < 3.20$. Data are not sufficient to hold that incomes differ by age.
25. **(a)** i) Goodness of fit, ii) Chi-square distribution
    (b) i) Analysis of Variance, ii) *F* distribution
    (c) i) Goodness of fit ii) Chi-square distrubution
    (d) i) *F* test ii) *F* distribution
    (e) i) Independence of Classification ii) Chi-square distribution

27. No problem. The goodness of fit tests do compare observed data to the theoretical distributions. But it is the differences that we measure, and these differences should be normally distributed if the distributions agree.

29. **(a)** $df = (k - 1) - m$ where k = number of groups, m = number of parameters that must be estimated.

(b) $df = (r - 1)(c - 1)$ where r = number of rows, c = number of columns

(c) $df = (k - 1)$ where k = number of groups

(d) $df = (T - k)$ where T = total number of observations, k = number of groups

31. **(a)** Variance is the SSD divided by the appropriate *df*.

(b) SSD_B is the *SSD* of group means ($\bar{x}$) from the grand mean ($\bar{\bar{x}}$), whereas SSD_W is the *SSD* of individual values from their own group means. Then these *SSD*s are summed over all groups (hence the $\Sigma\Sigma$ sign).

(c) Treatment variation is another term for SSD_B, so they are the same.

(d) Variation is *SSD* and variance is $SSD \div df$. **(e)** Variance and mean squared deviation (*MSD*) are the same.

33.

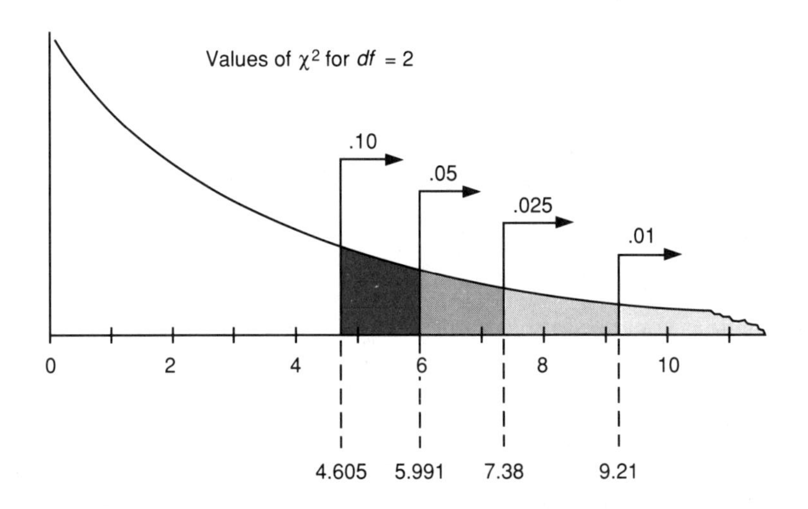

35. **(a)** 2.50, **(b)** Expected frequencies for number of errors of 0, 1, 2, 3, 4, 5, 6, and 7 are 14.8, 36.9, 46.2, 38.5, 24.0, 12.0, and 7.6, respectively. **(c)** Do not reject H_0: Data are Poisson distributed, because the chi-square calculated value of $4.361 <$ chi-square reject for .05, and 5*df* of 11.000.

(d) If μ were known in advance, the *df* would equal $7 - 1 = 6$ and reject value would be 12.592 instead of 11.000.

37. Reject H_0: Forecast is independent of source of employment because $30.575 > 16.812$

39. F_{calc} of $2.25 > F_{.05,\ 20,\ 20}$ of 2.12 so the two variances are different at the .05 level.

41. Do not reject equality of means because $.21 < F_{.05,\ 5,\ 24}$ of 2.62.

43. From a computer printout, the contingency table would have a calculated value of 35.198 with only one *df*, so the hypothesis of equal proportions across the employee classification is rejected at the .001 level. Sales are not independent of the number of employees.

CHAPTER 11

1. Regression involves estimating the value of a dependent variable from one or more independent variables. It describes the slope and intercept of the relationship. Correlation is a measure of the degree to which two or more variables are related.
3. **(a)** *Univariate* is one-variable data, and our best estimate of it is the mean. *Bivariate* is two-variable data. If the two variables are related, we can sometimes use one to help predict the other.
   (b) Simple regression uses only one independent variable to estimate the value of the dependent variable, whereas multiple regression uses two or more independent variables to predict the dependent variable.
   (c) Linear regression is a data relationship where the scatter diagram points lie in a fairly linear or straight line manner. A nonlinear regression is one where the relationship between the two variables follows some form of curve or nonlinear pattern.
5. The general form of the regression model is $Y = f(X, \varepsilon)$. This is a theoretical expression which says that the Y value is dependent upon the X value and some error term. The simple linear model $Y_c = a + b X$ does not include an error term. Although the form we use does not make specific mention of the error, we measure and calculate error in the form of the standard error of estimate, s_e, as part of the analysis.
7. **(a)** nonlinear, **(b)** no apparent relationship, **(c)** linear.
9. **(a)** The relationship has a positive slope and appears to be linear. **(b)** $Y_c = 3.69 + .36 X$
   (c) $6,200
11. $Y_c = 4.62 + .79 X$
13. **(a)** $Y_c = 2.09 + 3.29 (15) = 51.44$, say 51 **(b)** $Y_c = 2.09 + 3.29 (50) = 166.59$, say 167.
15. **(a)** along (parallel to) the Y-axis, **(b)** divide ΣX by n and ΣY by n.
   (c) The $\Sigma(Y - Y_c)$ is always equal to zero because Y_c is on the "mean" line that is the mathematical center of the data points. The summation of deviations around it is zero.
   (d) We use $(n - 2)$ because two degrees of freedom are lost in estimating the constants a and b.
17. **(a)** The standard error of estimate is the standard deviation of individual points about the regression line, so it is not a standard error in the way we used it before as being $s \div \sqrt{n}$.
   (b) The standard error of estimate differs from other standard deviations in that the mean of the distributions differs for each X value. Thus the deviations must be computed around a "variable" or moving mean. Also, instead of dividing by $(n - 1)$ in the denominator, we must divide by $(n - 2)$ because two *df* are lost.
19. Using the Σ values, plus the a and b values from the equation, we get $s_e = 6.28$.
21. $a = 4.62$, and $s_e = 3.56$ **23.** Interval is from 228 to 252 cases.
25. **(a)** $s_e = .7230$, **(b)** $s_e = .7332$ (difference due to rounding),
   (c) The 95% prediction limits are from $6,395 to $9,647, but values may vary due to rounding.
   (d) $s_{IND} = s_e (1.0644) = .7770$, so interval is from $6,165 to $9,877.
   (e) Any differences between (a) and (b) are due to rounding. For (c) the approximate limits are $6,395 to $9,647, with a range of $3,252, and for (d) the exact limits are $6,165 to $9,877, with a range of $3,712. The exact limits are about $460 broader in order to compensate for the intercept and slope errors.
27. **(a)** From computer output we have $Y_c = 22.095 + .325 X$ **(b)** $s_e = 19.25$
   (c) Individual prediction limits are 17.845 to 123.92, so would be rounded to from 18 to 124 nuclear reactors for an individual country with 150 million people.
   (d) Mean confidence limits are 51.7 to 89.9, so would be rounded to from 52 to 90 nuclear reactors on average for countries with 150 million people.
29. **(a)** After calculating $s_b = .4396$, we do not reject H_0: $\beta = 0$ because $3.87 < 4.303$.
   (b) Slope is not significant, so no confidence interval is established around it.

31. **(a)** With regression, we assume an independent/dependent relationship whereas with correlation both variables have equality.
 (b) Regression is concerned with the nature of the relationship, and is quantified in terms of the slope, b, and the intercept, a, of the regression line. Correlation is concerned with the closeness or strength of the relationship, and it is quantified in terms of the correlation coefficient.
 (c) With regression we assume a normal distribution of points in the Y direction around the regression line. With correlation we assume that joint normal distributions exist in both the X and Y directions.
33. **(a)** The coefficient of correlation, r, is simply the square root of the coefficient of determination, r^2.
 (b) The coefficient of determination is more intuitively understandable because it tells the amount of variation that is explained by the regression line relative to the total.
35. **(a)** $r^2 = .769$ **(b)** This means that approximately 77% of the variation in the employment level in this community is associated with freightcar shipments.
37. $r = -.9003$
39. **(a)** $r = -.94$, **(b)** Reject H_0: $\rho = 0$ between hardness and strength because $7.793 > 3.355$.
41. **(a)** The terms *dependent* and *independent* are used to identify the Y and X variables, respectively.
 (b) A positive relationship is one where increases in Y are associated with increases in X. A negative relationship exists if Y decreases as X increases. When the slope, b, is positive, r is given a positive sign. If b is negative, r is given a negative sign.
 (c) High correlation coefficients indicate a close association of the variables, but while they may suggest a possible causal relationship, they should never be interpreted as proving causation.
 (d) s_x measures variation around x, s_y measures variation around y, and s_e measures variation around Y_c.
43. Scatter diagram should show linear relationship between sales taxes collected, X, and units of M sold, Y.
45. **(a)** 1804, **(b)** 476, **(c)** 1.9181, **(d)** 7.557, **(e)** 14.16, **(f)** 2.57,
 (g) s_e is the standard error of estimate.
47. Do not reject H_0: $\beta = 2.000$ because $.69 < 3.355$
49.

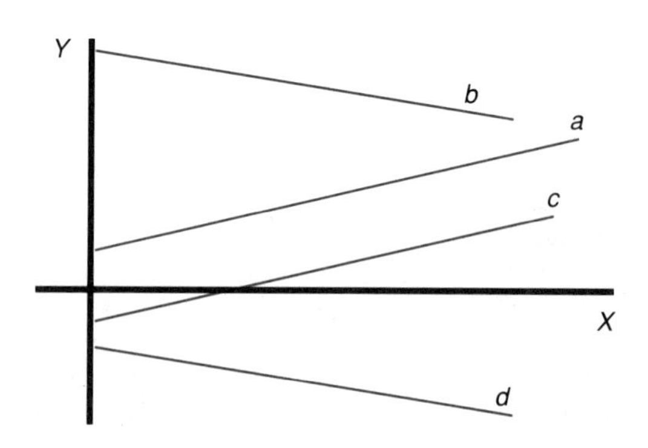

51. **(a)** .971, **(b)** .985 53. **(a)** $t_c = 4.32$, which is > reject value of approximately 1.99
 (b) Reject H_0: $\rho = 0$ and conclude the correlation is significantly different from zero.
55. **(a)** A computer printout of the correlation results is:

Corr. Coeff. X_1: EPSC Y_1: EPS1

Count:	Covariance:	Correlation:	R-squared:
30	1.64	.938	.881

A standard test of the H_0: $\rho = 0$ against H_1: $\rho \neq 0$ at the $\alpha = .05$ (assumed since none is specified) yields $t_c = 14.32$. The conclusion is to reject because $14.32 > t_{\alpha/2,\ n-2df} = t_{.05/2,\ 28} = 2.048$. The correlation is very strong and would have come about by chance less than 5% of the time.

57. (a) A scatter diagram of the data as drawn by computer would show a positively sloped relationship. The resultant regression equation is: $Y_c = 382.907 + 7.953X$

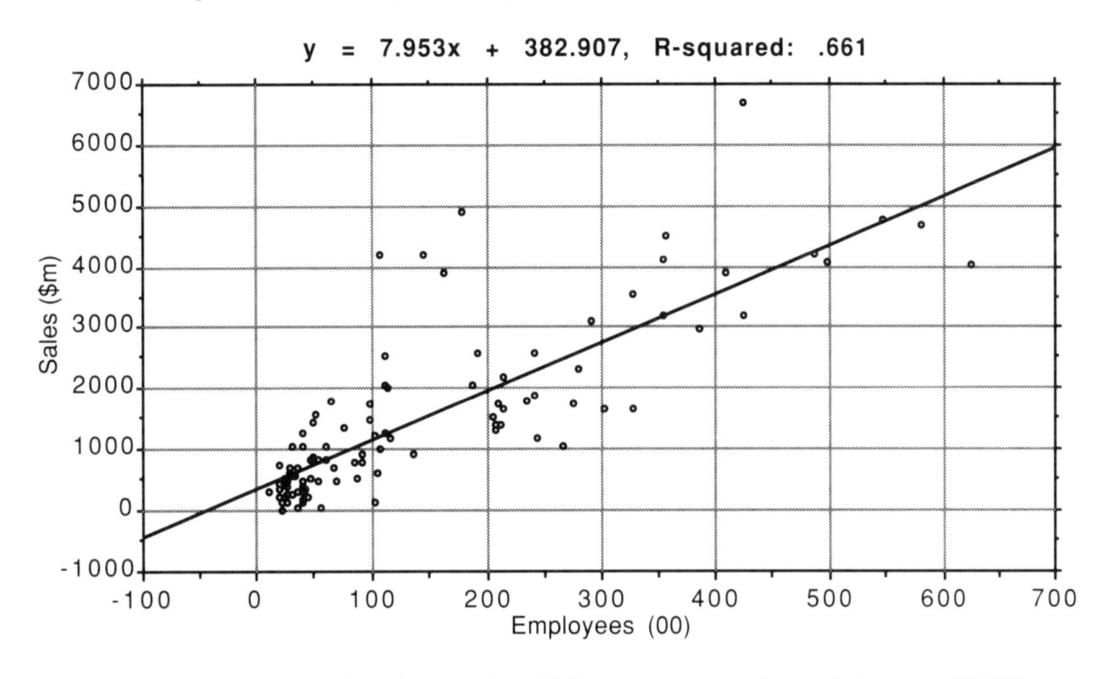

(b) $Y_c = 382.907 + 7.953(150) = \$1{,}575.86$ m. **(c)** From computer printout below $s_e = 827.589$

Simple Regression X_1 : Employees (00) Y_1 : Sales ($m)

DF:	R:	R-squared:	Adj. R-squared:	Std. Error:
99	.813	.661	.657	827.589

Analysis of Variance Table

Source	DF:	Sum Squares:	Mean Square:	F-test:
REGRESSION	1	130649113.665	130649113.665	190.756
RESIDUAL	98	67120511.325	684903.177	p = .0001
TOTAL	99	197769624.99		

(d) Per the computer output below, the slope is 7.953. This signifies that for each 100 additional employees, the average sales are increased by $7.953 m (or $79,530 per employee). The slope has a calculated t value of 13.811 and is significant at the .0001 level.

Simple Regression **X_1: Employees (00)** **Y_1 : Sales ($m)**

	Value:	Std. Err.:	Std. Value:	t-Value:	Probability
INTERCEPT	382.907				
SLOPE	7.953	.576	.813	13.811	.0001

(e) From the computer output shown above, $r^2 = .661$, so we can estimate that 66.1% of the variability in sales is associated with the number of employees.

CHAPTER 12

1. Multiple regression entails the same assumptions and limitations as simple regression, but instead of incorporating only one independent variable in the model we incorporate two or more.

3. **(a)** There is no theoretical limit, but practical considerations will be a limit. **(b)** 7

5. **(a)** $Y_c = 35.48 - 2.70X_1 + .85X_2 + 120.70X_3$ **(b)** 269.58

7. **(a)** A computer solution to the problem shows the equation is $Y_c = 2.782 + .288\ X_1 + .147\ X_2$

Regression Coefficients Table

Parameter:	Value:	Std. Err.:	t -Value:	Partial F:
INTERCEPT	2.782	1.975	1.409	
X 1	.288	.17	1.696	2.876
X 2	.147	.304	.483	.234

(b) When $X_1 = 7$ and $X_2 = 9$, $Y_c = 2.782 + .288(7) + .147(9) = 6.121$, or $6,121.

9. **(a)** s_e is the standard deviation of points around the multiple regression hyperplane. It is a measure of the unexplained, or error, deviation.

(b) The df for s_e are the number of data points in the sample, n, minus the number of independent variables, k, minus 1 (dependent variable).

(c) The s_e value is multiplied by the t or Z value for the level of prediction accuracy desired. This product is then one half of the prediction interval.

11. In ANOVA, the F_{calc} was the ratio of explained MSD to the residual, or error MSD. A large F value suggested that there was a difference in the means. In regression, we refer to the explained variation as the sum of squares due to regression (SSR) and the residual as the error sum of squares (SSE). It is still a ratio of explained to error, except with regression the explanation is the regression equation (or model), whereas with ANOVA it is due to the difference in the means of the groups involved.

13. **(a)** Both items are the same. They are simply different ways of identifying the explained variation.

(b) Both items are the same. They are different terms for the total variation.

(c) The first term, $\Sigma(Y - \bar{y})^2$ is the total variation, whereas the second, $[b_1\Sigma(X_1 - \bar{x}_1)(Y - \bar{y}) + b_2\Sigma(X_2 - \bar{x}_2)(Y - \bar{y})]$, is the explained variation. The difference between them is the unexplained variation—which is the numerator in the s_e equation.

15. Multicollinearity is the condition that occurs when some independent variables are highly correlated with each other. When this happens, the influence of one independent variable overlaps that of another, and it is not clear what proportion of the effect is explained by any given independent variable. One way of reducing multicollinearity is to delete one of more of the intercorrelated variables from the model. Another is to use a mathematical approach to reduce the effect.
17. One use is to help select highly correlated variables that tend to be of most value in predicting Y. A second is to identify multicollinearity in the data.
19. **(a)** 1.932, **(b)** 12.804, **(c)** $F_{.05,\ 6,\ 15} = 2.79$
21. **(a)** .6220, **(b)** $SSE = 1.160$, $SSR = 6.173$,
    (c) Do not reject H_0 that regression is not significant because $7.98 < 9.55$.
    (d) .842,
    (e) $R^2_A = .7367$, so the values are different. With a small sample of only two independent variables and only 6 observations, the R^2 value is unlikely to be very accurate.
23. **(a)** 1.893, **(b)** .921, **(c)** 43.794,
    (d) Reject H_0 that regression is not significant at the .05 level because $43.794 > 3.06$, which is $F_{.05,4,15}$.
25. CL of $\beta_4 = -.206$ to $-.014$.
27. **(a)** The r values of 1.00 are reporting a correlation of each variable with itself, which must be 1.00.
    (b) The apparent multicollinearity problem is the X_1, X_4 and X_4, X_1 correlation of .85. The X_3Y value of .74 is not a problem because that is the correlation between a predictor X_3 and a Y variable.
29. Using the computer printout data in the previous problem:
    (a) From the Analysis of Variance Table, $F_{calc} = 32.74$. This is significant at the .0001 level, so the overall regression is significant at the .05 level as well.
    (b) For this problem with $n = 20$ observations and $k = 4$ independent variables, the reject value for t is $t_{\alpha,n-k-1} = t_{.05/2,20-4-1} = t_{.05/2,15} = 2.131$. From the Regression Coefficients Table, the absolute value of two variables exceeds this. The significant variables are X_2 Temp °F with a t value of 2.339, and X_3 Snow Depth with a t value of 4.061.
    (c) Yes. Multicollinearity can be confirmed by checking the simple correlation between any of the predictor variables. If the computer program being used delivers a correlation table, it would reveal this quite clearly, as shown below.

Correlation Table

Variables	Y	X_1	X_2	X_3	X_4
Y	1.000	−.744	.874	.879	−.812
X_1	−.744	1.000	−.810	−.672	.686
X_2	.874	−.810	1.000	.717	−.853
X_3	.879	−.672	.717	1.000	−.699
X_4	−.812	.686	−.853	−.699	1.000

31. Computers are essential because (1) the initial (simultaneous) solution of several equations can be extremely time consuming and requires significant manipulation and (2) computers can assist substantially in the subsequent calculations and modeling activities. The job is too big to be done by hand calculation.
33. An example of a computer printout for this problem is as shown:

Multiple - Y : Y   Two X variables

DF:	R-squared:	Std. Err.:	Coef. Var.:
5	.874	.852	15.033

Analysis of Variance Table

Source	DF:	Sum Squares:	Mean Square:	F-test:
REGRESSION	2	15.156	7.578	10.443
RESIDUAL	3	2.177	.726	.025 < p ≤ .05
TOTAL	5	17.333		

Regression Coefficients Table

Parameter:	Value:	Std. Err.:	T-Value:
INTERCEPT	2.992	1.856	1.612
X1	.486	.197	2.47
X2	-.030	.09	-.331

(a) From the computer calculation, the standard error is $s_e = .852$.
(b) $R_2 = .874$, **(c)** $R_A^2 = .790$. An adjustment must be made because the sample size is small and the number of predictor variables is large relative to the number of observations, that is, $2 \div 6$ or 33%.
(d) The F_{calc} for the regression is 10.443, which is significant at the .05 level, but not at the .01 level.
(e) The .10 reject value for the t-test of the regression coefficients is $t_{.10/2,\ 6-2-1} = 2.353$. At this .10 level the calculated t value for X_1 (of 2.470) is significant, but the $-.331$ for X_2 is not. At the .05 level, the t reject value is 3.182, so neither coefficient is significant. This is not a very strong model.

35. **(a)** 2.261, **(b)** .863, **(c)** $s_{b_1} = .379$, $s_{b_2} = .106$,
(d) The t_{reject} value is $t_{.05/2,9} = 2.262$ so the X_1 coefficient (of $-.686$) is not significant but the X_2 value (of .252) is significant. **(e)** 6 to 16 errors

37. **(a)** $Y_c = 6153.473 + 8.896X_1 + 1.353X_2 - 40.577X_3$ **(b)** $s_e = 229.736$,
(c) Yes. The F_{calc} value of 36.669 is significant at the .0001 level so the regression is highly significant.
(d) $Y_c = 6153.473 + 8.896(110) + 1.353(375) - 40.577(20) = \$6,827.87$,
(e) $R^2 = .887$, so 88.7% of the variation in product cost is explained by the three variables.
(f) The reject value for $t_{\alpha/2,n-k-1} = t_{.05/2,\ 18-3-1} = 2.145$. From the regression coefficients table, the absolute values of the coefficients for assembly steps $X_1 = 3.244$ and batch size $X_3 = -3.834$ both exceed this and are significant.

39. **(a)** $Y_{PL} = 6,628.87 \pm 2.145(229.736) = \$6,335$ to $\$7,321$
(b) $Y_c = 6497.676 + 10.902X_1 - 44.626X_3$, $s_e = 228.399$, $R^2 = .88$
(c) $Y_{PL} = 6,804.38 \pm 2.131(228.399) = \$6,317.66$ to $\$7,291.10$
(d) The two estimates are close. Note that the standard errors of estimate are essentially the same and the coefficient of multiple correlation is almost as high. The X_2 variable really did not add much to the model and it had a high cross correlation (multicollinearity) with X_1 so the model is better without it.

41. A stepwise multiple regression program (using an F value of 4 to enter) confirms that share price is the only significant parameter in the model, so the resultant equation is $Y_c = .319 + .032\ X_3$.

STEP NO. 1 Stepwise Regression Y1 :Div ($) 5 X variables

Variables in Equation

Parameter:	Value:	Std. Err.:	Std. Value:	F to Remove:
INTERCEPT	.319			
Sh Price ($)	.032	.006	.497	32.098

Variables Not in Equation

Parameter:	Par. Corr:	F to Enter:
OP (yrs)	.164	2.67
AG (yrs)	.186	3.46
Exp (%)	-.122	1.469
Ad Exp ($m)	-.017	.029

CHAPTER 13

1. A time series is a set of successive observations of a variable at regular intervals in time, whereas an index number is the ratio of a summary measure of data in a given time period, to a similar measure of comparable data in a base period. Both involve the use of descriptive data from different points in time.

3. **(a)** intercept, a, and slope, b,
(b) One method would be to draw a trend line through the plot, measure its slope and intercept, and use that to write the equation $Y_t = a + bX$.

5. **(a)** 770 verdicts, **(b)** $Y_t = 770 + 60\ X$ ($1990 = 0$, X = yrs, Y = number verdicts/yr),
(c) $Y_t = 770 + 5\ X$ (July 1, $1990 = 0$, X = months, Y = number verdicts/yr)

7. **(a)** $Y_t = .38 + 5.93\ X$ ($1980 = 0$, X = yrs, Y = number of robots $\times$ 100),
(b) This means that the trend value for the number of robots was 38 in 1980 (when $X = 0$) and it increases by 5.93(100) or at 593 per year.
(c) 8,933 robots installed.

9. **(a)** Yes, a linear trend does appear to fit the data.
(b) $Y_t = 5.46 + 1.00\ X$ ($1975 = 0$, X = yrs, Y = \$ billion),
(c) $Y_t = 20.462 + 1.00\ X$ ($1990 = 0$, X = yrs, Y = \$ billion),
(d) 30.462 or $30.5 billion.
(e) Although the data do have a strong linear trend, projecting a trend 10 years into the future is very risky. A myriad of social, political, or economic factors could cause the trend to shift, and the projection is very likely to be wrong.

11. Seasonal effects are recurring and predictable increases and decreases that occur at regular times throughout a time series.
(a) Toy sales are likely to be higher in December (at Christmas time).
(b) Subway traffic is heavier during the "rush" hours when commuters are going to and from work. This would be a daily and/or hourly seasonal.
(c) The workload would probably tend to be heaviest at the end of the year when the "books are closed" or prior to when income taxes must be filed.

13. A specific seasonal is the preliminary seasonal value obtained when the actual Y value is divided by the moving average centered on the given period. In equation form it is $(TCSI) \div (TC)$ which equals (SI). The specific seasonals might differ for the same month, not because the trend is different, but because of the variation in actual values. This is largely due to the irregular component which has not yet been removed.
15. Given the trend and seasonal index, the steps are: (1) Compute the seasonally adjusted value $[(TCSI) \div S = (TCI)]$ (2) Divide by trend to isolate the cyclical and irregular $[(TCI) \div T = (CI)]$ (3) Smooth out the irregular by averaging the (CI) values (e.g., by using a 3 or 5 period moving average). This yields the cyclical index. (4) Divide (CI) by C to remove the cyclical and leave the irregular index.
17. The specific seasonals beginning with 3rd quarter 1986 are 96.55, 111.36, 113.83, 68.85, 89.96, 111.63, 130.80, and 66.67.
19. The format for computing the REVISED VALUES OF CYCLICAL & IRREGULAR COMPONENTS is shown below, along with calculation of the first value (2nd Qtr 1982) and last value (3rd Qtr 1988).

Year &	Quarter	(1) Actual Units Y	(2) Trend Value T	(3) Seasonal Index S	(4) Seasonally Adjusted $\frac{Y}{S} = \frac{TCSI(100)}{S} = TCI$	(5) C & I (4)÷(2) CI	(6) 3-Period MA of (5) C	(7) Irregular (5)÷(6) I
1982	1	290	266.85	110.08	263.44	98.72		
	2	235	271.00	84.25	278.93	102.93	97.28	105.81
1988	3	310	374.75	96.72	320.51	85.53	90.21	94.81
	4	460	378.90	108.95	422.21	111.43		

Other irregular values are:

For 1982:		105.81,	98.28,	98.88	**For 1983:**	95.51,	99.82,	100.60, 98.09
For 1984:	98.65,	106.75,	98.97,	94.54	**For 1985:**	98.35,	108.93,	96.18, 103.92
For 1986:	93.10,	107.24,	97.36,	100.78	**For 1987:**	105.91,	86.32,	100.02, 98.19
For 1988:	132.44,	83.14,	94.81					

21. Using the data of problem 20, the first value corresponds with July 1987 where the centered 12-period moving average is 4.46, and the ratio of actual to moving average is 22.43. Remaining months of 1987 are 65.45, 187.83, 120.00, 57.60, and 74.42. For 1988 the twelve months are 89.55, 67.61, 130.72, 144.00, 139.35, 60.38, 74.07, 99.41, 134.83, 106.67, 81.82, and 68.18. The first six months of 1989 are 92.31, 129.34, 193.29, 107.69, 100.60, and 82.29.
23. Calculations yield modified totals of 268.0, 434.4, 314.0, and 616.0 for the four quarters, respectively. After a correction factor of .9802 is applied to the modified means, the final seasonal indexes are 65.67, 106.44, 76.94, and 150.94 for the four quarters, respectively.
25. Cyclical and irregular components for the first five months of 1987 are as shown below. When the remaining values are computed, they suggest a cyclical effect is present. (A 5-period moving average would make the cycle even more apparent.) Column (7) evidences some irregular effect in the data because the numbers deviate substantially from 100.

ANALYSIS OF CYCLICAL & IRREGULAR COMPONENTS

MODEL Y = 3.1 + .2*X* (Dec 1986 = 0. *X* = mo, *Y* = units sold)

Year &	Month	(1) Actual Units *Y*	(2) Trend Value *T*	(3) Seasonal Index *S*	(4) Seasonally Adjusted $\frac{Y}{S} = \frac{TCSI(100)}{S} = TCI$	(5) *C* & *I* (4)÷(2) *CI*	(6) 3-Period *MA* of (5) *C*	(7) Irregular (5)÷(6) *I*
1987	1	4	3.3	89.8	4.45	134.98		
	2	2	3.5	97.3	2.06	58.73	92.72	63.34
	3	5	3.7	160.0	3.12	87.46	95.86	88.10
	4	7	3.9	124.3	5.63	144.40	117.45	122.95
	5	6	4.1	118.5	5.06	123.49	122.28	100.99
	6	3	4.3	70.5	4.26	98.96	89.68	110.35

27. In a composite *price* index, the base year quantities serve as weights, whereas in a composite *quantity* index, the base year prices serve as weights.

29. The simple price, quantity, and value indexes for 1981 are 100.0. For other years they are:

Year	1981	1982	1983	1984	1985	1986	1987
Price	65	61	57	49	45	41	36
Price Index	100.0	93.8	87.7	75.4	69.2	63.1	55.4
Quantity (000)	.2	.5	.9	2.0	4.1	7.7	13.0
Quantity Index	100.0	250.0	450.0	1000.0	2050.0	3850.0	6500.0
Value (*PxQ*)	13.0	30.5	51.3	98.0	184.5	315.7	468.0
Value Index	100.0	234.6	394.6	753.8	1419.2	2428.5	3600.0

31. The price index for food items with year 5 as the base can be calculated in this format:

ITEM	Base Q_0	Price per Unit ($) Yr 1	Yr 2	Yr 3	Yr 4	Yr 5	Price × Base Yr Quantity Yr 1	Yr 2	Yr 3	Yr 4	Yr 5
A	14	.30	.33	.36	.36	.39					
B	9	.25	.24	.30	.32	.30					
C	16	.20	.25	.28	.32	.30					
D	1	2.00	2.40	2.50	2.50	2.60					
Totals		- - - - Not Required - - -									
Price Index							100.0	113.1	126.4	133.4	133.6

33. **(a)** Price index can be calculated in the standard tabular format:

ITEM	Base Q_3	Price per Unit ($) Yr 1	Yr 2	Yr 3	Yr 4	Yr 5	Price × Base Yr Quantity Yr 1	Yr 2	Yr 3	Yr 4	Yr 5
A	18	775	650	545	425	390					
B	30	600	395	320	275	295					
C	143	385	240	200	185	175					
D	10	—	560	425	310	365					
Price Index							166.5	121.5	100.0	87.0	85.2

(b) The quantity index can be calculated in similar format with the result as shown:

	Yr 1	Yr 2	Yr 3	Yr 4	Yr 5
Quantity Index	20.8	36.7	100.0	139.7	161.2

(c) The value index can be calculated in similar format. (Year 1 and 2 totals shown below for reference.)

ITEM	Price per Unit ($) Yr 1	Yr 2	Yr 3	Yr 4	Yr 5	Number of Packages Sold Yr 1	Yr 2	Yr 3	Yr 4	Yr 5
Totals of each year Price × Quantity, or $(\Sigma P_n Q_n)$						19,920	23,400			
Value Index $[(\Sigma P_n Q_n) \div 52{,}260]$ (100)						38.1	44.8	100.0	122.4	137.1

(d) Prices have dropped substantially and volume has increased substantially. The value index shows that the volume increase has more than offset the price drop, for the total value shows a significant increase.

35. **(a)** LASPEYRES PRICE INDEX = [16,565/23,400] (100) = 70.9
(b) PAASCHES PRICE INDEX = [71,665/103,072] (100) = 69.4
(c) The two indexes are unusually close. This is because as prices dropped, consumption increased to compensate for the decline in prices. As a result, both indexes yield about the same conclusion, i.e., that the price index for these software items has dropped by about 30%.

37. Seasonal effects occur at regular intervals with a relatively predictable level of magnitude. Cyclical factors cannot always be identified and are not as consistent and predictable as seasonal factors.

39. **(a)** The sequence is (1) seasonal, (2) trend, (3) irregular (leaving cyclical), and (4) cyclical (leaving irregular).
(b) by division,
(c) The seasonal is first removed by smoothing in order to calculate the seasonal index. Later the irregular is removed by smoothing.

41. **(a)** It is difficult to say whether a linear trend is present from the graph, but the latter values are generally larger than earlier values, and a trend could very well be present.
(b) From projecting a trend it appears the year 20 expenses could be in the neighborhood of $50,000

43. **(a)** slope: $b = 3.50$, intercept: $a = 24.5$, $Y = 24.5 + 3.50\,X$ (19X0 = 0, X = yrs, Y = housing starts),
(b) $Y = 24.5 + 3.50\,(9) = 56$ housing starts.

45. Trend Line is $Y_t = 56.000 + 1.788\,X$ (1890 = 0, X = Yrs, Y = Population). The year 2000 is (2,000 − 1890) = 110 years distant from 1890. $Y_t = 56.000 + 1.788\,(110) = 252.68$, or 253 million. This is lower than the Census Bureau estimate of 267 million. (The growth rate is not linear.)

47. A graph of the data shows an upward (possibly linear) trend.

(a) Specific seasonals, along with the modified totals for Q_1 and Q_2, are shown below:

Year	Q_1	Q_2	Q_3	Q_4
1982	—	—	96.6	102.7
1983	103.9	97.2	97.3	102.1
1984	104.1	97.6	95.8	99.2
1985	104.9	99.2	98.4	103.6
1986	101.5	96.6	98.8	104.0
1987	101.0	97.0	97.2	101.2
1988	103.7	97.6	—	—
Modified Total	413.2	389.4		
(b) *Final Indexes* are: (*M*)(.9990)	103.2	97.2	97.3	102.3

(c) The trend, cyclical, and irregular factors for 1983 and 1984 are as shown below:

Year &	Quarter	(1) Actual Units Y	(2) Trend Value T	(3) Seasonal Index S	(4) Seasonally Adjusted $\frac{Y}{S}=\frac{TCSI(100)}{S}=TCI$	(5) C & I (4)÷(2) CI	(6) 3-Period *MA* of (5) C	(7) Irregular (5)÷(6) I
1983	1	290	305.7	103.2	281.0	91.9		
	2	280	308.0	97.2	288.1	93.5	93.3	98.5
	3	285	310.4	97.3	292.9	94.4	94.9	98.5
	4	310	312.7	102.3	303.0	96.9	96.6	97.7
1984	1	320	315.1	103.2	310.0	98.4	98.1	98.8
	2	305	317.4	97.2	313.8	98.9	99.0	99.4
	3	310	319.7	97.3	318.6	99.7	99.6	99.3
	4	330	322.1	102.3	322.6	100.2	100.5	99.2

Other year values for the final (irregular) calculation are as follows:

	1985	1986	1987	1988	1989
Irregular Q_1	99.3	94.6	102.9	103.7	99.2
Q_2	100.8	98.4	100.0	100.5	100.2
Q_3	100.6	100.5	100.1	101.1	100.5
Q_4	97.5	101.9	102.9	100.4	

(d) Yes, it appears to be fairly strong, rising from 93.3 to 109.7 and back down to 95.8. However, more data should be examined to confirm the exact nature of the cyclical effect.

49. Judgment samples are used in constructing index numbers, whereas probability samples are used when inferences are to be made.

51. The simple price, quantity, and value indexes are as follows:

Year	1982	1983	1984	1985
Price Index	100.0	96.7	87.9	93.1
Quantity Index	100.0	160.0	260.0	340.0
Value Index	100.0	154.7	228.4	316.5

53. Using 1986 as the base year, the composite value indexes for 1986, 1987, 1988, and 1989 are 100.0, 116.1, 128.6, and 136.9, respectively.

55.

Year	1985	1986	1987	1988	1989
(a) Deflated inventory bal. ($000)	311.8	304.5	309.3	318.8	321.7
(b) Same (with 1987 = base)	92.0	98.1	100.0	102.6	106.8

57. November 15, 1992 is 7 yrs times 4 quarters/yr = 28 quarters away.

Trend: $Y_t = 280.0 + 6.2\,(28) = 453.6$ or $453,600.

Trend, seasonal, & cyclical: $Y = 453.6\,(1.051)\,(1.18) = 562.5$

Therefore the forecast is $562,500.

CHAPTER 14

1. Parametric methods rely upon the use of parameters such as μ and σ and typically require some limiting assumptions about the population. Nonparametric methods do not assume any underlying distribution parameters and generally require only minimal assumptions about the population under study.

3. **(a)** the two-sample, or paired data sign test, **(b)** the runs test

5. The hypotheses are:

H_0: the sportswear items are randomly arriving from A and B

H_1: the sportswear items are not randomly arriving from A and B.

Using the runs test at $\alpha = .25$, we reject if $Z_{calc} < -1.15$ or > 1.15. There are 10 runs in the data, and $\mu_r = 12.92$ and $\sigma_r = 2.38$. Reject hypothesis that observations are random because Z_{calc} of $-1.23 < -1.15$.

7. The hypotheses are: H_0: $\mu = 10$ lbs and H_1: $\mu < 10$ lbs. Use sign test with the binomial distribution for $n = 14$ and $p = .50$. Reject if $P(\geq 4$ negative signs$) < .15$. Do not reject because probability of getting 4 or more negative signs is .9712. There is no basis for questioning this hypothesis.

9. The hypotheses are: H_0: $\pi = .50$ and H_1: $\pi \neq .50$. Use the paired difference sign test, and let $p =$ proportion of + signs. For .08 level test, reject if $Z_{calc} < -1.75$, or if $Z_{calc} > 1.75$. $Z_{calc} = 1.964$, so we reject H_0 and conclude that the expectations of the brokers differ depending upon whether a Republican or Democrat is appointed to the position.

11. **(a)** The rank sum test is the nonparametric equivalent to the small sample t-test and the corresponding large sample normal distribution test of the difference in means.

(b) None, except that the variables are continuous and the samples independent.

13. First, arrange each group into an array and assign ranks to the elements in the two groups. The hypotheses are: H_0: $\mu_A = \mu_P$ and H_1: $\mu_A \neq \mu_P$. Using the rank sum test, we reject if $Z_{calc} < -1.96$ or > 1.96. $Z_{calc} = -.69$, so the conclusion is do not reject H_0 of equal means. The sums of the ranks are not sufficiently different to reject the hypothesis at the .05 level.

15. First, arrange each group into an array and assign ranks to the elements in the two groups. The hypotheses are: H_0: $\mu_M = \mu_F$ and H_1: $\mu_M > \mu_F$. Using the rank sum test, we reject if $Z_{calc} > 2.33$. $Z_{calc} = 4.22$, so we reject the H_0 of equal means. The sums of the ranks are significantly different, allowing us to conclude, at the .01 level, that the lung cancer rate among men is significantly greater than among women in these countries. **17.** $-.7536$

19. **(a)** .6818,

(b) Hypotheses are: H_0: $\rho_r = 0$, H_1: $\rho_r \neq 0$, and reject value is $t_{.10/2,8} = 1.860$. Reject H_0 of no correlation because $2.63 > 1.860$.

21. Using the runs test at $\alpha = .05$, we reject if $Z_{calc} < -1.96$ or > 1.96. There are 11 runs in the data, and $\mu_r = 15.93$ and $\sigma_r = 2.68$. Do not reject hypothesis that selections are random because $-1.96 < Z_{calc}$ of $-1.84 < +1.96$.

23. Using the sign test with the binomial distribution for $n = 13$ and $\pi = .50$, we reject the H_0: average is 3.1% because the probability of getting 10 or more positive signs is .0461, which is less than the α limit of .05.

25. Using the sign test with the normal distribution, we reject the H_0: $\pi = .50$ because Z_{calc} of $3.41 > 2.33$. Opinions of passengers are changed as a result of viewing the film.

27. Using the rank sum test, do not reject H_0: $\mu_{US} = \mu_F$ because $-2.24 < Z_{calc}$ of $-1.18 < +2.24$.

29. Both tests use ranks and are nonparametric. The rank sum test is for testing whether the means of two populations are different and corresponds to the t test and normal distribution test of the difference in means. The test of a rank correlation coefficient is used to determine whether two population variables are closely associated with each other, so it is a test of association or relationship.

31. Using the rank sum test, reject the H_0: $\mu_R = \mu_C$ because Z_{calc} of $-3.62 < -1.96$.

Index

Absolute frequency distribution, 31, 34
Acceptance region, 268–269, 271, 274, 278–286, 299
Accuracy in sampling, 9–10
Addition rule, 115, 117–121, 137
Adjusted R^2, 442, 451
Alpha (α), definition of, 266
Alternative hypothesis, 258–259, 264–265, 271–275, 299, 346–347, 350, 356, 362, 363, 368, 402, 412, 435–439, 517–518, 520–523, 527–528, 530, 536
Analysis of variance (ANOVA), 342–344, 358, 363–369, 371–373, 435, 437
  assumptions of, 364
  definition of, 363
  logic of, 364
  table, 367–368
  two-way, 368–369
Analysis of variance for regression, 435–437, 446–447
Analytical tables, 28
ANOVA. *See* Analysis of variance.
Answers to odd-numbered problems, Appendix H
Appendix tables, 545–573
Approximate prediction intervals, 397–398, 434–435
A priori probability, 103
Area Charts, 42–43
Arithmetic mean. *See* mean.
Array, 28, 57, 219
  definition of, 27
Bar Charts, 42–43, 46
Bayes, Reverend Thomas, 124
Bayes' Rule, 119, 124
Bayes' Theorem, definition of, 124
Bayesian decision theory, 100, 101, 124–128
Before and After tests, 318–322
Bernoulli process, 156
Beta (β). *See* also regression coefficients.
  computation of, 290–294
  definition of, 266
Beta coefficients, 439
Between-group variation, 364–367
Bias, 9, 200, 230, 236
Binomial distribution, 146, 147, 155–162, 167, 183, 189, 246, 262, 324, 351, 521–522
  conditions, 156
  graph, 161
  probability equation, 158
  tables, 160, Appendix C
Bivariate data, 382, 394, 415
BMDP, 424, 449
Box-Jenkins, 501
Bureau of the Census Catalog, 13

Cases
  Chapter 1 Statistics In the Supermarket, 17
  Chapter 2 Midland Distribution Company, 46–49
  Chapter 3 Pfafner, Jeffrey and Jones Investment Company, 91–93
  Chapter 4 Alliance National Bank, 138–139
  Chapter 5 Detroit Lakes Power Company, 190–192
  Chapter 6 Debate at the City Council, 231–232
  Chapter 7 Northwestern Grain Cooperative, 253–254
  Chapter 8 International Printing and Publishing Company, 300–301
  Chapter 9 Transcontintental Communications Company, 336–337
  Chapter 10 Aerospace Industries, 373–374
  Chapter 11 Kathy Dubois, Legislative Assistant, 416–417
  Chapter 12 Forecasting at Sun Belt Equipment Company, 452–453
  Chapter 13 Winter Coats Inventory at Fashionworld, 505–507
  Chapter 14 Globe Personnel Services - We'll Find You a Job, 538–540
Cause and Effect, 390, 413, 415
Census, 7, 235
Central limit theorem, 198, 214–216, 237–238, 261–263
  definition of, 215
Central tendency, 54–57, 77–80, 89, 214
  measures of, 61–69
Charts, 42–43, 46, 48
Chart/graph software, 45, 46
Chebyshev's Inequality, 146, 187–188

statement of, 187
Chi-square, 342–358, 371–372
Chi-square distribution, 344–346, 348
definition of, 345
tables, Appendix F
Class
boundaries, 32–33, 78–79
interval 30, 32–33, 78–79
limits, 32–33
midpoint, 33, 37, 38, 77–78
Classical probability, 103–104, 137
Classical time series. *See* Time Series.
Cluster sampling, 205, 207–208, 230
Coefficient of correlation, 406–412, 415, 422, 533–536
definition of, 407
inference about the, 411–412, 535–536
Coefficient of determination, 407–412, 427, 440–442
Coefficient of multiple
correlation, R, 442
determination, R, 422, 423, 429, 440–442, 451
definition of, 440
Coefficient of variation, 54, 56, 85–87, 91–93
definition of, 86
Collectively exhaustive events, 107, 108
definition of, 108
Combinations, 100, 133–135, 137, 158–159, 211
Comparing the mean, median, and mode, 83, 89
Complement, 107–108, 109
rule, 115–116, 119, 120, 137
Components of a time series, 461, 503
Composite index numbers, 486–487, 490–495, 504
Computers (and microcomputers), 25, 449
software programs, 16, 44–46, 189, 287, 358, 424, 428, 449
use in statistical analysis, 3, 12, 15–16, 20, 23, 44–46, 47, 88, 137, 390, 422, 423, 424, 428–430, 436–437, 449, 463, 501
Conclusions, 11, 17
Conditional probability, 100, 101, 114–115, 393–394
definition of, 114
Confidence coefficient. *See* Confidence level.
Confidence interval, 234, 236–257, 278, 400, 439–440
definition of, 236
estimate of correlation coefficient, 412
estimate of difference between means, 313, 322
estimate of difference between proportions, 327
estimate of mean, 239–241, 252, 400
estimate of slope, 402–403, 439–440
estimate of proportion, 246–249, 252
Confidence level, 238, 239, 252, 254
Confidence limit. *See* Confidence interval.
Consistent estimator, 236
definition of, 236
Constant, 27
Consumer Price Index, 496–497
Contingency table, 354–355, 357–358, 372
Continuity correction factor, 182–184, 357
Continuous data, 27, 32–33, 46, 148
Continuous variables, 27, 32–33, 153, 169
definition of, 27
Continuous probability distribution, 151–154, 169, 221
definition of, 151
Corporate statistical database, 20–23 (*Also* Appendix A)
Correlation, 380, 406–413, 415–417, 423, 442–444, 533–535
Correlation analysis, 380, 383, 404–413, 415, 440–444, 501, 533–536
definition of, 383
Correlation coefficient, 410–412, 415, 443–444, 533–536
Correlation matrix, 443–444, 451
Counting procedures, 120–135
County and City Data Book, 13
Covariation, 433–434, 441
Criteria for solution, 10–11, 16
Cross-classified tables, 121, 354–357
Cumulative frequency distribution, 34–35, 38–40, 46, 47–48
definition of, 34
Curvilinear relationship, 383–384
Cyclical variation, 458, 459, 461–462, 471–472, 480–483, 502, 503
definition of, 461

Data, 3, 46, 57
analysis and testing, 10–11, 17, 260–261, 343
array, 27, 57
bivariate, 382, 415
census and sample, 7, 156, 267
collection of, 10–11, 17
continuous, 27, 46, 151
continuum of, 6
definition of, 26
discrete, 27, 46, 151
external 11, 12–13
grouped, 56–57, 77–83
internal, 11, 12
multivariate, 424–425
nominal, 515
point, 26
primary, definition of, 12
rank-ordered, 515
raw, 27, 30, 57, 261
secondary, definition of, 12
set, 26

sources and types of, 12–15, 20, 199
ungrouped, 56–57, 71–76
univariate, 382
Database, 1, 15–16, 200, 449
Deciles, 65
Decision
making, 3–6, 16, 197
problem, 11, 381
rule, 296, 329–331
support system (DSS), 12, 16
Decision tree analysis, 127–128
Deduction, 5, 16

Degrees of freedom, 222–225, 311, 348, 351, 354–355, 360–361, 367, 433, 436–437
Deming, W. Edwards, 2
Density function, 169
Dependence, 120–121, 317, 321
Dependent variable(s), 382, 423, 424, 428, 440
Descriptive statistics, 4, 11, 57
Deseasonalization, 478–479
Difference, tests of, 308. *See* also tests of differences.
Discrete data, 27, 32, 46
Discrete probability distribution, 151–154, 155–162, 164–167
definition of, 151
Discrete variables, 27, 148, 153, 169
definition of, 27
Dispersion, 54–55, 71–76, 80–82, 90, 215–216
measures of, 56, 71–76, 80–82, 90
Distribution-free test, 354, 516, 519

Econometric forecasts, 501
Efficient estimator, 214, 236
definition of, 236
Error, 8–10, 243–245, 265–268, 331, 385, 398–399, 408–409, 415, 425, 433–434. *See also* Sampling error, Type I and II errors.
Error variance, 367–368
Estimate, 7–8, 236–238
interval. *See* Interval estimate.
point. *See* Point estimate.
Estimated mean, 239–241, 390
Estimated regression coefficient, 387–391, 426–430, 438–440
Estimation, 234–257, 278, 313, 322, 327, 439–440, 466
error, 243–245, 266–268, 290–294, 299, 397–400
logic of, 237–238
Estimation using the
regression line, 393–403, 434–435
time series, 465–466
Estimators, 240
consistent, 236
efficient, 214, 236
unbiased, 214, 236
Event, 102, 105, 107–108, 150–152
definition of, 105
Expected frequencies, 346–347, 349, 355–356, 372
computation of, 347, 349–350, 355
Expected value, 149
of sample means, 214
Experiment, defintion of, 199
Explained variation, 408–410, 433–437, 440–442
Exponential distribution, 152

F-distribution, 342–344, 360–369, 372, 435–437
definition of, 360
F table, 361, Appendix G
F-test for regression, 435–437, 451
Factorials, 130
Federal Reserve, 13, 381
Finite population, 209, 218–219, 248–249
Finite population correction factor, 218–219, 230, 245, 249, 252
Fisher's Ideal Index, 495
Forecasting, 458, 459, 501–502, 504
Frame, 200
Freehand estimation, 463
Frequency distributions, 30–35, 46, 47, 57, 78, 81, 217
absolute, 31, 34, 38–39, 78
construction of, 30–33
cumulative, 34–35, 38–40, 47–48
definition of, 30
relative, 34–35, 38–39
of sample means, 211–213, 217
Frequency polygon, 25, 37-38
definition of, 37
Frequency table, 31, 32, 109–110
Fundamental rule, 100, 131

Game theory, 6
Gauss, Karl, 171
Gaussian distribution, 171
Geometric mean, 69, 89
Goodness-of-fit-test, 346–351, 372
Gossett, W. S., 221
Grand mean, 364–367
Graph constructions, 37–39, 42–43, 45, 161
Graphic presentation of data, 36–40, 44–46, 161, 170
Grouped data, 30–35, 77–83

Harmonic mean, 68, 89
Histogram, 25, 36–37
definition of, 36
Hypergeometric, 147, 152, 153
Hyperplanes, 424, 427, 451
Hypothesis, 260–265, 299, 309, 346, 347, 350, 356, 362–363, 368, 402, 412, 438–439, 517–518, 520–523, 527–528, 530, 536
definition of, 260
Hypothesis testing, 258–301, 306–337
of correlation coefficient, 411–412, 525–526
definition of, 260

logic of testing, 261–263, 308–310
method of, 260, 309–310
procedure for, 275, 312
of regression line slope, 401–402, 438–439
of several means, 363
of several proportions, 346
two-sample tests, 306–337
types of tests, 260–261, 264–265, 309
of variances, 362

Independence, statistical, 108, 117, 120–121, 137, 150–152, 308
Independent variable, 382, 424, 428–429, 440–441
Index numbers, 85, 458–460, 486–504
composite, 486–487, 490–495, 504
definition of, 460
purpose of, 486
simple, 488–490
types of, 486–487
Induction, 5, 16, 237–238
Industrial Production Index, 497
Inferential
procedures, 11, 221, 238, 261–263, 275, 308–310, 358, 394, 401–403, 411–412, 433–440, 465–466, 520–528
statistics, 5
Infinite population, 151–152, 209, 217
Information, 11, 25
Intercept, 387, 389, 426–430, 463–465
Interquartile range, 72, 90
Intersection of events, 108–109, 116–121
Interval estimate, 8, 234–237
definition of, 8, 236
Intervals, number of, 32
Introduction to statistics, 1–23
Irregular variation, 458, 459, 461–462, 471–472, 480–483, 503
definition of, 461

Joint probability, 100, 101, 114
definition of, 114
Judgment sampling, 200, 204–205

Kurtosis, 54, 82–83

Laspeyres Index, 458, 494–495
Leading indicators, 452–453
Least-squares method, 386–390, 393–394, 463–466
Level of significance (α), 268–269, 275, 287, 308, 312
definition of, 268
Line diagrams, 42–43
Line of best fit, 386–391, 393–403
Location, measures of, 56

Mail questionnaire, 13, 14, 16, 200
Management information system (MIS), 12, 16
Management Science, 3–4
Marginal probability, 100, 101, 112–113
definition of, 113
Mathematical methods, 4, 6
Mean, 54–55, 77–78, 89, 181, 186–187
of binomial distribution, 162, 181
calculation of, 8, 61–64, 77–78, 186–187
definition of, 62
estimate of, 234–257
of normal distribution, 170–173
of sampling distribution, 212–213
of uniform distribution, 153–154
Mean absolute deviation (MAD), 54–55, 71, 72–73, 76, 90
Mean square deviation (MSD), 367, 435
Measures of central tendency. *See* Central tendency.
Measures of dispersion. *See* Dispersion.
Measures of variability. *See* Dispersion.
Median, 39–40, 64–65, 78–79, 89
definition of, 64
Microcomputer. *See* Computer.
Minitab, 16, 424, 449
Mode, 54–55, 68, 80, 89
definition of, 68
Model
ANOVA, 363–367
multiplicative, 462
regression, 384–385, 424–425, 427
time series, 462
Modified mean, 477–478
Monthly Labor Review, 13
Moving average, 458, 473–474, 475–477, 481–482, 503
Multicollinearity, 422, 443–444, 451
definition of, 443
Multinomial distribution, 153, 344
Multiple choices, 100, 132, 135, 137
Multiple correlation, 423, 442–444
Multiple regression, 422–440, 449–453
Multiple regression and correlation analysis, 422–455
Multiple regression equations, 427-430
Multiple regression model, 424–430
Multiplication rule, 116–117, 119–121, 137
Multiplicative model, 462
Multivariate analysis, 422, 423–424, 449
Mutually exclusive events, 107–

108, 110, 137
definition of, 108

Negatively skewed distribution, 83
Nonlinear, 424–425, 428, 469
Nonparametric methods, 514–540
advantages and disadvantages of, 516–517
definition of, 516
Nonparametric tests, 354, 514–540
Normal distribution, 146, 147, 169–184, 189, 190–192, 214–216, 221, 237–238, 272, 309, 344–345, 351, 357–358, 394, 397–400, 407, 415, 427, 518–519, 522–523
approximation to binomial, 180–184, 226
characteristics, 170
equation, 169
density function, 169
importance of, 171
in regression, 393–400, 407, 415, 427
Table, 174–180, Appendix B
Normal equations, 387, 389, 427, 430
Null hypothesis, 258–260, 264–265, 271–275, 299, 308–309, 312, 346, 347, 350, 356, 362, 363, 368, 372, 402, 412, 435–439, 517–518, 520–523, 527–528, 530, 536
definition of, 264

Observation, 13, 199
Ogive, 38–40
definition of, 38
One-sample hypothesis test, 261
One-tailed test, 271, 273–274, 278–279, 280–281, 283, 284, 286–287, 320–321, 362
definition of, 273
Open-ended class interval, 32
Operating characteristic curve, 258, 297–299

P-value (in hypothesis testing), 287
Paasches Index, 458, 494–495
Paired difference test, 317–322
Paired samples test, 522–524
Parameters, 8, 153–154, 156, 164, 169, 200, 236, 239, 260
definition of, 8
Percentile, 67
Permutations, 100, 132–133, 135, 137
Personal Interview, 13, 14, 16, 204–205
Pie Charts, 42–43
Point estimate, 7–8, 62, 234, 236, 390–391, 426
definition of, 7, 236
Poisson distribution, 146, 147, 164–167, 189, 190–192, 348–351, 372
approximation to binomial, 167
conditions, 164
fit of data to a, 348–351
tables and graphs, Appendix D, 166
Poisson equation, 165
Pooled variance, 311, 315, 332–333
Population, 4–5, 8, 74, 156, 198, 209, 215–216, 236–238, 260, 308–309, 521
definition of, 4
parameter, 8, 200, 230, 236, 260
proportion, 156–162, 236, 262–265, 308
regression line, 385, 425
Population arithmetic mean, 61, 236
Positively skewed distribution, 83
Posterior probability, 124–128
Power curve, 258, 294–297, 299, 300
computation of, 295–296
definition of, 295
Precision, 243–245, 250–251, 398 399
Prediction intervals, 396–400, 422, 434–435
Price index, 487, 488, 489, 491–494, 503–504
Prior probability . See Probability, prior.
Probability , 100–144
a priori, 103
axioms, 115
Bayesian, 100, 101, 119, 121, 124–128, 137
classical, 103–104, 137
conditional, 100, 101, 114–115, 393–394
definition of, 102–103
general equation for, 104
joint, 100, 101, 114
marginal, 100, 101, 112–113
posterior, 124–128
prior, 4, 124–128
relative frequency, 103, 137
samples, 200, 205–208
subjective, 103, 104, 137
table, 121
versus confidence, 239
Probability distributions, 146–195. *See also* Binomial, Normal, Poisson, *t*.
definition of, 149
Probability rules, 100, 101, 115–121, 137
Probability sampling, 200, 202, 205–208, 331
definition of, 205
Probability tree, 127
Problem, identification of, 10–11, 16
Producer Price Index, 497
Proportion, 236–237, 252, 324–327
standard error of, 226–228, 230, 236, 252, 310
tests of. *See* Tests of proportions.

Quality control, 329–331, 333

Quantiles, definition of, 65
Quantity index, 487, 489, 492, 494, 503–504
Quartiles, 40, 65–67

Random number table, 202–203
Random sampling, 202–203, 206–208, 238, 267, 344. *See also* Probability sampling.
 definition of, 206
Random variable, 146, 148–149, 239, 267, 407
 definition of, 148
 test of randomness, 516–519
Range, 32, 54–55, 71, 72, 80, 90
 definition of, 72
Rank correlation, 514, 516, 533–536, 538
 definition of, 534
Rank correlation coefficient, 534–536
 computation of, 534–535
 testing of, 535–536
Rank sum test, 514, 516, 527–531, 538
 definition of, 527
 logic of, 527–528
Ratio-to-moving average method, 475–478
 steps in the, 475
Raw data, 27, 30
 definition of, 27
Real class limits, (class boundaries), 33
Reference tables, 28
Region of acceptance, 268–269, 271, 274, 278, 279, 281, 282, 285, 286, 291–294, 298, 320–321, 331, 531
Region of rejection, 268–269, 271, 274, 531. *See also* region of acceptance.
Regression. *See also* regression analysis.
 assumptions, 394
 coefficients, 422, 426–430, 437–440
 curvilinear, 383
 definition of, 383
 linear and nonlinear, 383
 multiple, 422–453
 simple and multiple, 383, 423–425
 sum of squares, 435–437, 442, 451
Regression analysis, 380–403, 406–407, 415–417, 423–440, 501
Regression equation, 385, 389–391, 425–430, 451
Regression line, 385–387, 389–391, 415, 417, 427
Regression model, 384–385, 393, 423, 427, 435, 451
Rejection region. *See* Region of rejection.
Relationships
 curvilinear, 384
 linear, 384
 negative linear, 384
 positive linear, 384
Relative dispersion, 85–87
Relative frequency
 distribution, 34
 probabilities, 103–104, 137
Residual variation, 433–437
Revision of probabilities, 125–128
Rhind Mathematical Papyrus, 147
Rounding conventions, 59
Run, definition of, 517
Runs test, 514, 516, 517–519, 538

Sample, 6, 74, 80, 151, 157–161, 227, 235–238, 267, 516, 520, 528
 convenience, 204–205
 data, 199
 definition of, 4
 large, 247
 purpose, 204–205
 quota, 204–205
 random, 205–208, 267
 size, 11, 243–245, 311, 331, 368
 small. *See* Small samples.
 statistic, 8, 200, 230
 use of, 7, 235, 267
Sample arithmetic mean, 61–64, 236
Sample coefficient of correlation, 410–412, 442
Sample proportion, 156, 236
Sample regression line, 385, 425
Sample space, 101, 105–107, 149–151
 definition of, 105
Sampling, 196
 bias, 200, 230
 cluster, 205, 207–208, 230
 design, 199–200
 errors in, 200–201, 267
 importance of, 198
 with replacement, 217–218
 without replacement, 218–219
 simple random. *See* Simple random sampling.
 stratified, 205, 207, 230
 systematic, 205, 206, 230
Sampling distributions, 160, 196, 209–219, 230, 237–238, 247, 262–263
 of differences, 309
 formation of, 211
 of the mean, 211–213, 237–238, 261–262
 of proportions, 213, 247, 262–263
 of *t*, 221
Sampling error, 8–9, 200–201, 243–245, 250–251, 331
 definition of, 9
Sampling fraction, 218, 241, 245, 248–249, 252
SAS, 16, 424, 449
Scatter diagram, 383–384, 429
 definition of, 383
Seasonable adjusted data, 478–479
Seasonal index, 458, 459, 471–479, 502, 503
 definition of, 472
 using the, 478–479
Seasonal variation, 209–210, 458, 459, 461–462, 481–483

definition of, 461
Set, 101, 105, 108–110, 137
Shifting,
time series, 467–468
index numbers, 495–496
Sign test, 514, 516, 520–524, 538
one-sample, 520–522, 538
paired-sample, 522–524, 538
Signature of an equation, 466, 503
Significance level, 263, 268–269, 308
definition of, 268
Significant digits, 58–61
Simple random sampling, 205, 206
definition of, 206
Simple regression, 383
Simultaneous solution, 430
Skewness, 54, 82–83, 156
of line, 387, 390–391, 401–403, 426–430, 463–468
Slope of line, 387, 390–391, 401–403, 426-430, 463–468
Small frequency adjustment, 350–351, 442
Small samples, 240–241, 246–247, 276, 311, 314–315, 397–398, 442, 528
Sources of data, 12–13
Spearman rank correlation coefficient, 534
Specific seasonals, 477–478
SPSS, 16, 424, 449
Standard deviation, 54, 55, 71, 72, 73–76, 80–82, 90, 162, 181–182, 186–188, 212–213, 228, 344
definition of, 74
of regression. *See* Standard error of estimate.
Standard error
of differences, 310–311, 318–320, 325, 332–333
of the means, 216–219, 225, 228, 230, 236–238, 248, 252, 310, 329–331, 332
of proportion, 226–228, 230, 236–238, 248, 252, 310, 332
of sampling distribution, 217–218, 238
Standard error of estimate, 394–399, 422, 433–435, 437, 451
definition of, 395, 433
Standard error of the slope, 401–403, 437, 438–440
Standard normal probability distribution, 169–184, 344–345
Standardized coefficients, 439
Stated class limits, 33
Statistic, 8, 236
Statistical Abstract of the United States, 13, 496
Statistical independence. *See* Independence, statistical.
Statistical inference, 5, 214–216, 221–225, 237–239, 261–263, 308–310
Statistical significance, 263, 308, 313, 332
Statistical tables, 25, 28
Statistical techniques, 6, 57
Statistical terminology, 7–10
Statistics, 1, 8, 16, 235
definition of, 4
descriptive, 4, 57
inferential, 5
reason for studying, 1, 2–3, 235
Statview, 436, 449
Steps in
constructing a frequency distribution, 31–33
a goodness of fit test, 346
regression, 386
a statistical study, 10–11
a test of hypothesis, 260, 275, 312, 334–335, 346, 355–357
a test of independence of classification, 355
a time series forecast, 502
Stepwise regression, 422, 444, 450
Stratified sampling, 205, 207, 230
Student's *t*-distribution. *See* *t*-distribution.
definition of, 221
Subjective probability, 103, 104, 137
definition of, 104
Summary statistics, 56, 89–90
Summation notation, 59
Survey of Current Business, 13, 496
Surveys and field studies, 13–15
Systematic sampling, 205, 206, 230

t-distribution, 221–225, 235, 240–241, 252, 281–283, 309, 311, 314–315, 319–322, 397–398, 401–403, 415, 438–440
t-tables, 222–225, Appendix E
Table of random digits, 202–203
Tables, analytical and reference, 28
Tally table, 30–31, 32
Target population, 199–200
The Wall Street Journal, 381, 478
Test criteria, 271–276, 309–310
Test of independence of classification, 354–358, 372
Test procedure, 275, 355–357
Test statistic, 275–276, 309–310, 332, 345–346, 355, 361–363, 365, 401–402, 435–437, 518, 529
Testing hypotheses. *See* Hypothesis testing.
Tests
of correlation coefficient, 412
of differences, 308–337
of means, 261–262, 271–274, 276, 277–283, 309–310, 363–369, 521
of proportions, 262–263, 276, 284–287, 309–310, 324–327, 346-351, 354–357, 523
of regression model, 435–437
of slope of regression line, 401–402, 438–439

of variances, 362–363
Theorem 1, 214
Time, 460
Time series, 458–483, 501–503
components, 461–462
definition of, 461
equation, 462
Treatment variance, 367–368
Tree diagrams, 100, 127–128
Trend, 458–459, 475, 502, 503
definition, 461
estimation of, 461–469
nonlinear, 469
Two-sample hypothesis test, 261, 306–337
Two-tailed test, 271–273, 277–278, 281–283, 284–285, 295, 309, 318–320, 362, 530–531
definition of, 271
Type I error, 258, 265–268, 269, 291–294, 299, 331
definition of, 265
Type II error, 258, 259, 265–268, 290–294, 299, 331, 524, 529
definition of, 265
Types of probability, 112–115

Unbiased, 74, 217, 225
estimator, 214, 217, 236
Unconditional probability, 117
Unexplained error, 424
Unexplained variation, 395, 408–411, 433–437, 440–442
Ungrouped data, 56–66, 71–76
Uniform distribution, 147, 152–154, 346–348
Union of events, 108–109
Univariate data, 382, 394

Value index, 487, 488, 489, 490, 492, 503–504
Variables, discrete and continuous, 27
Variability, in sampling, 209–210, 382, 407
Variance, 54, 71, 73–75, 80–81, 90, 153–154, 311, 317, 363–369, 394
between-column, 367
inference about population, 358, 362–363
within-column, 367
Variation, 364–367, 395, 408–411
Venn diagram, 101, 108–110

Weighted mean, 54, 61–64
definition of, 62
Wilcoxon rank sum tests, 527
Within-group variation, 364–367

$\overline{X}$-bar ($\overline{X}$), 8

Yates continuity correction, 357

Z-table. *See* normal distribution.